W0082109

Multiscale Phenomena in Materials—Experiments and Modeling

MATERIALS RESEARCH SOCIETY
SYMPOSIUM PROCEEDINGS VOLUME 578

Multiscale Phenomena in Materials—Experiments and Modeling

Symposium held November 30–December 2, 1999, Boston, Massachusetts, U.S.A.

EDITORS:

I.M. Robertson
University of Illinois
Urbana, Illinois, U.S.A.

D.H. Lassila
Lawrence Livermore National Laboratory
Livermore, California, U.S.A.

B. Devincre
CNRS-ONERA
Chatillon, France

R. Phillips
Brown University
Providence, Rhode Island, U.S.A.

Materials Research Society
Warrendale, Pennsylvania

Single article reprints from this publication are available through
University Microfilms Inc., 300 North Zeeb Road, Ann Arbor, Michigan 48106

CODEN: MRSPDH

Copyright 2000 by Materials Research Society.
All rights reserved.

This book has been registered with Copyright Clearance Center, Inc. For further information, please contact the Copyright Clearance Center, Salem, Massachusetts.

Published by:

Materials Research Society
506 Keystone Drive
Warrendale, PA 15086
Telephone (724) 779-3003
Fax (724) 779-8313
Web site: http://www.mrs.org/

Library of Congress Cataloging-in-Publication Data

Multiscale phenomena in materials—experiments and modeling : symposium held
    November 30–December 2, 1999, Boston, Massachusetts, U.S.A. / editors, I.M. Robertson,
    D.H. Lassila, B. Devincre, R. Phillips
        p.cm.—(Materials Research Society symposium proceedings,
        ISSN 0272-9172 ; v. 578)
        Includes bibliographical references and indexes.
        ISBN 1-55899-486-6
        1. Materials—Computer simulation—Congresses. 2. Continuum mechanics—Computer
    simulation—Congresses. 3. Deformations (Mechanics)—Mathematical models—Congresses.
    4. Plasticity—Mathematical models—Congresses. 5. Dislocations in crystals—Mathematical
    models—Congresses. I. Robertson, I.M. II. Lassila, D.H. III. Devincre, B. IV. Phillips, R.
    V. Materials Research Society symposium proceedings ; v. 578
TA405 .M888 2000
620.1'1—dc21   00-035475

Manufactured in the United States of America

CONTENTS

PLASTICITY AT THE CONTINUUM LENGTH SCALE

NON-LOCAL PLASTIC THEORY AND DISLOCATION PHENOMENA

*Invited Paper

DISLOCATION DYNAMICS— EXPERIMENTS AND SIMULATION

*Invited Paper

DISLOCATION CORE PROPERTIES AND EFFECTS

FRACTURE AND CRACK PROPAGATION

*Invited Paper

*Invited Paper

MICROSTRUCTURAL MODELING FOR INDUSTRIAL METALS PROCESSING

PREFACE

Since the development of quantum mechanics, great strides have been made in understanding materials phenomena based on "first principles" associated with atomic and sub-atomic structure. Perhaps the most notable examples are in the fields of microelectronics and nuclear energy. In other fields, the connections between the fundamental nature of atomic structure and important properties are problematic because in addition to atomic structure, other structural aspects of the material at longer length scales are important. This is the case when considering the mechanical behavior of materials. Although general relationships between composition of materials and their mechanical properties have been established, a detailed understanding of the effects of the composition and microstructure based on first principles is lacking. For example, it is well known that a metal's mechanical properties such as yield strength can be grain size dependent, however, in many instances fundamental understanding of this phenomenon has been illusive.

In recent years there has been increasing interest in using what are termed "multiscale modeling" approaches to understand the effects of composition and microstructure on mechanical behavior. It is now quite common to find publications that report on related phenomena using atomistic, microscale (*dislocation dynamics*) and mesoscale/continuum computer simulations. With this enhanced interest in multiscale phenomena emerges a need for experimental work that validates and challenges the predictive capabilities of modeling and simulations.

The main purpose of this Symposium A, "Multiscale Phenomena in Materials— Experiments and Modeling," held November 30–December 2 at the 1999 MRS Fall Meeting in Boston, Massachusetts, is to bring together experimentalists and individuals working in the areas of modeling and simulations. To this end, we structured the sessions such that talks would be principally at a given length—scale and modeling and experimental subject matter was interspersed. Selected papers contributed by the symposium participants are presented with the same goal in mind. We hope that the reader will benefit from the varied nature of the papers and reflect on the many possibilities of making contact between experiments and simulations.

In addition to the papers from this symposium, selected papers from Symposium C, "Microstructural Modeling For Industrial Metals Processing," are included in the last section of this proceedings.

I.M. Robertson
D.H. Lassila
B. Devincre
R. Phillips

February 2000

ACKNOWLEDGMENTS

The organizers would like to thank the invited speakers, the session chairs, and the symposium assistants for helping to make this a successful symposium. We are grateful to Lawrence Livermore National Laboratory for financial support. Several papers from Symposium C, "Microstructural Modeling for Industrial Metals Processing," appear as part of this proceedings, and we would like to recognize the efforts of Armand J. Beaudoin, Jr., Paul R. Dawson, Warren J. Poole, and Hugh Shercliff in organizing that symposium.

Finally, we would like to thank our respective institutions for their support of our work on this symposium:

> University of Illinois (I.M. Robertson)
> Lawrence Livermore National Laboratory (D.H. Lassila)
> CNRS-ONERA (B. Devincre)
> Brown University (R. Phillips)

MATERIALS RESEARCH SOCIETY SYMPOSIUM PROCEEDINGS

MATERIALS RESEARCH SOCIETY SYMPOSIUM PROCEEDINGS

Prior Materials Research Society Symposium Proceedings available by contacting Materials Research Society

Plasticity at the Continuum Length Scale

SYMMETRY INVESTIGATION OF TEXTURED POLYCRYSTAL PROPERTIES (INVITED)

P. J. Maudlin*, J. F. Bingert*, G. T. Gray III*, and R. K. Garrett, Jr.**
*Los Alamos National Laboratory, Los Alamos, NM 87545
**Naval Surface Warfare Center, Indian Head Division, Indian Head, MD 20604-5035

ABSTRACT

Tantalum plate and rod materials that demonstrate mild-to-strong anisotropic plastic flow during large deformation are analyzed in terms of tensorial property symmetry. Texture interrogations of these materials reveal duplex orientation components that have implications with regard to the symmetry realized during plastic deformation; specifically these materials show less symmetry than one would expect from a cursory examination of the texture. Mesoscale polycrystal simulations are performed to probe a general shape for the yield surface function based on a discrete orientation distribution representation of the material texture and previously established single-crystal deformation modes. The yield surface shape is mathematically represented in terms of second and higher-order tensors. A plastic compliance analysis is presented and applied to graphically map the deformation symmetry contained in these tensors for both ideal and real materials. Compliance results are shown to be consistent with finite element simulations of r-value specimens loaded in uniaxial tension.

INTRODUCTION

A practical understanding of the deformation symmetry exhibited by various metallic polycrystals can occasionally be rather elusive. A detailed knowledge of s ͺle-crystal deformation mechanisms and the orientation distribution function (ODF) of cry 'als (texture) may not be quantitative enough in order to predict the symmetry of bulk plastic flow even when the material is subjected to uniform loading. In other words, given a symmetrically loaded cubic polycrystal with apparent orthotropy in texture via cursory pole figure inspection, the deformation can exhibit a remarkable lack-of-symmetry.

If we assume that plastic deformation of a subject material can be represented in terms of a potential surface (or yield surface), and further assume that the shape of such a surface can be accurately captured with some smooth function f, then an understanding of the deformation symmetry can be obtained via mathematical interrogation of the tensorial terms of f. Generally, characterization of the tensors contained in f could be conducted either experimentally as discussed by Hill [1], or with polycrystal simulations [2] which would require as input the appropriate deformation mechanisms and the measured texture. Spectral analysis can then be employed to investigate the symmetry of these tensors, either by solving for and analyzing the eigensystem analytically as, for example, Zuo and Hjelmstad [3] presented for the fourth-order elastic stiffness tensor, or by forming a lower-order tensor via contraction with appropriate vectors analogous to the acoustic problem (see Hearmann [4] and also Johnson [5]).

In this effort an approach coined *plastic compliance analysis*, which is akin to the acoustic problem, is applied to deduce the symmetry of tensors representing both ideal materials, e.g., isotropic, hexagonal, orthotropic, etc. symmetries, and real polycrystals, i.e., rolled and forged tantalum stock. Finite element simulations of r-value tests are performed and the results used to validate and demonstrate the utility of the described compliance analysis approach.

3

PLASTIC COMPLIANCE ANALYSIS

Consider the classical associated flow law [6] typically used for evolving the plastic part of the rate-of-deformation tensor $\underline{\underline{D}}$ in a rate-dependent, current-configuration plasticity formulation:

$$\underline{\underline{D}}^{\mathrm{p}} = \dot{\lambda}\frac{\partial f}{\partial \underline{\underline{\sigma}}}$$

(1)

Equation (1) employs a Cauchy stress $\underline{\underline{\sigma}}$ as the stress measure and $\underline{\underline{D}}$ as the appropriate work conjugate rate-of-strain measure. The quantity $\dot{\lambda}$ in Eq. (1) is a time-dependent scalar and f is a general but smooth yield function (plastic potential). In the presented analysis we assume an absence of rigid body rotation and use *unrotated* tensors. Tensor *order* is denoted here by the number of underbars.

Taylor expansion of f, assuming f depends only on the deviator of $\underline{\underline{\sigma}}$ (denoted by $\underline{\underline{s}}$), gives:

$$f = f_0 + \frac{\partial f}{\partial \underline{\underline{s}}}\bigg|_0 : \left(\underline{\underline{s}} - \underline{\underline{s}}_0\right) + \frac{1}{2}\left(\underline{\underline{s}} - \underline{\underline{s}}_0\right) : \frac{\partial^2 f}{\partial \underline{\underline{s}}^2}\bigg|_0 : \left(\underline{\underline{s}} - \underline{\underline{s}}_0\right) + \frac{1}{6}\left(\underline{\underline{s}} - \underline{\underline{s}}_0\right) : \frac{\partial^3 f}{\partial \underline{\underline{s}}^3}\bigg|_0 : \left(\underline{\underline{s}} - \underline{\underline{s}}_0\right) : \left(\underline{\underline{s}} - \underline{\underline{s}}_0\right)$$

$$+ \frac{1}{24}\left(\underline{\underline{s}} - \underline{\underline{s}}_0\right) : \left(\underline{\underline{s}} - \underline{\underline{s}}_0\right) : \frac{\partial^4 f}{\partial \underline{\underline{s}}^4}\bigg|_0 : \left(\underline{\underline{s}} - \underline{\underline{s}}_0\right) : \left(\underline{\underline{s}} - \underline{\underline{s}}_0\right) + \ldots$$

(2)

Under the assumption of a stress-free initial state, defining second- through eighth-order tensors respectively:

$$\underline{\underline{\gamma}} \equiv \frac{\partial f}{\partial \underline{\underline{s}}}\bigg|_0 \quad , \quad \underline{\underline{\alpha}} \equiv \frac{\partial^2 f}{\partial \underline{\underline{s}}^2}\bigg|_0 \quad , \quad \underline{\underline{\delta}} \equiv \frac{\partial^3 f}{\partial \underline{\underline{s}}^3}\bigg|_0 \quad \text{and} \quad \underline{\underline{\beta}} \equiv \frac{\partial^4 f}{\partial \underline{\underline{s}}^4}\bigg|_0$$

(3)

and truncating any higher-order terms from the above Taylor expansion produces a *quartic* yield function (designated here with a superscript):

$$f^{(4)} = f_0 + \underline{\underline{\gamma}} : \underline{\underline{s}} + \frac{1}{2}\underline{\underline{s}} : \underline{\underline{\alpha}} : \underline{\underline{s}} + \frac{1}{6}\underline{\underline{s}} : \underline{\underline{s}} : \underline{\underline{\delta}} : \underline{\underline{s}} + \frac{1}{24}\underline{\underline{s}} : \underline{\underline{s}} : \underline{\underline{\beta}} : \underline{\underline{s}} : \underline{\underline{s}}$$

(4)

If for some material-of-interest sign independence (inversion symmetry) in the plastic deformation mechanisms is appropriate, i.e., $f^{(4)}\left(\underline{\underline{s}}\right) = f^{(4)}\left(-\underline{\underline{s}}\right)$, then the following even form of the yield function would apply:

$$f^{(4)} = f_0 + \frac{1}{2}\underline{\underline{s}} : \underline{\underline{\alpha}} : \underline{\underline{s}} + \frac{1}{24}\underline{\underline{s}} : \underline{\underline{s}} : \underline{\underline{\beta}} : \underline{\underline{s}} : \underline{\underline{s}}$$

(5)

Note that the effects of kinematic hardening and sign-dependent deformation can be accommodated with the first- and third-order terms of Eq. (4) (e.g., see Tsai and Wu [9]), although here we proceed with the quartic form given by Eq. (5).

Further truncation of the fourth-order term in Eq. (5), which assumes that a second-order plastic potential function is sufficiently accurate, results in the more familiar *quadratic* function valid for general anisotropic materials restricted only by inversion symmetry:

$$f^{(2)} = f_0 + \frac{1}{2}\underline{\underline{s}} : \underline{\underline{\alpha}} : \underline{\underline{s}}$$

(6)

The yield function shape tensor denoted by α (the appearance of tensors in the text of order four or higher are written without underbars for convenience) and is nominally fourth-order with major and minor symmetries. This tensor would appear to contain twenty-one independent coefficients, however, since α is a deviatoric tensor, there are six associated constraint equations that reduce the maximum number of independent constants contained in α to fifteen. For a material-of-interest that has orthotropic or higher symmetry, α simplifies to the classic quadratic

4

expression proposed by Hill [1]. Hill's quadratic yield function contains a maximum of six independent constants, and is given here in Voigt-Mandel reduced components form:

$$VM\left(\underline{\underline{\alpha}}\right)_{\text{Hill}} = \begin{pmatrix} G+H & -H & -G & 0 & 0 & 0 \\ & F+H & -F & 0 & 0 & 0 \\ & & F+G & 0 & 0 & 0 \\ & & & L & 0 & 0 \\ & & & & M & 0 \\ & & & & & N \end{pmatrix} \tag{7}$$

Taking the deviatoric stress gradient of $f^{(4)}$ as indicated in Eq. (1) gives:

$$\frac{\partial f^{(4)}}{\partial \underline{s}} = \underline{\underline{\alpha}} : \underline{s} + \frac{1}{6}\underline{s} : \underline{\underline{\beta}} : \underline{s} : \underline{s} \tag{8}$$

and the resulting higher-order associated flow law becomes:

$$\underline{D}^{p} = \dot{\lambda}\left(\underline{\underline{\alpha}} : \underline{s} + \frac{1}{6}\underline{s} : \underline{\underline{\beta}} : \underline{s} : \underline{s}\right) \tag{9}$$

Again, if we assume material behavior that is adequately described by $f^{(2)}$, then the associated flow law reduces to:

$$\underline{D}^{p} = \dot{\lambda}\,\underline{\underline{\alpha}} : \underline{s} \tag{10}$$

Consistent with the assumption of plastic incompressibility for metals (i.e., $\underline{D}^{p}$ is traceless), it follows that α is a deviatoric tensor and $\underline{s}$ can be replaced by $\underline{\sigma}$ in Eq. (10) without any loss of interpretation (contraction with α extracts only the deviatoric part of $\underline{\sigma}$):

$$\underline{D}^{p} = \dot{\lambda}\,\underline{\underline{\alpha}} : \underline{\sigma} \tag{11}$$

This evolution equation can be expressed in *compliance* form by defining a plastic compliance tensor $\underline{\underline{K}}^{p} \equiv \dot{\lambda}\,\underline{\underline{\alpha}}$ that absorbs the associated flow scalar. Thus, we have a plastic compliance relationship that is analogous to the elastic compliance form of Hooke's Law:

$$\underline{D}^{p} = \underline{\underline{K}}^{p} : \underline{\sigma} \quad \Leftrightarrow \quad \underline{\varepsilon}^{e} = \underline{\underline{K}} : \underline{\sigma} \tag{12}$$

where K^{p} has dimensions of inverse viscosity ($\text{Pa}^{-1}\text{s}^{-1}$) as compared to the dimension of K which is inverse stress (Pa^{-1}), and $\underline{\varepsilon}^{e}$ is elastic strain.

Recall the definition of Lagrangian strain-rate $\dot{\underline{E}}$ that is formally related to the rate-of-deformation $\underline{D}$ via:

$$\dot{\underline{E}} = \underline{F}^{T} \cdot \underline{D} \cdot \underline{F} \tag{13}$$

where $\underline{F}$ is the deformation gradient. Partitioning $\underline{D}$ into elastic and plastic parts (consistent with a multiplicative decomposition of the deformation gradient [7], i.e., $\underline{F} = \underline{F}^{e} \cdot \underline{F}^{p}$) and inserting this partition into Eq. (13) gives:

$$\dot{\underline{E}} = \underline{F}^{T} \cdot \underline{D}^{e} \cdot \underline{F} + \underline{F}^{T} \cdot \underline{D}^{p} \cdot \underline{F} \tag{14}$$

Equation (14) suggests a partition of $\dot{\underline{E}}$ into elastic and plastic parts such that the plastic part can be identified as:

$$\dot{\underline{E}}^{p} = \underline{F}^{T} \cdot \underline{D}^{p} \cdot \underline{F} \tag{15}$$

Now, if we substitute the relationship between Cauchy and second Piola-Kirchhoff stress, i.e., $\underline{\underline{\sigma}} = \frac{1}{J}\underline{F} \cdot \underline{\sigma}^{pk2} \cdot \underline{F}^{T}$, into the left-hand-equation of Eq. (12), replace $\underline{D}^{p}$ in same with Eq. (15),

5

and move the resulting deformation gradients to the right-hand side, than an *initial configuration* version of the associated flow law is realized:

$$\underline{\dot{\underline{E}}}^P = \frac{1}{J}\left(\underline{\underline{F}}^T \cdot \underline{\underline{F}}^T \cdot \underline{\underline{K}}^P \cdot \underline{\underline{F}} \cdot \underline{\underline{F}}\right):\underline{\sigma}^{pk2} \tag{16}$$

where J is the Jacobian ρ_o/ρ (initial to current density ratio). Since we have already restricted our analysis to unrotated tensors, $\underline{\underline{F}}$ in Eq. (16) can be replaced via polar decomposition with the right-hand stretch $\underline{\underline{U}}$ to give:

$$\underline{\dot{\underline{E}}}^P = \frac{1}{J}\left(\underline{\underline{U}}^2 \cdot \underline{\underline{K}}^P \cdot \underline{\underline{U}}^2\right):\underline{\sigma}^{pk2} \tag{17}$$

If we further confine our analysis to small strain ($\underline{\underline{U}} \cong \underline{\underline{I}}$ and $J \cong 1$), then Eq. (17) simplifies to:

$$\underline{\dot{\underline{E}}}^P = \underline{\underline{K}}^P : \underline{\sigma}^{pk2} \tag{18}$$

Following the strain-vector formulation presented by Malvern [8], $\underline{\dot{\underline{E}}}$ contracted by an arbitrary unit vector $\underline{n}$ pointing in the direction of an initial configuration *fiber element* of differential length dS can be interpreted as a *unit relative displacement rate vector*:

$$\frac{d\underline{\dot{u}}}{dS} = \underline{\dot{\underline{E}}} \cdot \underline{n} \tag{19}$$

Again using our small-strain assumption for a strain-rate partition into elastic and plastic parts,

$$\frac{d\underline{\dot{u}}^e}{dS} + \frac{d\underline{\dot{u}}^P}{dS} = \left(\underline{\dot{\underline{E}}}^e + \underline{\dot{\underline{E}}}^P\right) \cdot \underline{n} \tag{20}$$

we select the plastic part and interpret the differential $d\underline{\dot{u}}^P$ as the plastic part of the displacement rate:

$$\frac{d\underline{\dot{u}}^P}{dS} = \underline{\dot{\underline{E}}}^P \cdot \underline{n} \tag{21}$$

Next we replace the second Piola-Kirchhoff stress in Eq. (18) with a dyadic product involving a plane normal $\underline{m}$ and a traction $\underline{t}$:

$$\underline{\sigma}^{pk2} \rightarrow \underline{m} \otimes \underline{t} \tag{22}$$

where $\underline{m}$ and $\underline{t}$ are considered here to be unit vectors (implying normalized stress) of arbitrary direction. Note that only the symmetric part of $\underline{m} \otimes \underline{t}$ is extracted when contracted with the symmetric compliance tensor $\underline{\underline{K}}^P$ in Eq. (18), thus side-stepping the stress symmetry issue of Eq. (22). We also replace $\underline{\dot{\underline{E}}}^P$ in Eq. (18) with an eigenvalue decomposition to attain the relationship:

$$\sum_{k=1}^{3}\xi_k\, \underline{p}_k \otimes \underline{p}_k = \underline{m} \cdot \underline{\underline{K}}^P \cdot \underline{t} \tag{23}$$

Introducing a second-order symmetric compliance tensor $\underline{\underline{C}}^P \equiv \underline{m} \cdot \underline{\underline{K}}^P \cdot \underline{t}$, we contract both sides of Eq. (23) with the eigenvector $\underline{p}_i$ and re-arrange to formally uncover the eigenvalue problem:

$$\left(\underline{\underline{C}}^P - \xi_i\underline{\underline{I}}\right) \cdot \underline{p}_i = \underline{0} \quad , \quad \text{no sum on i} \tag{24}$$

The eigenvalue in Eq. (24) is interpreted as a *principal unit relative plastic strain rate* in the direction of the corresponding eigenvector, consistent with a principal axis representation of $\underline{\dot{\underline{E}}}^P$. This becomes obvious after inserting the eigenvalue decomposition of $\underline{\dot{\underline{E}}}^P$ into Eq. (21) and noting the result when the fiber direction $\underline{n}$ is coincident with one of the eigenvectors $\underline{p}_i$:

$$\frac{d\underline{\dot{u}}^P}{dS} = \sum_{k=1}^{3}\left(\xi_k\, \underline{p}_k \otimes \underline{p}_k\right) \cdot \underline{n} = \sum_{k=1}^{3}\left(\xi_k\, \underline{p}_k \otimes \underline{p}_k\right) \cdot \underline{p}_i = \xi_i\, \underline{p}_i \quad \text{no sum on i,} \quad i = 1, 2, 3 \tag{25}$$

We now reinterpreted the unit relative plastic displacement rate $d\underline{\dot{u}}^p/dS$ of Eq. (21) as a plastic straining (strain-rate) vector [8]:

$$\underline{\dot{\varepsilon}} = \sum_{k=1}^{3} \left(\xi_k \, \underline{p}_k \otimes \underline{p}_k \right) \cdot \underline{n} \tag{26}$$

where the projection of this vector in the initial fiber direction $\underline{n}$ is the extension rate:

$$\dot{\varepsilon}_n = \sum_{k=1}^{3} \underline{n} \cdot \left(\xi_k \, \underline{p}_k \otimes \underline{p}_k \right) \cdot \underline{n} = \sum_{k=1}^{3} \xi_k \left(\underline{p}_k \cdot \underline{n} \right)^2 \tag{27}$$

and the corresponding orthogonal projection is the lateral shear rate:

$$\dot{\varepsilon}_s = \sqrt{|\underline{\dot{\varepsilon}}|^2 - \dot{\varepsilon}_n^2} \tag{28}$$

When $\underline{n}$ assumes an eigenvector direction, i.e., $\underline{p}_k \cdot \underline{n} = 1$, Eq. (27) indicates that the extension rate is identical to the associated eigenvalue; if the eigenvectors are coincident with the cartesian basis vectors, which is true for all materials of orthotropic or higher symmetry, then the extension rate is equal to the appropriate eigenvalue on the cartesian axes.

As an example of how one might use these *plastic compliance* definitions to interrogate the symmetry of yield function tensors, let

$$\underline{n} = \underline{\underline{Q}} \cdot \underline{n}_0 \quad \text{where} \quad \underline{\underline{Q}} = \begin{pmatrix} \cos\theta & -\sin\theta & 0 \\ \sin\theta & \cos\theta & 0 \\ 0 & 0 & 1 \end{pmatrix} \tag{29}$$

where the rotation tensor $\underline{\underline{Q}}$ contains some arbitrary plane rotation. Insertion of Eq. (29) into Eqs. (27) and (28), and using the solution of eigensystem Eq. (24), gives $\dot{\varepsilon}_n$ and $\dot{\varepsilon}_s$, respectively, in a form advantageous for graphical representation, i.e., polar plots. These plots represent the projections $\dot{\varepsilon}_n$ and $\dot{\varepsilon}_s$ as functions of θ, and the compliance contours portrayed reflect the symmetry inherent in α and/or β, as numerically illustrated in the next section.

ANALYSIS OF SHAPE TENSORS OF IDEAL SYMMETRY

Before proceeding with numerical examples that interrogate the symmetry of several ideal yield function tensors, a Voigt-Mandel change-of-basis is applied for convenience. In this reduced-components notation the Piola-Kirchhoff stress and Lagrangian strain-rate tensors become six-dimensional vectors:

$$\underline{\sigma}^{pk2} = VM\left(\underline{\underline{\sigma}}^{pk2}\right) = \left(\sigma_{11}^{pk2} \quad \sigma_{22}^{pk2} \quad \sigma_{33}^{pk2} \quad \sqrt{2}\sigma_{23}^{pk2} \quad \sqrt{2}\sigma_{31}^{pk2} \quad \sqrt{2}\sigma_{12}^{pk2} \right)^T$$

$$\underline{\dot{E}}^P = VM\left(\underline{\underline{\dot{E}}}^P\right) = \left(\dot{E}_{11}^P \quad \dot{E}_{22}^P \quad \dot{E}_{33}^P \quad \sqrt{2}\dot{E}_{23}^P \quad \sqrt{2}\dot{E}_{31}^P \quad \sqrt{2}\dot{E}_{12}^P \right)^T \tag{30a,b}$$

and the tensor α becomes second-order and six-dimensional:

$$\underline{\underline{\alpha}} = VM\left(\underline{\underline{\underline{\alpha}}}\right) = \begin{pmatrix} \alpha_{1111} & \alpha_{1122} & \alpha_{1133} & \sqrt{2}\alpha_{1123} & \sqrt{2}\alpha_{1131} & \sqrt{2}\alpha_{1112} \\ & \alpha_{2222} & \alpha_{2233} & \sqrt{2}\alpha_{2223} & \sqrt{2}\alpha_{2231} & \sqrt{2}\alpha_{2212} \\ & & \alpha_{3333} & \sqrt{2}\alpha_{3323} & \sqrt{2}\alpha_{3331} & \sqrt{2}\alpha_{3312} \\ & & & 2\alpha_{2323} & 2\alpha_{2331} & 2\alpha_{2312} \\ & & & & 2\alpha_{3131} & 2\alpha_{3112} \\ & & & & & 2\alpha_{1212} \end{pmatrix} \tag{30c}$$

The more awkward Voigt-Mandel transformation of the eight-order tensor β produces a fourth-order six-dimensional tensor:

$$\underset{\equiv}{\beta} = \begin{pmatrix} \beta_{1111mnop} & \beta_{1122mnop} & \beta_{1133mnop} & \sqrt{2}\beta_{1123mnop} & \sqrt{2}\beta_{1131mnop} & \sqrt{2}\beta_{1112mnop} \\ \beta_{2211mnop} & \beta_{2222mnop} & \beta_{2233mnop} & \sqrt{2}\beta_{2223mnop} & \sqrt{2}\beta_{2231mnop} & \sqrt{2}\beta_{2212mnop} \\ \beta_{3311mnop} & \beta_{3322mnop} & \beta_{3333mnop} & \sqrt{2}\beta_{3323mnop} & \sqrt{2}\beta_{3331mnop} & \sqrt{2}\beta_{3312mnop} \\ \sqrt{2}\beta_{2311mnop} & \sqrt{2}\beta_{2322mnop} & \sqrt{2}\beta_{2333mnop} & 2\beta_{2323mnop} & 2\beta_{2331mnop} & 2\beta_{2312mnop} \\ \sqrt{2}\beta_{3111mnop} & \sqrt{2}\beta_{3122mnop} & \sqrt{2}\beta_{3133mnop} & 2\beta_{3123mnop} & 2\beta_{3131mnop} & 2\beta_{3112mnop} \\ \sqrt{2}\beta_{1211mnop} & \sqrt{2}\beta_{1222mnop} & \sqrt{2}\beta_{1233mnop} & 2\beta_{1223mnop} & 2\beta_{1231mnop} & 2\beta_{1212mnop} \end{pmatrix}_{KL}$$

(30d)

where in this reduced subscripts notation, the indices K and L assume the values 1 through 6 where each value for K or L implies the corresponding double index values mn or op = 11, 22, 33, 23, 31, 12, respectively. Note also in Eqs. (30) that the number of underbars conveniently discriminates between 3D and 6D tensors, e.g., $\underset{\equiv}{\alpha}$ is the 3D shape tensor and $\underset{\equiv}{\alpha}$ is the 6D reduced components shape; thus, the Voigt-Mandel transformation of Eq. (9) can be written:

$$\underline{D}^p = \dot{\lambda}\left(\underset{\equiv}{\alpha} \cdot \underline{s} + \frac{1}{6}\underline{s} \cdot \underset{\equiv}{\beta} \cdot \underline{s} \cdot \underline{s}\right)$$

(31)

Proceeding with an isotropic example of the construction of the compliance rates $\dot{\varepsilon}_n$ and $\dot{\varepsilon}_s$, we first select the von Mises shape tensor (one independent constant):

$$\underset{\equiv}{\alpha}_{von\,Mises} = \begin{pmatrix} 2 & -1 & -1 & 0 & 0 & 0 \\ & 2 & -1 & 0 & 0 & 0 \\ & & 2 & 0 & 0 & 0 \\ & & & 3 & 0 & 0 \\ & & & & 3 & 0 \\ & & & & & 3 \end{pmatrix}$$

(32)

and specify a constant loading of uniaxial tension along the 3-axis:

$$\underline{m} = (0, 0, 1)^T \quad \text{and} \quad \underline{t} = (0, 0, 1)^T$$

(33a)

Here the dyad involving $\underline{m}$ and $\underline{t}$ results in the familiar uniaxial stress:

$$\underset{\equiv}{\sigma}^{pk2} = \begin{pmatrix} 0 & 0 & 0 \\ 0 & 0 & 0 \\ 0 & 0 & 1 \end{pmatrix} \quad \text{or} \quad \underline{\sigma}^{pk2} = (0 \ 0 \ 1 \ 0 \ 0 \ 0)^T$$

(33b)

We further specify the initial orientation of fiber dS to be $\underline{n}_o = (1, 0, 0)^T$, which we use for all the compliance analyses presented below, and solve eigensystem Eq. (24) with the von Mises shape given by Eq. (32) (in conjunction with setting λ to $1\,Pa^{-1}s^{-1}$). This eigensolution indicates apparent pure normal deformation: the set of eigenvalues is $\xi_i = \{-1, -1, 2\}$ and the corresponding set of eigenvectors $\underline{p}_i = \{(1,0,0)^T, (0,1,0)^T, (0,0,1)^T\}$ are coincident with the cartesian basis vectors; note the eigenvalues sum to zero thus satisfying plastic incompressibility.

As a subsequent aid for interpreting the compliance polar plots, Fig. 1 presents results from a dynamic finite element (FE) simulation of a r-value test [1]. These simulations model a L/D = 8 square bar with a 6.350 mm cross-section using hexahedral single-integration-point finite elements. Constant velocity boundary conditions (± 10 m/s) were applied to the ends of the test specimen for a period of 1 ms. The rate-dependent constitutive description given in Ref. [2] was utilized for these FE calculations.

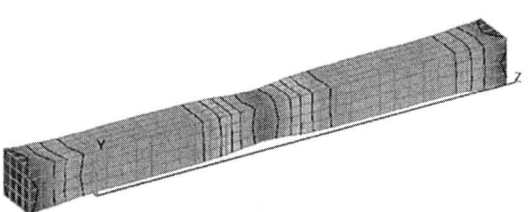

Figure 1: FE simulated geometry showing an initially square rod (r-value specimen) loaded in a state of uniaxial tension. Contours of equivalent plastic strain indicate that necking has occurred at rod center for this von Mises case after about 70% equivalent strain.

Computing the plastic extension rate using the above von Mises eigensolution, and plotting $\dot{\varepsilon}_n$ as a function of θ results in the polar plot of Fig. 2a. The extension rate in the 1-2 plane for this tensile loading is of course negative as indicated by the eigenvalues above, but is illustrated as a positive radius in Fig. 2a by plotting $-\dot{\varepsilon}_n$. Note the directional uniformity of $\dot{\varepsilon}_n$ in this figure, i.e., $\dot{\varepsilon}_n$ is independent of fiber orientation $\underline{n}$ as one might expect for an isotropic material. Also, inspection of Eqs. (26) and (27) indicates that $\dot{\varepsilon}_s$ has zero value for this case of a von Mises material and uniaxial stress loading. The center cross-section of the simulated r-value specimen is shown in Fig. 2b, indicating a r-value, i.e., $r \cong ln(y/y_0)/ln(x/x_0)$, of unity and velocity magnitude contours tending to 3-axis symmetry.

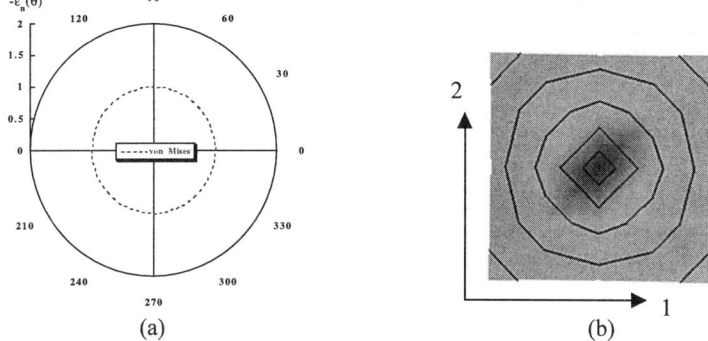

(a)   (b)

Figure 2: Plasticity results for a von Mises material: a) polar plot of the plastic extension rate in the $x_1 - x_2$ plane, and b) center cross-section of an initially square r-value specimen showing contours of the velocity vector magnitude.

Consider a shape tensor possessing hexagonal symmetry (three independent constants) whose initial axis-of-symmetry has alignment with the 3-axis (i = 3, least compliant direction), and is then transformed by rotation to align the symmetry axis with first the 1-axis (i = 1) and then the 2-axis (i = 2):

$$
\underline{\underline{\alpha}}_{Hex,i} = \left\{ \begin{pmatrix} 2 & -1 & -1 & 0 & 0 & 0 \\ & 2.5 & -1.5 & 0 & 0 & 0 \\ & & 2.5 & 0 & 0 & 0 \\ & & & 4 & 0 & 0 \\ & & & & 1 & 0 \\ & & & & & 1 \end{pmatrix}, \begin{pmatrix} 2.5 & -1 & -1.5 & 0 & 0 & 0 \\ & 2 & -1 & 0 & 0 & 0 \\ & & 2.5 & 0 & 0 & 0 \\ & & & 1 & 0 & 0 \\ & & & & 4 & 0 \\ & & & & & 1 \end{pmatrix}, \begin{pmatrix} 2.5 & -1.5 & -1 & 0 & 0 & 0 \\ & 2.5 & -1 & 0 & 0 & 0 \\ & & 2 & 0 & 0 & 0 \\ & & & 1 & 0 & 0 \\ & & & & 1 & 0 \\ & & & & & 4 \end{pmatrix} \right\} \quad \text{for } i = 1, 2, 3 \quad (34a, b, c)
$$

Solution of the Eq. (24) eigensystem for Eq. (34c) contracted with Eq. (33a) gives eigenvalues and eigenvectors that have von Mises values reflecting in-plane isotropy. Computing the plastic extension rate for each of the above three orientations, we obtain the results compared in Fig. 3a. This comparison of hexagonal compliances is physically intuitive: the 3-axis symmetry case shows in-plane isotropy in terms of compliance, the 1-axis symmetry case Eq. (34a) shows an elliptical plastic response with maximum compliance realized along the 2-axis, and the 2-axis symmetry case Eq. (34b) also shows an elliptical plastic response with maximum compliance realized along the 1-axis. The aspected center cross-section of the r-value specimen is shown in Fig. 3b for the 2-axis symmetry case, indicating a r-value of 0.671.

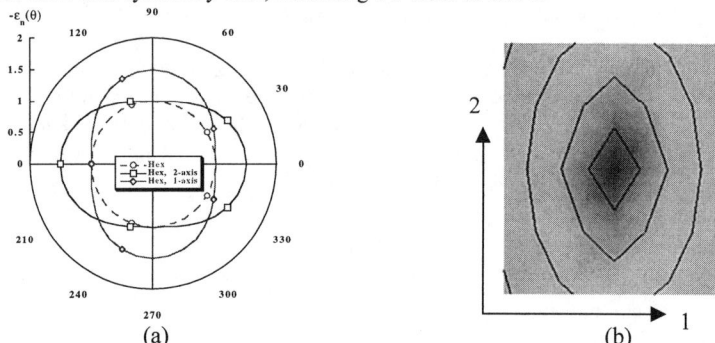

(a)   (b)

Figure 3: Plasticity results for a material possessing hexagonal symmetry: a) polar plot of the plastic extension rate in the $x_1 - x_2$ plane with the axis of symmetry initially aligned with the 3-axis, then aligned with the 2-axis, and then aligned with the 1-axis, and b) center cross-section of an initially square r-value specimen showing contours of the magnitude of the velocity vector for the 2-axis symmetry case.

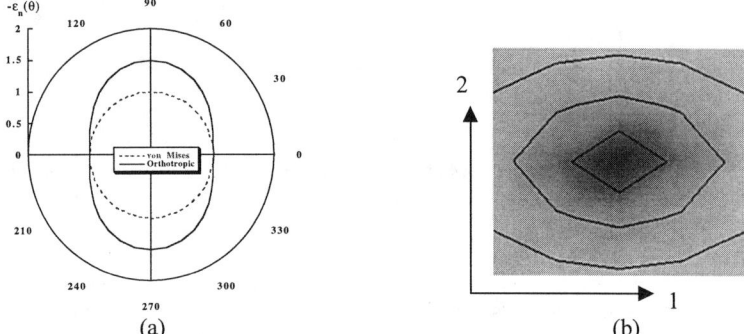

(a)   (b)

Figure 4: Plasticity results for a material possessing orthotropic symmetry: a) polar plot of the plastic extension rate in the $x_1 - x_2$ plane, where the 1-direction is the least compliant direction, and b) center cross-section of an initially square r-value specimen showing contours of the magnitude of the velocity vector.

Consider a shape tensor possessing orthotropic symmetry (six independent constants) where the least-compliant direction for this material is the 1-direction:

$$\underset{=}{\alpha}_{\text{Orthotropic}} = \begin{pmatrix} 2 & -1 & -1 & 0 & 0 & 0 \\ & 2.5 & -1.5 & 0 & 0 & 0 \\ & & 2.5 & 0 & 0 & 0 \\ & & & 1 & 0 & 0 \\ & & & & 2 & 0 \\ & & & & & 3 \end{pmatrix} \tag{35}$$

Solution of the Eq. (24) eigensystem using Eq. (35) contracted with Eq. (33a) gives $\xi_i = \left\{-1, -\frac{3}{2}, \frac{5}{2}\right\}$ and $\underset{\sim}{p}_i = \left\{(1,0,0)^T, (0,1,0)^T, (0,0,1)^T\right\}$. Computing the plastic extension rate for this orthotropic tensor, we obtain the results presented in Fig. 4a. This comparison indicates that the 2-direction is 50% more compliant than the von Mises material, and the compliance in the 1-direction is identical with von Mises. The aspected center cross-section of the simulated r-value specimen is shown in Fig. 4b, indicating a r-value of 1.49.

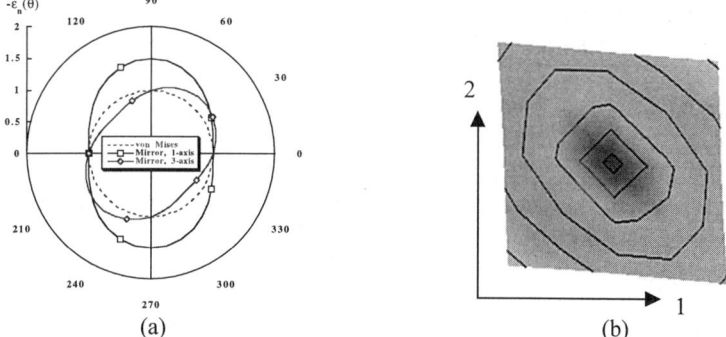

(a)   (b)

Figure 5: Plasticity results for a material possessing mirror plane symmetry: a) polar plot of the plastic extension rate in the $x_1 - x_2$ plane where the mirror normal is either in the 1- or 3-direction, and b) center cross-section of an initially square r-value specimen for the 3-axis mirrored case showing contours of the magnitude of the velocity vector.

Consider a shape tensor possessing mirror-plane symmetry (nine independent constants) where the mirror-plane normal for the base case is aligned with the 1-axis, and a second case featuring rotation of the shape tensor about the 2-axis by 90° that produces 3-axis mirror plane symmetry:

$$\underset{=}{\alpha}_{\text{mirror},i} = \left\{ \begin{pmatrix} 2 & -1 & -1 & 0.25 & 0 & 0 \\ & 2.5 & -1.5 & 0.5 & 0 & 0 \\ & & 2.5 & -0.75 & 0 & 0 \\ & & & 1 & 0 & 0 \\ & & & & 1.5 & 0.1 \\ & & & & & 2 \end{pmatrix}, \begin{pmatrix} 2.5 & -1.5 & -1 & 0 & 0 & 0.75 \\ & 2.5 & -1 & 0 & 0 & -0.5 \\ & & 2 & 0 & 0 & -0.25 \\ & & & 2 & -0.1 & 0 \\ & & & & 1.5 & 0 \\ & & & & & 1 \end{pmatrix} \right\} \quad \text{for } i = 1, 3 \ (36\text{a,b})$$

Solution of the Eq. (24) eigensystem for Eq. (36a) contracted with Eq. (33a) gives $\xi_i = \left\{-1, -1.57, 2.57\right\}$ and $\underset{\sim}{p}_i = \left\{(1,0,0)^T, (0,0.992,0.129)^T, (0,-0.129,0.992)^T\right\}$. Computing the plastic extension rate for the Eq. (36) shapes, we obtain the results presented in Fig. 5a. This comparison indicates that the 1-axis mirrored material has a compliance response similar to the orthotropic material above, and the 3-axis mirrored material has a *rotated* compliance response with maximum compliance at 45°. Note that the compliance of the 3-axis mirrored material matches the von Mises values at the 1- and 2- axis intercepts. Again, these responses can be understood if

11

one considers the resultant contraction of uniaxial stress in the 3-direction, Eq. (33b), with each of the two shape tensors given by Eqs. (36). The skewed cross-section of the simulated r-value specimen for the 3-axis mirrored case is shown in Fig. 5b, indicating a r-value of unity and enhanced compliance at the 45° angle.

SYMMETRY ANALYSIS OF REAL POLYCRYSTALS

Tantalum in the forms of rolled and recrystallized plate and forged and round-rolled rod are investigated for deformation symmetry in this section, using yield function tensors characterized from polycrystal simulations as described in some detail in Refs. [2] and [10]. These simulations use textures measured via x-ray pole figure analysis from the as-received metal in conjunction with experimentally established single-crystal deformation modes, i.e., two slip modes ($\{110\}\langle111\rangle, \{112\}\langle111\rangle$) for both the plate [2] and rod [11] where the critical resolved shear stresses are set equal for the plate and biased by 20% favoring $\{110\}\langle111\rangle$ slip for the rod.

The measured sample texture for the tantalum plate is presented in terms of the rolling plane-normal (3-axis) ODF-recalculated pole figures given in Fig. 6. This texture is primarily comprised of two three-fold $\{111\}\langle uvw \rangle$ components within the rolling plane with a background of $\langle111\rangle$ fiber normal to the rolling plane.

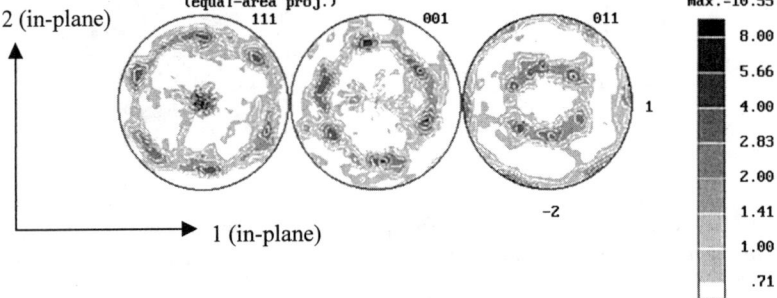

Figure 6: Recalculated pole figures (through-thickness direction (3-axis) normal) for the rolled and recrystallized Ta plate. Intensity scale in multiples of random distribution.

Characterized shape tensors for the tantalum plate for a base case where the 3-axis corresponds to through-thickness direction (least compliant) and for a second case that has experienced a 90° rotation about the 1-axis are, respectively:

$$\underline{\underline{\alpha}}_{\text{Ta Plate}} = \left\{ \begin{pmatrix} 2.22 & -1.23 & -0.99 & 0.19 & 0.12 & -0.06 \\ & 2.04 & -0.81 & -0.02 & -0.14 & 0.11 \\ & & 1.80 & -0.17 & 0.02 & -0.05 \\ & & & 4.23 & 0.06 & -0.06 \\ & & & & 3.96 & -0.33 \\ & & & & & 3.33 \end{pmatrix}, \begin{pmatrix} 2.22 & -0.99 & -1.23 & -0.19 & -0.06 & -0.12 \\ & 1.80 & -0.81 & 0.17 & -0.05 & -0.02 \\ & & 2.04 & 0.02 & 0.11 & 0.14 \\ & & & 4.23 & 0.06 & 0.06 \\ & & & & 3.33 & 0.33 \\ & & & & & 3.96 \end{pmatrix} \right\} \quad (37a,b)$$

Solution of the Eq. (24) eigensystem using Eq. (37a) contracted with Eq. (33a) gives $\xi_i = \{-0.99, -0.81, 1.80\}$ and $\underline{p}_i = \left\{ (0.982, 0.190, 0.002)^T, (-0.190, 0.982, 0.047)^T, (0.007, -0.047, 0.999)^T \right\}$. Computing the plastic extension rates based on this eigensolution for the rolling plane ($x_1 - x_2$ plane) and in the orthogonal $x_1 - x_3$ plane (corresponding to Eq. (37b)), we obtain the results presented in Fig. 7a. This comparison indicates that the rolling plane is nearly isotropic with a slight tilt (a few degrees), and that the 2-direction is somewhat less compliant consistent with the direction of rolling. The orthogonal-plane case shows large compliance at the 170° angle, and produces the skewed cross-section of the simulated r-value specimen given in Fig. 7b, indicating a r-value of 0.662.

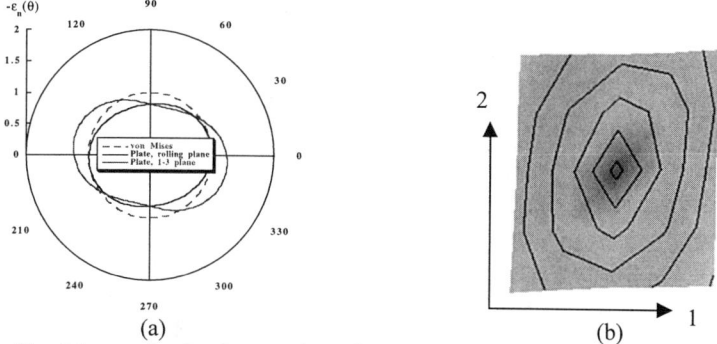

(a)   (b)

Figure 7: Plasticity results for the tantalum plate: a) polar plot of the plastic extension rate in the $x_1 - x_2$ plane (rolling plane) and in the $x_1 - x_3$ plane, and b) center cross-section of an initially square r-value specimen for the latter case ($x_1 - x_3$ plane) showing contours of the magnitude of the velocity vector.

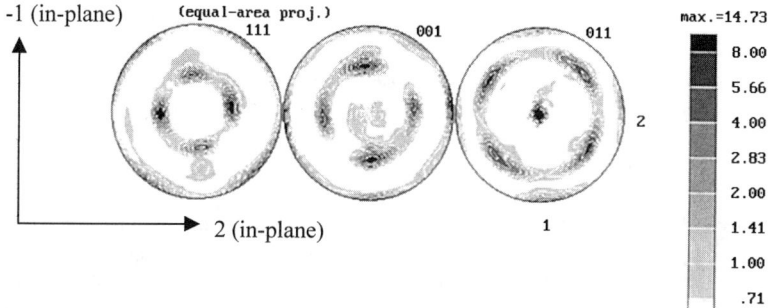

Figure 8: Recalculated pole figures (rod axis (3-axis) normal) for the forged and round-rolled Ta rod. Intensity scale in multiples of random distribution.

The measured sample texture for the tantalum rod is presented in terms of the rod axis-normal (3-axis) ODF-recalculated pole figures given in Fig. 8. At first glance this texture appears to primarily consist of a single $\{110\}\langle uvw \rangle$ orientation. Further inspection reveals two 2-fold $\{110\}$ components of unequal strength, namely the $\{110\}\langle \bar{1}10 \rangle$ (stronger) and $\{110\}\langle 001 \rangle$, based on an arbitrarily chosen vertical reference direction. This texture results from the lack of axisymmetry in the manufacturing process, and produces implications concerning the symmetry of subsequent dynamic plastic deformation. The characterized shape tensor for the tantalum rod where the 3-direction (least compliant) is coincident with the rod-axis is:

$$\underset{\approx}{\alpha}_{\text{Ta Rod}} = \begin{pmatrix} 2.18 & -1.18 & -1.00 & 0.43 & 0.63 & 0.28 \\ & 2.24 & -1.06 & -0.47 & -0.48 & 0.32 \\ & & 2.05 & 0.04 & -0.15 & -0.60 \\ & & & 4.57 & -0.64 & -0.41 \\ & & & & 4.89 & -0.08 \\ & & & & & 4.90 \end{pmatrix} \tag{38}$$

13

Solution of the Eq. (24) eigensystem using Eq. (38) contracted with Eq. (33a) gives $\xi_i = \{-0.61, -1.45, 2.06\}$ and $\underline{p}_i = \{(0.730, -0.683, 0.038)^T, (0.683, 0.730, 0.015)^T, (-0.038, 0.015, 0.999)^T\}$. Computing the plastic extension rates based on this eigensolution in the plane perpendicular to the rod axis (i.e., the $x_1 - x_2$ plane), we obtain the results presented in Fig. 9a. This polar plot indicates that the most compliant direction is at $52.5°$ and further demonstrates minimum compliance (even less than von Mises) at $142.5°$. A highly skewed cross-section of the simulated r-value specimen for this material is shown in Fig. 9b, indicating an aspect ratio near unity and large compliance at the $45°$ angle consistent with the polar plot. Considerably less symmetry is observed than what would be presumed from a cursory examination of the texture in Fig. 8.

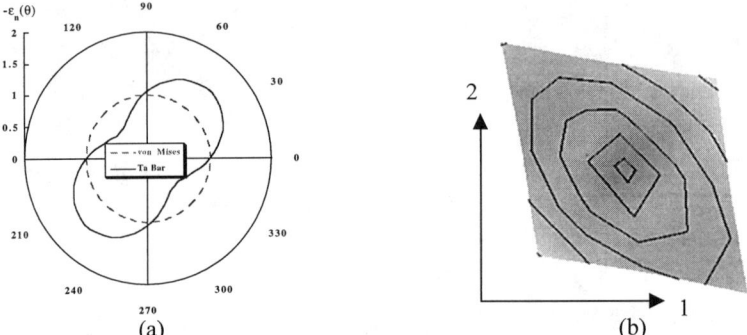

(a)   (b)

Figure 9: Plasticity results for the tantalum rod: a) polar plot of the plastic extension rate in the $x_1 - x_2$ plane (perpendicular to the rod axis), and b) highly skewed center cross-section of an initially square r-value specimen showing contours of the magnitude of the velocity vector.

SUMMARY

Analytical yield functions expressed in terms of shape tensors are interrogated for symmetry using a plastic compliance analysis based on an initial configuration associated flow law. This analysis employs spectral decomposition of the plastic part of the Lagrangian strain-rate tensor and forms a plastic compliance eigenvalue problem analogous to the acoustic problem. Plastic extension rates constructed from the eigensolution are numerically illustrated using polar plots to reveal the inherent symmetry of selected yield function tensors representing both ideal and real materials. The presented plastic compliance technique was successfully applied to two tantalum polycrystals to reveal the low-symmetry of deformation associated with these duplex-texture materials.

REFERENCES

1. R. Hill, *The Mathematical Theory of Plasticity*, Oxford University Press, London, 1950, pp. 315-340.
2. P. J. Maudlin, J. F. Bingert, J. W. House, and S. R. Chen, I. J. Plasticity, **15**, p. 139-166 (1999).
3. Q. H. Zuo and K. D. Hjelmstad, J. Acoust. Soc. Am., **103**, Number 4, p. 1727-1733 (1998).
4. R.F.S. Hearman, *Applied Anisotropic Elasticity*, Oxford University Press, London, 1961, pp. 68-89.
5. J. N. Johnson, J. App. Phys., **42**, Number 13, p. 5522-5530 (1971).
6. Y. C. Fung, *Foundations of Solid Mechanics*, Prentice-Hall, NJ, 1965, pp. 142-145.
7. D. Peric and D. R. J. Owen, Rep. Prog. Phys., **61**, p. 1495-1574 (1998).
8. L. E. Malvern, *Introduction to the Mechanics of a Continuous Medium*, Prentice-Hall, NJ, 1969, pp. 123-132.
9. S. W. Tsai and E. M. Wu, J. Composite Materials, **5**, p. 58-80 (1971).
10. P. J. Maudlin, S. I. Wright, U. F. Kocks and M. S. Sahota, Acta mater., **44**, Number 10, p. 4027-4032 (1996).
11. J. F. Bingert, Los Alamos National Laboratory, work in progress.

STRAIN TENSORS AT THE ATOMIC SCALE

M.F. HORSTEMEYER*, M.I. BASKES**
*Sandia National Laboratories, MS9405, Livermore, CA 94550, mfhorst@sandia.gov
**Los Alamos National Laboratory, MST-8, Los Alamos, NM 87545, baskes@lanl.gov

ABSTRACT

Almansi and Green strain tensors are developed for use in large deformation molecular dynamics/statics simulations that employ Embedded Atom Method (EAM) potentials for metals. The strain tensors are formulated with respect to the deformation gradient. A scalar potential function is used with a weighting function that is dependent upon a cutoff radius for the deformation gradient. For a homogeneous or inhomogeneous deformation, a cutoff distance of one lattice parameter can be used to approximate local strain level. Inhomogeneous deformation reveals different results for Almansi and Green strain tensors indicating that the small strain assumption cannot be used to determine large atomic strains.

INTRODUCTION

Studies employing continuum quantities, such as a strain tensor, at the atomic level must be performed in order to bridge length scales from the discrete atomic level to the macroscopic continuum scale. Mott and Argon (1995) developed a strain tensor for Lennard-Jones potentials for use in polymers. Falk (1999) developed a strain tensor for glass with no crystalline order using Lennard-Jones potentials. Their formulations focus on employing a small strain tensor in the context of a pair-potential. In this paper, we develop a method to determine the deformation gradient of a deformed material that can undergo multiaxial deformation states to large strain levels by employing a many-bodied potential in the same spirit as the EAM potential employs it. From this point, we can determine any strain measure without limitation of scale size or strain amount. In particular, we develop forms for the Almansi strain tensor and Green strain tensor. The Eulerian Almansi strain tensor is developed with current (deformed) configuration coordinates, and the Lagrangian Green strain tensor is developed with reference (undeformed) configuration coordinates. At small strains, these two strain measures equal each other. Small strain is defined when the deformation gradient components are small compared to unity. The Green strain is more suitably used in elasticity since there is usually an undeformed state to which the body is elastically unloaded. The Almansi strain is more suitably used for large strain measures because of the large amount of dislocations that are present that distort the lattice.

Strain Formulation

In a rectangular Cartesian coordinate system, we employ an incremental deformation path that considers a displacement of an atom from the reference configuration, $\underline{x}$, to the current configuration, $\underline{x}^*$, in which a small perturbation, $d\underline{x}$ and $d\underline{x}^*$, is assumed in each configuration. As the limit $d\underline{x} \rightarrow 0$, we get

$$d\underline{x}^* = \frac{\partial \underline{x}^*}{\partial \underline{x}} d\underline{x} \,, \tag{1}$$

in which the deformation gradient, $\underline{F}$, is defined as

$$\underline{F} = \frac{\partial \underline{x}^*}{\partial \underline{x}} \,. \tag{2}$$

Now assume an incremental form and allow m to be a counter for the atom number and N to be the time step counter, we rewrite equations (1)-(2) as

$$\left(\Delta\underline{x}^m\right)^N = \underline{F}\left(\Delta\underline{x}^m\right)^{N-1} \tag{3}$$

and given in indicial notion as

$$\left(\Delta x_i^{\,m}\right)^N = F_{ij}\left(\Delta x_j^{\,m}\right)^{N-1} \tag{4}$$

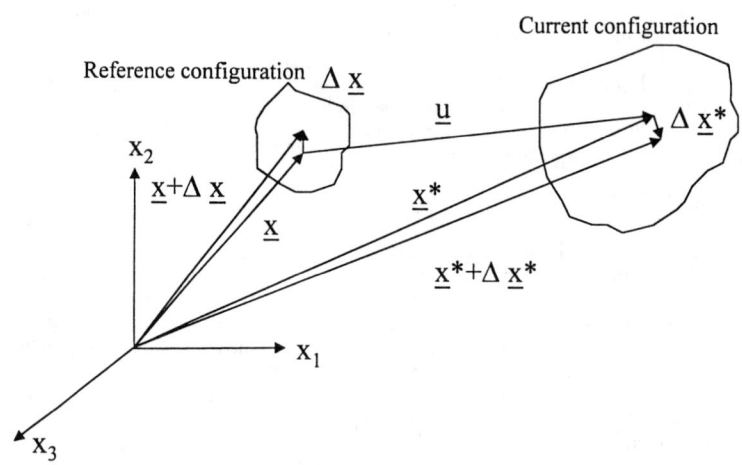

Figure 1. Schematic of reference and current configuration with respect to updated displacement, $\underline{u}$.

where i,j=1,3. For ease of differentiation let $b_i^{\,m} = \left(\Delta x_i^{\,m}\right)^N$ and $c_j^{\,m} = \left(\Delta x_j^{\,m}\right)^{N-1}$, so then we can write

$$b_i^{\,m} - F_{ij}^{\,m}c_j^{\,m} = a_i^{\,m} \tag{5}$$

where a is a residual to be minimized, and m is not related to contravariant and covariant notation symbols. A potential scalar function can be written

$$\phi^m = \left(b_i^{\,m} - F_{ij}^{\,m}c_j^{\,m}\right)\left(b_k^{\,m} - F_{kl}^{\,m}c_l^{\,m}\right)\delta_{ik}. \tag{6}$$

With the summation on all the atoms and with a weighting function we get

$$\phi = \sum_m\left(b_i^{\,m} - F_{ij}^{\,m}c_j^{\,m}\right)\left(b_k^{\,m} - F_{kl}^{\,m}c_l^{\,m}\right)\delta_{ik}W(m) \tag{7}$$

in which $W(m)=rc^2-dist^2$ is used for the weighting function. Here, rc is the cutoff radius, and *dist* is the distance from the atom of interest with respect to the atom in question. The cutoff radius was varied to analyze the nonlocality of the deformation gradient. Figure 2 shows how the weighting function changes as a function of cutoff radius.

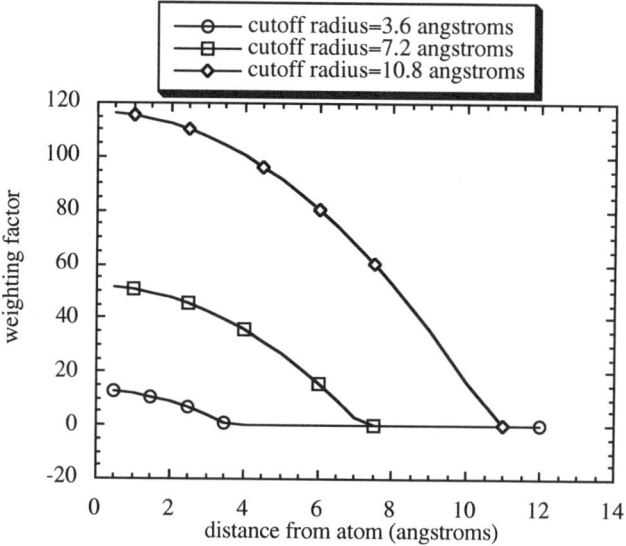

Figure 2. Weighting factor as a function of distance from the atom in question for different cutoff radii to determine the deformation gradient.

Now we minimize the potential by taking the derivative with respect to the deformation gradient

$$\frac{\partial \phi}{\partial F_{st}} = \sum_m \left(b_i{}^m - F_{ij}{}^m c_j{}^m \right) \left(-\frac{\partial F_{kl}{}^m}{\partial F_{st}{}^m} \right) c_l{}^m \delta_{ik} W(m) + \left(b_k{}^m - F_{kl}{}^m c_l{}^m \right) \left(-\frac{\partial F_{ij}{}^m}{\partial F_{st}{}^m} \right) c_j{}^m \delta_{ik} W(m) = 0_{st}.$$

(8)

By reducing terms we get

$$\frac{\partial \phi}{\partial F_{st}} = -\sum_m a_i{}^m c_l{}^m \Pi_{klst} \delta_{ik} W(m) - \sum_m a_k{}^m c_j{}^m \Pi_{ijst} \delta_{ik} W(m) = -\sum_m a_i c_l \Pi_{ilst} W(m) - \sum_m a_k c_j \Pi_{kjst} W(m) = 0_{st}$$

(9)

Now substitute equation (5) into equation (9) to get

$$\sum_m \left(b_i{}^m - F_{ij}{}^m c_j{}^m \right) c_l{}^m W(m) = 0_{il}.$$

(10)

By rearranging terms, we get

$$\sum_m b_i{}^m c_l{}^m W(m) = \sum_m F_{ij}{}^m c_j{}^m c_l{}^m W(m).$$

(11)

Solving for the deformation gradient, we get

$$F_{ij}^{\ m} = \sum_m b_i^{\ m} c_l^{\ m} W(m) (\sum_m c_j^{\ m} c_l^{\ m} W(m))^{-1}.$$ (12)

Now replacing $b_i^{\ m} = \left(\Delta x_i^{\ m}\right)^N$ and $c_j^{\ m} = \left(\Delta x_j^{\ m}\right)^{N-1}$, we get

$$F_{ij}^{\ m} = \sum_m \left(\Delta x_i^{\ m}\right)^N \left(\Delta x_l^{\ m}\right)^{N-1} W(m) \left(\sum_m \left(\Delta x_j^{\ m}\right)^{N-1} \left(\Delta x_l^{\ m}\right)^{N-1} W(m) \right)^{-1}$$ (13)

Once the deformation gradient, $\underline{F}$, is determined, we can now solve for the Lagrangian Green strain tensor, $\underline{E}$, with respect to reference coordinates as

$$\underline{E} = \left(\underline{F}^T \underline{F} - \underline{1}\right) = \frac{1}{2}\left(\frac{\partial \underline{x}^*}{\partial \underline{x}} \frac{\partial \underline{x}^*}{\partial \underline{x}} - \underline{1}\right)$$ (14)

and the Eulerian Almansi strain tensor, E*, with respect to current coordinates as

$$\underline{E}^* = \frac{1}{2}\left(\underline{1} - \left(\underline{F}^{-1}\right)^T \underline{F}^{-1}\right) = \frac{1}{2}\left(\underline{1} - \frac{\partial \underline{x}}{\partial \underline{x}^*} \frac{\partial \underline{x}}{\partial \underline{x}^*}\right).$$ (15)

The Almansi strain is often related to the true strain, and the Green strain is often related to the engineering strain. However, upon close examination of definition of the true strain and engineering strain, respectively, we can observe the difference,

$$E^{true} = \ln\left(\frac{l_f}{l_i}\right)$$ (16)

$$E^{eng} = \left(\frac{l_f - l_i}{l_i}\right)$$ (17)

in which l_f is the final length and l_i is the initial length.

RESULTS

To illustrate results from the strain formulation, we perform molecular dynamics simulations under simple shear (torsion) on a small and large set of atoms (486 and 9546 atoms) using EAM potentials (Daw and Baskes, 1984; Angelo et al., 1995; Baskes et al., 1997) for nickel up to 50% strain. In these simulations, the direction of the simple shear loading was in the x-direction along the y-face. Free edges without boundary conditions existed on the x-faces. The dimensions of the geometry were two units in the x-direction per one unit in the y-direction. Several lattice parameter distances were assumed in the z-direction with periodicity also assumed in the z-direction. Horstemeyer and Baskes (1999) explain in greater detail the rationale for relating these conditions to continuum level simple shear formulations. In Figure 3 we show the calculated value for the Green shear strain and Almansi shear strain for an atom that is near the center of the block of atoms (#207) and at the edge of the applied displacement (#77). The calculated strain values are added to the total values from each increment. We also show the average of the Green shear strain and Almansi shear strain for the small block of atoms (486) in Figure 3. For "small" strains less than 15%, the central, edge, and average strain for both strain measures are almost equal. For larger strains, a significant deviation in the local atomic strains exists due to the nucleation of dislocations. The average Almansi shear strain is somewhat greater than the Green shear strain.

Figure 4 shows the various strain values as in Figure 3 but with a larger number of atoms (9546 atoms). In Figure 4, the center atom is #4693 and the edge atom is 4885. As in the small atom case, the center atom appears to correlate more closely with the average values. Also, we see that the average Almansi shear strain is somewhat greater than the average Green shear strain. In the small and large blocks of atoms, the cutoff distance was 3.6 Å, slightly larger than the lattice parameter for nickel (3.52 Å). In both Figures 3 and 4 the atoms near the edge experience a surface effect and as such do not track as close to the average result as the atoms in the center of block of material.

Note that the average Green strain and average Almansi strain values are not in one-to-one correspondence with the applied shear strain. Though not shown here because of space limitations, the same correlation is observed when finite element simulations are performed under the same boundary conditions. The finite element simulations reveal that a strain gradient exists in the corner of the specimens. As such, strain accomodation within the material occurs such that a slightly lower aggregate response arises when an average of the local strain values is performed.

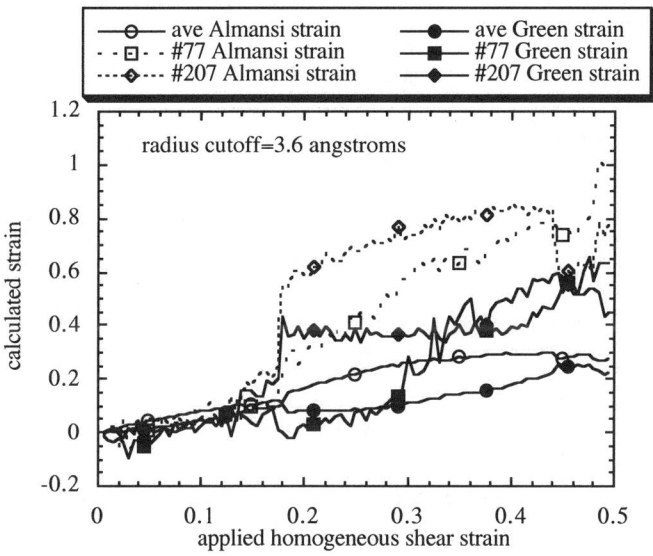

Figure 3. Calculated strains versus the applied shear strain for the 486 atom specimen.

CONCLUSIONS

Strain tensors are developed for atomistic simulations in the undeformed and deformed configurations based upon a deformation gradient. The cutoff distance for the weighting function needs only to include the first nearest neighbors although the formulation is not limited to that assumption.

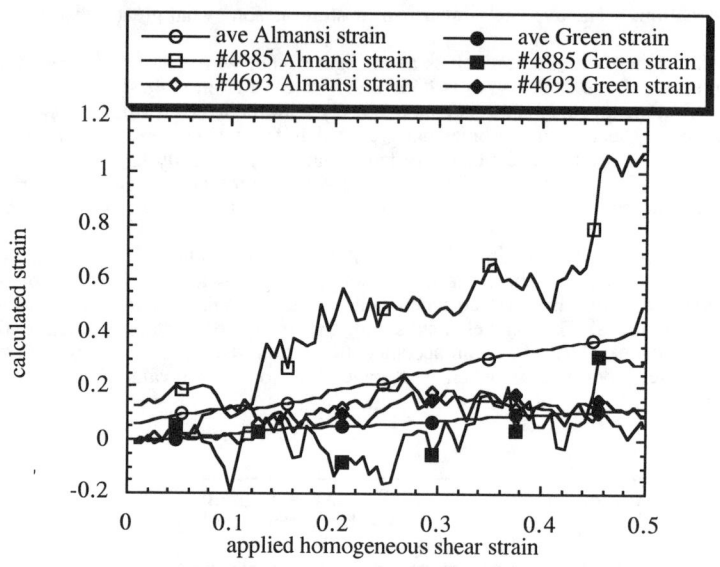

Figure 4. Calculated strains versus applied shear strain for the 9546 atoms specimen.

ACKNOWLEDGMENTS

This work has been sponsored by the U.S. Department of Energy, Sandia National Laboratories under contract DE-AC04-94AL85000.

REFERECES

1. Mott, P.H., Argon, A.S., and Suter, U.W., J. Comp. Physics, V101, No.1, p. 140 (1992).
2. Falk, M.L., Phys. Rev B., Vol. 60, No. 10, p. 7062 (1999).
3. Daw, M.S. and Baskes, M.I., Phys. Rev., Vol B29, p. 6443 (1984)
4. Angelo, J.E., Moody, N.R., and Baskes, M.I., J. Modell. Simul. Mater. Sci. Eng., Vol. 3, p. 289 (1995).
5. Baskes, M.I., Sha, X., Angelo, J.E., and Moody, N.R., J. Modell. Simul. Mater. Sci. Eng., Vol. 5, p. 651 (1997)
6. Horstemeyer, M.F. and Baskes, M.I., J. Engring. Matls, Techn., V121, p. 114 (1999)

EXPERIMENTAL STUDY OF MOBILE DISLOCATION DENSITIES AND VELOCITIES THROUGH TRANSIENT MECHANICAL TESTS

B. LO PICCOLO*, J.L. MARTIN* and J. BONNEVILLE**
*Physics Department, Ecole Polytechnique Fédérale, CH 1015 Lausanne, Switzerland
**Laboratoire de Métallurgie Physique, Université de Poitiers, SP2MI, Bd 3-Téléport 2 – B.P.
179, F 86960 Futuroscope Cedex

ABSTRACT

The use of transient mechanical tests such as repeated load relaxations and repeated creep tests provide some insight on the respective contribution of mobile dislocation densities and velocities to the plastic strain-rate and also on the hardening mechanisms. Improvements of the second technique are reported as well as results on Ni_3Al polycrystals. Other materials (e.g. Cu, TiAl) are considered for comparison. Correlations are found between dislocation exhaustion rates, work-hardening coefficients and the amplitude of yield point at reloading after the relaxations.

INTRODUCTION

The Orowan equation which stipulates that the plastic strain-rate $\dot{\varepsilon}_p$ is proportional to the mobile dislocation density ρ_m and to their average velocity v can be used to correlate mechanical properties and dislocation mechanisms. However, any conventional mechanical test fails to separate the respective contributions of v and ρ_m to this parameter. In other words, the question remains open on whether the crystal contains few mobile dislocations moving at high velocities or many dislocations with low velocities, for given deformation conditions.

On the experimental side, dislocation velocities have been measured mostly using etch pit experiments in a variety of crystals (see e.g. [1]), seldom by dynamic observations in the electron microscope (see e.g. [2]). However, mobile dislocation densities are a poorly documented parameter.

However, the importance of this parameter for a complete understanding of the deformation rate has recently stimulated some numerical simulations. The latter focus on mobile dislocation densities, on multiplication processes in some selected materials (see e.g. [3] for BCC crystals, [4] for Cu), and new relations are proposed which describe the variation of the mobile dislocation density as a function of stress e.g. in Si [5], a covalent material with a low initial defect density. These relations need to be verified experimentally.

We have developed repeated transient mechanical tests [6] which aim at separating the respective contribution of ρ_m and v to $\dot{\varepsilon}$. The principles of such tests are recalled here, together with the main assumptions about their interpretations. Some recent results are presented to illustrate which kind of information about ρ_m is accessible and to assess the validity of our interpretation.

THE REPEATED TRANSIENT TECHNIQUES

The first one consists of successive relaxations [6]. The sample is loaded at constant strain-rate and at the desired stress σ_M, it is allowed to relax for a given time interval Δt (30 s in the present experiments). The total amount of relaxed stress is $\Delta\sigma_1$. Subsequently the sample is reloaded up to σ_M and allowed to relax again during Δt, with a total relaxed stress $\Delta\sigma_2$. The procedure is repeated several times and then the constant strain-rate deformation starts anew. A yield point at reloading is usually observed, of amplitude $\Delta\sigma$, the nature of which is discussed below. At low enough temperatures, the amounts of relaxed stress $\Delta\sigma_1, \Delta\sigma_2 \dots \Delta\sigma_n$ are observed to decrease. Therefore the relaxations in a series are slowing down due to mobile dislocation exhaustion and the subsequent hardening: each relaxation in the series starts with a lower mobile dislocation density as compared to the previous one and a smaller effective stress, although σ_M is the same. It is thought that no dislocation multiplication and no change in internal stress take place between two relaxations because of the quasi-elastic reloading conditions. It is therefore possible to get some estimations of the strain-hardening as well as of the mobile dislocation exhaustion-rate which are characteristic of a relaxation, by comparing the $\Delta\sigma_i$ values along a series. The present technique has been applied successfully to a variety of crystals (see e.g. [7] for γTiAl polycrystals).

21

A second technique of repeated creep tests has been developed more recently. At σ_M, the load is kept constant while the specimen creeps with duration Δt by an amount $\Delta\varepsilon_1$. In the original method [8], the load was then lowered, so as to allow the plastic strain incurred in the specimen to recover. Then a second creep test was started during Δt with a creep-strain $\Delta\varepsilon_2$ etc. Since $\Delta\varepsilon_2$ is usually too small to be safely measured, the procedure has been modified [9]: the stress is increased by a small amount $\Delta\sigma'$ between two creep tests. $\Delta\sigma'$ has to be large enough to ensure a measurable strain during the subsequent creep transient, and small enough for reloading quasi-elastically, in view of a more simple interpretation.

In the case of logarithmic relaxation or creep transients, which are most commonly observed at low enough temperatures, the assumptions and equations used for the interpretation have been described in detail [10, 11]. They are summarized below. The Orowan equation is used:

$$\dot{\varepsilon}_p = \alpha \rho_m b \, v \qquad (1)$$

where α is a geometric coefficient and b the Burgers vector. v is thermally activated i.e.

$$v = vd \, \exp\left(-\frac{\Delta G(\sigma^*, T)}{kT}\right) \qquad (2)$$

where v is the vibration frequency of the dislocation line, and d the mean free path of the dislocations between two obstacles. The activation energy ΔG can be developed with respect to the effective stress σ^*, which is the difference between the applied stress σ and the long range internal stress σ_i:

$$\Delta G = \Delta G_o - \sigma^* V \qquad (3)$$

where V is the physical activation volume, a meaningful quantity related to the area swept by the dislocation during the activation process.

On the macroscopic point of view, the so-called "machine equation" is:

$$\dot{\sigma} = M\left(\dot{\varepsilon}_t - \dot{\varepsilon}_p\right)$$

where $\dot{\sigma}$ is the stress-rate, $\dot{\varepsilon}_t$ the total strain-rate and M the specimen-machine modulus. In a single relaxation test, the stress decrease can be expressed [12] as a function of time by:

$$\sigma - \sigma_M = -\frac{kT}{V_a} \ln\left(1 + \frac{t}{C_r}\right) \qquad (4)$$

V_a (which has the dimension of a volume) and the time constant C_r can be determined by fitting relation (4) with the experimental curve. The relation between V_a and V can be written [10, 11]:

$$V_a = \Omega V \qquad (5) \qquad \text{with}$$
$$\Omega = \left(1 + \frac{K}{M}\right)(1 + \beta) \qquad (6)$$

Ω being a structural parameter, K the strain-hardening coefficient during relaxation and β an exhaustion coefficient in the expression of ρ_m:

$$\rho_m = \rho_{mo} \exp\left[\frac{\beta V}{kT} \Delta\sigma^*\right] \qquad (7)$$

ρ_{mo} corresponds to the onset of relaxation. In (7), the exponential variation of ρ_m with the change in effective stress $\Delta\sigma^*$ during relaxation is compatible with (4).

The technique of repeated relaxations allows the determination of Ω from the values $\Delta\sigma_1$, $\Delta\sigma_2$ ...[6, 10] and then V (relation 5) and β (relation 6) with reasonable assumptions on K [6, 10]. An exhaustion parameter of the mobile dislocations $\Delta\rho_m/\rho_{mo}$ can be obtained using (7), where $\Delta\rho_m$ is the decrease of density which corresponds to a decline of one order of magnitude of the initial relaxation rate [6] (or a time interval equal to $9C_r$). The stress-dependence of the dislocation velocity is also known (relations 2 and 3).

Similar considerations apply to creep transients. Along a creep curve, the strain $\Delta\varepsilon_p$ is:

$$\Delta\varepsilon_p = \frac{kT}{MV_c} \ln\left(1 + \frac{t}{C_c}\right) \qquad (8)$$

where V_c has the dimensions of a volume and C_c is a time constant. Both parameters can be determined by fitting relation (8) with the creep curve. The experiment shows that V_a is constant and C_r changes for relaxation, while both V_c and C_c are varying for creep [9].

From two successive creep tests (numbered respectively n and n+1 in the series), V_c^n, V_c^{n+1} can be obtained (relation 8), which yields V as a function of $\Delta\sigma'$, $\Delta\varepsilon_n$ and $\Delta\varepsilon_{n+1}$ [9, 11].

More experimental details are given elsewhere [6, 7, 9].

RESULTS AND DISCUSSION

It has been checked (fig. 1) that the values of V along a stress-strain curve of a $Ni_{75}Al_{25}$ polycrystal are quite comparable when measured with both types of techniques [9, 10]. This emphasizes the reliability of both transient tests and confirms that the dislocation mobility mechanism is the same.

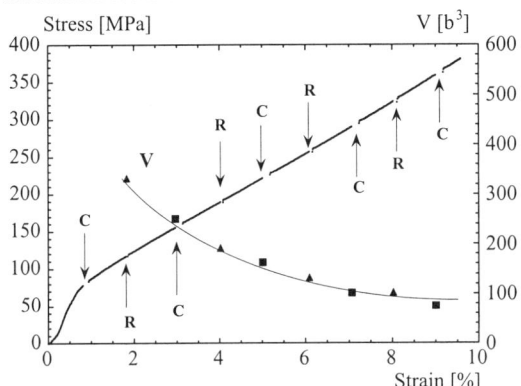

Fig. 1: Comparison of activation volume values V along a deformation curve, determined respectively by repeated relaxation R and repeated creep transients C. (The two first creep tests in the series are considered.) $Ni_{75}Al_{25}$ polycrystal. T=300K.

The dislocation exhaustion rates $\Delta\rho_m/\rho_{mo}$ have been measured through relaxations, for different types of materials [10] and compared with the work-hardening coefficient θ. Along a stress-strain curve both parameters $\Delta\rho_m/\rho_{mo}$ and θ exhibit parallel trends (in single crystals of $Ni_3(Al,Hf)$ as well as of Cu [10]). When different materials are compared, high values of θ/G (G=shear modulus) correspond to elevated dislocation exhaustion rates ($Ni_3(Al,Hf)$ single crystals and TiAl polycrystals), while θ/G and $\Delta\rho_m/\rho_{mo}$ are both low for Cu single crystals oriented in double glide. This corresponds to different work-hardening mechanisms in both classes of materials [13]. The importance of dislocation exhaustion in Ni_3Al has already been evidenced experimentally [14].

Another type of check of the validity of the above determined $\Delta\rho_m/\rho_{mo}$ parameter, is to study the yield point at reloading after the relaxations. Its amplitude $\Delta\sigma$ has been measured in a $Ni_{76}Al_{24}$ polycrystal, along the stress-strain curve, at different temperatures. Corresponding to a plastic strain of 3%, the $\Delta\sigma$ values, normalized to the stress σ at reloading ($\sigma \sim \sigma_M$), are plotted on fig. 2, as a function of temperature. On fig. 3, the mobile dislocation exhaustion rates measured by repeated relaxations prior to the yield point are also plotted as a function of temperature. Similar trends of variation of $\Delta\sigma/\sigma$ and $\Delta\rho_m/\rho_{mo}$ are observed, by comparing fig. 2 and 3. Both parameters are constant between 300 and 400K, reach a maximum value at around 500K and then decrease to low values around 700-800K. This can be understood by considering that $\Delta\sigma$ is the stress increment necessary to multiply fresh dislocations after the mobile ones have been exhausted by Kear-Wilsdorf lock formation during the preceding relaxations. The larger the exhaustion parameter, the higher $\Delta\sigma$. The good agreement between the data of fig. 2 and 3 supports the assumptions made for the interpretation of the transients. In addition, the peak temperature of about 500K, corresponds rather well to the peak temperature of θ ($T_{p\theta}$) measured at the same strain [9], which is much smaller than the peak temperature for yield stress (larger than 800K). The proposed interpretation is as follows: Below 500K, dislocation exhaustion dominates and increases with T since the locking cross-slip process is thermally activated. Above 500K, screw dislocation locking is in competition with yielding of the weakest locks. Therefore the net exhaustion-rate $\Delta\rho_m/\rho_{mo}$ is lower as well as $\Delta\sigma$ and θ values. All three parameters decrease with temperature because the stress increases and consequently the barrier unlocking frequency. This agrees with the predictions of a recent 3 D computer simulation of plastic flow in Ni₃Al [15]. It has been proposed that the locks which yield first are incomplete ones because of the right orders of magnitude of the corresponding stresses which were estimated [16] using antiphase boundary energies, measured by transmission electron microscopy [17] [18]. The parallel trends of variation of $\Delta\sigma/\sigma$ (fig. 2) and $\Delta\rho_m/\rho_{m0}$ (fig. 3), emphasize the validity of the technique used to measure mobile dislocation exhaustion.

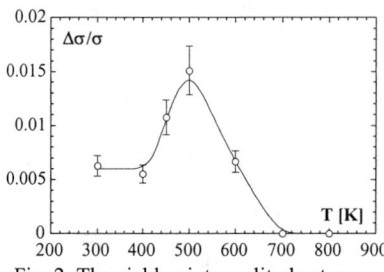

Fig. 2: The yield point amplitude at re-loading $\Delta\sigma$, normalized to the stress σ, after 6 successive relaxations as a function of temperature. $\varepsilon_p =3.10^{-2}$. (Ni₇₆Al₂₄).

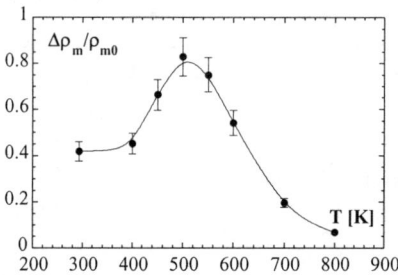

Fig. 3: Mobile dislocation exhaustion-rate $\Delta\rho_m/\rho_{m0}$, as a function of temperature. Same material and conditions as fig. 2.

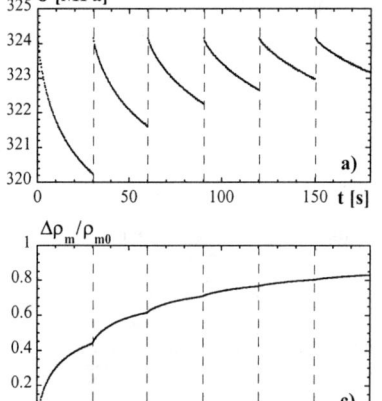

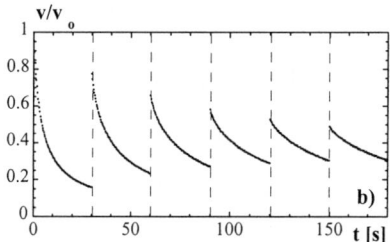

Fig. 4: An experiment of 6 successive relaxations Ni₇₅Al₂₅ polycrystal at 300K. **a)** Experimental curve. **b)** Computed dislocation velocities normalized to v_o. **c)** Mobile dislocation exhaustion parameter $\Delta\rho_m/\rho_{mo}$, (v_o corresponds to the onset of the first relaxation).

After describing the experimental technique and justifying the related assumptions by the above results, the question about the Orowan equation is now examined. Fig. 4 illustrates an example of successive relaxations in a binary Ni₃Al polycrystal. The variation of v/v_o, and $\Delta\rho_m/\rho_{mo}$, estimated assuming K=θ, are also represented. v/v_o is computed considering (2) and (3), while the repeated relaxations yield V. If the onset and the end of the first relaxation are compared ($\Delta t = 30$ s), $\Delta\rho_m/\rho_{mo}$ increases from 0 to 0.41 (fig. 4c) i.e. ρ_m/ρ_{mo} decreases from 1 to 0.59, v/v_o decreases from 1 to 0.15 (fig. 4b) and $\dot\varepsilon_p / \dot\varepsilon_{po}$ decreases from 1 to 0.09 (from (1) or by comparing the $\dot\sigma$ values on fig. 4a). It can be concluded that, under the above conditions, a strong decrease of the plastic strain-rate (down to 9% of its onset value) is accounted for by a moderate reduction of ρ_m (down to 59% of its onset value) and an important decrease of v (down to 15% of its original magnitude). These quantitative considerations bring some answer to the relative contributions of v and ρ_m to $\dot\varepsilon_p$, in the present conditions. Work is in progress and a number of materials under a variety of deformation conditions are being investigated to estimate the respective role of ρ_m and v in the Orowan equation.

CONCLUSIONS

The technique of repeated creep tests has been improved and successfully applied to Ni$_3$Al polycrystals. In particular, it allows determination of a microscopic activation volume, which characterizes the stress dependence of the dislocation velocity. The values of this parameter are in good agreement with those measured through repeated relaxations, a well-established and reliable procedure. Using the latter technique, mobile dislocation exhaustion rates are measured, and found to be closely related to those of the work-hardening coefficient. Ni$_3$Al exhibits high values of both parameters because of the efficiency of cross-slip exhaustion leading to Kear-Wilsdorf lock formation. Consequently the amplitude of the yield point at reloading after relaxation is found to be closely related to the mobile dislocation exhaustion rate, as the temperature changes. At last, it has been shown that these methods allow clarification of the respective contribution to the strain-rate of the mobile dislocation density and velocity in the frame of the Orowan equation. An interpretation is proposed of the decrease in work-hardening rate with temperature.

All the above results provide a body of evidence which supports the validity of the experiments of repeated transients, together with the assumptions used for their interpretation.

ACKNOWLEDGEMENTS

The financial support of Fonds National Suisse de la Recherche Scientifique is gratefully acknowledged.

REFERENCES

1. J.J. Gilman, Austral. J. Phys. 13, 327 (1960).
2. A. Couret and D. Caillard, Acta Metall. 3, 1455 (1985).
3. M. Tang, B. Devincre and L.P. Kubin, Modelling Simul. Mater. Sci. Eng., 7, 893 (1999).
4. D. Gomez-Garcia, B. Devincre and L.P. Kubin, this conference.
5. A. Moulin, M. Condat and L.P. Kubin, Acta Mater. 47, 2879 (1999).
6. P. Spätig, PhD thesis n° 1407, Ecole Plytechnique Fédérale, Lausanne (1995).
7. J. Bonneville, B. Viguier and P. Spätig, Scripta Mater. 36, 275 (1997).
8. A. Orlova, J. Bonneville and P. Spätig, Mat. Sci. Engng, A 191, 85 (1995).
9. B. Lo Piccolo, PhD thesis n° 2044, Ecole Polytechnique Fédérale, Lausanne (1999).
10. J.L. Martin, B. Matterstock, P. Spätig and J. Bonneville, Proceedings of the 20th Risø Int. Symp. on Mater. Sci., Edit. J.B. Bilde-Sorensen, J.V. Carstensen, W. Hansen, D. Juul Jensen, T. Leffers, W. Pantleon, O.B. Pedersen and G. Winther, Roskilde, pp. 103-121 (1999).
11. J.L. Martin, B. Lo Piccolo and J. Bonneville, to appear in Intermetallics, 2000.
12. F. Guiu and P.L. Pratt, Phys. Stat. Sol. 6, 111 (1964).
13. B. Matterstock, J.L. Martin, J. Bonneville and T. Kruml, Mater. Res. Symp. Proc. Vol. 552 (1999), Materials Research Society p. KK5.17.1-6.
14. K.J. Hemker, M.J. Mills and W.D. Nix, Acta metall. mater., 39, 1901 (1991).
15. B. Devincre, P. Veyssière and G. Saada, Phil. Mag. 79, 1609 (1999)
16. D. Caillard and G. Molénat, Proc. of the 20th Risø Int. Symp. on Mater. Sci. Ibidem, p 1-23.
17. D.M. Dimiduk, A.W. Thompson and J.C. Williams, Phil. Mag. A67, 675 (1993).
18. T. Kruml, J.L. Martin and J. Bonneville, Phil. Mag. accepted for publication in 2000.

FINITE ELEMENT SIMULATIONS OF THE DEFORMATION OF BCC AGGREGATES USING A CRYSTAL PLASTICITY MODEL - LOCAL ORIENTATION EFFECTS

D. ARIZMENDI*, J. L. RAPHANEL**
* LPMTM-CNRS, Université Paris Nord, 93430 Villetaneuse, FRANCE
**LMS-CNRS, Ecole polytechnique, 91128 Palaiseau Cedex, FRANCE.
raphanel@lms.polytechnique.fr

ABSTRACT

One has simulated by a finite element computation, the tensile plastic deformation of bcc aggregates with the same initial crystallographic texture but a different spatial distribution of orientations. The computations show fairly large differences between the simulations in terms of axial strain fields: one aggregate presents an homogeneous field, while another one localizes the strain in a narrow band. This "macroscopic effect" is then analyzed at the smaller scale of the crystallite by comparing local re-orientations, glide activity and stress levels. The influence of neighboring grains and of the "local" texture is thus shown. This is an effect that polycrystalline models (Taylor or self-consistent models) cannot account for.

INTRODUCTION

The response of crystalline aggregates to plastic deformation may seem at first homogeneous at a macroscopic scale, but subsequently, in most cases, it becomes heterogeneous, which may lead to localization and loss of ductility. The apparent homogeneity is the result of an average over local heterogeneities and it disappears when local differences become too large. It is thus of interest to design structures and models which can provide data on what actually happens at a smaller scale and, in a first approach, one may consider the scale of the grain (or crystallite) and its immediate neighbors.

Experiments have been conducted on special crystals, called multicrystals [1][2], made of a single layer of grains. They have provided a lot of interesting data on the local intragranular strain field and misorientations which showed the influence of initial crystalline orientations and geometry on the mechanical response. Some experiments have also been made on actual polycrystals [3], but they remain limited to surface observations.

Our approach has thus been to create a virtual three dimensional aggregate with enough crystallites to behave more in a polycrystalline fashion than in a single crystal one, but not too many grains so that local analysis and computing times remain tractable. A finite element code is then used in order to simulate a plastic deformation, in our case, an uni-axial tensile test. The code includes crystal plasticity: the plastic deformation results from dislocation glide on a well defined set of slip systems; the hardening rules are related to the physics of glide and take account of the evolutions of the densities of dislocations on the different slip systems. The parameters of the model have been evaluated using experimental data from the literature for iron single crystals [4]. The model is then validated by comparisons of the simulations with experiments on iron single crystals done in our laboratory and taken from the literature. One thus gains confidence into the predictability of the model and applies it to idealized multicrystalline structures. Keeping the same geometry, grain shapes and sizes, one changes the spatial distribution of initial orientations. One analyses both the

Mat. Res. Soc. Symp. Proc. Vol. 578 © 2000 Materials Research Society

character of the macroscopic response (its "homogeneity") and the local behavior of some chosen initial orientations (their stress levels, plastic glide and re-orientation).

MODELING

The mechanical modeling follows the framework developed by Teodosiu et al. [5]. The mechanical behavior is taken as elastoviscoplastic, with anisotropic elasticity and a power-law viscoplastic relation. The case treated here concerns α iron or mild steel at room temperature, so that the viscosity effects are very small, and the permanent deformation is essentially due to instantaneous plasticity. In bcc crystals this deformation is related to dislocation glide on well defined crystallographic slip systems. These systems have in common a glide direction of type $< 111 >$ and planes whose normals have been found to be of type $\{110\}$ or $\{112\}$ (in some instances $\{123\}$ planes have also been observed, but not predominantly and they shall be left out of our model). An asymmetry of glide on the $\{112\}$ planes has been reported very early for bcc crystals [6] and later explained [7]. One thus needs to differentiate between easy (or twinning) and uneasy (or antitwinning) glide directions for the same $\{112\} < 111 >$ system. The crystal plasticity model accounts for 12 $\{110\} < 111 >$ systems and 24 oriented $\{112\} < 111 >$ systems. They are characterized by their initial critical shear stress.

An original part of the model is to include a hardening rule based on the physics of glide for cubic crystals. The model is inspired by the works of Mecking and Kocks [8], and Tabourot et al. [9][10] for fcc crystals, but is adapted for bcc crystals at an intermediate temperature. This requires consideration of the critical shear stress on a system (s) as made of two components:

$$\tau_c^{(s)} = \tau^{*(s)}(\dot{\epsilon}, T) + \tau_\mu^{(s)}(\rho^{(1)}, \ldots, \rho^{(n)}) \tag{1}$$

where the first component is related to lattice friction, and is in our case taken as constant and the second is an athermal term (similar to the one considered for fcc metals), linked to dislocation interactions and expressed in terms of dislocation densities:

$$\tau_\mu^{(s)} = \mu b \left(\sum_{u=1}^{u=n} a^{(su)} \rho^{(u)} \right)^{1/2} \tag{2}$$

where μ is the shear modulus, b the amplitude of the Burgers vector and the $a^{(su)}$ are interaction coefficients between slip systems s and u [11][12]. It remains to express an evolution rule for dislocation densities in terms of glide magnitude [13][14]:

$$\frac{d\rho^{(s)}}{d\gamma^{(s)}} = \frac{1}{b\Lambda^{(s)}} - f(\dot{\epsilon}, T)\rho^{(s)}, \tag{3}$$

where $\Lambda^{(s)}$ is the mean free path of dislocations for the glide system s and $f(\dot{\epsilon}, T)$ characterizes the restoration process. Among other things, $\Lambda^{(s)}$ depends on the grain size and in our model it will be considered constant.

SIMULATIONS

The sample

One has designed a sample in the shape of a tensile specimen, with three layers of parallelepipedic grains. Each grain is elongated in the tensile direction. Each layer is translated with respect to its neighbor, so that the corner points are not aligned along the z axis and as few grains as possible meet at a corner point. This morphology is chosen in order to represent, albeit crudely, a cold-rolled mild steel sheet subjected to a uniaxial tension test along its previous rolling direction. To push the comparison further, one has selected initial orientations from a set of measured preferred orientations of a cold-rolled mild steel sample. The exact geometry and grain shapes are shown on figure 1, each grain is divided into 6 elements, except for some grains on the side of the intermediate layer which have only 3 elements.

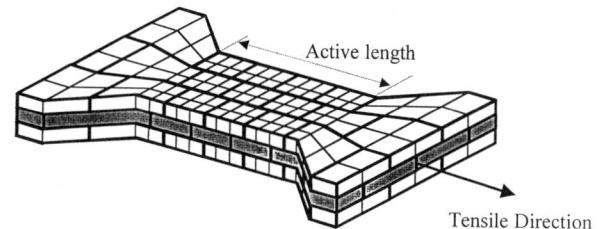

Figure 1: Sketch of the model aggregate, showing the grains and elements. The distribution of orientations is modified in the intermediate layer only (shaded grey).

The active length and width of the sample are respectively 12mm and 6mm and its thickness 3mm. One applies boundary conditions that are idealized conditions, aiming at simulating a tensile test at constant cross-head velocity, with a clamped sample (i.e. no rotations are allowed at the grips). For highly anisotropic materials, this induces shear strains and stresses in the sample.

Several simulations have been performed, where the only change is a different distribution of initial orientations among the grains belonging to the active length of the middle layer. In other terms, the initial morphological and crystallographic textures of the samples have been kept identical for each test.

The parameters of the model

One must distinguish three classes of input parameters.
• The first one describes the test and sample : temperature (T=295K), cross-head velocity ($v = 10^{-3}s^{-1}$), sample dimensions, grain size ;
• The second one consists of material constants for α iron: Burgers vector (b=2.48 10^{-10}m), shear modulus ($\mu = 74.2\ 10^3$MPa), viscosity exponent ($n = 50$), reference glide velocity ($\dot{\gamma}_0 = 10^{-3}s^{-1}$), mean free path ($\Lambda^{(s)} = 8.37\ 10^{-6}$m);
• Lastly, one gives the set of parameters identified on single crystal tests from the literature [4], which provide the best fit for the hardening description that we have chosen to use: the initial density of dislocations (the same on all the systems), $\rho_0^{(s)} = 4.16\ 10^6$m/m^2; the effective shear stresses for each family of slip system (they are assumed equal in a first approximation), $\tau^* = 13.5$ MPa; the interaction coefficients $a^{(su)}$ (seven values corresponding to the various cases of self-hardening and latent hardening, ranging from $a_0 = 0.07$ for $\{110\} < 111 >$ self-hardening, to $a_2'' = 6.3a_0$ for latent hardening between two non-collinear systems $\{112\} < 111 >$ gliding in antitwinning direction.

The computations

The main characteristics of the computations are their use of large transformations, with an updated Lagrangian scheme. The configuration and internal variables are updated explicitly, except for the slip rates which are updated with a forward gradient procedure. Very small time increments are thus required. This allows for a progressive local lattice re-orientation and a precise account of glide activity during the course of the deformation. The finite element mesh is made of 324 quadratic hexahedral elements (20 nodes per element, and under-integration over 8 Gauss points). The boundary conditions are zero force on the free surfaces, and displacement conditions on the two end surfaces, which simulates a clamped tensile specimen deformed at constant cross-head velocity. At each step, the new configuration is known as well as the strain field. The stresses, dislocation densities, local orientations, glide increments on each slip system are evaluated and updated at each Gauss point.

RESULTS

Owing to space limitations, some supporting curves and figures will not appear here, but shall be found in a future reference by the same authors or in a PhD dissertation [15], furthermore, we shall limit the discussion to only two of the several simulations conducted in the project, chosen as exhibiting the largest differences. The first one (labeled H) corresponds to a random distribution of initial orientations taken from a set of orientations which represents the crystallographic texture of a cold-rolled mild steel sheet. The second one (labeled L) differs only by the spatial arrangement of the same orientations in the active zone of the intermediate layer of grains. More explicitly, 13 grains have been assigned two different permutations of the same set of initial orientations, the 54 other grains remaining unchanged. One emphasizes the relatively small differences between the samples of simulations H and L.

Although the overall responses in simulation H and L, as seen on the curves axial force-axial displacement (or tensile stress-axial displacement) , are essentially the same and very similar to a typical polycrystalline, a local approach reveals greater differences. At a macroscopic axial strain of 0.25, the axial strain field is almost homogeneous for H ($E_{11}^{max} \approx 1.5 E_{11}^{macro}$), both at an intragranular and intergranular level, while it is inhomogeneous for L ($E_{11}^{max} \approx 2.3 E_{11}^{macro}$). The axial strain is concentrated in a narrow band near one end of the sample for simulation L. The tendency to develop a band of localized deformation is present in the two extreme layers of grains, in H the distribution of orientations acts as a moderating factor while in L it is the opposite and the strain concentration extends across the three layers of grains. The total dislocation densities follow the same trend and are larger in L than H, in the band of strain concentration. Although this band is mostly a region of single glide, its magnitude is such that the generation of dislocations on the active system is very large.

The glide activity appears very different for simulation H, where many systems have been activated in the intermediate layer (up to the 5 necessary systems, in many grains) and for simulation L, where the situation of single glide predominates, except for the grains near the free edge of the layer where one finds again up to 5 active systems.

In terms of stresses, the results are less clear-cut. One verifies that the maximum stress reached in a grain is closely related to the mode of plastic deformation. For the same local axial strain, a zone in single glide has a lower axial stress than a zone in double or multiple glide. Large stress gradients appear both for H and L across some grains. The anisotropy of the crystalline structure and the imposed boundary conditions generate a local state of stress which is not uniaxial, some regions of a grain may thus be in compression rather than tension. The grain boundaries appear as regions of local stress concentrations. It is not possible , however, given the small number of grains considered in these samples to clearly correlate a stress level with a particular orientation.

The local misorientation or reorientation of the grains are different in each simulation. The inverse pole figures for the seven internal grains of the active zone in the intermediate layer (figure 2) show a scatter of orientations for simulation H, whereas for simulation L, a more acute texture appears with convergence towards some preferred orientations, One notices that within each grain a fairly large misorientation takes place, which will contribute to spread the crystallographic texture. These inverse pole figures imply that the deformation textures for H and L are going to present some differences which could not have been predicted by a model such as those now currently used for deformation texture calculation, the Taylor models or the self consistent models.

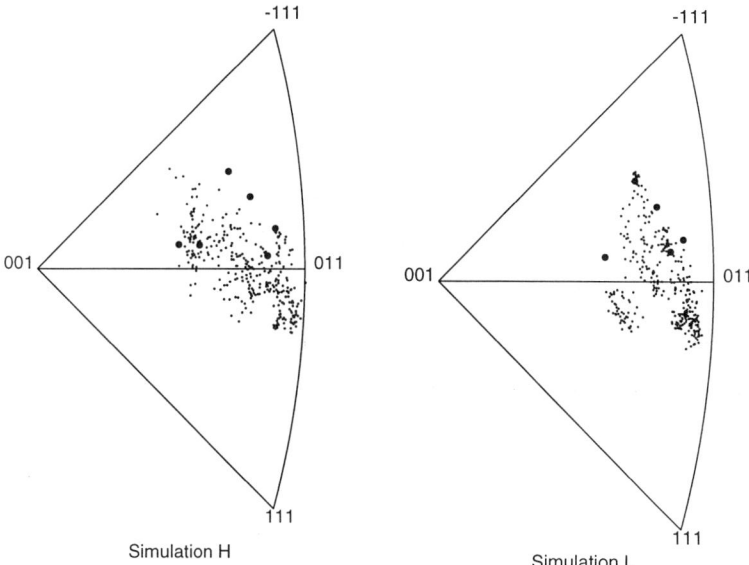

Figure 2: Inverse pole figures showing the position of the tensile axis, for the internal grains of the intermediate layer of the deformed samples, at an average strain of 0.25 (the large dots represent the initial position of the axis for each grain, take note that initial orientations may be the same for 2 grains). Each small dot corresponds to an integration point

CONCLUSION

A slight difference in the local distribution of orientations may produce large macroscopic effects, more particularly in terms of the homogeneity of the plastic response of the polycrystalline aggregate. One may argue that with polycrystals made of an almost infinite number of grains, the statistical averaging is enough to cancel out this effect. However, if one considers thin layers or if one has reason to believe that the orientations are not randomly distributed, but that there exists spatial correlations, the effect may well be significant.

The finite element code that we have developed and applied to bcc aggregates is a useful tool to understand and predict the development of local heterogeneities and to follow the evolution of local internal variables such as dislocation densities. Its three dimensional nature gives insight into the behavior of crystallites completely surrounded by neighbors, whereas most experiments and observations are limited to surface or near surface grains. The very large amount of collected data requires the use of adequate tools for quantitative comparisons, and a fairly good idea of the level at which the structure should be analyzed, from an individual Gauss point, to an element or a grain.

Such a method does not aim at replacing the polycrystalline models such as the self-consistent schemes, but at improving the analysis of local interactions and it may thus help to redefine the representative element at the microscopic scale in instances where one sees that a crystallite cannot be considered alone but rather with its neighbors.

REFERENCES

[1] C. Rey, P. Mussot, and A. Zaoui. pages 867–874. Trans. Japan Inst. of Metals, 1986.

[2] F. Delaire, J.R. Raphanel, and C. Rey. *Acta Materialia, to be published*, 2000.

[3] R. Becker and S. Panchanadeeswaran. *Acta Metall. et Mat.*, 43 No7:2701–2719, 1995.

[4] W. Spitzig and A. S. Keh. *Acta Metallurgica*, 18:611–622, 1970.

[5] J. L. Raphanel, C. Teodosiu, and L. Tabourot. In C. Teodosiu, J.L. Raphanel, and F. Sidoroff, editors, *Large Plastic Deformations : Fundamental Aspects and Applications to Metal Forming*, pages 153–168. A.A. Balkema, 1993.

[6] G. I. Taylor. *Proceedings of the Royal Society of London*, A118:1–, 1928.

[7] J. W. Christian. *Metallurgical Transactions*, 14A:1237–1256, 1983.

[8] H. Mecking and U.F. Kocks. *Acta Metallurgica*, 29:1865–1875, 1981.

[9] L. Tabourot, M. Fivel, and E. Rauch. *Materials Sci. and Eng.*, A234-236:639–642, 1997.

[10] L. Tabourot, M. Fivel, and E. Rauch. pages 511–516. RisøNational Laboratory, Roskilde, Denmark, 1998.

[11] P. Franciosi. *Acta Metallurgica*, 31:1331–1342, 1983.

[12] P. Franciosi. *Acta Metallurgica*, 33:1601–1612, 1985.

[13] E. Kröner and C. Teodosiu. In A. Sawczuk, editor, *Problems of plasticity*, pages 49–82, Warsaw, 1972. Noordhoff Insternational Publishing, Leyden.

[14] C. Teodosiu. volume 2, pages 837–876. NBS Spec. Pub. 317, 1970.

[15] D. Arizmendi. PhD thesis, Université Paris Nord, 93430 Villetaneuse, FRANCE, 1999.

THE EFFECT OF MATERIAL INHOMOGENEITY ON
SERRATED PLASTIC FLOWS

X. LI and W. TONG
Department of Mechanical Engineering, Yale University, New Haven, CT 06520-8284
wei.tong@yale.edu

ABSTRACT

Serrated plastic flows or propagating Portevin-Le Chatelier (PLC) bands in uniaxial tensile aluminum alloy sheet samples were analyzed by a 3D finite element model. A local dynamic strain aging viscoplasticity model motivated by dislocation dynamics was implemented in an ABAQUS user material subroutine. When the sheet metal was assumed to be spatially homogeneous, individual plastic deformation bands inclined about 50° with regard to the axial tensile loading were predicted by the model to propagate along the thin sheet sample, similar to the experimentally observed behavior of the solid solution strengthened alloy AA5182-O. When the sheet metal was assumed to be inhomogeneous by making a few material elements slightly harder or softer, the model predicted a propagating single neck formed by double shear bands, similar to the behavior of the precipitation strengthened alloy AA6111-T4.

INTRODUCTION

Plastic deformation of metals and alloys can be unstable under certain thermal-mechanical loading conditions. The so-called Portevin-Le Chatelier (PLC) plastic deformation bands, often observed in aluminum alloys at ambient temperature, are one of macroscopically manifested unstable or serrated plastic flows. Understanding of the origin of PLC bands and the effects of unstable plastic flows on surface finish and formability of aluminum sheet metals is of both scientific and industrial significance. Recently, a robust and accurate plastic strain mapping technique was developed and applied to study the temporal-spatial characteristics of the PLC bands in aluminum alloys [1]. By directly monitoring the plastic strain rate distribution during the course of uniaxial tension of two aluminum alloys AA5182-O and AA6111-T4, the detailed behavior of PLC bands could be identified. The band-like deformation regions in AA5182-O were found to be inclined to the loading axis with an angle about 50 degrees and multiple bands appeared across the entire gage section (see Fig. 1a). On the other hand, the localized deformation regions in AA6111-T4 developed eventually into a single band or neck and the macroscopic orientation of the band is almost perpendicular to the loading axis (see Fig. 1b). It was speculated that these two distinctive dynamic behaviors of local plastic deformation in these two alloys may be related to the two different strengthening mechanisms, solid solution (AA5182-O) and precipitation strengthening (AA6111-T4) mechanisms [1].

In this study, a numerical simulation of the plastic deformation of aluminum alloys is given with an aim to clarify the mechanisms leading to the two distinctively different PLC bands. A local dynamic strain aging viscoplasticity model will first be described. The motivation of the model based on dislocation dynamics will be outlined. The procedure in implementing the model in a user-defined material subroutine and in analyzing uniaxial tension of thin sheet samples in the nonlinear finite element programs ABAQUS will be given briefly. Results of the numerical simulations will be presented and conclusions will be offered based on the study. Additional research efforts towards further understanding of propagating plastic deformation bands in aluminum alloys will be discussed.

33

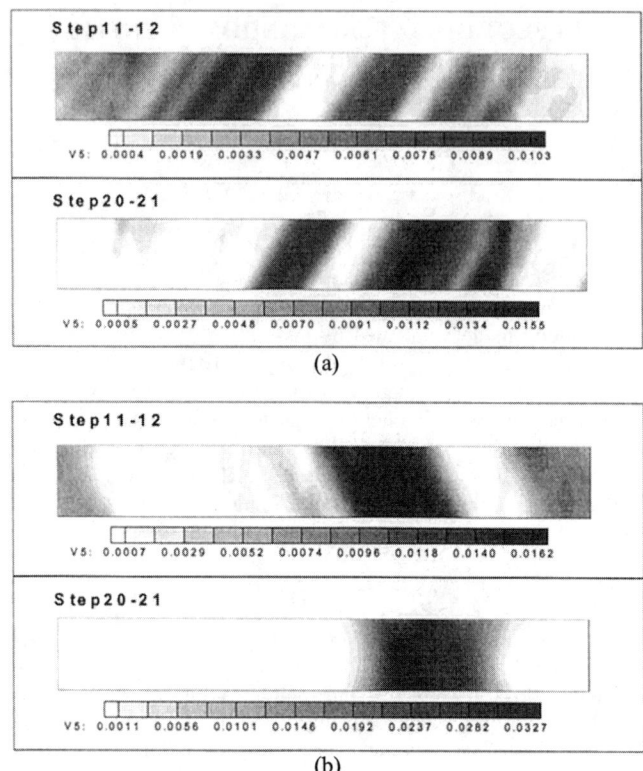

Fig.1 Experimentally measured plastic strain increment maps of two aluminum alloys under uniaxial tension loading: (a) AA5182-O; (b) AA6111-T4.

MATERIAL MODEL

Production of dislocations and their motion and interaction are the predominant physical mechanism of the plastic deformation of metals and alloys. For polycrystalline materials, the flow stress σ_f and the plastic strain rate $\dot{\varepsilon}_p$ are related to the immobile (forest) and mobile dislocation densities ρ_f and ρ_m respectively

$$\sigma_f = \alpha\mu\sqrt{\rho_f}, \quad \dot{\varepsilon}_p = b\rho_m\bar{v}, \tag{1}$$

where b is the Burgers vector of the crystal, μ is the shear modulus of the material, and $\bar{v}$ is the average velocity of the mobile dislocations. Description of the evolution of both immobile and mobile dislocation densities as well as the dependence of the average velocity of the mobile dislocations on the applied stress and the dislocation densities is needed to complete the model. For simplicity, a local evolution equation may be prescribed for each type of dislocations, namely,

$$\dot{\rho}_f = \dot{\rho}_w + \dot{\rho}_p - \dot{\rho}_a - \dot{\rho}_c - \dot{\rho}_r \quad \text{and} \quad \dot{\rho}_m = \dot{\rho}_n + \dot{\rho}_c + \dot{\rho}_r - \dot{\rho}_p - \dot{\rho}_w, \tag{2}$$

where $\dot{\rho}_w$ is the production rate of the dislocation cells, $\dot{\rho}_p$ the pinning rate of mobile dislocations by impurity or solute atoms, $\dot{\rho}_a$ is the annihilation rate of forest dislocations, $\dot{\rho}_c$ the rate of forest dislocations climbing over cell walls, $\dot{\rho}_r$ the release rate of the pinned dislocations, and $\dot{\rho}_n$ the generation of new mobile dislocations (due to Frank-Read source and others).

The model given by Eqs.(1)-(2) can be simplified greatly by assuming that the immobile dislocation density can be related to the equivalent inelastic strain and the mobile dislocation density can be related to the inelastic strain and an aging time t_a which simulate the interaction between the mobile dislocations and the solute atoms in aluminum alloys [2]. The material is assumed to be elastic-viscoplastic. The elasticity is in a classical hypoelastic formulation and the plasticity model consists of two components: a standard rate-dependent plastic model and a dynamic strain aging component. The plasticity model can be expressed as

$$\dot{\varepsilon}_p = \dot{\varepsilon}_0 \exp(\frac{\sigma - \sigma_f}{S} - P_1 C_s') \tag{3}$$

or

$$\sigma = \sigma_f + S\ln(\frac{\dot{\varepsilon}_p}{\dot{\varepsilon}_0}) + SP_1 C_s' \tag{4}$$

where σ is the equivalent von Mises stress, ε_p is the equivalent plastic strain, the dot denotes the rate of change with respect to time t, $\dot{\varepsilon}_0$ is the reference strain rate, and σ_f is the flow stress of the material at the reference strain rate, which has the form of

$$\sigma_f = C_{f1} + C_{f2}[1 - \exp(-\frac{\varepsilon_p}{C_{f3}})] \tag{5}$$

The strain rate sensitivity S is a function of the equivalent plastic strain,

$$S = C_{s1} + C_{s2}\sqrt{\varepsilon_p} \tag{6}$$

The value of C_s' is determined by the aging time t_a and the equivalent plastic strain

$$C_s' = 1 - \exp(-\frac{P_2 \varepsilon_p^{\alpha} t_a^{n}}{k_2^{n}}) \tag{7}$$

The aging time is related to the waiting time t_w by a differential equation

$$\dot{t}_a = 1 - \frac{t_a}{t_w} \tag{8}$$

The waiting time is a function of both the equivalent plastic strain and equivalent plastic strain rate.

$$t_w = \frac{C_{w1} + C_{w2}\varepsilon_p^{\beta}}{\dot{\varepsilon}_p} \tag{9}$$

The equations (3)-(9) together complete a dynamic strain aging material model. The parameter C_{f1}, C_{f2}, C_{f3}, C_{s1}, C_{s2}, C_{w1}, C_{w2}, P_1, n, k_2, α, β are the model constants which can be determined empirically [2].

FINITE ELEMENT SIMULATION

The material model was implemented in a user material routine in the nonlinear finite element program ABAQUS [3]. Since the model will simulate the serration of the load-force curve, attention is paid to the integration scheme used in the material routine. In our study, the implicit integration of combined bisection and Newton-Raphson methods was used for solving both equivalent plastic strain rate and the aging time [4]. J-2 flow rule is also assumed in this study, so that the equivalent stress is equal to the material flow stress. The 3-D 8-node incompatible solid element was used in this study. The mesh number was 30x5x2, representing a gage section of 20mm x 4.6mm x 1mm. With the origin of the coordinate system fixed at the lower-left corner of the mesh, the boundary conditions used in this study can be written as following,

$$u(0,y) = 0; u(L,y) = d$$
$$v(0,W) = v(L,W) = 0$$
$$w(0,y) = w(L,y) = 0$$

where u, v, w are the displacement in x-, y- and z-direction, respectively, L is the length of the specimen, W denotes the width of the specimen. The constants in the material model are chosen as in Table 1.

To study the effect of the randomly distributed inhomogeneity on the propagating PLC strain band patterns, simulations have been carried out using ABAQUS for both spatially homogeneous and inhomogeneous materials. In order to initiate the PLC bands in the homogeneous material, a stress disturbance of 10 MPa was added to the lower-left corner of the specimen at the very beginning of the loading process. The inhomogeneous material was modeled by assigning different values of C_{fl} at some randomly picked elements. The values vary between 145 MPa to 155 MPa.

Parameter	Value	Parameter	Value
P_1	18.0	n	1/3
P_2	0.62	C_{f1}	150.0
α	0.44	C_{f2}	205.0
β	0.68	C_{f3}	0.15
k_2	2.16e-3	C_{s1}	0.41
C_{w1}	7.2e-5	C_{s2}	2.912
C_{w2}	4.32e-3	$\dot{\varepsilon}_0$	2.2955E-7

Table 1: List of the material model constants used in the simulations

RESULTS

The global engineering stress-strain curves of rectangular thin sheet samples from the two simulations are plotted in Fig. 2(a) and Fig. 3(a). The curves show a smooth part at the early loading stage, even though an initial stress disturbance was introduced in both cases. At the strain level around 8%, both curves start to show serrations, indicating the PLC bands were initiated. Fig. 2(b) and Fig. 3(b) shows the contour plots of the equivalent plastic strain rate at selected loading stages, indicating the dynamic nature of the PLC band patterns. For simulation without the material inhomogeneity, the PLC bands were found always to incline from the axial direction. At the earlier deformation stage shortly after the onset of the PLC bands, two inclined bands symmetrically oriented were seen to move away from each other. In the later times, only one inclined band is dominant, moving along the specimen (Fig. 2b). In contrast, a double-band formed neck was observed throughout the entire loading process when the material inhomogeneity is introduced in the finite element analysis (Fig. 3b).

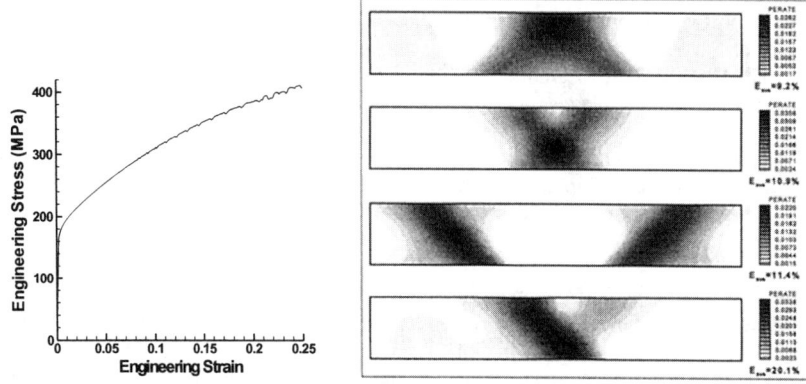

Fig. 2 Stress-strain curve (a) and plastic strain rate distribution (b) of a homogeneous material.

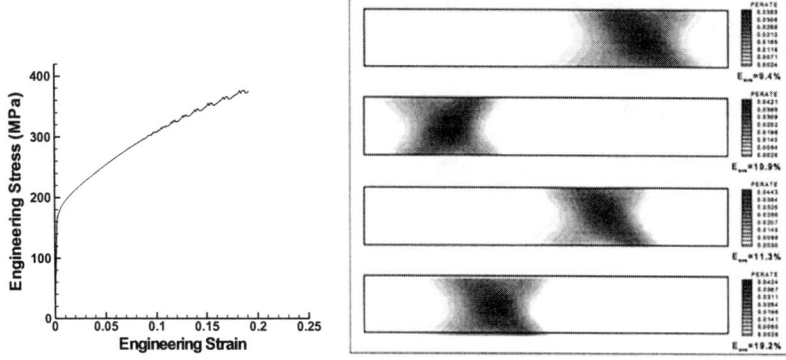

Fig. 3 Stress-strain curve (a) and plastic strain rate distribution (b) of an inhomogeneous material.

CONCLUSIONS

The local dynamic strain aging viscoplasticity model implemented in the 3D finite element analysis was able to produce qualitatively similar behaviors of PLC bands in aluminum alloys. The spatial coupling of the PLC band strain patterns may be the result of interaction of the local material instability and the global structural constraint. From the second 3D finite element simulation case, it was found that the randomly distributed material inhomogeneity initiates a double-band neck moving along a uniaxial tensile sheet specimen. The detailed spatial-temporal characteristics of the PLC bands may thus be affected by the spatial heterogeneity of the material.

The microstructural differences between AA5182-O and AA6111-T4 may provide the rationale behind the two 3D finite element simulations that explain the two different behaviors of the PLC bands in the two materials. AA6111-T4 alloy is a precipitate strengthened material, and the precipitates have in general a much larger size than the strengthening "particles", the solute atoms, as in the AA5182-O aluminum alloy. This will make the AA6111-T4 alloy much more heterogeneous and its plastic deformation much more inhomogeneous at the length scale comparable to the size of precipitates. Unlike the aluminum alloy AA6111-T4, the alloy AA5182-O is strengthened mainly by the solute Mg atoms. The solute atoms are small in comparison with the precipitates in AA6111-T4 and are much more uniformly distributed throughout the material via solid diffusion process.

The spatial-temporal characteristics of the PLC band strain patterns can be affected by many factors, such as the specimen geometry, the force and displacement boundary conditions and the microstructure of materials. The 3D finite element analysis presented in this study provides for the first time one of possible explanations for a propagating neck formed by double shear bands in AA6111-T4 tensile specimens. Without turning to any nonlocal plasticity models, the spatial coupling of the PLC bands is interpreted in terms of local material instabilities due to dynamic strain aging, global structural constraints, some mesoscale material inhomogeneities (consequently, a naturally introduced length scale). Micromechanics experiments will be needed to confirm the heterogeneous nature of the plastic deformation at the mesoscale and even smaller scales in AA5182-O and AA6111-T4 and a more physically based micromechanics model of dynamic strain aging should be formulated for aluminum alloys.

ACKNOWLEDGMENTS

The research reported here has been supported by NSF (Drs. K. Chong and Davies), Lawrence Livermore National Labs (Dr. W.E. King) and Alcoa Technical Center (Drs. O Richmond, H. Weiland, L.G. Hector, and M. Li). Their financial and technical assistance is gratefully acknowledged.

REFERENCES

1. X. Li and W. Tong in *Multiscale Modeling of Materials* (Mater. Res. Soc. Proc. **538**, Pittsburgh, PA, 1999), p.179-184.

2. P. G. McCormick and C. P. Ling, *Acta Metall. Mater.* Vol. 43, No. 5, pp.1969-1977 (1995).

3. *ABAQUS/Standard User's Manual*, version 5.8, Hibbitt, Karlsson & Sorensen, Inc. (1997).

4. R. Becker, *The Integration of Material, Process and Product Design*, Zabaras et al. (eds), pp. 77-84 (1999).

A MODEL FOR CALCULATING SUBSTRATE CURVATURE DURING COALESCENCE OF PT ISLANDS ON AN AMORPHOUS SUBSTRATE

M. A. PHILLIPS, V. RAMASWAMY, B. M. CLEMENS and W. D. NIX
Department of Materials Science & Engineering, Stanford University, Stanford CA 94305

ABSTRACT

Previous work using *in-situ* curvature measurement has shown a correlation between stress and microstructure during the early stages of thin film growth. The model presented here can be used to predict the curvature change of the substrate during part of this growth process. Curvature, and thus film stress, is measured *in-situ* during growth of sputter-deposited Pt on amorphous substrates. The average film stress is observed to be slightly compressive initially, followed by a change towards a tensile maximum, after which the stress becomes compressive again. Plan view TEM micrographs of Pt films of thicknesses up to 35 Å show the evolution of microstructure from isolated islands to a coalesced film. This evidence suggests that the tensile regime is due to island coalescence. The model calculates the curvature induced in a substrate during the tensile excursion associated with island coalescence, where discontinuous islands are modeled as a series of cracks in an otherwise continuous film. Quantitative measurements of island size and areal fraction covered are extracted from the TEM micrographs and used to predict the curvature during coalescence. The predicted stresses are shown to compare favorably with the measured stresses.

INTRODUCTION

Understanding the relationship between film stress and the corresponding microstructure at various stages of growth allows accurate prediction and control of film microstructure. Film stress, obtained from substrate curvature is measured during growth of sputter-deposited Pt on amorphous substrates, is observed to be slightly compressive at thicknesses less than 8 Å. This is followed by a change to tension in thicker films, leading to a tensile maximum at about 35 Å, after which the stress becomes compressive again. Such stress behavior is commonly observed during the deposition of high mobility metal films by evaporation [1].

As discussed by Nix and Clemens [2], the development of tensile stresses in thin films has been associated with the coalescence of crystallites during film deposition, and the tensile stress maximum then marks the completion of film coalescence [3]. It can be shown that when surfaces of the neighboring isolated crystallites come into close proximity the crystallites spontaneously snap together producing both grain boundaries and elastic strains in the film. While the model for crystallite coalescence does provide a rationale for the maximum tensile stresses that can be created in a film, it does not describe the gradual change in curvature as the crystallites coalesce to form islands and as the islands grow to cover the surface of the substrate. In this paper we present a simple quantitative model for tensile stress that develops due to crystallite coalescence, taking account of island size and areal coverage of the substrate. We use a model developed by Hutchinson and co-workers [4] for describing curvature changes associated with cracking in residually stressed films to determine the curvature when the film is discontinuous.

EXPERIMENT

During the *in-situ* stress measurement, Pt layers of varying thickness were deposited onto 110 μm thick glass coverslips. The samples were deposited in a UHV chamber, at room

Mat. Res. Soc. Symp. Proc. Vol. 578 © 2000 Materials Research Society

temperature and at an Ar pressure of 3 mtorr. The Pt deposition rate, measured by an oscillating quartz crystal rate monitor, was about 1 Å/second. Stress behavior during growth was obtained from *in-situ* substrate curvature measurement, using a multiple parallel laser beam technique [5]. The frequency of data acquisition permitted observation of curvature change with sub-monolayer sensitivity.

TEM samples for *ex-situ* investigation were prepared by depositing Pt on coated TEM grids (grid = 200 mesh). The coating is a 40 nm thick SiO_2 film providing a growth surface similar to glass slides, and completely eliminates TEM sample preparation. A Philips CM20 (200 keV FEG gun) and JEOL 4000 (400 keV) were used to obtain the TEM images shown in this paper.

RESULTS

Figure 1 shows curvature, K, versus nominal thickness, $\bar{t}_f$ for Pt deposited on the glass coverslips. The nominal thickness is a measure of the amount of material that has been deposited. As the material does not deposit uniformly, but forms discrete islands in the early stages, the average height of the islands is greater than the nominal thickness. The average island height, t_f, is given by,

$$t_f = \frac{\bar{t}_f}{A_A}$$

where A_A is the areal fraction of substrate covered by Pt.

The change in curvature during growth of Pt on glass is compressive at thicknesses less than 8 Å. This is followed by a change towards a tensile maximum at 35 Å, after which the stress becomes compressive again. These regions have been respectively associated with island nucleation and growth, coalescence, and peening. In the present paper we focus on the tensile regime observed during coalescence.

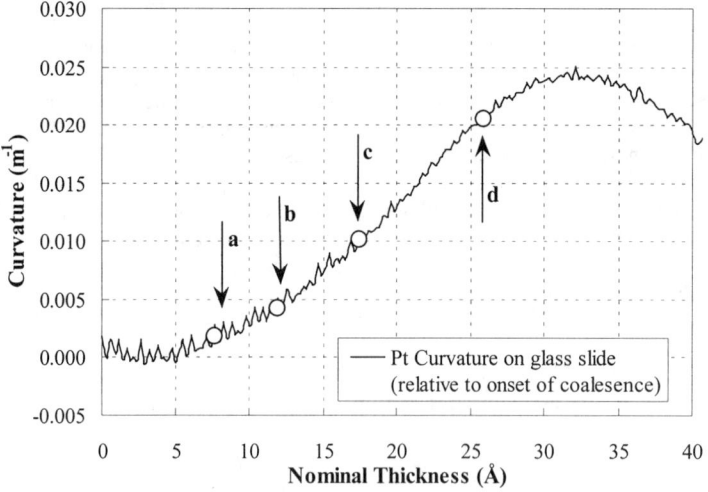

Figure 1: *In-situ* curvature versus nominal thickness for Pt deposited on SiO_2 substrate. Points labeled a,b,c,d represent thicknesses where TEM samples were collected and respectively correspond to images a,b,c,d in Figure 2.

The points labeled **a**, **b**, **c** and **d** on Figure 1 indicate the nominal thicknesses at which the TEM images in Figure 2 where collected. In all the TEM images the SiO_2 substrate appears as a light background, and locations where Pt has been deposited appear darker, due to electron density contrast effects between Pt and the lighter Si and O in the electron beam.

Figure 2(a) shows a sample with largely isolated Pt islands on the SiO_2 substrate, corresponding to the thickness in Figure 1 where the tensile behavior begins. At greater thicknesses, the Pt islands are much larger and coalescence has begun (Figure 2(b) & (c)). Note that these images correspond to points in Figure 1 well into the tensile regime. Figure 2(d) shows a TEM image for a thicker sample, where coalescence is approaching completion, but a number of channels remain between the islands.

Dark Field TEM images and High Resolution TEM images have revealed that the islands are polycrystalline and consist of a number of smaller crystallites which have coalesced together. Figure 2(e) shows a Dark Field Image of the 18 Å Pt film. Note that the bright crystals are much smaller than the size of the islands in Figure 2(c). Figure 2(f) shows a TEM image for a 12 Å Pt film on SiO_2. It is also clear from this image that the larger islands consist of smaller crystallites of different orientations.

MODEL

In the tensile regime, when the film is not yet continuous, the curvature change will not have reached the maximum value, both because the islands are all of finite size and because the substrate is not yet completely covered by the film. Thus two effects must be taken into account to model the gradual change in curvature during coalescence. These effects are easiest to understand if we think first of a continuous film of thickness t_f with tensile stress σ_f producing a substrate curvature K_o. Making the film discontinuous with islands of size s can be modeled by introducing cracks in the films with spacing s, as shown in Figure 3. The presence of cracks, which act like the edges of islands, causes curvature to relax from K_o to K_1. Using the Stoney equation;

$$K_o = \frac{1}{\bar{E}_s} \frac{6 t_f \sigma}{t_s^2}$$

where $\bar{E}_s$ is the biaxial elastic modulus of the substrate, σ is the stress in a fully coalesced film, t_f and t_s are the film and substrate thicknesses, respectively.

The change in curvature due to the introduction of cracks in the film is given by,

$$\Delta K = K_1 - K_o, \tag{1}$$

where ΔK is negative. Based on the work of Hutchinson and Xia [4], Hutchinson has developed a formula for the change in substrate curvature that occurs when cracks are introduced into a residually stressed film. For plane strain, the relative change in curvature is;

$$\frac{\Delta K}{K_o} = -\left(\frac{2l}{s}\right)\tanh\left(\frac{s}{2l}\right), \tag{2}$$

where l is a characteristic length proportional to the film thickness, t_f,

$$l = \frac{\pi}{2} g(\alpha, \beta) t_f$$

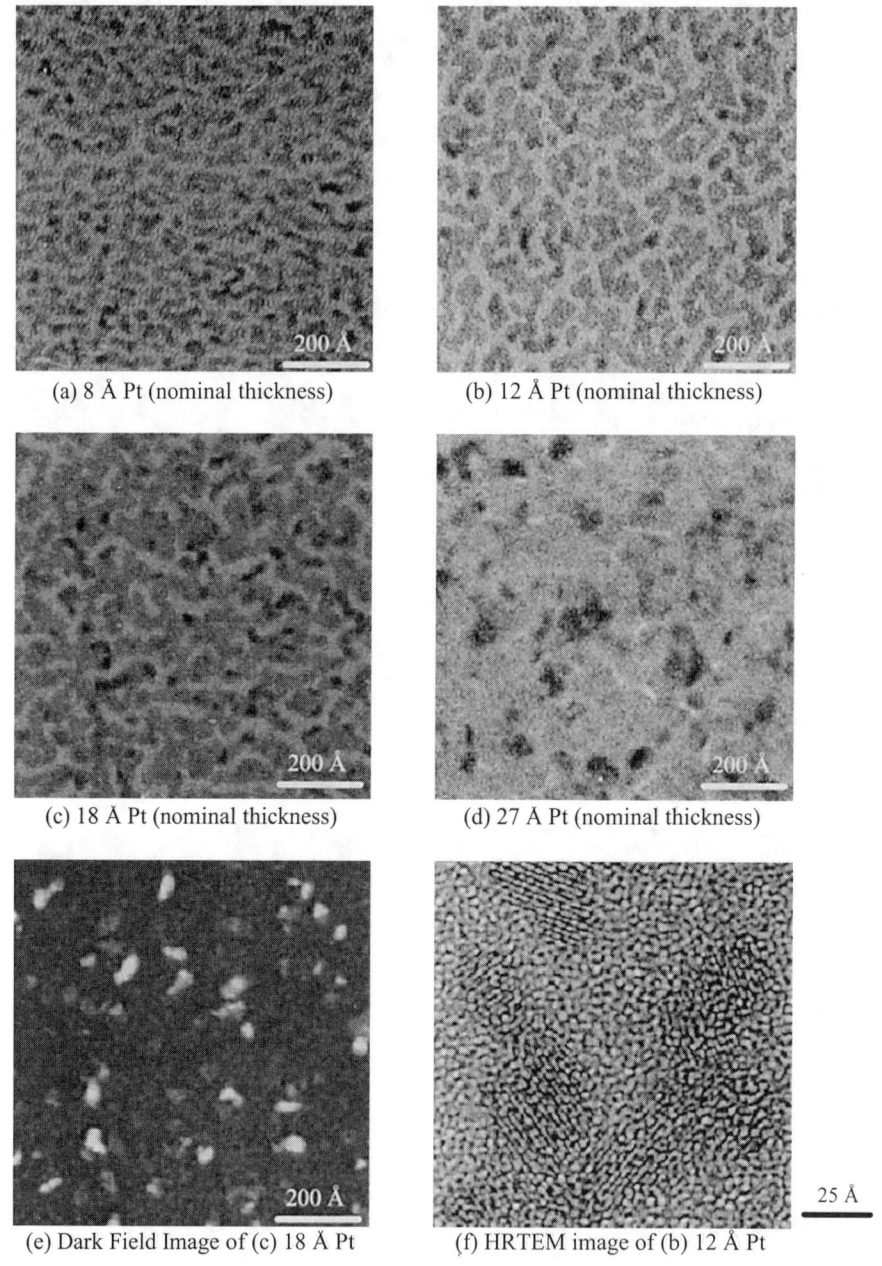

Figure 2: TEM images of Pt deposited on SiO₂ substrates, showing; (a) discrete islands, (b) & (c) partial coalescence and (d) almost complete coalescence; (e) Dark Field and (f) HRTEM, show additional information collected from the TEM

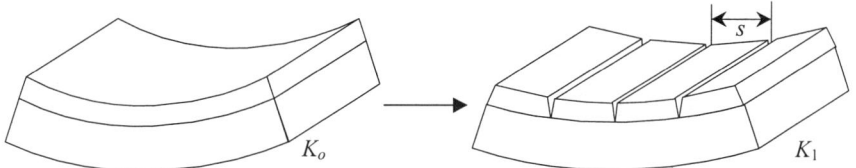

Figure 3: Illustration of reduced curvature by introduction of cracks

Here $g(\alpha,\beta)$ is a function of the Dunders parameters (elastic mismatch between film and substrate), and are given in table form by Beuth [6]. Using material properties for SiO_2 and Pt, $\alpha = 0.43$, $\beta = 0.2$ and $g(\alpha,\beta) = 1.89$.

Less than 100% of the substrate is covered in the early stages of film growth, so the curvature of the substrate is expected to be less than that for a continuous film. For the case of plane strain the contribution to curvature from a partially covered substrate is simply proportional to A_A. The curvature induced in the substrate by coalesced islands occupying a limited areal fraction is given by;

$$K = A_A K_1.$$ (3)

Combining eqns (1), (2) and (3) we obtain a final expression for the change in curvature relative to the maximum curvature:

$$\frac{K}{K_o} = A_A \left[1 - \left(\frac{2l}{s} \right) \tanh \left(\frac{s}{2l} \right) \right]$$ (4)

Comparison with Experiment

To compare the model predictions with experiment, we determine the following parameters; areal coverage, A_A and mean "crack spacing," s. Areal fraction is obtained from the TEM images using point counting methods. The crack spacing in the model is estimated as the *reciprocal of one half of island edge length per unit area*. A_A and s are plotted in Figure 4.

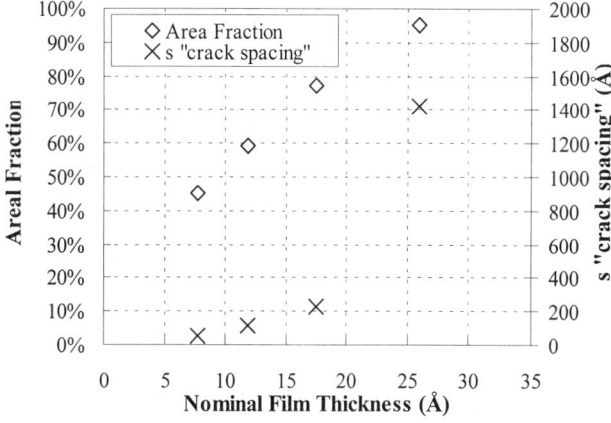

Figure 4: Areal fraction covered by Pt and island "crack spacing" vs nominal thickness

Values for $\bar{t}_f$, s, and A_A were used in equation (4) along with material property parameters. The modeled result and curvature data (normalized to the tensile maximum) are plotted in Figure 5, which shows that the model does closely follow the behavior of the data during the tensile excursion.

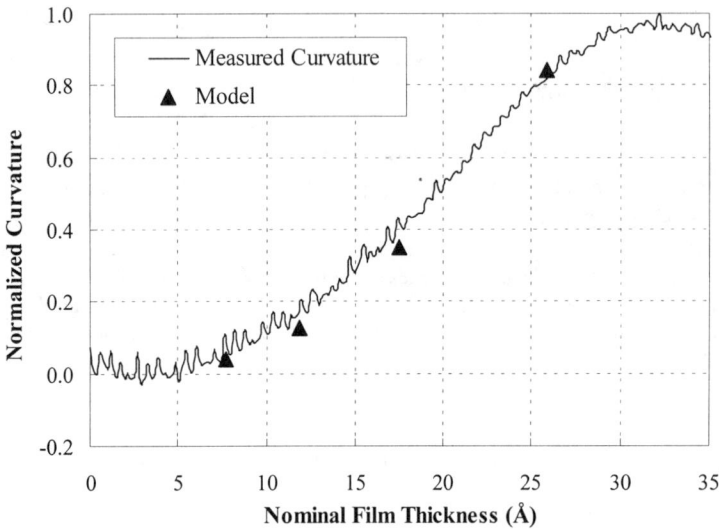

Figure 5:   Measured and modeled curvature versus nominal thickness of Pt deposited on SiO_2

CONCLUSIONS

We have developed a model to explain the change in curvature in the tensile direction observed *in-situ* during sputter deposition of Pt films on amorphous SiO_2 substrates. The curvature change is a result of crystallite coalescence. Transmission Electron Microscopy was used to measure areal fraction and island size, and with these parameters we are able to predict the magnitude of the curvature developed in the substrate during the island coalescence.

ACKNOWLEDGEMENTS

The authors gratefully acknowledge funding from National Science Foundation through Grant No. DMR-9408552, and the assistance of Gerhard Dehm at the Max-Planck-Institut für Metallforschung, Stuttgart, Germany for assistance with preparation of TEM images.

REFERENCES

1   R. Abermann and R. Koch, Thin Solid Films **129**, 71-8 (1985).
2   W. D. Nix and B. M. Clemens, Journal of Materials Research **14**, 3467-3473 (1999).
3   R. Abermann, R. Kramer, and J. Maser, Thin Solid Films **52**, 215-29 (1978).
4   J. W. Hutchinson, Harvard University, (private communication).
5   V. Ramaswamy, B. M. Clemens, and W. D. Nix, (Mater. Res. Soc. Proc. **528**, San Francisco, CA, 1998), pp. 161-8
6   J. L. Beuth, Jr., International Journal of Solids and Structures **29**, 1657-75 (1992).

THE DETERMINATION OF EVOLVING MICROSTRUCTURE
USING CONSTITUTIVE RELATIONSHIPS

B.J. DIAK, S. SAIMOTO
Department of Materials and Metallurgical Engineering, Nicol Hall, Queen's University, Kingston, Ontario, K7L 3N6, Canada

ABSTRACT

The evolution of the constitutive parameter, the mean slip distance, λ, is monitored during tensile deformation of 3.2 μm grain size aluminum at 200 K. Transmission electron microscopy (TEM) confirms that the grain size, D, sets an upper limit to λ.

INTRODUCTION

An ideal constitutive relation describing plastic deformation should integrally manifest the microplastic processes using microstructural parameters giving rise to the macroscopic phenomenon of the stress-strain, σ-ε, behaviour. However, the complexity of the evolving microstructure, which has been modeled by so-called internal state variable models, relies on experimental parameters for "curve fitting" purposes, and has been criticized for not representing *any physical features of the microstructure* [1]. By extending a kinematic microscopic model of slip [2], Diak et al. [3] have shown that the correlation between the σ-ε behaviour and the microplastic processes can be described by the activation volume, V, mean slip distance, λ, and mean slip velocity, $\dot\lambda$: V is inversely proportional to the flow stress, σ_f, which is the difference between the total stress, σ, and the stress for inception of dislocation activity, σ_o (at 0.02 % offset strain); λ describes the operation of a slip packet from initiation to termination and is inversely proportional to the product of σ_f and the work hardening coefficient, $d\sigma/d\varepsilon$; and $\dot\lambda$ is a power law function of σ_f, which incorporates the mobile dislocation density. Thus the functional relationship, $\sigma = f(\lambda, \dot\lambda, T)$, can be evolved and the parameters λ and $\dot\lambda$ determined from the experimental conditions as extensively discussed [3]. Through this development the parameters calculated from the mechanical data correlate very well to the model except that the magnitude of λ is about one order too large. Thus the model, which inherently assumes that all generated dislocations are stored, has to be relaxed by invoking the dynamic annihilation of dislocations. The primary evidence for such an effect is the well-known phenomenon that the stored work is only about 5 % of the mechanical work expended [2]. Therefore, the real mean slip distance is theoretically tractable from dynamic measurement of the σ-ε behaviour as [3]

$$\bar\lambda = \left(\frac{n}{2A\alpha}\right)\left(\frac{b\mu^2}{2}\frac{d\gamma}{\tau d\tau}\right) = \left(\frac{nM^3}{2A\alpha}\right)\left(\frac{b\mu^2}{2}\frac{d\varepsilon}{\sigma_f d\sigma}\right) = \left(\frac{nM^3}{2A\alpha}\right)\lambda \tag{1}$$

where n = number of successive loops from the same source, A = annihilation factor, α = strength parameter, **b** = Burgers vector, μ = shear modulus, and τ and γ are the shear stress and strain, respectively, calculated from σ_f and ε using the Taylor factor, M.

Tabata et al.'s [4] in situ observation of dislocation glide within cell interiors suggests that $\bar\lambda$ could correlate to the cell size, d_c. Field and Weilland [1] have used TEM to quantitatively assess d_c in commercial purity aluminum with 100 μm grain size, D, compressed to small strains between room temperature and 523 K. They observe a more rapid decrease in mean free path for

45

<111> oriented grains versus <922> ones, with values for the mean free path approaching 0.4 μm at 0.02 strain. In the early stages of cell formation, when the structure is open, it is possible for a dislocation to glide through more than one cell [5]. Argon and Haasen [6] have shown that Stage III work hardening is typified by a decreasing d_c with progressing work hardening, as observed in aluminum [7]. The cell size evolution is dependent on D, with smaller D's characterized by a smaller d_c for a given strain reported for copper [8], and nickel [9], but less conclusive behaviour observed for 70 μm aluminum [7]. More recently, Chu and Morris, Jr. [10] reported the tensile work hardening behaviour of recrystallized (D = 100 μm) and recovered (d_c = 2.2 μm) 99.94 wt.% aluminum at 77 K, but do not relate the measured dσ/dε, to d_c.

One way to illustrate that our proposed model is compatible to the observed evolution of microstructure is to demonstrate that $\bar{\lambda}$ does not exceed D. This requires a polycrystalline material with D of a few micrometers. Figure 1 illustrates the low temperature (a) σ–ε and (b) λ–σ_f behaviour of 3.2 μm aluminum, as reported previously [3]. The behaviour is characterized by a temperature independent portion before the yield point elongation (YPE) followed by a temperature dependent part (Fig. 1(b)). The λ behaviour suggests that during the pre-YPE strain most dislocations are stored, whereas beyond the YPE, λ jumps up by an order of magnitude suggesting that dynamic annihilation has been initiated. Furthermore, the Cottrell-Stokes law [12] is obeyed in both regions [3] indicating that indeed the dislocation-forest interactions govern σ_f.

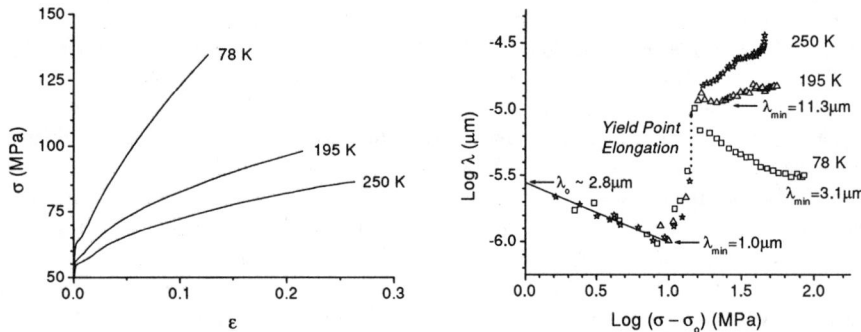

Fig. 1. Low temperature dependence of (a) the stress - strain behaviour for the ultra-fine grain sized aluminum, and (b) the corresponding mean slip distance, λ, versus flow stress representation. The nominal strain rate was 3.2×10^{-5}/s (after Diak et al. [3]).

The objective of this study is to quantify the scale of the grain and cell structure of the recrystallized and deformed material using TEM, and compare to the dynamically measured constitutive parameter, λ. These measurements will help in estimating the overall value of the combined terms in Eqn. (1) relating $\bar{\lambda}$ to λ. This report concentrates on the work hardening behaviour at 195 K, which exhibits a slightly increasing, but relatively constant λ after the YPE.

EXPERIMENTAL PROCEDURE

Material and Microstructure

Details regarding the preparation of the ultra-fine grain size microstructure from the 99.98wt.% aluminum alloy used for tensile testing have been reported earlier [11]. After thermomechanical processing, a reduced gauge section of 2.85 mm diameter and 32 mm length was prepared from the dynamically recrystallized rods by chemically etching 2.94 mm diameter swaged rods. The specimens were then annealed at 230°C for 100 hours to obtain a stable microstructure referred to as Rx/Ann. The grain size remains very uniform up to 275°C [11].

Mechanical Testing

Using the expected work hardening behaviour at 195 K as a guide (Fig. 2), continuous tensile tests were performed on a high resolution, cryostat-type, screw-driven tensometer by immersion in a mixture of ethanol and dry ice. The nominal strain rate was 3.2×10^{-5}/s. To assess the evolving microstructure, different tensile specimens were deformed past 0.02% yield to: (1) a state as close as possible to the minimum value of λ before the macroscopic yield point and unloaded; and (2) a state past the YPE at which λ appears to stabilize, and unloaded.

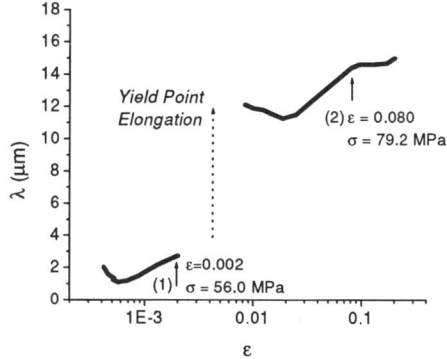

Fig. 2. λ versus ε at 195 K. The arrows indicate the work hardened states (1) and (2) at which the tests were unloaded.

Electron Metallography

TEM was conducted on tensile specimens in the Rx/Ann condition, and after deformation at 200 K to 0.002 and 0.080 plastic strain. TEM specimens were prepared at room temperature from the tensile rods by spark erosion of 0.30 mm thick discs perpendicular to the tensile axis. The discs were ground down to ~ 0.20 mm on emery paper using a hand-held tripod polisher followed by standard jet polishing for aluminum. All imaging was performed using a Philips CM20. Grain, and cell, d_c, sizes were determined by the Heyn lineal intercept technique on images captured from foils oriented perpendicular to the beam, and are reported as the mean lineal intercept without assuming any volume filling shape. Three intercept lines were constructed from the center of the image every 60° and at least 120 intercepts were counted for each specimen. Dislocation structures were imaged by tilting the foil to obtain a low index diffraction condition and represent about 2/3 of the total population. Electron channeling contrast (ECC) images obtained in a JSM 840 scanning electron microscopy (SEM) were used to check the neighboring orientation differences in the TEM foil of the Rx/Ann material.

RESULTS

As-recrystallized and Annealed

An ECC image of the Rx/Ann microstructure from a TEM foil is shown in Fig.3 (the dark region to the right is lost signal due to the hole in the foil). The average mean lineal intercept for this material is 3.2 μm, and x-ray peak scans and pole figure analysis reveal the microstructure to have a strong cube fibre texture. Figure 4(a) shows a bright field TEM image of the microstructure, and (b) the corresponding selected area diffraction pattern from the same region shows clustering of spots along the ring rather than a uniform distribution. This observation corresponds to the observed cube fibre texture.

Micro-Strain Region

After small deformation to 0.002 strain the material exhibits evidence of dislocation activity (debris trails) in the grain interiors. Figure 5 (a) taken from a thicker part of the foil shows a stabilized array. Also observed were regions of tangles and loops both in the grain interior (Fig. 5(b)), and near grain boundaries. The mean lineal intercept length for the cell structure is 1.6 μm.

Macro-Strain Region

The microstructure after 0.080 strain is

Fig. 3. SEM electron channeling contrast image of TEM foil taken from the Rx/Ann cross-section of the tensile rod.

characterized by further subdivision of the grains to about 1.2 μm (Fig. 6 (a)). The most obvious change from the micro-strain region is that the sub-grain boundaries are tangled walls. The sharper boundaries as indicated by the arrows appear to be the original higher angle grain

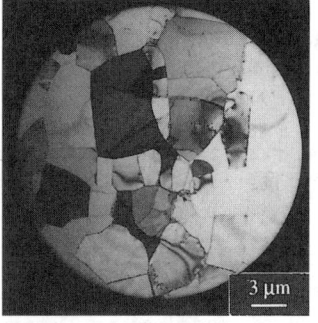

Fig. 4. (a) Selected area bright field image of Rx/Ann material, and (b) the corresponding selected area diffraction pattern.

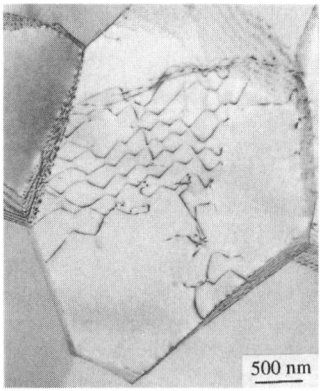

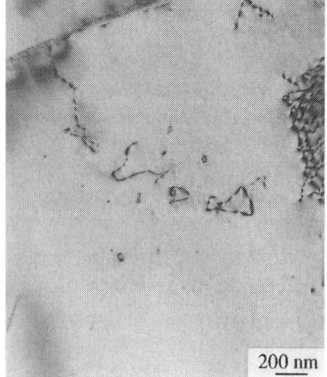

Fig. 5. (a) Brightfield TEM image of stabilized network in grain interior of 0.002 strain material. (b) A different area of the foil reveals dipole and loop remnants in the interior of a grain.

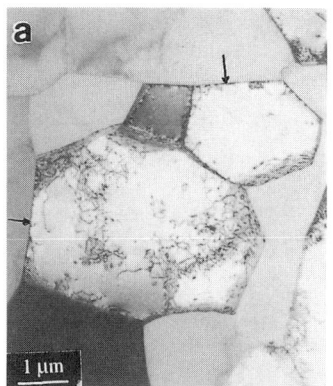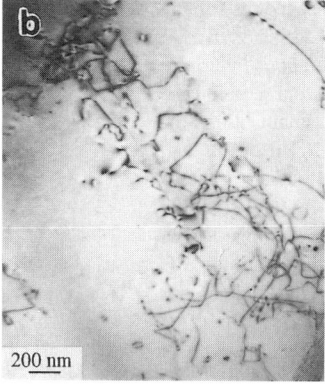

Fig. 6. Sample deformed to 0.080 strain. (a) Tangled wall cell structure subdivides a grain. (b) Higher magnification reveals deformation loops trapped in and around the cell walls.

boundaries. The expected increases in misorientation between cells results in greater image contrast. At higher magnification (Fig. 6(b)) the tangled cell walls are seen to consist of dislocation loops both inside and outside the walls.

DISCUSSION

Table 1: Variables describing the property / microstructure evolution during tensile testing.			
T= 200 K, $\dot{\varepsilon}_o$ = 3.2x10⁻⁵/s, σ_o = 43.3 MPa, b = 2.86 Å, μ = 27.4 GPa			
σ (MPa)	ε (%)	λ (μm)	D or d_c (μm)
43.3	0.0	~2.8 (extrapolated)	D = 3.2
48.9	0.06	1.0 (minimum)	3.2 > d_c >1.6
56.0	0.2	2.8	d_c = 1.6
79.2	8.0	14.4	d_c = 1.2

Table 1 summarizes the macroscopic and microstructural variables of interest tractable from this study. The results show that in the micro-strain region, λ is comparable to d_c, but less than D, and increases by an order of magnitude beyond 0.002 strain. In the micro-strain region, Fig. 5(a) shows evidence of dislocation passage in the grains, but the tangled walls are not visible. This does not mean that dislocation density within the grains did not exist during micro-straining, because an unstabilized array will disappear during the foil thinning process [13]. Thus our contention that λ decreases as the forest dislocation density builds up during micro-strain is not refuted. On the other hand, tangled cell walls are stable during thin foil preparation and subsequent in situ deformation [14], as seen in Fig. 6 where the crystallite interior has been sub-divided by tangled walls and cells. However, as the tangled walls build up and cells form, λ increases. Thus λ must be greater than d_c with an upper limit of D. Figure 7 plots the measured d_c values vs. λ, and clearly indicates that λ exceeds d_c after λ_{min}, and exceeds both d_c and D at the largest strain. As mentioned in the introduction, the over-prediction of λ results, because the model assumes no recovery/annihilation and hence $\mu/(d\tau/d\gamma)$ is about two orders of magnitude too large. Saimoto et al. [15] used this functional relationship to show that it reflects the evolution of the internal structure, even at elevated temperatures and high strains where obvious recovery and annihilation are possible. In comparing the dislocation densities in tangled walls to those geometrically necessary to account for the imposed strain, McLean [16] found that the required densities were about five times larger than those observed, and therefore he postulated that either annihilation or passage through the tangled wall had occurred. Invoking the annihilation factor of Eqn. (1), such that $\bar{\lambda} = \lambda/10$ [3], shifts the data into the grain size limited area (Fig. 7). After the YPE the data suggests that $\bar{\lambda}$ could exceed d_c. Göttler [17] has observed slip lines formed during Stage III deformation with lengths exceeding d_c.

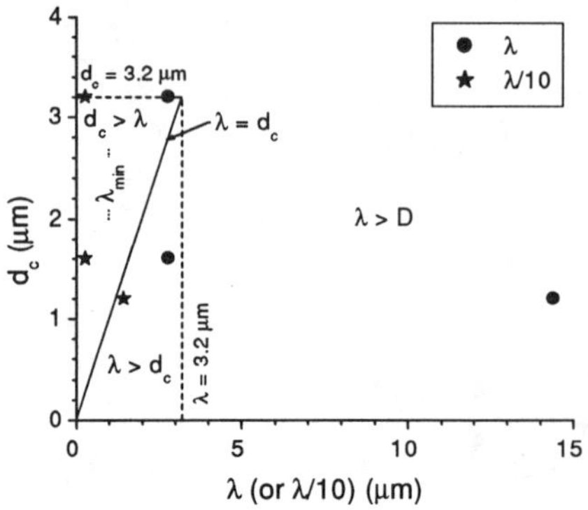

Fig. 7. Observed relationship between d_c and λ (or $\lambda/10$). The dotted box indicates the region limited by the grain size, D = 3.2 μm..

CONCLUSIONS

The dynamic and TEM observations indicate that the λ is some multiple of d_c. Since λ cannot exceed the grain size of 3.2 μm, the use of a parameter A to account for dynamic annihilation was invoked so that $\bar{\lambda} = \lambda/10$.

ACKNOWLEDGEMENTS

We would like to thank Alcan K.R.D.C. for supplying the aluminum alloy and jet polishing the TEM samples. We would also like to acknowledge Dr. G.J.C. Carpenter of CANMET for assistance with the TEM work.

REFERENCES

1. D. Field and H. Weilland, Mat. Sci. Forum, **157-162,** 1181 (1994).
2. F.R.N. Nabarro, Z.S. Basinski and D.B. Holt, Adv. Phys., **13,** 193 (1964).
3. B.J. Diak, K.R. Upadhyaya and S. Saimoto, Prog. Mat. Sci. **43,** 223 (1998).
4. T. Tabata, H. Fugita, M.-A. Hiroaka and S. Miyake, Phil. Mag., **46,** 801 (1982)
5. C. Schwink and E. Göttler, Acta Metall., **24,** 173 (1976).
6. A.S. Argon and P. Haasen, Acta Metall. Mater., **41,** 3289 (1993).
7. D. Sil and S.K.. Varma, Met. Trans., **24A,** 1153 (1993).
8. J.J. Gracio and J.V. Fernandes and J.H. Schmitt, Mat. Sci. Eng., **A118,** 97 (1989).
9. J.G. Rao and S.K. Varma, Met. Trans., **24A,** 2559 (1993).
10. D. Chu and J.W. Morris, Jr., Acta Mater., **44,** 2599 (1996).
11. H. Jin, J. Li, B.J. Diak and S. Saimoto, in *4th Intl. Conf. On Recrystallization and Related Phenomena*, edited by T. Sakai and H.G. Suzuki (JIM, 1999) 277-282.
12. A.H. Cottrell and R.J. Stokes, Proc. R. Soc. Lond. A, **233,** 17 (1955).
13. E.A Faulkner and R.K. Ham, Phil. Mag., **7,** 279 (1962).
14. S. Saimoto, H. Saka, T. Imura, Scripta Metall., **11,** 615 (1977).
15. S. Saimoto, S. Sang and L.R. Morris, Acta Metall., **29,** 215 (1981).
16. D. McLean, Can. Journ. Phys., **45,** 973 (1967).
17. E. Göttler, Phil. Mag., **28,** 1057 (1973).

Non-Local Plastic Theory and Dislocation Phenomena

THE STRAIN GRADIENT EFFECTS
IN MICRO-INDENTATION HARDNESS EXPERIMENTS

Z. Xue[*], Y. Huang[*], H. Gao[**], W.D. Nix[***]
[*]Dept. of Mechanical & Industrial Eng., Univ. of Illinois, Urbana, IL 61801
[**]Division of Mechanics and Computation, Stanford University, Stanford, CA 94305
[***]Dept. of Materials Science and Engineering, Stanford University, Stanford, CA 94305

ABSTRACT

Micro-indentation experiments have shown very strong size dependence of the indentation hardness, a phenomenon that cannot be explained by classical plasticity theories. A theory of mechanism-based strain gradient (MSG) plasticity has been developed based on the Taylor model in dislocation theories, and is intended for materials and structures whose dimension controlling plastic deformation falls roughly between 0.1 to 10 microns. The MSG plasticity theory is used in the present study to investigate the size effects observed in micro-indentation experiments. It is shown that the theory can indeed reproduce micro-indentation experimental data, thus providing an important self-consistent check of the MSG plasticity theory.

INTRODUCTION

Micro-indentation and nano-indentation experiments have repeatedly shown that the hardness of metallic materials at micro-scale can be significantly different from that of bulk materials. For example, the hardness at the micron or sub-micron depth of indentation can double or even triple that at large depth of indentation[1-7]. Classical plasticity theories have no internal constitutive length parameters and fail to predict this size dependence. Over this 1~10 micron scale, however, there are still hundreds or even thousands of dislocations such that there should be a continuum theory of plasticity that can describe the collective behavior of dislocations at the micron scale.

The phenomenological theory of strain gradient plasticity[8-10] represents one such theory intended for applications to materials whose dimension controlling plastic deformation falls roughly within a range from a tenth of a micron to ten microns. The strain gradients are introduced in the constitutive law, and the internal constitutive length parameters scaling the strain gradient range from sub-microns to microns from fitting the micro-scale experiments.

The dislocation model developed by Nix and Gao[11] for micro-indentation experiments has shed some lights on both the internal material lengths in strain gradient plasticity and the experimental law needed to advance a mechanism-based theory of strain gradient plasticity. They started from the Taylor relation[12] between shear strength and dislocation density, and derived a fundamental law among the flow stress σ, strain ε, and strain gradient η as

$$\sigma = \sigma_{ref}\sqrt{f^2(\varepsilon) + l\eta} \,, \tag{1}$$

where σ_{ref} is a reference stress in uniaxial tension such that the uniaxial stress-strain law can be written as $\sigma = \sigma_{ref}f(\varepsilon)$, η is the effective strain gradient, and l is identified as the internal constitutive length in strain gradient plasticity, given by

$$l = 18\alpha^2\left(\mu/\sigma_{ref}\right)^2 b \,. \tag{2}$$

53

Here μ is the shear modulus, b is the Burgers vector, and α is an empirical constant in the Taylor relation[12] and is on the order of one[13]. For micro-indentation hardness experiments, the fundamental law (1) predicts a linear dependence of the square of the hardness, H^2, on the inverse of indentation depth, $1/h$, which agrees remarkably well with the hardness data for single crystal and cold worked polycrystalline copper[7], as well as with single crystal silver[5] and polycrystalline copper[6].

Motivated by this remarkable agreement, Gao, Huang, Nix and Hutchinson[14,15] have developed a theory of mechanism-based strain gradient (MSG) plasticity based on the Taylor model in dislocation theory. The theory is established from a multi-scale, hierarchical framework to link the micro-scale (e.g., $10\sim100$nm), at which dislocation interactions are governed by the Taylor model, to the meso-scale (e.g., $1\sim10$ microns), at which the plasticity theory is formulated. On the micro-scale, the strain gradient term is treated as a measure of the density of geometrically necessary dislocations[16] whose accumulation increases the flow stress strictly following the Taylor model. On the meso-scale, the constitutive equations are obtained by averaging micro-scale plasticity laws over a representative cell. As the characteristic length associated with deformation becomes much larger than microns, MSG plasticity degenerates naturally to the classical plasticity theories.

The theory of MSG plasticity[14,15] and the associated finite element method[17] are used in this paper to investigate the micro-indentation experiments. The MSG plasticity theory is presented next. The hardness predicted by MSG plasticity is then compared with the micro-indentation hardness experiments in order to valid this continuum plasticity theory at the micron and sub-micron scales.

MECHANISM-BASED STRAIN GRADIENT (MSG) PLASTICITY

The theory of MSG plasticity[11,14,15] is briefly described in this paper. Its starting point is the Taylor relation[12] between the shear strength τ and the total dislocation density ρ_T in a material,

$$\tau = \alpha\mu b\sqrt{\rho_T} = \alpha\mu b\sqrt{\rho_S + \rho_G}, \tag{3}$$

where μ is the shear modulus, b is the Burgers vector, α is an empirical constant on the order of one in the Taylor relation[12], ρ_G is the density of geometrically necessary dislocations and is related to the effective strain gradient η as[16,17]

$$\rho_G = 2\eta/b, \tag{4}$$

and ρ_S is the density of statistically stored dislocations, which can be determined from the uniaxial stress-strain law $\sigma = \sigma_{ref} f(\varepsilon)$ since there are no strain gradients in uniaxial tension and therefore no geometrically necessary dislocations[11,17],

$$\sigma_{ref} f(\varepsilon) = \sigma = 3\tau = 3\alpha\mu b\sqrt{\rho_S}. \tag{5}$$

The substitution of (4) and (5) into (3) yields the fundamental law (1) and the internal constitutive length (2) for strain gradient plasticity. Gao, Huang, Nix and Hutchinson have generalized the flow stress in (1) to a three-dimensional constitutive law for mechanism-based strain gradient plasticity theory[14,15].

INDENTATION MODEL

The indentation model of Begley and Hutchinson[18] is adopted in the present study. A conical, frictionless indenter with the half-angle of $72°$ (~ Vickers indenter) is indented into the solid. For a given indentation load, the finite element method based on MSG plasticity predicts the contact area. The effect of pile-up or sink-in may influence the micro-indentation hardness, and is fully accounted for in the indentation model. The effect of indenter tip radius on micro-indentation hardness is also discussed.

The micro-indentation hardness is defined in the usual way as the mean pressure exerted by the indenter at the maximum load. The maximum load can be obtained straightforwardly from the experiments, while the contact area between the indenter and the indented material is usually determined from the contact compliance[19,20], or from direct SEM observations[7] if there is pile-up or sink-in. For strain-hardened materials and metallic glasses that exhibit a low strain-hardening rate, the volume of material displaced by the indentation pushes out to the sides of the indenter and forms a pile-up of material, making the projected contact area larger than the cross-sectional area of the indenter at that depth. For well-annealed soft metals that exhibit a high strain-hardening rate, the displaced volume is accommodated mainly by the displacements in the elastic far-field, producing what is called a sink-in effect and making the projected contact area less than the cross-sectional area of the indenter at that depth. The effect of pile-up or sink-in, if not accounted for, leads to errors in the absolute measurement of micro-indentation hardness. Recently, McElhaney, Vlassak and Nix[7] have used both indentation loads and displacements and direct scanning electron microscopy (SEM) images of the impressions left by large indentations to develop a new technique for determining the indenter shape and the actual cross-sectional area of the indenter. This new method has properly accounted for the effect of pile-up and sink-in on the contact area. For copper, the error in the contact area is more than 20% if pile-up or sink-in is not accounted for. However, after pile-up or sink-in is taken into account, the corrected micro-indentation hardness still shows a strong dependence on the depth of indentation. For example, the hardness at the depth of 2 microns is only $1/3$ of that for 0.16 micron depth. Therefore, even though pile-up and sink-in have significant effect on micro-indentation hardness, they are not the reason of size-dependent indentation hardness.

The radius of the indenter tip may also have significant effect on micro-indentation hardness, particularly for very shallow indentation. If the tip radius becomes comparable to the depth of indentation, it will also contribute to the size-dependent micro-indentation hardness observed in experiments. McElhaney, Vlassak and Nix[7], however, have used very sharp indenters whose tip radii are much less than sub-microns. This is clearly observed from their SEM micrographs of indentation, which show that, over the scale of one micron, the indenter tips are still very sharp such that the tip radii must be about 100nm. The indenter tip radii have also been determined by measuring the elastic contact (with the displacement in the range of $0 \sim 10$nm) on flat fused quartz and observed a Hertz-type force-displacement relation[21]. It is confirmed that the radii of sharp indenter tips are indeed about 100nm. Furthermore, the indentation displacement at which these sharp indenters become self-similar is less than 25nm. This is much less than even the smallest depth of indentation reported in McElhaney, Vlassak, and Nix's experiment[7]. Therefore, the indenter tip radius has essentially no effect on the observed depth dependence of the hardness.

The indenter is modeled as rigid and axisymmetric such that the finite element analysis is two-dimensional. The finite element formulation is established from the principle of virtual work for higher-order continuum theories, and is identical to that given by Begley and Hutchinson[18] except that its constitutive law is replaced by MSG plasticity. Three different finite elements[17], namely the triangular C_1 element, the hybrid element, and the high-order isoparametric element, are used in the finite element analysis in order to make sure that there is no dependence on the elements. The elements are implemented in the ABAQUS finite element program through the USER-ELEMENT interface, and are validated by comparing with the analytical solutions for void growth in MSG plasticity[15]. For a large depth of indentation such that the strain gradient effects become negligible, these elements predict the same indentation hardness with the conventional elements in ABAQUS (without strain gradients), which serves another validation of the elements.

NUMERICAL RESULTS

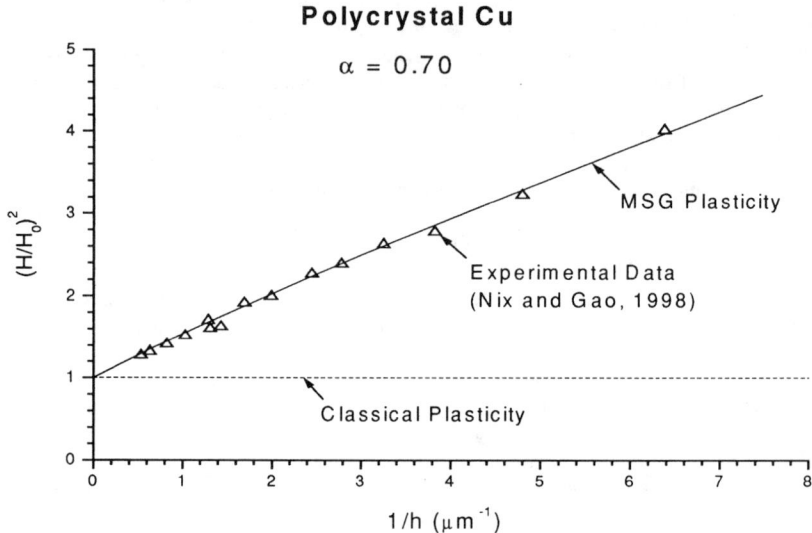

Figure 1. Depth dependence of the hardness of polycrystalline copper; the solid line is the hardness predicted by MSG plasticity; the triangles are experimental data[7,11]; H is the micro-indentation hardness, $H_0 = 834 MPa$ is the conventional indentation hardness (i.e., for large depth of indentation), h is the depth of indentation, the uniaxial stress-strain curve $\sigma = 408\varepsilon^{0.3}$ MPa, shear modulus $\mu = 45$ GPa, and the Burgers vector $b = 0.255$ nm.

The indentation hardness predicted by the finite element method for MSG plasticity is compared with the experimental measured hardness for polycrystalline and single crystal copper[7]. The uniaxial stress-strain data for polycrystalline and single crystal copper are $\sigma = 408\varepsilon^{0.3}$ MPa and $\sigma = 283\varepsilon^{0.3}$ MPa, respectively[22]. The plastic work hardening exponent 0.3 is consistent with independent experiments for polycrystalline copper[9]. The shear modulus $\mu = 45$ GPa, and the Burgers vector $b = 0.255$ nm. For a large depth of indentation (>>microns), at which the strain gradient effects become negligible, the finite element method predicts the conventional indentation hardness of $H_0 = 834$ MPa for polycrystalline copper and

$H_0 = 581$ MPa for single crystal copper, which agree well with the experimentally measured hardness for large depth of indentation[7,11].

The empirical coefficient α in the Taylor model, which is on the order of one[12,13], is determined by fitting micro-indentation hardness data over the micron and sub-micron depth of indentation. Figures 1 and 2 present the micro-indentation hardness predicted by MSG plasticity, $(H/H_0)^2$, versus the inverse of indentation depth, $1/h$, for polycrystalline and single crystal copper, respectively. Here H is the micro-indentation hardness, H_0 is the conventional indentation hardness (for a large depth of indentation), h is the depth of indentation, and $\alpha = 0.70$ for polycrystalline copper and $\alpha = 1.1$ for single crystal copper. The experimental data[7], as plotted by Nix and Gao[11], are also presented for comparison. It is clearly observed that the numerically predicted hardness by MSG plasticity agree very well with the experimentally measured micro-indentation hardness over a wide range of indentation depth, from one tenth of a micron to several microns. The coefficient α estimated from the experimental data also has the correct order of magnitude (of one). Moreover, the numerical results based on MSG plasticity give straight lines in Figs. 1 and 2, consistent with the estimate based on dislocation models[11].

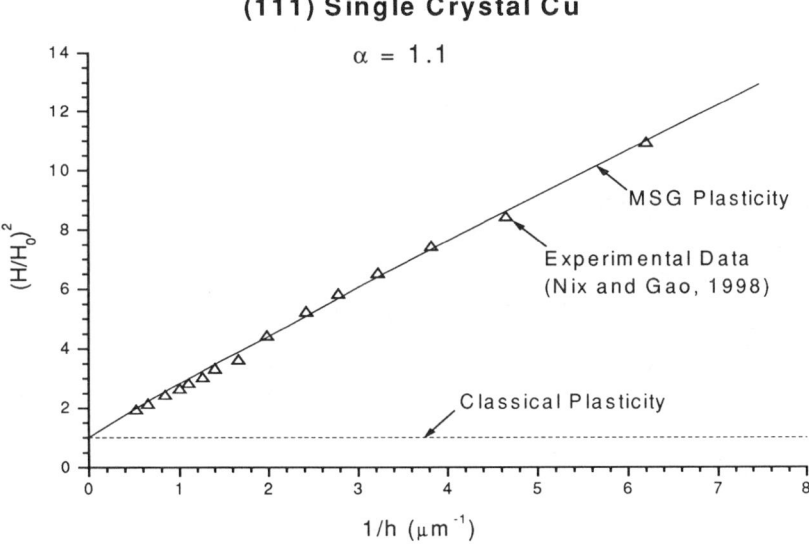

Figure 2. Depth dependence of the hardness of (111) single crystal copper; the solid line is the hardness predicted by MSG plasticity; the triangles are experimental data[7,11]; H is the micro-indentation hardness, $H_0 = 581$ MPa is the conventional indentation hardness (i.e., for large depth of indentation), h is the depth of indentation, the uniaxial stress-strain curve $\sigma = 283\varepsilon^{0.3}$ MPa, shear modulus $\mu = 45$ GPa, and the Burgers vector $b = 0.255$ nm.

CONCLUSIONS

The agreements between the predicted hardness by MSG plasticity and the experimentally measured micro-indentation hardness in Figs. 1 and 2 provide a validation of the theory of MSG plasticity. Both the numerical analysis and the micro-indentation experiments show a linear relation between the square of indentation hardness, H^2, and the inverse of indentation depth, $1/h$. Moreover, the coefficient α determined from the experimental data is indeed on the order of one, consistent with the Taylor model[12,13]. This validates MSG plasticity both as a new mechanics theory and as an engineering analysis tool for advanced materials technology at the micro-scale.

ACKNOWLEDGMENTS

YH and HG acknowledge the helpful discussions with Dr. J.W. Hutchinson. The work of YH was supported by the NSF through CMS-98-96285 and by the NSF of China. The work of HG was supported by the NSF through Young Investigator Award MSS-93-58093. The work of WDN was supported by the DOE through DE-FG03-89ER45387.

REFERENCES

1. W.D. Nix, Met. Trans. **20A**, 2217 (1989).
2. M.S. De Guzman, G. Neubauer, P. Filnn, and W.D. Nix, Mater. Res. Symp. Proc. **308**, 613 (1993).
3. N.A. Stelmeshenko, M.G. Walls, L.M. Brown, and Y.V. Milman, Acta Metall. Mater. **41**, 2855 (1993).
4. M. Atkinson, J. Mater. Res. **10**, 2908 (1995).
5. Q. Ma and D.R. Clarke, J. Mater. Res. **10**, 853 (1995).
6. W.J. Poole, M.F. Ashby, and N.A. Fleck, Scripta Metall. Mater. **34**, 559 (1996).
7. K.W. McElhaney, J.J. Vlassak, and W.D. Nix, J. Mater. Res. **13**, 1300 (1998).
8. N.A. Fleck and J.W. Hutchinson, J. Mech. Phys. Solids **41**, 1825 (1993).
9. N.A. Fleck, G.M. Muller, M.F. Ashby, and J.W. Hutchinson, J. W., Acta Metall. Mater. **42**, 475 (1994).
10. N.A. Fleck and J.W. Hutchinson, Adv. Appl. Mech. **33**, 295 (1997).
11. W.D. Nix and H. Gao, J. Mech. Phys. Solids **46**, 411 (1998).
12. G.I. Taylor, J. Inst. Metals **62**, 307 (1938).
13. W.D. Nix and J.C. Gibeling, in *Metals/Materials Technology Series 8313-004* (ASM, Metals Park, OH, 1985).
14. H. Gao, Y. Huang, W.D. Nix, and J.W. Hutchinson, J. Mech. Phys. Solids **47**, 1239 (1999).
15. Y. Huang, H. Gao, W.D. Nix, and J.W. Hutchinson, J. Mech. Phys. Solids **48**, 99 (1999).
16. M.F. Ashby, Phil. Mag. **21**, 399 (1970).
17. Z. Xue, *A STUDY OF MICRO-INDENTATION HARDNESS TESTS BY MECHANISM-BASED STRAIN GRADIENT PLASTICITY*, MS thesis, Univ. of Illinois, Uabana, IL (1999).
18. M.R. Begley and J.W. Hutchinson, J. Mech. Phys. Solids **46**, 2049 (1998).
19. M.F. Doerner and W.D. Nix, J. mater. Res. **1**, 601 (1986).
20. W.C. Oliver and G.M. Pharr, J. mater. Res. **7**, 1564 (1992).
21. Unpublished work of W.D. Nix and coworkers (1999).
22. D. McLean, *Mechanical Properties of Metals* (John Wiley and Sons, New York, 1962).

EFFECT OF PRECIPITATE MORPHOLOGY ON THE GRADIENT-DEPENDENT BEHAVIOUR OF TWO-PHASE SINGLE CRYSTALS

E. P. BUSSO, F. T. MEISSONNIER and N. P. O'DOWD
Mechanical Engineering Department, Imperial College, London SW7 2BX, UK

ABSTRACT

In this work a recently proposed gradient and rate-dependent crystallographic formulation is used to investigate the macroscopic behaviour of a precipitated single crystal. It relies on strain gradient concepts to account for the additional strengthening mechanism caused by presence of interfacial and geometrically necessary dislocations (GNDs). The total slip resistance is assumed to be due to a mixed population of mobile and sessile forest obstacles arising from both statistically stored dislocations (SSDs) and GNDs. The non-local crystallographic theory is implemented numerically into the finite element method. It requires the calculation of the slip rate gradients at the element level to determine the evolutionary behaviour of the GND densities, and a fully implicit numerical algorithm within a large strain kinematics framework and non-isothermal conditions. The effects of the relevant microstructural features (precipitate size, morphology and volume fraction) and deformation gradient-related length scales on the macroscopic behaviour is investigated and compared with experimental results.

INTRODUCTION

During high temperature deformation of single crystal superalloys, the initially cuboidal precipitates undergo morphological and volume fraction changes which strongly affect the single crystal mechanical properties [1]. In this work, the combined effects of the precipitate size, morphology and volume fraction on the macroscopic stress-strain behaviour are investigated using a micro-macro continuum mechanics approach.

A recently proposed gradient dependent crystallographic formulation [2][3] is used to describe the behaviour of the soft matrix of a precipitated single crystal. The finite element implementation of the non-local crystallographic model includes the calculation of the slip rate gradients at the element level to determine the evolutionary behaviour of the GND densities, and a fully implicit numerical algorithm within a large strain kinematics framework to update the local stresses and internal slip system variables [4]. The numerical procedure is then applied to investigate the influence of the different microstructural length scales introduced by changes in the precipitate morphology in two phase γ/γ' single crystals. The capability of the proposed integration procedure is shown through three-dimensional unit-cell deformation studies. Three typical precipitate morphologies are considered, viz. cuboidal, plate-like and the limiting case corresponding to the coalescence of two neighbouring precipitates. A comparison between the gradient independent and gradient dependent predicted macroscopic responses is also presented.

GRADIENT-DEPENDENT CRYSTALLOGRAPHIC FORMULATION

The hyperelastic crystallographic formulation used in this work accounts for finite strain kinematics and non-isothermal conditions. It relies on the multiplicative decomposition of the total deformation gradient, $\mathbf{F}$, into a thermal component, $\mathbf{F}^\theta$, which represents the deformation of the crystal lattice due to temperature changes, an inelastic component, $\mathbf{F}^p$, associated with pure slip while the lattice remains undistorted and unrotated but expanded due to temperature changes, and an elastic component, $\mathbf{F}^e$, which accounts for the elastic stretching and rigid-body rotations.

From the kinematics of dislocation motion, the time rate of change of the inelastic deformation gradient is given by [5],

$$\dot{\mathbf{F}}^p = \left(\sum_{\alpha=1}^{n_\alpha} \dot{\gamma}^\alpha \mathbf{P}^\alpha \right) \mathbf{F}^p, \quad \text{with} \quad \mathbf{P}^\alpha \equiv \mathbf{m}^\alpha \otimes \mathbf{n}^\alpha , \tag{1}$$

59

where, $\mathbf{m}^\alpha$ and $\mathbf{n}^\alpha$ are the slip direction and the slip plane normal unit vectors on the slip system α.

The constitutive stress-strain relation is defined as,

$$\mathbf{T}^* = \mathcal{L} : \mathbf{E}^e , \quad \text{with} \quad \mathbf{E}^e = \frac{1}{2}\left(\mathbf{F}^{eT}\mathbf{F}^e - \mathbf{1}\right) , \tag{2}$$

where $\mathbf{E}^e$ is the elastic Green-Lagrange tensorial strain measure, $\mathcal{L}$, the anisotropic elastic moduli, $\mathbf{1}$, the second order identity tensor, and, $\mathbf{T}^*$, the work conjugate of $\mathbf{E}^e$. From the basic relation between the Cauchy stress tensor σ and $\mathbf{T}^*$, it can be shown that,

$$\sigma = \det\left(\mathbf{F}^e\right)^{-1}\mathbf{F}^e\,\mathbf{T}^*\,\mathbf{F}^{eT} . \tag{3}$$

The shear strain rate, in its most general form is given by,

$$\dot{\gamma}^\alpha = \dot{\gamma}_o \exp\left[-\frac{F_o}{k\theta}\left\langle 1 - \left\langle \frac{|\tau^\alpha| - S^\alpha \mu/\mu_0}{\hat{\tau}_0 \mu/\mu_0}\right\rangle^p\right\rangle^q\right] sgn(\tau^\alpha) , \tag{4}$$

where τ^α is the resolved shear stress, S^α, the total slip resistance, θ the absolute temperature, μ, μ_0 the shear moduli at θ and 0 K, respectively, and F_0, $\hat{\tau}_o$, p, q and $\dot{\gamma}_0$ are material parameters.

In Eq. 4, contributions to the overall slip resistance are assumed to be due to both the SSDs and the GNDs acting as obstacles to dislocation motion. Three types of GNDs per slip system are considered: a set of screw dislocations parallel to the slip direction $\mathbf{m}^\alpha$ and two sets of edge components oriented parallel to $\mathbf{n}^\alpha$ and $\mathbf{t}^\alpha = \mathbf{m}^\alpha \times \mathbf{n}^\alpha$, respectively. The corresponding total slip resistance is defined as the mean square of the sum of the contributions from the above dislocation populations:

$$S^\alpha \equiv \left[(S_S^\alpha)^2 + (S_{G_s}^\alpha)^2 + (S_{G_{et}}^\alpha)^2 + (S_{G_{en}}^\alpha)^2\right]^{1/2} , \tag{5}$$

where, S_S^α, denotes the slip resistance due to the SSDs, and $S_{G_s}^\alpha$, $S_{G_{et}}^\alpha$ and $S_{G_{en}}^\alpha$ those due to the screw and *edge* GND components, respectively.

The evolutionary relation for the slip resistance caused by statistically-stored forest obstacles is given by,

$$\dot{S}_S^\alpha = \sum_{\beta=1}^{n_\alpha} \delta_S^{\alpha\beta}\left[h_s - d_D(S_S^\beta - S_{S0}^\beta)\right]|\dot{\gamma}^\beta| , \tag{6}$$

where, h_S and d_D, are material constants associated with hardening and dynamic recovery processes, respectively, S_{S0}^α is the initial value of S_S^α, and, $\delta_S^{\alpha\beta}$, the cross-hardening interaction matrix (e.g. $\delta_S^{\alpha\beta} = \delta_{\alpha\beta}$, the Kronecker delta, for self-hardening).

The evolutionary law for the slip resistance contribution arising from the GNDs is given by [3],

$$\dot{S}_{G_i}^\alpha = \frac{\left(\lambda_{G_i}\mu b_G^\alpha\right)^2}{2S_{G_i}^\alpha}\sum_{\beta=1}^{n_\alpha}\delta_S^{\alpha\beta}\dot{\rho}_{G_i}^\beta, \quad \text{for } i = \{s, et, en\} , \tag{7}$$

where, $\dot{\rho}_{G_i}^\alpha$, represents the time rate of change of each set of GND densities (for $i = s, et, en$), λ_{G_i}, are statistical coefficients which account for the deviation from regular spatial arrangements of the GND populations, and, b_G^α, is the magnitude of the Burger's vector of the individual GNDs.

The determination of the density evolution of the GNDs in Eq. 7 is determined from Nye's dislocation tensor in terms of the spatial gradient of the slip rate, $\dot{\gamma}^\alpha$ [3][6],

$$b_G^\alpha\left(\dot{\rho}_{G_s}^\alpha\mathbf{m}^\alpha + \dot{\rho}_{G_{et}}^\alpha\mathbf{t}^\alpha + \dot{\rho}_{G_{en}}^\alpha\mathbf{n}^\alpha\right) = \mathbf{curl}\left(\dot{\gamma}^\alpha\mathbf{n}^\alpha\mathbf{F}^p\mathbf{F}^\theta\right) . \tag{8}$$

NUMERICAL INTEGRATION PROCEDURE

The strain-gradient crystallographic formulation has been implemented numerically into a finite element code [7]. The implementation required the development of (i) a fully implicit user-defined material subroutine for large strains to update the stresses, solution-dependent variables, and material jacobian at each integration point, and (ii) a user-defined element for a 3-dimensional isoparametric solid with reduced integration to determine the slip rate gradients required to incrementally update the GND densities via Eq. 8.

In non-linear finite element methods, an estimate of the incremental displacement field which satisfies the displacement and traction boundary conditions is used to compute the deformation gradient at each material point $(\mathbf{F}_{n+1})$ at the end of a generic time increment n, $\Delta t = t_{n+1} - t_n$. (Hereafter, the subscripts n and $(n+1)$ will identify the variables valued at the beginning and end of the time increment, respectively). First, the known initial conditions at t_n, namely $\mathbf{F}_n$, $\boldsymbol{\sigma}_n$, and the solution-dependent variables associated with the inelastic deformation and current state, $\{\mathbf{F}_n^p, S_{S\,n}^\alpha, S_{Gi\,n}^\alpha\}$, are recovered within the material subroutine. An implicit time-integration procedure is then used to solve the simultaneous incremental non-linear equations associated with the crystallographic formulation. This procedure explicitly incorporates the effects of the GND densities and relies on a Newton-Raphson scheme with a single level of iteration to update both the stress and internal state variables.

The solution of the implicit incremental problem leads to the update of the inelastic deformation gradient and slip resistances, $\{\mathbf{F}_{n+1}^p, S_{S\,n+1}^\alpha, S_{Gi\,n+1}^\alpha\}$, and the new lattice orientation, $\{\mathbf{F}_{n+1}^e \mathbf{m}^\alpha, \mathbf{n}^\alpha \mathbf{F}_{n+1}^{e-1}\}$. Furthermore, the jacobian matrix required to iteratively solve the global equilibrium equations,

$$ \mathcal{C} \equiv \frac{\partial \boldsymbol{\sigma}_{n+1}}{\partial \boldsymbol{\epsilon}_{n+1}} \; , \tag{9} $$

is explicitly computed from the updated variables. Full details of the numerical implementation are given in [4].

To determine the evolution of the GND densities, the slip rate gradient terms given by the components of $\mathbf{curl}\left(\dot{\gamma}^\alpha \mathbf{n}^\alpha \mathbf{F}^p \mathbf{F}^\theta\right)$ in Eq. 8 must be calculated within the 3-dimensional 20-node elements. The interpolation of the slip rate gradients is based on linear shape functions associated with an 8-node element and full (2x2x2) Gauss integration. In this way, it is possible to capture up to linear spatial variations of slip rate gradients within the element.

RESULTS AND DISCUSSION

The study of the gradient dependent behaviour of a precipitated single crystal (SC) was conducted on a commercial superalloy, i.e. CMSX4, and relied on detailed unit cell FE calculations to account explicitly for the features such as precipitate size, morphology and volume fraction.

From the typical microstructure of the two-phase SC, the following geometrical parameters can be identified in an evolving γ/γ' microstructure (see Fig. 1): the precipitate width, depth and height, l, w and h, respectively, and the corresponding γ-channel widths, d_l, d_w, and d_h. Here, the precipitate volume fraction is related to the ratio $(l\,h\,w)/(d_l\,d_h\,d_w)$. For a given precipitate volume fraction, the dimensionless length scale which describes the scale-dependent behaviour of the two-phase single crystal is given by the ratio between the width of the deformation gradient region intrinsically associated with the presence of GNDs in the vicinity of the γ/γ' interface, and the corresponding γ-channel width, i.e. l_G/d_l in Fig. 1.

In the FE unit cell analyses, a uniform displacement was applied on the periodic boundaries normal to the $< 001 >$ crystallographic axis to simulate uniaxial loading under constant true strain rate and isothermal conditions. In the unit cell computations, the γ-phase is described by the gradient-dependent model whereas the precipitate is assumed to remain elastic throughout the deformation. Details of the calibration of the two phases are given in [3].

The effect of precipitate size on the macroscopic behaviour of the precipitated SC was first evaluated by varying the precipitate size from 0.13 μm to 1.8 μm while maintaining the precipitate volume fraction constant at 68%. This range was based on the statistical variations of the precipitate size from the mean value experimentally measured in CMSX4.

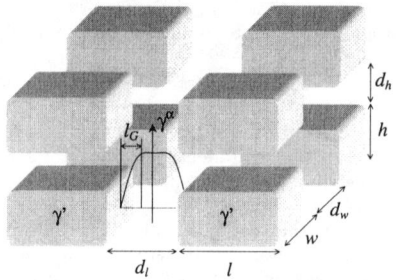

Fig 1. Schematics of the accumulated inelastic strain profile in a generic slip system (α) across the soft γ-phase channel of width d_l, showing the length scale (e.g. l_G) associated with the slip gradient

Figure 2 shows the predicted steady-state $< 001 >$ flow stress dependency on the precipitate size for an applied $< 001 >$ strain rate of 10^{-3} s^{-1} at 850°C. Here, l has been normalised by the mean precipitate size ($l_{mean} = 0.52$ μm), and the steady-state flow stress, σ_{001}, by that corresponding to the mean precipitate size ($\sigma(l_{mean}) = 1040$ MPa).

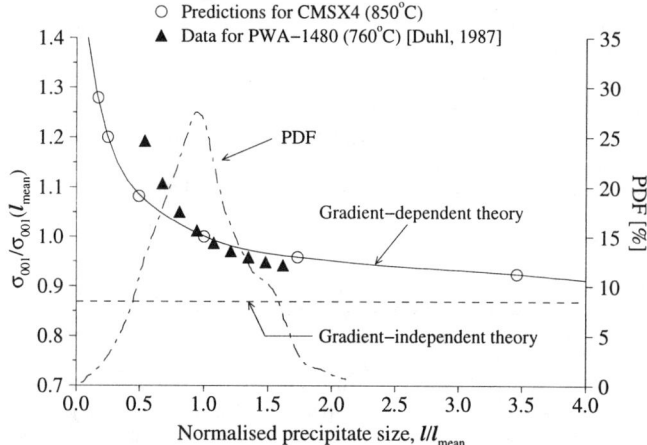

Fig 2. Predicted <001> steady-state flow stress normalised by the value corresponding to the mean precipitate size ($l_{mean} = 0.52\mu$m). Also shown are data for another high precipitate volume fraction superalloy [8] .

It can be seen that the predicted trend compares well with the data reported for another high precipitate volume fraction SC [8]. As expected, the gradient-dependent predictions gradually approach the gradient-independent limit as the precipitate size increases. Conversely, as the precipitate size decreases, the local density of GNDs strongly dominates the local resistance to slip hence the flow stress dramatically increases. The implications of this size effect are illustrated with the probability density function of the precipitate size seen in Fig. 2. It is shown that in a material such as CMSX4, where $0.01 < l/l_{mean} < 2.0$, the corresponding flow stress varies from 0.95 to 1.35.

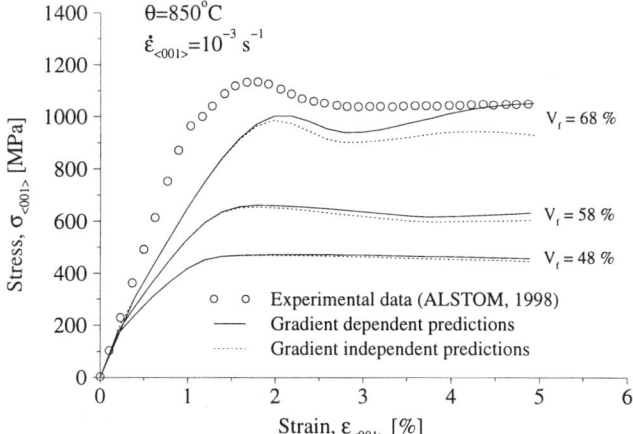

Fig 3. Effect of precipitate volume fraction on the $< 001 >$ monotonic response predicted by the FE unit cell calculations for CMSX4 with $\dot{\epsilon}_{<001>} = 10^{-3}\ s^{-1}$ at 850°C. A comparison between the scale-dependent and scale-invariant computations is given

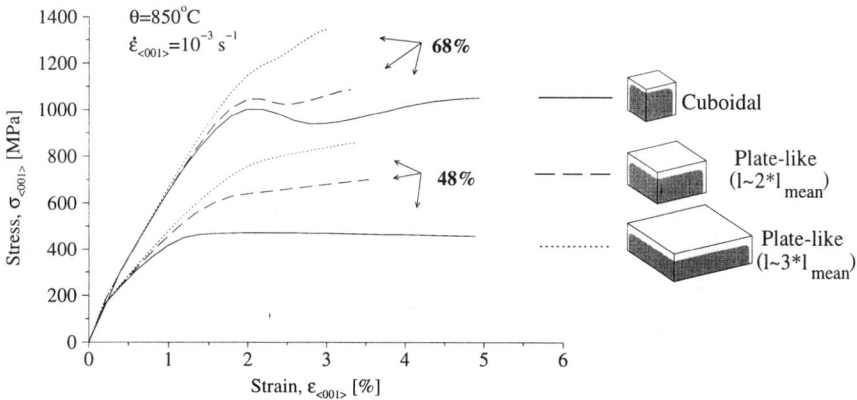

Fig 4. Gradient dependent predictions of the effects of precipitate volume fraction and morphology on the predicted $< 001 >$ tensile response with $\dot{\epsilon}_{<001>} = 10^{-3}\ s^{-1}$ and 850°C

Figure 3 shows the effect of the cuboidal precipitate volume fraction on the predicted < 001 > monotonic response of CMSX4. The gradient independent predictions are consistent with those reported in Busso et al. [9], that is the macroscopic flow stress decreases with decreasing volume fraction and the strain gradient effects becomes more important at high volume fractions.

It is important to point out that the macroscopic softening behaviour seen for $V_{\gamma'}=0.68$ and observed in $\gamma - \gamma'$ single crystals immediately after yielding, often explained as being due to the restructuring of the initial dislocation structure, can in fact be accounted for by the activation of additional slip systems caused by large lattice rotations. This issue has been studied in detail in Busso et al. [9].

The effect of the precipitate morphology is shown in Fig. 4. Here, three different precipitates are considered: cuboidal and plate-like precipitates with in-plane aspect ratios of twice and three times the mean precipitate size. It is seen that a rafted microstructure exhibits a significant strengthening effect when loaded in the out-of-plane direction. This effect is also predicted at the lowest volume fraction of 48%. However, the maximum stress value obtained for a volume fraction of 48% is considerably lower than the one obtained for the unrafted 68%-precipitate.

SYUMMARY

A gradient and rate dependent crystallographic formulation has been applied to investigate the effect of precipitate size, morphology and volume fraction on the overall cubic stress-strain response of a γ/γ' Ni-base superalloy. Unit cell based computations were relied upon to establish an explicit link between the precipitate morphology at the microscopic level with the macroscopic behaviour of an equivalent homogeneous single crystal.

ACKNOWLEDGEMENTS

The authors are grateful to the Engineering and Physical Science Research Council (UK) for its financial support (Grant GR/K73688) and ABB ALSTOM POWER (UK) for the provision of the CMSX4 test data. The ABAQUS program was provided under academic license by HKS Inc, Rhode Island.

REFERENCES

1. Pollock, T. M. and Argon, A. S. Directional coarsening in nickel-base single crystals with high volume fractions of coherent precipitates. *Acta Metall. Mater.*, **42**, pp. 1859–1874 (1994).

2. Busso, E. P. and McClintock, F. A. Mechanisms of cyclic degradation of NiAl Single crystals at high temperatures. *Acta Metall. Mater.*, **42**, p. 3263 (1994).

3. Busso, E. P., Meissonnier, F. T. and O'Dowd, N. P. Gradient-dependent deformation of precipitated single crystals. *J. Mech. Phys. Solids.*, (Submitted) (1999).

4. Meissonnier, F. T., Busso, E. P. and O'Dowd, N. P. Finite element implementation of a generalised non-local rate-dependent crystallographic formulation for finite strains. *Int. J. Plasticity.*, (Submitted) (1999).

5. Asaro, R. J. and Rice, J. R. Strain localization in ductile single crystals. *J. Mech. Phys. Solids*, **25**, pp. 309–338 (1977).

6. Dai, H. and Parks, D. M. Geometrically-necessary dislocation density and scale-dependent crystal plasticity. *Proc. of Sixth International Symposium on Plasticity.* A. Khan ed., Gordon & Breach Publ., pp. 17-18 (1997).

7. ABAQUS, HKS Inc., Providence, Rhode Island, USA, 1999.

8. Duhl, D. N. Directionally solidified superalloys. *Superalloys II: high temperature materials for aerospace and industrial power.* Sims et al. (eds.), pp. 189-214 (1987).

9. Busso, E. P., Meissonnier, F. T., O'Dowd, N. P. and Nouailhas, D. Length scale effects on the geometric softening of precipitated single crystals. *J. de Physique IV*, **8**, pp. 55-61 (1998).

EXPERIMENTAL ANALYSIS AND CRYSTALLOGRAPHIC MODEL OF PLASTIC DEFORMATION AFTER A CHANGE OF LOADING PATH IN MILD STEEL POLYCRYSTALS

T. HOC, C. REY

Laboratoire de Mécanique des Sols, Structures et Matériaux, UMR 8579, Ecole Centrale Paris, Grande Voie des Vignes, 92295 Châtenay-Malabry cedex, France

ABSTRACT

Strain localization in mild steel submitted to a sequential loading paths is investigated at macroscopic, mesoscopic and microscopic scales. The experimental results demonstrate that the morphology of the localization and the nominal load-displacement curves depend on the microstructural anisotropy. A crystalline model using a finite element code is proposed. The anisotropy is described by a hardening matrix whose terms correspond to dislocation-dislocation interactions and depend on the evolution of the dislocation densities on the activated slip systems during the sequential tests. The strain localization predicted by this model fits with the experimental observation and allows us to assume that localization is correlated to the saturation on the activated slip systems.

INTRODUCTION

In polycrystals, strain localization, i.e., sets of narrow bands carrying a large amount of plastic deformation, corresponds to an instable phenomenon, frequently observed during cold forming processes [1,2,3,4]. In order to simulate such processes and their effect on localization, changes in loading path were extensively investigated at macroscopic and microscopic scales [5,6,7], but there is relatively little information available on the evolution on the local strain field during localization [8].

The aim of this paper is to analyse microstructural effects on strain localization after changes in loading path. For this purpose, experimental analysis at macroscopic, mesoscopic and microscopic scales have been performed. On the basis of these experimental results, a numerical crystalline model using the Abaqus Finite element code is proposed to predict strain localization

EXPERIMENTAL RESULTS

Mild steel sheets (0.9 mm thick) obtained by cold rolling and then annealing were used in this investigation. The polycrystal had an average grain size of $20\mu m$ and an initial crystallographic texture close to a $< 111 >$ fiber texture.

Rectangular samples (250 mm×50 mm×0.9 mm) were first strained in plane tension in the rolling direction (RD) at room temperature. The experiments were carried out with strain rates of $10^{-3}s^{-1}$ up to the macroscopic strain values of about 0.18. Then, (50 mm×10 mm×0.7 mm) samples for uniaxial tension, were machined from the pre-strained sheets parallel (Rolling Direction) and orthogonal (Transverse Direction) to the rolling direction. The samples were deformed in-situ in a scanning electron microscope in order to determine simultaneously the local strain field via microextensometry of microgrids and local grain orientation via Electron Back Scattering Diffraction technique (EBSD).

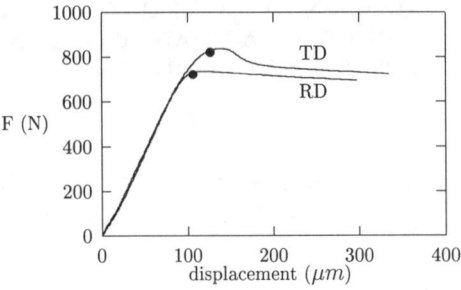

Figure 1: Load-elongation curves in the rolling and the transverse direction after prestraining.

Figure 1 shows typical nominal load-displacement curves for the prestrained samples tested in uniaxial tension along the rolling direction and the transverse direction. This figure shows that prestrained material presents a sharp elastoplastic transition in the case of the rolling direction and a parabolic elastoplastic transition in the case of the transverse direction. In the both tests, localization occurs for the maximum of the load, at 54° to the tensile axis. The morphology of localization depends strongly on the direction of the second loading. In fact, two symmetrical diffuse macrobands are obtained for the rolling direction while two different macrobands (the second one narrower than the first one) are observed for the transverse direction (see figure 2).

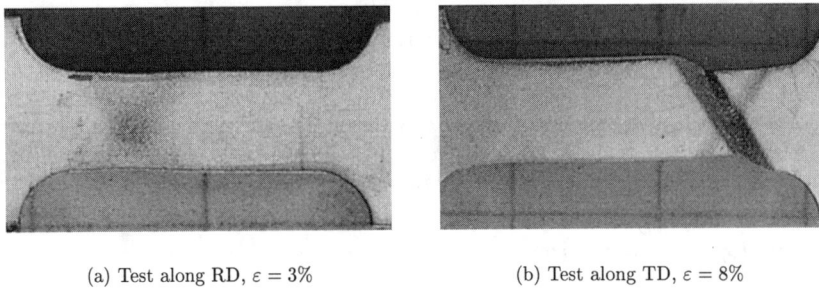

(a) Test along RD, $\varepsilon = 3\%$ (b) Test along TD, $\varepsilon = 8\%$

Figure 2: Form of the localization for rolling and transverse direction after prestraining. Initial width of sample = 3mm

This anisotropic behavior can be linked to the microstructure of dislocations. Microstructures corresponding to the end of the first loading and to the second loading in the transverse direction are respectively given on figures 3 a) and b). Two main features can be noticed, in the prestrained material, the microstructure consists in two families of walls forming mainly parallelepipedic cells with a set of walls at about 90° to the tensile axis presenting a better definition than the other. In the macrobands (second loading), the first dislocation arrangement (corresponding to the preloading) is replaced by a new set of walls. These observations are similar to Schmitt's results [9].

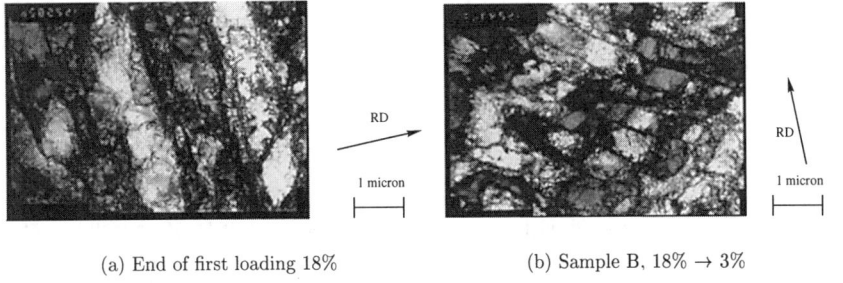

(a) End of first loading 18%   (b) Sample B, 18% → 3%

Figure 3: Morphology of the localization for rolling and transverse direction after pre-strained. Initial width of sample = 3mm

At an intermediate scale, called mesoscopic scale, local strain field evolution is determined during the second loading by using microgrids (surface : $1mm^2$, mesh : $5\mu m \times 5\mu m$) laid on the samples. The result of the computation of the local Green-Lagrange component E_{22}, shows that the whole plastic deformation is concentrated within the macrobands, whereas the deformation outside (matrix) is close to zero (elastic deformation). Moreover, a regular pattern of two families of small parallel bands of localized deformation appears within the macrobands. These small bands correspond to coarse slip bands observed at the surface of the samples [10].

CRYSTAL MODEL

All these results lead us to propose a numerical model which takes into account crystallographic plasticity and where anisotropy is described in terms of dislocation densities on each slip system. For this purpose, a viscoplastic power law for the glide on the twenty four slip systems, usually used in BCC structure [11], $\{110\} < 111 >$ and $\{112\} < 111 >$ is assumed. The applied resolved shear stress τ^s, the critical shear stress τ_c^s and the shear strain rate $\dot{\gamma}^s$ are related by :

$$\dot{\gamma}^s = \dot{\gamma}_o \left| \frac{\tau^s - \tau_c^s}{\tau_0} \right|^n \text{sgn}(\tau^s) \quad \text{if} \quad |\tau^s| > \tau_c^s \quad \text{and} \quad \dot{\gamma}^s = 0 \quad \text{otherwise} \tag{1}$$

where $\dot{\gamma}_o$ is a reference strain rate, and τ_o is the friction stress. In this formulation, the conditions of activation of slip system (s) in a grain are given by Schmid's law. In the phenomenological approach used here, critical resolved stress is linearly related to slip rate through a hardening matrix. The components of this (24x24) hardening matrix depend on short-range interactions between two families of dislocations and is not a constant. Moreover, the anisotropy of the hardening matrix increases with the number of activated slip systems. The evolution law of dislocations densities based on Orowan's relation and annihilation process of dislocation dipoles is given by :

$$\dot{\rho}^s = \frac{1}{b} \left(\frac{1}{L^s} - G_c \rho^s \right) |\dot{\gamma}^s| \quad \text{with} \quad L^s = K \left(\sum_{u \neq s} \rho^u \right)^{-\frac{1}{2}} \tag{2}$$

Plane	$\{110\} \cap \{110\}$	$\{110\} \cap \{112\}$	$\{112\} \cap \{112\}$
Same	0.2		0.26
Colinear	0.2	0.21	0.26
No colinear	0.23	0.2205	0.299

Table 1: Matrix interaction coefficients, a^{su}

where b is the magnitude of the Burgers vector, G_c is a parameter proportional to the characteristic length associated with the annihilation process of dislocation dipoles, L^s is the mean free path of system (s), and ρ^u is the total dislocation density on latent systems (u) in a grain. In this formulation, K is a material parameter and only dislocation-dislocation interactions (forest interaction) are taken into account.

The critical shear stress on system (s) is related to the evolution of the dislocation density through the relation :

$$\tau_c^s = \tau_o + \mu\, b \left(\sum_u a^{su}\, \rho^u \right)^{\frac{1}{2}} \qquad (3)$$

where a^{su} are constants characterizing the different kinds of interaction between two families of dislocations (s) and (u). In this paper, six different coefficients a^{su} are considered (see table 1).

The values of the parameters are determined from mechanical tests (plane tension tests, tensile tests, cyclic shear tests) and crystallographic texture, using an automatic identification procedure given by the software SiDoLo and a polycrystalline model [12,13]. The computed parameters, given in tables 1 and 2, are close of those proposed in the literature.

$\tau_0 (MPa)$	$\rho_0 (m^{-2})$	G_c (nm)	K	$\dot{\gamma}_0\ (s^{-1})$	n
20	64.10^9	10	20	1.14	15

Table 2: Different parameters of the model

MODELING RESULTS

In order to simulate qualitatively the strain localization observed after prestraining, an experimental surface S_e of $190 \times 90\mu m^2$, corresponding to a layer of 114 grains, was considered. This surface with a ($5\mu m \times 5\mu m$ mesh) microgrid is presented in figure 4 a. The orientation of the 114 grains was previously determined by using the EBSD technique.

This surface is the base of a finite element elementary motif (FEEM) constituted by a set of 8 node brick elements. Each element corresponds to the orientation of an actual grain (see figure 4 b).

The FEEM is 8 times juxtaposed and boundary conditions of a uniaxial test are applied to this sample, for rolling and transverse directions, respectively. The density of dislocations and the orientations of 114 grains computed at the end of the first loading are introduced in the finite element code at the beginning of the second loading. Figures 5 and 6 show maps of longitudinal deformation for this two samples. In the two cases, strain localizations are obtained with a morphology which corresponds qualitatively to the experimental results.

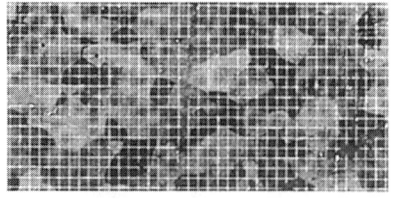 

(a) Experimental surface S_e (b) FEEM with individual grain orientation

Figure 4: Actual aggregate and finite element elementary motif

CONCLUSION

Localization threshold and morphology (diffuse necking or macrobands) depend on the loading path, i.e., the anisotropy evolution. In the model, anisotropy is taken into account through a hardening matrix described in terms of dislocation-dislocation interactions and of dislocation densities whose evolution laws can describe saturation phenomena. It is worth noting that localization occurs without the introduction of defects in the FE code. Strain localizations predicted by this model fit qualitatively experimental results. In the BCC structure, where parallel slip planes can be activated in adjacent grains, localization can be the result of the saturation of dislocation densities on the activated slip systems.

REFERENCES

1 R.J. Asaro. *Acta metal.*, 27:445–453, 1979.

2 D. Pierce. *J. Mech. Phys. Solids*, 31:133–153, 1983.

3 H. Deve, S. Harren, C. McCullough, and R.J. Asaro. *Acta Metal.*, 36(2):341–365, 1988.

4 E.F. Rauch and S. Thuillier. *Archives of Metallurgy*, 38:167–177, 1993.

5 J.H. Schmitt, E.L. Shen, and J.L. Raphanel. *Int. Jour. of Plasticity*, 10(5):535–551, 1994.

6 R.H. Wagoner and J.H. Laukonis. *Metal. Trans. A*, 14:1487, 1983.

7 A. Korbel and P. Martin. *Acta Metal.*, 36(9):2575–2586, 1988.

8 C. Rey and P. Viaris. *Mater. Sci. Eng. A*, A234-23:1007, 1997.

9 J.V. Fernandes and J.H. Schmitt. *Phil. Mag.*, 48(6):841–870, 1983.

10 T. Hoc, C. Rey and P. Viaris de Lesegno *Scripta Mat.* to be published.

11 S. Thuillier *Thèse Institut National Polytechnique de Grenoble*, 6 novembre 1992, France.

12 P. Pilvin *Int. Seminar Mécamat "the inelastic Behaviour of Solids", Oytana c. et al ed. Besancon*, September 1988, p155.

13 T. Hoc *Thèse Ecole Centrale Paris*, 8 september 1999, France.

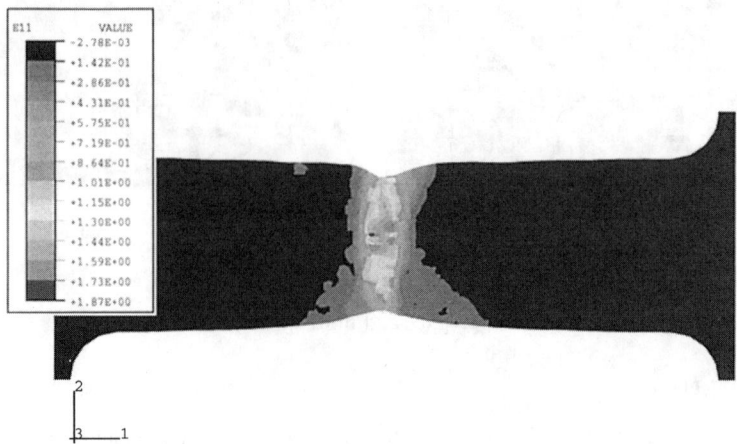

Figure 5: Longitudinal deformation for reloading in the rolling direction. $\varepsilon = 12.8\%$

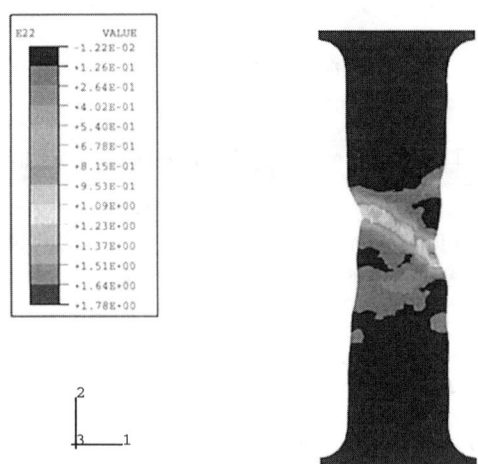

Figure 6: Longitudinal deformation for reloading in the transverse direction $\varepsilon = 15.7\%$

COMPLEMENTARY EXPERIMENTAL TECHNIQUES FOR MULTI-SCALE MODELING OF PLASTICITY

L. E. Levine, G. G. Long, and D. R. Black, Materials Science and Engineering Laboratory, National Institute of Standards and Technology, Gaithersburg, MD 20899

ABSTRACT

Some recently-developed experimental techniques, such as *in situ* ultra-small-angle X-ray scattering (USAXS), have demonstrated a capability for measuring aspects of dislocation structure evolution that are inaccessible to other experimental methods. However, no single technique can provide the entire range of information required by theoretical and computational researchers. It is only through the synergy of several experimental techniques (such as USAXS, transmission electron microscopy, and X-ray diffraction imaging) that much of the required quantitative information can be obtained. Ultimately, the development of additional new experimental techniques will also be required.

INTRODUCTION

The recent surge of activity in the fundamental science of plasticity has been driven primarily by the revolution in computer hardware and the subsequent development of "dislocation dynamics" codes[1–3] that simulate the behavior of many interacting dislocations in three dimensions. Phenomena such as cross-slip and kink nucleation that occur at smaller length- and time-scales are handled through a "rules" based approach. New theoretical advances are also taking place, based largely upon modern developments in statistical physics.[4–7] Unfortunately, progress in these areas is being hampered by a paucity of experimental data on key facets of dislocation structure evolution. Such data is required both as a guide to the development of theoretical and computational models as well as the ultimate test of their validity.

The plastic deformation of metals involves phenomena that interact over a wide range of length- and time-scales. The largest length scale of importance is the macroscopic size and shape of the sample, which can strongly affect its mechanical behavior through a variety of mechanisms, ranging from changes in the statistical distribution of defects to simple mechanical instabilities. At the next smaller length-scale, given by the grain size, texture plays a major role. As slip occurs in a given grain, the resulting deformation produces inhomogeneous stresses on the surrounding grains. These stresses often result in grain rotations and, at elevated temperatures, grain boundary sliding. At this same length scale, surface roughening can take place through several mechanisms, including grain rotations and the still incompletely understood process of the Portevin-Le Chatelier[8–10] effect. At the sub-grain level, the interaction of dislocations with the grain boundaries can be extremely complicated,[11] and the transport of strain across these boundaries depends strongly upon the characteristics of the specific boundary. Because of the large number of grain boundary misorientations in a polycrystalline sample, some type of dislocations-based statistical theory is called for, but no such theory has yet been proposed. Within the grains, the dislocations interact through long-range stress fields to produce three dimensional (3D) dislocation structures that impede the motion of mobile dislocations during straining. The process is complicated, with a rich variety of dislocation structures forming and changing in response to changes in extrinsic parameters such as the applied stress distribution, sample temperature, applied strain, and strain rate.[12–15] At the level of individual dislocation interactions, the simple picture of two straight dislocations cutting through each other has been shown to be incorrect.[16–18] Instead, two approaching dislocations often form complicated looping arrangements that interact in a complex fashion. Such behavior can complicate the use of rules in 3D dislocation dynamics simulations. At the atomic scale, the dislocation core structure can greatly affect the nucleation of cross slip and kink pairs, which can play important roles in work hardening. Additional atomic-level phenomena include short-range dislocation-dislocation interactions as well as interactions between dislocations and point defects such as vacancies and interstitials.

73

The above list, while far from exhaustive, is intended to underline the processes that must be taken into account in any multi-scale model of plasticity. Since the effects of small length-scale processes often propagate all the way to the macroscopic level, a fundamental understanding of plastic deformation can only be attained through some form of multi-scale modeling.

What kinds of experimental information are required at these various length scales? Since the size and shape of the sample can be considered as model inputs, we will only consider the smaller length-scales. At the grain level, modelers can use information on the sizes, shapes, and orientations of the grains, as well as 3D maps of the internal stresses within the sample. Most of this information can be obtained using optical and scanning electron microscopy of etched sample surfaces, cross-sectional transmission electron microscopy (TEM), orientation imaging microscopy, and neutron diffraction residual stress measurements. Surface roughening can be measured by means of profilometry, optical and electron microscopy, and scanning probe microscopy. Residual stresses are more difficult to measure in single- and few-crystal specimens, but recent work using X-ray microbeams at the Advanced Photon Source promises to make such measurements possible.[19] In principle, residual stresses in large single crystals could be measured using neutron diffraction, and work on developing the necessary experimental techniques is currently in progress.[20]

Skipping down to the atomic level, most of the phenomena listed above cannot be measured directly using any existing or projected experimental techniques. In some cases, *ab initio* simulations can take the place of experimental measurements, but such simulations can only handle very small system sizes. In general, *ab initio* calculations cannot be used for many dislocation studies, since dislocations are extended 3D structures, incorporating large numbers of atoms. Multi-ion interatomic potentials have been developed[21−23] for simulating such structures. Here, the many-body angular forces are accounted for through explicit three- and four-ion potentials. Parameters for the potentials are determined through comparison with *ab initio* calculations and with high-resolution TEM (HRTEM) measurements of selected grain boundary structures.[24] HRTEM can also be used to measure directly the atomic spreading of dislocation cores,[25] but such experiments are extremely difficult in materials with high stacking-fault energy, and surface effects must be taken into account in their interpretation.[26]

At the next larger length scale where the curving shape of the individual dislocations becomes important (e.g., for dislocation-dislocation interactions), simpler potentials, such as embedded atom method and Lennard-Jones potentials, are used to handle the large numbers of atoms required.[16,27] Since these potentials are much less accurate than those discussed above, the simulation results often provide qualitative rather than quantitative information. Discrete dislocation dynamics calculations are also used to model dislocation behavior at this length scale. The experimental technique that is most often employed to investigate phenomena at this length scale is, again, TEM. A good example of the usefulness of TEM measurements to the modeling of interacting dislocations appears in a paper by K. W. Schwarz,[28] where cross-sectional TEM pictures of a graded SiGe layer grown on Si(100)[29] are compared with discrete dislocation dynamics simulations. The dislocation structures are shown to result from the interaction of Frank-Read sources on intersecting glide planes. An example of a useful parameter that could be measured using *in situ* TEM is the critical breaking angle, ϕ_c, for a dislocation junction of a specific type. This parameter could be incorporated into the "rules" used by dislocation dynamics codes. Attempts have been made to estimate ϕ_c,[16,30] but experimental measurements are required for a more accurate determination.

TEM can also be used effectively for investigations of the interactions of dislocations with various interfaces. As an example, T. Foecke and D. vanHeerden recently conducted an *in situ* TEM study of deformation and fracture mechanisms in single crystal Cu/Ni nanolaminates.[31] The resulting images of the Orowan bows and pileups can be readily interpreted to give quantitative values for the local stress fields. When interesting events such as the break-away of dislocations from pinning points occur, these local stress measurements can provide quantitative values for a variety of critical parameters. Such measurements could be used both as parameters in multi-scale modeling programs and as tests for model predictions.

As indicated above, significant quantitative experimental input can be obtained for both large and small length scales. This is not generally true, however, for the intermediate length scale of dislocation pattern formation within individual grains and single crystals. The experimental probe that is most useful at smaller length scales, TEM, has inherent limitations that curtail its usefulness at the intermediate length scale. The primary difficulty is that TEM can only be used on foils with thicknesses less than about 0.3 μm. Since the formation of dislocation cells is a bulk process, meaningful *in situ* experiments generally require sample thicknesses of at least 50 μm. TEM can be used *ex situ* to examine sample cross-sections, but it has been demonstrated[32-36] that dislocation structures can be altered significantly during the required thinning process. This is true particularly for materials with a low Peierl's stress, such as high-purity fcc metals. *Ex situ* measurements also suffer from the inability to examine the dislocation structures under the applied load. Even if the sample unloading and thinning effects were negligible, reconstructing the evolution of large-scale, three-dimensional, inhomogeneous structures from isolated cross sections is a daunting task. Nevertheless, several comprehensive *ex situ* TEM studies have been attempted and many important insights into the general work hardening process have resulted.[35-41] Finally, it should be emphasized that, in spite of some limitations, TEM can be used to provide crucial data concerning dislocation structure evolution. The experimentalist must simply be aware of these limitations and work around them. Examples of how TEM can be used effectively to obtain quantitative results from work hardened fcc metals will be discussed in the next section.

Other experimental techniques that can provide useful results are etch-pit studies,[42] ultrasonic measurements,[43] slip-line studies,[44] resistivity measurements,[45] positron annihilation spectroscopy,[46,47] diffraction peak profiling,[48-52] diffraction imaging,[53-55] and ultrasmall-angle X-ray scattering (USAXS).[56-58] By themselves, however, no one of these techniques can provide a coherent and quantitative picture of the deformation process. Such information can only be obtained through a coordinated use of complementary experimental techniques. One particularly promising set of complementary techniques is TEM, ultrasmall-angle X-ray scattering, and high-resolution X-ray diffraction imaging. The remainder of this paper will consider what parameters are required for modeling at this intermediate length scale and will explore how these three experimental techniques can be used together to obtain much of the required quantitative information.

COMPLEMENTARY TECHNIQUES

The information that we consider to be most useful to modelers and theorists is listed in question form below. Whenever possible, these data should be obtained *in situ* as a function of the changing extrinsic parameters.

1) What is the spatial distribution of dislocation walls (e.g., cell sizes and shapes in stage III)?

2) What is the dislocation density distribution across the walls?

3) Does a significant fraction of the component dislocations occupy ordered positions? If so, what percentage and how are they ordered?

4) What is the net Burgers vector (misorientation) of the walls?

5) What is the total dislocation line length for the various active slip systems?

6) What do the dislocation-wall stress fields look like?

7) How are the dislocation structures tied to the lattice?

Rather than addressing these items in order, it is more coherent to examine each experimental technique in turn. No one experimental technique can answer completely any of the above questions. Together, the techniques provide pieces of the puzzle that must be painstakingly assembled. Thus, a simple description of how to answer the above questions is not possible. Nevertheless, it will be shown that using several experimental techniques together can provide some quantitative information on all of the above topics.

Transmission Electron Microscopy

Ex situ TEM has been used extensively to explore the spatial distribution of dislocation walls (item 1, above) in late stage II and stage III.[35–41,51] In spite of the limitations discussed in the Introduction, this type of study usually produces valid results even for samples having a low Peierl's stress. The primary reason for this success is that late stage dislocation walls generally contain a significant number of dislocations with uncompensated Burgers vectors, producing rotational misorientations, θ_M, between the crystal regions separated by the walls.[59] Although significant numbers of geometrically "unnecessary" dislocations may be lost during the sample thinning process, the overall arrangement of the walls remains mostly intact. Irradiating the sample with fast neutrons[60] prior to thinning can help prevent the loss of geometrically unnecessary dislocations by producing vacancies and interstitials that can act as pinning points for the dislocations. Such processing allows dislocation wall studies to be conducted on samples at smaller strains.

Since the angular misorientation across cell walls (item 4) is unlikely to be greatly affected by sample thinning, the misorientation distribution can be measured directly by TEM.[61] In the most comprehensive study of this type conducted to date, Hughes *et al.* report that the misorientation distributions in rolled Al samples have a peaked shape and that the average misorientation increases with increasing strain according to a simple power law. The only drawbacks to this technique are that the samples must be examined *ex situ*, which doesn't allow the misorientation across specific walls to be measured as a function of strain, and that very small rotations (less than approximately 0.1°) cannot be measured.

The dislocation-wall stress fields (item 6) are very difficult to measure directly. We know of only one attempt to measure this using TEM.[60] Mughrabi took a Cu single crystal and fatigued it, producing the characteristic persistent slip band (PSB) walls. While the sample was still under an applied stress, it was irradiated with fast neutrons and then thinned parallel to the primary slip planes. TEM images of curved dislocations between adjacent bands allowed the original local stresses to be calculated. A plot of the measured stress as a function of position between adjacent PSB walls showed a sharp fall-off in the stress level with increasing distance from the walls. These findings are in agreement with those from complementary experiments using diffraction peak profiling.[51,52]

Finally, TEM is the only technique that can shed light on the important question of how the dislocation structures are tied to the lattice (item 7). There is considerable disagreement on this question. Some prominent researchers in the field take the position that walls are strongly held in place by Lomer-Cottrell locks, and that such locks form the superstructure that allows the observed dislocation walls to exist. The primary experimental evidence supporting this viewpoint is a TEM study[62] looking at the recovery behavior of dislocation structures in pre-strained single-crystal samples of Cu and Ni. Upon heating, the dislocation structures displayed two distinct types of annihilation behavior. The first occurs slowly over several minutes, and is attributed to the slow climb and resulting annihilation of dipoles. The second mechanism occurs much more rapidly and is attributed to thermally activated penetration of pinning obstacles. These obstacles are considered to be the hard locks holding the walls in place. Other researchers consider Lomer-Cottrell locks to be incidental objects with no real significance. Their view is that walls are probably held in place by tangling which occurs through multiple cross-slip events. Experimental evidence supporting this viewpoint is the observation that the number of Lomer-Cottrell locks reported in TEM studies of dislocation walls is too small to be significant. Experimental evidence is therefore sparse on both sides of this argument, and a focussed TEM study of this issue would be welcome.

TEM clearly has an indispensable role to play in acquiring the quantitative data listed in the beginning of this section. The next sub-section will discuss the role of high-resolution diffraction imaging which complements TEM in several important ways.

High-Resolution X-ray Diffraction Imaging

High-resolution X-ray diffraction imaging (also known as X-ray topography) is a non-destructive characterization technique that images the defect microstructure of single crystals with a maximum resolution of ≈ 1 μm.[53–55] During a measurement, the sample is illuminated by an X-ray beam and an image of the diffracted beam is recorded. Image contrast results from local deviations from perfect long-range atomic order such as the strain field of a dislocation. The use of monochromatic X rays at a modern synchrotron source provides many advantages over other X-ray sources, and all of the discussion on diffraction imaging will be restricted to monochromatic synchrotron X rays. Also, although both reflection and transmission geometries are possible, transmission geometries provide the greatest information on internal defect structures, and reflection measurements will not be discussed in this paper.

Unlike the radiographic X-ray beam, the transmitted diffracted beam is able to pass through mm-thick samples. Thus, images can be formed from Al samples with thicknesses of ≈ 1 mm, guaranteeing that the resulting dislocation structures exhibit "bulk" rather than thin film behavior.

The diffraction imaging experiments described in this section were conducted on a dogbone shaped single-crystal tensile specimen of high purity Al (atomic fraction 99.99+ %) with a thickness of 1 mm. The crystal was grown using a soft-mold process similar to that of T. H. Alden.[63] Multiple samples with identical crystallographic orientation were produced by slicing the as-grown crystals in an acid saw. Samples used for the USAXS experiments described in the next subsection were produced in the same fashion. For a detailed description of the sample preparation procedure, the reader is referred to a paper by Levine, Long, and Thomson.[58]

For straining the samples *in situ*, we designed and built a computer-controlled low-profile tensile stage that is both small and light enough to be mounted on an X-ray goniometer.[64] It can deform a specimen by uniaxial tension *in situ* while simultaneously measuring the average longitudinal stress and the average linear strain.

Figure 1 shows a sequence of $\bar{2}00$ diffraction images (10 keV photon energy) of a sample that was deformed *in situ* to a maximum strain of 1.9 %. With a standard deviation of approximately $1.0°$, the tensile axis for this sample was [0.03 0.85 -0.53]. Thus, the largest Schmid factors (denoting the largest resolved shear stresses) are 0.42 for the [1 0 1] ($\bar{1}\,\bar{1}\,1$) slip system and 0.37 for the [1 1 0] ($\bar{1}\,1\,\bar{1}$). Since these Schmid factors are very close, it is expected that slip will occur on both the ($\bar{1}\,\bar{1}\,1$) and ($\bar{1}\,1\,\bar{1}$) slip planes. While each image was acquired, the sample was held at just below the flow stress (≈ 0.99).

The sequence of images in Fig. 1 shows the development of dislocation walls from diffuse, intermittent structures at 0.85 % strain to sharply defined walls at 1.9 % strain. Increased diffracted intensity appears black in these images, so the dislocation walls are white; they are aligned primarily along the diagonal directions. The white lines extending from the lower left to the upper right are dislocation walls that are parallel to the ($\bar{1}\,\bar{1}\,1$), and the similar, though less prominent, features on the other diagonal are ($\bar{1}\,1\,\bar{1}$) dislocation walls. The ($\bar{1}\,\bar{1}\,1$) planes are clearly the dominant slip planes for this region of the sample.

Since these images show just a single projection of the actual structures, it is not possible to determine accurately the wall thicknesses. However, the walls have relatively sharp edges where they exit the sample surface, and the spacings between these exit points can be used to determine the spacings of adjacent walls. Taking into account the geometrical projection of the image, we obtain ($\bar{1}\,\bar{1}\,1$) wall spacings ranging from 9 μm to 20 μm for 1.9 % strain. The average spacing is 13 μm. More complete structural information could be obtained by imaging stereo pairs.[65]

As mentioned above, a diffraction image does not show dislocations directly; what forms the image are the local deviations from perfect atomic order. This distinction is important for proper interpretation of diffraction images. An excellent example is the pair of bright horizontal lines that appear in these images. Although at first glance these appear to be continuous lines, a closer look shows that the "lines" are actually a series of bright spots that line up with the ($\bar{1}\,\bar{1}\,1$) dislocation walls. Whenever the ($\bar{1}\,\bar{1}\,1$) walls intersect these lines, they become brighter, indicating that the local displacements from perfect long range order are increased. The most likely explanation for this behavior is that some linear barrier to dislocation motion is present, causing a local increase in dislocation density

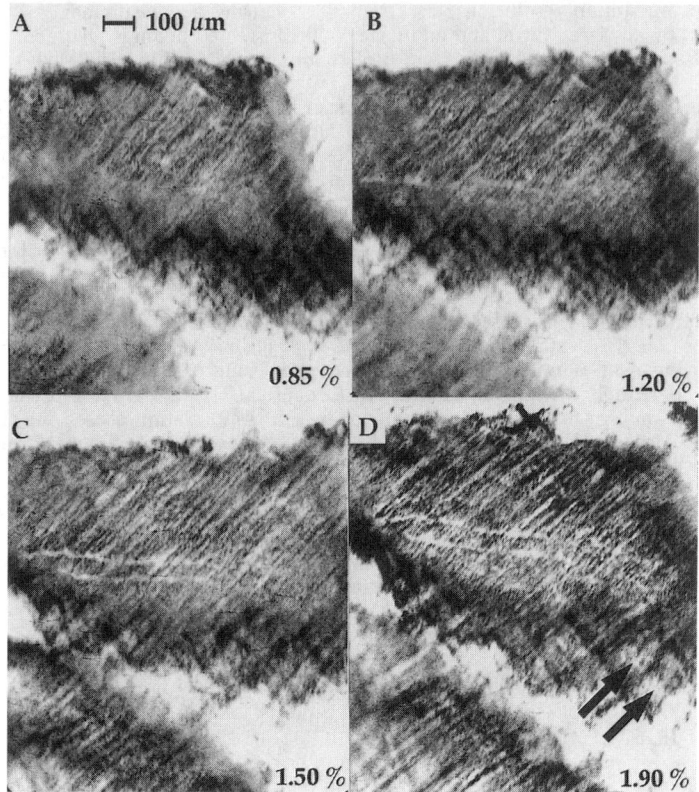

FIG. 1. Diffraction images from single crystal Al deformed *in situ* to strains of 0.85 %, 1.2 %, 1.5 %, and 1.9 % strain. The white lines are dislocation walls.

where the barrier intersects the dislocation walls. This explanation also accounts for the change in appearance of the "lines" in the sequence of images. As the strain increases, the number of dislocations that are affected by the barrier increases, resulting in a brightening of the lines. The barrier is probably a set of grown-in dislocations that lie on one of the remaining two (low resolved shear stress) slip planes. Further evidence is provided by the nearly uniform intensity and sharpness of the lines. This indicates that the depth of the linear barriers remains nearly constant. A horizontal orientation and a constant depth are consistent with dislocations on the (1 1 1) or (1 1̄ 1̄) planes, but not on the two principle slip planes.

Diffraction imaging is highly sensitive to sample orientation. As we will show, this sensitivity can be exploited to provide unique information on local misorientations (item 4). Figure 2 shows a calculated contour plot of the kinematic diffracted intensity from a square simple-cubic sample (200 atom edge length) containing 30 positive and 30 negative randomly positioned edge dislocations. The Burgers vectors all lie along the X-axis and the straight dislocation lines extend along the Z-axis which is assumed to be perfectly periodic. The plane of the figure is defined by the X and Y components of the scattering vector, $\mathbf{q}$, and the figure is centered on the expected position of the 100 reflection. The complicated lobe pattern of the reflection is caused by local atomic density (dilatation) and rotation fluctuations, which appear as deviations along the X and Y axes, respectively. If the

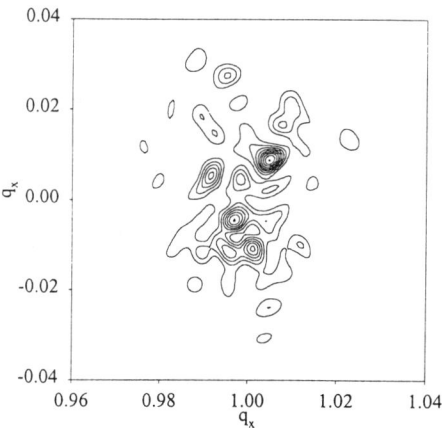

FIG. 2. Computer simulation of the 100 reflection from a 200 atom × 200 atom square sample containing 30 positive and 30 negative edge dislocations with random positions.

random collection of positive and negative dislocations were replaced by a tilt boundary, the complicated intensity distribution in the figure would reduce to two mirror image diffraction spots that are separated along the Y axis. The separation would be proportional to the misorientation across the tilt boundary.

Diffraction conditions are often described using a geometrical construction in reciprocal space called the Ewald sphere.[66] The Ewald sphere intersects the origin, and has a radius determined by the X-ray wavelength. Rotating the sample in the beam is equivalent to rotating the Ewald sphere in reciprocal space. Diffraction occurs when the Ewald sphere intersects a reciprocal lattice spot, such as the lobed 001 reflection in Fig. 2. The first lobes that are contacted by the Ewald sphere correspond to localized regions of the sample that, because of the fluctuations, are able to diffract first. These regions will be visible in a diffraction image and the adjacent, non-diffracting, regions will not. This explains why only portions of the sample are visible in Fig. 1. As the sample is rotated further, new lobes intersect the Ewald sphere, and a progression of local regions of the sample will be imaged. If the diffraction image is recorded continuously during a sample rotation (using an X-ray video system), the local fluctuations along one axis can be measured directly. By repeating this procedure using multiple reciprocal lattice spots and rotation axes, a misorientation map of the sample can, in principle, be constructed. In practice, it will likely prove impossible to obtain complete misorientation information from more than a small subset of the walls since only some of the dislocation walls will be visible (and identifiable) in all of the required image sequences.

A single rotation sequence from the sample shown in Fig. 1 was recorded on video tape. For a strain level of 1.9 %, corresponding to panel D, the image formation progressed in small distinct jumps, demonstrating that the misorientations result from geometrically necessary dislocations within the dislocation walls. The two arrows in Fig. 1D denote two regions that were compared as described above. The relative misorientation was measured to be $0.010° \pm 0.003°$. (All errors reported in this paper represent one standard deviation.) Since only a single rotation sequence was examined, this measured misorientation is just a projection of the true misorientation. Nevertheless, the general behavior of the image formation during rotation suggests that this measured rotation is a typical value for this sample at this strain level. The measurement also demonstrates diffraction imaging's capability of measuring misorientations that are much smaller than those which can be measured by TEM. For a tilt boundary in Al, a rotation of $0.010°$ corresponds to one uncompensated edge dislocation every 1.6 μm! Also, since diffraction imaging is a non-destructive probe that can be used *in situ*, the misorientations across specific walls could be followed as a function of strain in the system.

Diffraction imaging and TEM can be used together in a cooperative way to obtain information important to theory and modeling. Addressing item 1 at the beginning of this section, diffraction imaging can be used to study dislocation wall formation at the early stages of deformation. This can be accomplished *in situ* in thick specimens. At larger strains (greater than about 6 % to 8 %), the high density of dislocations seriously degrades the diffraction images. But at larger strains, the spatial distribution of dislocation walls can be studied *ex situ* by TEM on cross-sections. Similarly, misorientation studies (item 4) can be conducted using diffraction imaging at small strains and TEM at larger strains. For diffraction imaging studies using single crystals that are deformed by uniaxial tension, the corresponding TEM studies must also focus on such samples. Ideally, the samples for both studies would be cut from a single master crystal such that the tensile axis is the same in both sets of experiments.

Ultra-Small-Angle X-ray Scattering

USAXS can be used to provide quantitative, volume averaged, *in situ* information on dislocations and dislocation structures as they evolve during deformation. Some of the information obtainable from USAXS cannot be obtained using any other existing experimental technique. However, effective use of USAXS requires that it be used in conjunction with other experimental techniques such as TEM and X-ray diffraction imaging. This is because a quantitative interpretation of the USAXS data sometimes requires input from other techniques on parameters such as the orientations of the dislocation walls, and the orientations and Burgers vectors of the component dislocations.[56] In principle, all of this information could be extracted directly from an exhaustive set of USAXS data; in practice, the time required to acquire such a data set makes such an approach impractical.[58]

USAXS is particularly well suited for studies of dislocation structures over a broad range of length scales. Since the commissioning of the new USAXS facility on UNICAT sector 33 at the Advanced Photon Source,[67] the range of scattering vector that can now be probed in a single USAXS scan corresponds to length scales ranging from 10 Å up to 8 μm. Thus, a single scattering profile can provide information on the population of individual dislocations and dislocation dipoles, the organization of the dislocations into wall configurations, and the global structure of dislocation cells. The data presented in this paper, however, was obtained using the NIST USAXS facility at the National Synchrotron Light Source,[68] which only provides information on structures ranging in size from $\approx$ 200 Å to 1 μm. Thus, no information on dislocation cells was obtained.

As with X-ray diffraction imaging, USAXS can be used *in situ* on samples that are thick enough to exhibit bulk behavior. For Al, we typically use a 7 keV photon energy which makes it possible for us to avoid Bragg reflections while still allowing us to use samples with thicknesses of about 0.2 mm. The three samples described in this subsection were produced using the procedures described in the diffraction imaging subsection. The samples will be referred to as samples G, J, and P. Samples G and J were sliced from the same original crystal and thus have the same tensile axis. The diffraction imaging sample described above was also sliced from this crystal, allowing direct comparisons to be made. Thus, the diffraction imaging and USAXS experiments from these crystallographically identical samples provide an example of how these techniques can be used in a complementary fashion.

At the length scales probed in these experiments, USAXS from dislocations arises from scattering off the long-range dilatation fields. Thus, only the edge components of the dislocations are detectable. USAXS is also highly sensitive to the orientation of the scattering vector, $\mathbf{q}$, with respect to the dislocation lines and Burgers vectors. Thus, if $\mathbf{q}$ is oriented perpendicular to a given slip plane, then the dislocations on this plane are oriented for maximum scattering contrast, and dislocations on the other slip planes will not contribute significantly. This effect allows direct separate measurements of the increase in dislocation density on the primary and *secondary* slip planes (item 5).

Figure 3 shows absolute-calibrated, slit-smeared USAXS data from sample G, which was deformed *in situ* by uniaxial tension. Data are shown from strains ranging from 0.4 % to 6.7 %. The increase in scattering visible between 0.001 Å^{-1} and approximately 0.01 Å^{-1}, is directly proportional to the dislocation density on a secondary slip plane. A plot of this integrated intensity as a function of strain shows that the dislocation density remains constant out to approximately 3.5 % strain, and then starts to increase rapidly.[57,58] Although this general behavior is not surprising, the USAXS measurements can provide quantitative information on when the increase starts and how rapidly the dislocation density increases.

The data below 0.001 Å^{-1} in Fig. 3 provide information on the dislocation density distribution across the walls (item 2). As a first approximation, we assumed that the dislocation density was roughly constant except at the wall edges, where we assumed that the dislocation density decreased monotonically to zero. This distribution was modeled as the convolution of a square function with a Gaussian. The standard deviation of the Gaussian is thus a measure of the width of this interface region. The resulting predicted functional form for the scattering closely matches the observed behavior below 0.001 Å^{-1}. Fig. 4 shows the extracted interface widths (perpendicular to the walls) from samples G and J as a function of increasing strain. The results from both samples are consistent and show that the interface width decreases linearly with strain. To obtain these results, it was necessary to know the orientation of the walls with respect to the scattering vector. The diffraction images in Fig. 1 showed that two sets of dislocation walls were present, parallel to the ($\bar{1}\,\bar{1}\,1$) and ($1\,1\,\bar{1}$) planes. Since q was oriented nearly symmetrically between these two planes, both sets of walls are consistent with the data and we are not able to distinguish between them. Additional information on the orientations of the component dislocations is required to distinguish between these two possibilities.

Figure 5 shows USAXS data from sample P, which was deformed along the [0.66 0.42 0.62] axis to 10 % strain. This sample was oriented for maximum scattering contrast of the primary slip plane which is the ($\bar{1}\,1\,\bar{1}$). As before, the data below 0.001 Å^{-1} reflects the dislocation distribution across the walls. The desmeared equivalent slope of -4 indicates that the dislocation walls are very sharp, with an average interface width less than approximately 400 Å. At larger q, the partially buried peaks visible on top of a desmeared equivalent slope of -2 demonstrate that a fraction of the primary dislocations occupy ordered positions. The underlying slope tells us that the component dislocations are individual dislocations rather than dislocation dipoles. The positions of the peaks tell us that the dislocations are all of the same sign, with an apparent average spacing of 1830 Å $\pm$ 130 Å. Since this spacing is actually a projection, knowledge of the orientation of the walls is required to extract the true spacing. However, the strength of the peaks strongly suggests that the walls are aligned for maximum visibility, and thus are tilt boundaries. Making this assumption allows us to calculate the mean lattice rotation across the walls to be 0.090° $\pm$ 0.006°. The data also indicates that the distribution of misorientation angles has a peaked shape with a half width at half maximum of 0.03°, a result that is in qualitative agreement with the TEM study discussed previously.[61] Thus, USAXS is a third technique that can be used for measuring the distribution of misorientation angles in deformed samples (item 4). More important, however, is the demonstrated capability of conducting *in situ* measurements of the spatial correlations between dislocations (item 3). Such correlations have been observed in all of the samples that have been oriented for high dislocation scattering contrast. In some of these samples, the spatial correlations were observed to change with time during creep.[58]

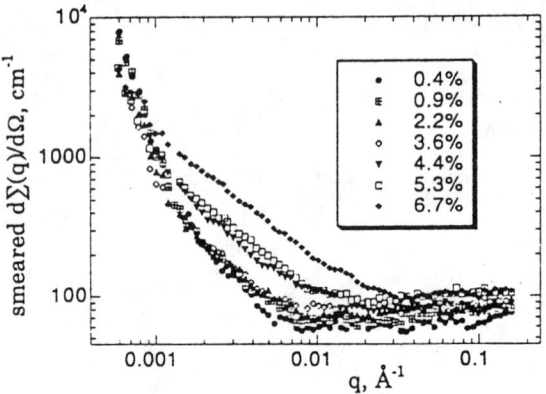

FIG. 3. Absolute calibrated, slit-smeared ultra-small-angle X-ray scattering data obtained *in situ* from sample G at strains ranging from 0.4 % to 6.7 %. The one sigma errors are given approximately by the symbol size.

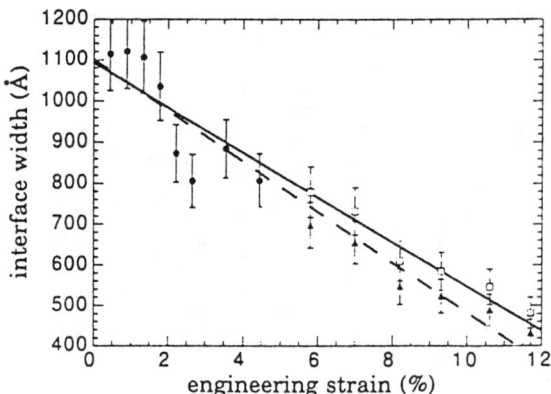

FIG. 4. Interface widths of dislocation wall boundaries in samples G (solid circles) and J. The squares and triangles are data from sample J assuming scattering from ($\bar{1}$ 1 $\bar{1}$) and ($\bar{1}$ $\bar{1}$ 1) planes, respectively. The lines are linear fits.

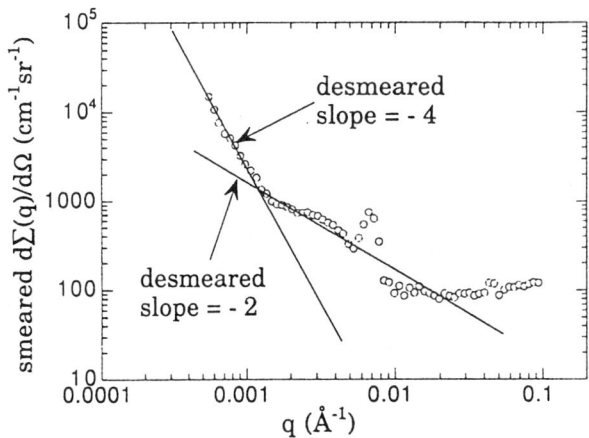

FIG. 5. USAXS data from sample P, strained *in situ* to 10 % strain. The peak structure is produced by correlations in the dislocation spacing. The one sigma errors are given approximately by the symbol size.

CONCLUSIONS

Obtaining quantitative information for modelers and theorists is most difficult at the intermediate length scale that includes the formation of dislocation structures in single grains. The experimental techniques discussed in the previous section, when used together, can provide quantitative information that can be used to answer in part the most pressing questions. Some of these questions can be answered almost completely. For example, measurements of the misorientations across dislocation walls can be made using TEM, diffraction imaging, and USAXS. When used together, these techniques can provide quantitative rotational information from the beginnings of deformation to sample failure. Other questions, such as those concerning the dislocation-wall stress fields and the importance of dislocation junction formation, are much more difficult to address, and only partial answers can be obtained at this time. In some cases, it is likely that new experimental techniques will be needed. Several research groups around the world are currently involved in developmental work, and improvements in our ability to supply information required for multi-scale modeling are inevitable.

The fundamental science of plasticity is burgeoning, with most of this activity concentrated in computer modeling. However, the multi-scale and multi-disciplinary nature of this work greatly increases the requirement for experimental input. The successful coupling of the relevant models and theories *must* include guidance and verification from experiments at all levels. As discussed in this paper, much of the required information can be obtained using currently available experimental techniques. Unfortunately, the number of experimentalists working in this field is far below what is required to support adequately the existing level of modeling activity. The greatest need is for experimentalists who are willing to work collaboratively using a variety of complementary techniques, and who will also interact directly with the theory and modeling communities.

ACKNOWLEDGMENTS

We thank H. E. Burdette for measuring the orientation of our single-crystal samples using Laue diffraction. The NSLS is operated under contract no. DE-AC02-76CH00016 with the U. S. Department of Energy.

REFERENCES

[1] L. P. Kubin and G. Canova, Scripta Metall., **27**, 957 (1992).

[2] H. M. Zbib, M. Rhee, and J. P. Hirth, in "Advances in Engineering Plasticity and Its Applications," Ed. T. Abe and T. Tsuta, Pergamon, Oxford, 15 (1996).

[3] K. W. Schwarz, J. Appl. Phys., **85**, 108 (1999).

[4] Michael Zaiser and Peter Hähner, Phil. Mag. Lett., **73**, 369 (1996).

[5] Peter Hähner, Karlheinz Bay, and Michael Zaiser, Phys. Rev. Lett., **81**, 2470 (1998).

[6] R. Thomson and L. E. Levine, Phys. Rev. Lett., **81**, 3884 (1998).

[7] B. Bakó and I. Groma, Phys. Rev. B, **60**, 122 (1999).

[8] A. Portevin and F. Le Chatelier, C. R. Acad. Sci. Paris, **176**, 507 (1923).

[9] F. Mertens, Scott V. Franklin, and M. Marder, Phys. Rev. Lett., **78**, 4502 (1997).

[10] P. Hähner, Acta Mater., **45**, 3695 (1997).

[11] A. P. Sutton and R. W. Balluffi, "Interfaces in Crystalline Materials," Clarendon Press, Oxford, 704 (1995).

[12] L. Kubin, in "Treatise in Materials Science and Technology," Vol. 6, Ed. R. W. Cahn, P. Haasen and E. L. Kramer, VCH-Weinberg (1993).

[13] J. Gil Sevillano, in "Treatise in Materials Science and Technology", Vol. 6, Ed. R. W. Cahn, P. Haasen and E. L. Kramer, VCH-Weinheim (1993).

[14] A. Argon, in "Physical Metallurgy", Ed. R. W. Cahn and P. Haasen, Pergamon, New York (1997).

[15] D. Kullman-Wilsdorf, Mat. Res. Innovat., **1**, 265 (1998).

[16] S. J. Zhou, D. L. Preston, P. S. Lomdahl, and D. M. Beazley, Science, **279**, 1525 (1998).

[17] D. Rodney and R. Phillips, Phys. Rev. Lett., **82**, 1704 (1999).

[18] L. K. Wickham, K. W. Schwarz, and J. S. Stolken, unpublished

[19] J. S. Chung and G. E. Ice, J. Appl. Cryst., **86**, 5249 (1999).

[20] L. E. Levine, A. J. Allen, and Thomas Gnaupel-Herold, unpublished.

[21] J. A. Moriarty, Phys. Rev. B, **42**, 1609 (1990).

[22] J. A. Moriarty, Phys. Rev. B, **49**, 12431 (1994).

[23] W. Xu and J. A. Moriarty, Phys. Rev. B, **54**, 6941 (1996).

[24] Geoffrey H. Campbell, Walter L. Wien, Wayne E. King, Stephen M. Foiles, and Manfred Ruhle, Ultramicroscopy, **51**, 247 (1993).

[25] Michael J. Mills and Pierre Stadelmann, Phil. Mag. A, **60**, 355 (1989).

[26] Michael J. Mills, Murray S. Daw, and Stephen M. Foiles, Ultramicroscopy, **56**, 79 (1994).

[27] Brad Lee Holian and Peter S. Lomdahl, Science, **280**, 2085 (1998).

[28] K. W. Schwarz, J. Appl. Phys., **85**, 120 (1999).

[29] F. K. LeGoues, unpublished.

[30] B. Devincre and L. P. Kubin, Mod. Simul. Mater. Sci. Eng., **2**, 559 (1994).

[31] T. Foecke and D. vanHeerden, in "Chemistry and Physics of Nanostructures and Related Non-Equilibrium Materials," Ed. E. Ma, B. Fultz, R. Shull, J. Morral, and P. Nash, TMS, Pittsburgh, 193, (1997).

[32] R. K. Ham, Phil. Mag **7**, 1177 (1962). 741 (1980).

[33] A. Seeger, *The Relation Between Structure and Mechanical Properties of Metals* **1**, H.M.S.O., London, p. 3 (1963).

[34] P. B. Hirsch, *The Relation Between Structure and Mechanical Properties of Metals* **1**, H.M.S.O., London, p. 39 (1963).

[35] Siegfried Mader, Alfred Seeger, and Hans-Martin Thieringer, J. Appl. Phys., **34**, 3376 (1963).

[36] Z. S. Basinski, Discuss. Faraday Soc. **38**, 93 (1964).

[37] J. W. Steeds, Proc. Roy. Soc. **A292**, 343 (1966).

[38] E. Göttler, Phil. Mag. **28**, 1057 (1973).

[39] F. Prinz and A. S. Argon, Phys. Stat. Sol. **A57**, 741 (1980).

[40] C. T. Young, T. J. Headley, J. L. Lytton, Mater. Sci. Eng. **81**, 391 (1986).

[41] D. A. Hughes, Acta Met. et Mater. **41**, 1421 (1996).

[42] F. W. Young, Jr., J. Appl. Phys. **32**, 192 (1961).

[43] G. A Alers and K. Salama, *Dislocation Dynamics*, Ed. A. R. Ronsenfield, G. T. Hahn, A. L. Bement, Jr., and R. I. Jaffee, McGraw-Hill, Inc., New York, 211 (1968).

[44] H. G. Van Bueren, *Imperfections in Crystals*, North-Holland Publishing Company, Amsterdam, Chapter VI (1961).

[45] H. G. Van Bueren, *Imperfections in Crystals*, North-Holland Publishing Company, Amsterdam, Chapter XV (1961).

[46] R. C. Reno, L. J. Swartzendruber, and L. H. Bennett, NDT International, October, 224 (1979).

[47] Eiji Hashimoto, Yoshitake Ueda, Nobuyuki Uematsu, Masayuki Iwami, and Takao Kino, J. Phys. Soc. Japan **61**, 3799 (1992).

[48] M. Wilkens, Phys. Status Solidi, A**2**, 359 (1970).

[49] M. A. Krivoglaz, *X-ray and Neutron Diffraction in Nonideal Crystals*, Springer, Berlin (1996).

[50] L. E. Levine and Robb Thomson, Acta Cryst., A**53**, 590 (1997).

[51] H. Mughrabi, T. Ungár, W. Kienle, and M. Wilkens, Phil. Mag. A**53**, 793 (1986).

[52] T. Ungár, Mat. Sci. For. **166-169**, 23 (1994).

[53] "Direct Observation of Imperfections in Crystals," Ed. by J. B. Newkirk and J. H. Wernick, Interscience Publishers, New York (1961).

[54] D. Keith Bowen and Brian K. Tanner, "High Resolution X-ray Diffractometry and Topography," Taylor & Francis, London (1998).

[55] Bruce Steiner, L. E. Levine, T. C. Cull and C. S. Ray, J. Non-Cryst. Sol. **204**, 13 (1996).

[56] Robb Thomson, L. E. Levine, and G. G. Long, Acta Cryst. A**55**, 433 (1999).

[57] G. G. Long, L. E. Levine, and Robb Thomson, J. Appl. Cryst., in press.

[58] L. E. Levine, G. G. Long, and Robb Thomson, submitted to Phys. Rev. B (1999).

[59] A. S. Argon and P. Haasen, Acta metall. mater. **41**, 3289 (1993).

[60] H. Mughrabi, in "Continuum Models of Discrete Systems 4," Ed. by O. Brulin and R. K. T. Hsieh, North-Holland Pub. Comp. (1981).

[61] D. A. Hughes, Q. Liu, D. C. Chrzan, and N. Hansen, Acta mater. **45**, 105 (1997).

[62] F. Prinz, A. S. Argon, and W. C. Moffett, Acta metall., **30**, 821 (1982).

[63] T. H. Alden, Rev. Sci. Instr., **31**, 897 (1960).

[64] L. E. Levine and R. J. Fields, NISTIR 5867, U. S. Department of Commerce, National Institute of Standards and Technology (1996).

[65] Kyoichi Haruta, J. Appl. Phys., **76**, 1789 (1965).

[66] B. E. Warren, "X-ray Diffraction," Dover, Mineola, N. Y., 19 (1990).

[67] G. G. Long, A. J. Allen, J. Ilavsky, P. R. Jemian, and P. Zschack, Abstracts for Synchrotron Radiation Instrumentation (SRI-XI), Stanford, CA, 44 (1999).

[68] G. G. Long, P. R. Jemian, J. R. Weertman, D. R. Black, H. E. Burdette, and R. Spal, J. Appl. Cryst., **24**, 30 (1991).

DISLOCATIONS AND INTERNAL STRESSES IN THIN FILMS: A DISCRETE-CONTINUUM SIMULATION

C. LEMARCHAND † , B. DEVINCRE ‡, L.P. KUBIN ‡ and J.L. CHABOCHE †
† ONERA, DMSE, BP 72, 92322 Chatillon Cedex, France
‡ Laboratoire d'Etude des Microstructures, CNRS-ONERA, BP 72, 92322 Chatillon, France

INTRODUCTION

The plasticity of thin films and layers is of considerable technological interest. For instance, the relaxation of internal stresses in semiconducting epitaxial layers has been the object of many studies [1, 2]. This relaxation is usually treated via the concept of critical thickness, the latter being defined as the maximum layer thickness below which dislocations cannot spontaneously move and relax the internal stresses. The various internal stresses present in epitaxial layers (e.g. the misfit and elastic incompatibility stresses at the film/substrate interface and the image force in a free-standing film) can be computed within a continuum frame. However, the way they influence the motion of a dislocation has not yet been computed, even in a approximate manner. An useful approximation that allows treating the boundary condition at the surface of a free-standing film consists of making use of the concept of image dislocation. Then, the critical stress for moving a dislocation in a free-standing film is the same as that of a capped layer of thickness twice that of the film. To date, models and dislocation dynamics (DD) simulations are available that involve several levels of approximation for the treatment of the dislocation/interface and dislocation/surface interactions [3-7] . For reasons that are not clearly understood, however, these models predict critical thicknesses that are systematically larger than the expected ones. The comparison with experiment is, in addition, made difficult because stresses have to be artificially introduced to replace the internal stresses and approximations have to be done to treat the image stresses. In the present work it is shown that it is now possible to fully account for the contribution of the various sources of internal stresses to the critical stress for the motion of a threading dislocation. This is performed numerically with the help of a hybrid code that combines a DD code for the treatment of the dislocation dynamics and a Finite Element (FE) code for the treatment of the boundary conditions. In what follows, several applications of this discrete-continuum model (DCM) to the study of dislocation motion in epitaxial layers are presented. The motion of a dislocation in a thin film is considered, including the image force and successively adding a misfit stress and an elastic incompatibility stress at the film/substrate interface.

NUMERICAL METHOD

The DD code used in the present work has been described in detail in several previous publications [8, 9]. It was recently applied to the problem of dislocation motion in a capped layer [7]. In short, it is based on a discretization of time, space and dislocation character. The elastic properties of dislocations (line tension, curvature, interactions) are fully accounted for within a linear elastic frame. Other properties, like mobility, core and contact properties are treated via local rules. In the present case, the motion of a single dislocation will be considered, with a mobility of the viscous drag type. In the DCM, the DD code is coupled to a FE code, ZeBuLon, jointly developed by ENSMP and ONERA. The coupling between the two codes and its validation have been described in [10]. In essence, the FE code treats the

Mat. Res. Soc. Symp. Proc. Vol. 578 © 2000 Materials Research Society

boundary value problem and cares of the conditions of local equilibrium in the considered representative volume element. The DD code cares of dislocation motion in the same volume and, hence, of the plastic strain. A full discussion of the hybrid DCM is presented in [11]. One important difference between conventional DD codes and the DCM code resides in the initial conditions. In order for the FE code to be able to properly account for the presence of a dislocation, the latter has to be introduced by a Volterra process. In practice, the Volterra cut is made by inserting from a free surface either a prismatic loop or a dislocation line ending at the surface. The dislocation line is then moved in place, usually in the centre of the crystal. Given this initial plastic shear, the FE part of the DCM code solves the boundary conditions at the free surfaces, the conditions of stress equilibrium in the volume and the stress field of the dislocation in the finite simulation box [10]. The DCM simulations are then started with this configuration. One may notice that the length scale of the DCM code is then provided by the elementary shear involved in the Volterra process, i.e., by the Burgers vector, b, of the dislocation.

In what follows, the elastic constants of the thin film are those of copper, either in the approximation of isotropic elasticity or within a full anisotropic treatement like in the last calculation presented here. The lattice parameter, from which the Burgers vectors used in what follows can be deduced, is also that of copper (a = 0.362 nm). The representative volume element consists of a substrate and a copper thin film, with a (001) interface, as shown in fig. 1 b). In all the numerical experiments, the simulated volume is meshed with 1575 elements, 15x15x5 for the thin film and 15x15x2 for the substrate.

RESULTS

Image Force

In a first test, only the image force at the surface of the free-standing film is considered. The simplified geometry showed in figure 1 a) is adopted. An edge dislocation of Burgers vector $b = [010]$, moving in a (100) plane, is introduced through the $x = 0$ face and moved to the centre of the simulation cube, thus setting the initial conditions. This line, of edge character, is glissile in a plane perpendicular to the interface and deposits screw dislocations on it. This configuration presents little practical interest, since screw dislocations do not relax internal stresses, but it is suited for checking the capped layer approximation. Two film thicknesses were defined, H_f = 12.7 or $127nm$ ($50b$ or $500b$). To eliminate the misfit and elastic incompatibility stresses, both the substrate and the thin film are assumed to be made of the same isotropic material, copper. Thus, the only condition set at the interface is that it cannot be penetrated by the dislocation moving in the thin film. The simulated volume is then loaded at a constant shear rate by imposing the following constant displacement rates at the external surfaces: $(-\dot{u}, 0, 0)$ on $y = 0$ and $(\dot{u}, 0, 0)$ on $y = H$, $(0, -\dot{u}, 0)$ on $x = 0$ and $(0, \dot{u}, 0)$ on $x = H$, where $H = H_f + H_s$ is the total thickness of the film plus substarte.

Under load, the dislocation bows out and starts moving in an irreversible manner when a critical stress is reached. In the usual conditions of mesoscopic simulations, the critical value of an applied stress is easily defined, since the applied stress is uniform in the considered volume. It is not so within the DCM, as the stress field that is computed is the full solution of the boundary value problem. To derive the critical applied stress, a loading sequence is performed in the absence of the dislocation, thus defining a loading stress vs.

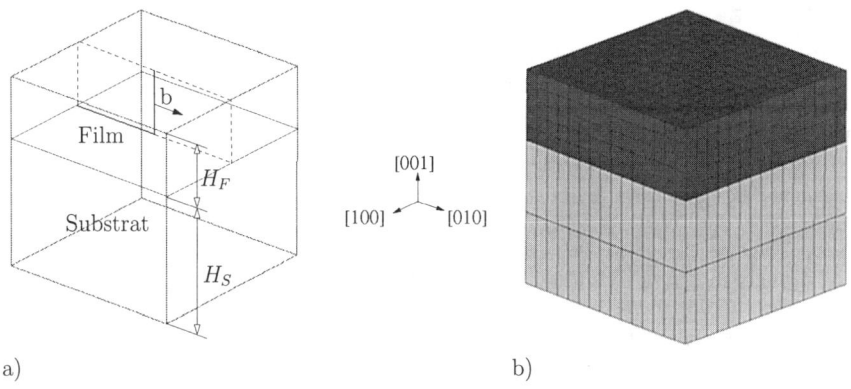

a)   b)

Figure 1: a) Initial configuration in the DD part of the code. The slip plane of the dislocation segment is materialized by the dashed lines. b) FE meshing of the representative volume element.

time curve. The critical stress is then defined as the value of the loading stress in the absence of dislocation at the time corresponding to the onset of irreversible dislocation motion in the full simulation.

The critical stresses obtained for the two thickness values are compared in figure 2 with the results yielded by the DD simulation alone (i.e., not coupled with a FE code). In the latter, the same geometry is used and the moving dislocation is simply confined between two linear boundaries representing the substrate, the film thickness being twice the thickness of the free-standing film. We see that the present results are in quite good agreement with those obtained within a simple two-dimensional approximation of the capped layer. This is not surprising since the configuration of an edge dislocation trailing a screw segment is the one for which the approximation of the image dislocation is the most accurate. The model of the capped layer should, of course, be less effective in the more realistic geometry considered below.

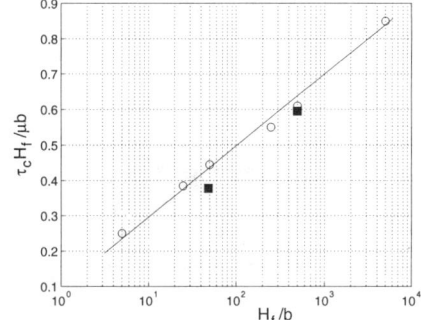

Figure 2: Critical stress for the motion on an edge dislocation in a thin film as a function of the film thickness (in reduced units). μ is the shear modulus of the thin film, H_f its thickness and b the magnitude of the Burgers vector of the mobile dislocation. The filled squares represent the two values yielded by the DCM. For comparison, the values yielded by the DD simulations of Gómez et al. [7] within the capped layer approximation (open circles) are also indicated. The full line is drawn as a guide to the eye.

Misfit Stress

In a second step, a misfit stress is introduced by mimicking a thermal dilatation between the film and the substrate. The misfit strain is adjusted to the value $\epsilon_0 = -0.3\%$, which is typical of real interfaces. The negative sign of the misfit strain indicates here that the thin film is submitted to compression stresses.

The numerical experiment is conducted in conditions that are now closer to those found in real layers. The threading dislocation introduced into the DCM is a screw segment of Burgers vector $a/2[101]$, which glides in a $(\bar{1}01)$ plane inclined with respect to the interface plane. Thus, the segment deposited at the interface is of edge character and its self-stress field strongly interacts with the misfit stress field. In this case, like in real layers, no loading is imposed. Then, the threading dislocation may or may not move under the effect of the misfit stresses, depending upon the film thickness and the value of the misfit strain. In the present case, the thickness is $H_f = 500b$, larger than the critical thickness, so that dislocation motion occurs.

The internal stress field induced by the misfit acts on the dislocation via the projection of the corresponding Peach-Koehler stress on its glide plane. This stress is written $\tau = \frac{1}{2}(\sigma_{33} - \sigma_{11})$ and the isovalues of 2τ in the absence of the threading dislocation are shown in figure 3. This stress field is positive in most of the slip area but changes sign near the free surface. Here and it what follows, this bending of the thin film is due to the boundary conditions adopted, according to which traction-free conditions are imposed to the faces of the substrate. Conditions involving rigid substrate faces are not considered here but can be implemented easily.

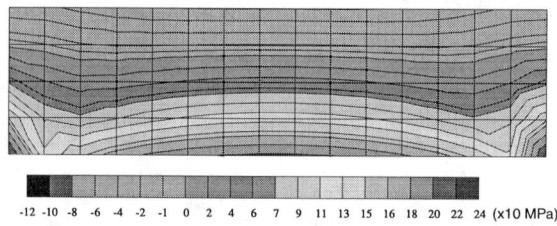

-12 -10 -8 -6 -4 -2 -1 0 2 4 6 7 9 11 13 15 16 18 20 22 24 (x10 MPa)

Figure 3: Isovalues of the stress $(\sigma_{33} - \sigma_{11})$ (in MPa) generated by a misfit strain of -0.3% and resolved in the slip plane of the threading dislocation (the latter has not been introduced at this step). The stresses are negatives near the free surface (top) and positive near the interface (bottom).

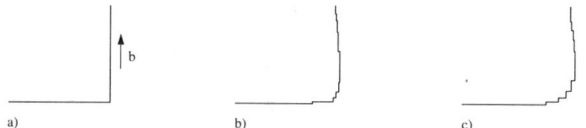

Figure 4: Successive shapes of the threading dislocation. a) Initial condition. b) During bowing out. c) During motion.

When the dislocation line is introduced, it bows out (*cf.* figure 4 b) and moves (*cf.* figure 4 c) under the effect of the misfit stress. As the interface is inclined with respect to the plane of figure 4, it follows that the dislocation line does not end perpendicular to the free surface. This feature occurs when the Burgers vector of the line is not perpendicular to the free surface. It is known to result from a compromise between two adverse effects: the minimization of the line length favors a normal angle whereas the dependence of the line energy on character favors a local screw orientation. The curvature of the moving dislocation is not uniform, which reflects the change in sign of the internal stress close to the free surface. As a result, it appears that the mobility of the dislocation segment is reduced close to the free surface. This effect cannot be accounted for within the approximation of the capped layer. The latter, thus, overestimates the global mobility of the dislocation or underestimates the critical stress value for its motion. To account for this discrepancy, an effective film thickness, smaller than the real one should be defined. Finally, the motion of the dislocation is caused by the attractive nature of the self-stress and misfit stress fields. In other words, this motion is favored because the edge segment deposited at the layer/substrate interface tends to relax the misift stress. Therefore, if the dislocations that spontaneously move in a thin film are such that they relax the misfit stresses, the hardening effect, with respect to the capped layer approximation, is always present whatever the sign of the misfit.

Full Calculation

In this last example, the geometry used is the same as in the previous section and all the possible sources of internal stresses are simultaneously accounted for. The image force at the free surface and the misfit stress are computed with the same numerical values ($H_f = 500b$, $\epsilon_0 = -0.3\%$) and for the same slip system as before. Elastic incompatibility stresses at the layer/substrate interface are introduced by assuming that the copper film is deposited on a silicon substrate. The elastic constants for the two materials are now the anisotropic ones. They essentially differ by the C_{12} modulus, 120 MPa in Cu and 64 MPa in Si.

The structure of the internal stress field is qualitatively similar to the one shown in figure 3. Under this field, the dislocation segment moves, achieving a shape that is also qualitatively similar to that of figure 4. An important difference, however, resides in the fact that in the present case, the dislocation stops before reaching the edge of the thin film. A comparison of the behaviour of the initially screw segment, with and without elastic incompatibility stresses, is shown in figure 5. In this graph, one of the two non-zero components of the plastic strain tensor ϵ^p is plotted as a function of time during the motion of the threading dislocation. One can see

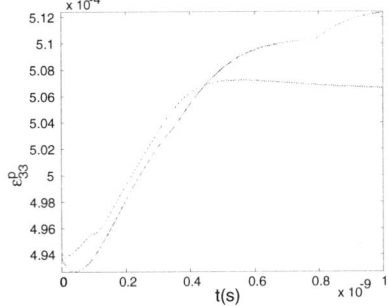

Figure 5: Time dependence of the plastic strain component ϵ^p_{33} during the motion of an initially screw threading dislocation. The full and dotted curves were obtained, respectively, without and with the elastic incompatibity field at the interface between the layer and the substrate. In the last case, the saturation of the plastic strain is due to the blocking of the dislocation. The non-zero initial strain values stem from the Volterra process by which the dislocation segment was introduced.

that the elastic incompatibility stress field at the interface induces an additional hardening, as the plastic strain produced by the threading dislocation saturates early in its presence.

DISCUSSION AND CONCLUDING REMARKS

These three examples illustrate the potential interest of the DCM for treating the motion of threading dislocations in a thin film, taking fully into account the complexity of the internal stress field. In the present case, the latter includes the self-stress of the dislocation in a bounded medium, the misfit stress and the elastic incompatibility stress at the internal interface, within an elastic treatment that can be either isotropic or anisotropic. It is worth mentioning that the film thickness, as governed by the DD component of the hybrid code, can be adjusted to values as low as a few Burgers vectors. As was shown here, the more realistic is the situation considered, given the film thickness, the smaller is the effective thickness to be considered. It is suggested that this may partly explain the present discrepancies between experiment and modeling. However, this result has to be confirmed with more realistic boundary conditions that the ones used in the present study. Finally, a wealth of different situations is expected to occur, depending upon the nature of the slip system considered, the orientation of the interphase, the nature of the materials considered and the possible presence of misfitting dislocations. An account of the most important cases realized in practice will be presented in a forthcoming publication.

REFERENCES

[1] P. Mooney. Strain relaxation and dislocations in SiGe/Si structures. *Mat. Sci. Engng.*, **R17**, 105-146, 1996.

[2] R. Beanland, D. Dunstand, and P. Goodhew. Plastic relaxation and relaxed buffer layers for semiconductor epitaxy. *Adv. Phys.*, **45**, 87-146, 1996.

[3] L. Freund. A criterion for arrest of a threading dislocation in a strained epitaxial layer due to an interface dislocation in its path. *J. Appl. Phys.*, **68**, 2073-2080, 1990.

[4] L. Freund. The mechanics of dislocations in strained-layer semiconductor materials. *Adv. Appl. Mechanics*, **30**, 1-66, 1994.

[5] K. Schwarz and Terzoff. Interaction of threading and misfit dislocations in a strained epitaxial layer. *Appl. Phys. Lett.*, **69**, 1220-1222, 1996.

[6] K. Schwarz. Simulation of dislocations on the mesoscopic scale. II: Application to a strained-layer relaxation. *J. Appl. Phys.*, **85**, 120-129, 1999.

[7] D. Gómez-García, B. Devincre, and L.P. Kubin. Dislocation dynamics in confined geometry. *Journal of Computer-Aided Materials Design*, 1999. In press.

[8] L.P. Kubin, G. Canova, M. Condat, B. Devincre, V. Pontikis, and Y. Brechet. Dislocations microstructures and plastic flow: a 3D simulation. *Solid State Phenom.*, **23-24**, 455-472, 1992.

[9] B. Devincre and L. P. Kubin. Mesoscopic simulations of dislocations and plasticity. *Mat. Sci. Engng*, **A234-236**, 8-14, 1997.

[10] C. Lemarchand, B. Devincre, L. P. Kubin, and J. L. Chaboche. Coupling of mesoscopic and macroscopic simulations of plastic deformation. In *MRS Symp. Proc.*, **538**, 63-68, 1999.

[11] C. Lemarchand. *De la dynamique des dislocations a la mecanique des milieux continus.* Doctoral Thesis, Universite Pierre et Marie Curie, 1999.

SYNCHROSHEAR VERSUS CONVENTIONAL SHEAR MECHANISMS IN TRANSITION-METAL DISILICIDES WITH THE C40 STRUCTURE

H. INUI and M. YAMAGUCHI

Department of Materials Science and Engineering, Kyoto University, Sakyo-ku, Kyoto 606-8501, Japan

ABSTRACT

The deformation behavior of (0001) $<1\bar{2}10>$ basal slip in single crystals of five different transition-metal disilicides with the C40 structure has been investigated in the temperature range from room temperature to 1500°C in compression. These disilicides are found to be classified into two groups depending on the onset temperature for plastic flow. The low-temperature group, which consists of VSi_2, $NbSi_2$ and $TaSi_2$, exhibits the onset temperature for plastic flow around 0.3 T/T_m (melting temperature) and deforms by a conventional shear mechanism. In contrast, the high temperature group, which consists of $CrSi_2$ and $Mo(Si,Al)_2$, exhibits the onset temperature around 0.6 T/T_m and deforms by a synchroshear mechanism. Factors affecting the deformation mechanism in these C40 disilicides are discussed in terms of directionality of atomic bonding and the relative stability of the C40 phase with respect to the $C11_b$ phase.

INTRODUCTION

Transition-metal disilicides with the $C11_b$ and C40 structures have been of great interest as structural materials to be used in oxidizing environments at temperatures higher than the upper limit for Ni-base superalloys [1-3]. These two structures are closely related to each other. The atomic arrangements on {110} in the tetragonal $C11_b$ structure is identical to that on (0001) in the hexagonal C40 structure and $<111>_{C11b}$ // $<1\bar{2}10>_{C40}$ if the axial ratio (c/a) of the $C11_b$ structure is idealistically $6^{1/2}$. As seen in Fig. 1, showing the $C11_b$ structure with $c/a=6^{1/2}$ projected normal to {110}, there are four equivalent positions, A~D for stacking {110} planes assuming the stacking is made in such a way that transition metals occupy saddle positions directly above Si-Si bonds [3]. The $C11_b$ and C40 structures differ from each other only in the sequence of the stacking of such atomic layers; the former structure is based on the AB stacking sequence while the latter structure is based on the ABC stacking sequence.

Four transition-metals, V, Cr, Nb and Ta, are known to form binary C40 disilicides with Si. Deformation experiments so far made on single crystals of these C40 disilicides [4-9] have indicated that basal slip is the only operative slip system and that while the three C40 disilicides, VSi_2, $NbSi_2$ and $TaSi_2$, deform at low temperatures such as room temperature, plastic flow is observed only above 700°C for $CrSi_2$. However, the reason why $CrSi_2$ behaves differently from the other disilicides has not been addressed. Of these disilicides, $CrSi_2$ exhibits lattice properties different from those of the other three C40 disilicides in many respects. For example, the melting temperature of $CrSi_2$ is quite low when compared to that of the other C40 disilicides and its electrical conductivity is semiconducting while the others exhibit a metallic conductivity [3]. In addition, the directionality in atomic bonding is recently reported to be the most significant for $CrSi_2$ of the four C40 disilicides when judged from their elastic constants [10].

In the present study, we investigate the deformation behavior of single crystals of not only these four binary C40 disilicides but also Al-added ternary C40 $Mo(Si,Al)_2$ in order to gain insight into deformation mechanisms of C40 disilicides.

Mat. Res. Soc. Symp. Proc. Vol. 578 © 2000 Materials Research Society

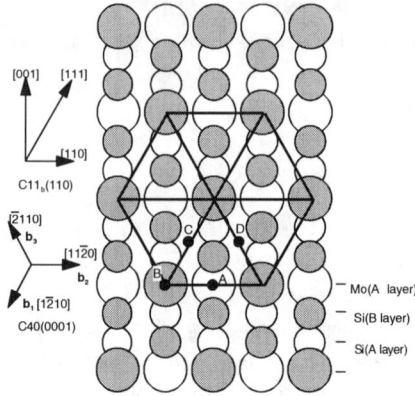

Fig. 1. Atomic arrangements on {110} planes in the C11$_b$ structure and (0001) planes in the C40 structure. The stacking positions of A-D and crystallographic directions in both structures are indicated while $\mathbf{b_1}$-$\mathbf{b_3}$ indicate Burgers vectors of 1/6<1$\bar{2}$10>-type partial dislocations in the C40 structure.

EXPERIMENTAL PROCEDURES

Single crystals of VSi$_2$, CrSi$_2$, NbSi$_2$, TaSi$_2$ and Mo(Si$_{0.85}$Al$_{0.15}$)$_2$ were grown with our ASGAL FZ-SS35W optical floating-zone furnace at a growth rate of 10 mmh^{-1} under an Ar gas flow. Specimens with dimensions 1.8 x 1.8 x 5 mm^3 for compression tests were cut from as-grown single crystals and then mechanically polished with diamond paste. The compression axis orientation investigated is [1$\bar{2}$12], which is inclined from the c-axis by about 45 degrees, to facilitate the operation of basal slip.

Compression tests were conducted on an Instron-type testing machine in vacuum in the temperature range from room temperature to 1500°C at a strain rate of 1x10^{-4} s^{-1}. Operative slip planes were determined by slip trace observations on two orthogonal surfaces of specimens by optical microscopy. Dislocation structures and dislocation Burgers vectors were examined by transmission electron microscopy (TEM) with a JEM-2000FX electron microscope operated at 200 kV. Dislocation core structures were examined by high-resolution transmission electron microscopy (HRTEM) with a JEM-4000EX electron microscope operated at 300 kV.

RESULTS

Deformation behavior

Critical resolved shear stresses (CRSS's) for basal slip in five different C40 disilicides are plotted in Fig. 2 as a function of temperature. As seen in Fig. 2, the onset temperature for plastic flow varies from disilicide to disilicide and ranges from 400°C for VSi$_2$ to 1100°C for Mo(Si,Al)$_2$. These onset temperatures for plastic flow observed presently in VSi$_2$, NbSi$_2$ and TaSi$_2$ are much higher than those (room temperature) reported for the corresponding disilicides by Umakoshi et al. [5]. This discrepancy is believed to be due to the difference in the mobile dislocation density in the crystals tested, as we discussed in our previous papers [7,9]. The values of CRSS for VSi$_2$, NbSi$_2$ and TaSi$_2$ strongly depend on deformation temperature in the low temperature range and exhibit a moderate anomalous increase with temperature in the intermediate temperature range and finally decrease with increasing temperature in the high temperature range. The extent

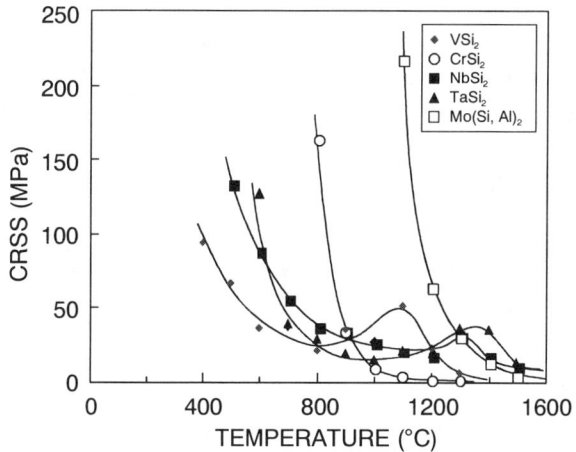

Fig. 2. Temperature dependence of CRSS for basal slip in five different C40 disilicides.

of the anomaly as well as the anomalous temperature range varies with disilicides. Since a serrated stress-strain behavior is usually observed in the anomalous temperature range, the anomaly is attributed to the so-called Portevin-Le Chatelier effect [5,7,9]. In contrast, the values of CRSS for $CrSi_2$ and $Mo(Si,Al)_2$ simply decrease with increasing deformation temperature without exhibiting any anomaly.

Dislocation dissociation

Contrast analysis has confirmed that dislocations carrying basal slip in all C40 disilicides have exclusively Burgers vectors of the $1/3<1\bar{2}10>$-type. At low temperatures, $1/3<1\bar{2}10>$ dislocations tend to align parallel to either $60°$ or screw orientations with the $60°$ orientations being dominant. In some areas, dislocations parallel to $30°$ and edge orientations are sometimes observed but their incidence is less frequent. Weak-beam analysis has confirmed that in all C40 disilicides, $1/3<1\bar{2}10>$ dislocations dissociate into two identical partials separated by a stacking fault (SF), as described below.

$$1/3<1\bar{2}10> \rightarrow 1/6<1\bar{2}10>+SF+1/6<1\bar{2}10>. \qquad (1)$$

The separation distance between two-coupled partial dislocations is found to be considerably larger for $CrSi_2$ and $Mo(Si,Al)_2$ than for the other three disilicides. Since the onset temperatures for the former two C40 disilicides are higher than those for the latter three, the present results are opposite to the general trend that the wider is the separation distance, the more deformable is the crystal. We thus made HRTEM examination of dislocation core structures in these C40 disilicides in order to see the origin of the opposite trend observed for these C40 disilicides.

Typical examples of HRTEM images of $1/3[1\bar{2}10]$ dislocations observed in $TaSi_2$ and $Mo(Si,Al)_2$ are shown in Fig. 3(a) and Fig. 4(a), respectively. In both images, the viewing direction is $[10\bar{1}0]$ and the dislocation imaged has a Burgers vector (**b**) of $1/3[1\bar{2}10]$ and is nearly of edge character having dissociated into two partials with **b**$=1/6[1\bar{2}10]$ with their approximate positions indicated by arrows. An enlargement of the fault region between the two-coupled partials in Fig. 3(a) and Fig. 4(a) is shown in Fig. 3(b) and Fig. 4(b), respectively. In order to reproduce the HRTEM image of Fig. 3(b) and Fig. 4(b), various models for the stacking sequence of the fault region are considered. The best matching between experimental and calculated

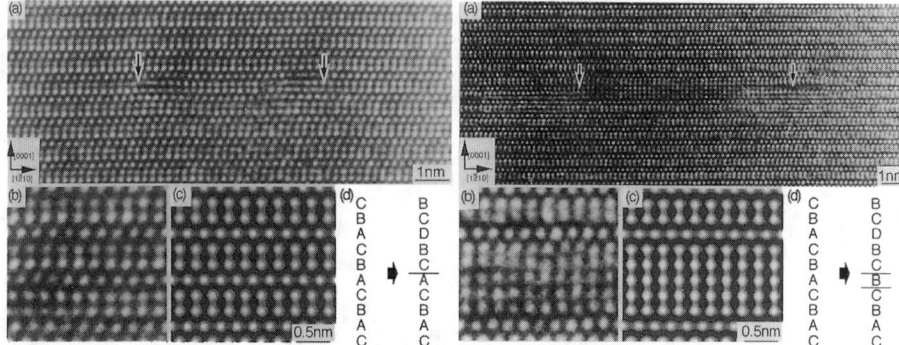

Fig. 3. (a) HRTEM image of an edge $1/3[1\bar{2}10]$ dislocation in a $[1\bar{2}12]$-oriented TaSi$_2$ single crystal deformed at 700°C viewed along $[10\bar{1}0]$ and an enlargement of the fault region in (a) is shown in (b), and (c) is an HRTEM image calculated based on the sequence shown in (d). The parameters used in the calculation are a crystal thickness of 8 nm and a defocus value of –85nm.

Fig. 4. (a) HRTEM image of an edge $1/3[1\bar{2}10]$ dislocation in a $[1\bar{2}12]$-oriented Mo(Si,Al)$_2$ single crystal deformed at 1100°C viewed along $[10\bar{1}0]$ and an enlargement of the fault region in (a) is shown in (b), and (c) is an HRTEM image calculated based on the sequence shown in (d). The parameters used in the calculation are a crystal thickness of 12 nm and a defocus value of -90nm.

images is obtained with the model shown in Fig. 3(d) and Fig. 4(d), and the calculated images based on these models are shown in Fig. 3(c) and Fig. 4(c) for TaSi$_2$ and Mo(Si,Al)$_2$, respectively. The indices of A-D in Fig. 3(d) and Fig. 4(d) denote positions of (0001) atomic layers in terms of the ABC stacking of the C40 structure (Fig. 1).

As seen in Fig. 3(d), a B atomic layer sitting on an A layer and all layers above it in TaSi$_2$ are displaced by the Burgers vector ($[1/6[1\bar{2}10]$) of the partial dislocation, undergoing the transition A $\rightarrow$ D, B $\rightarrow$ C, C $\rightarrow$ B, relative to the positions fixed on the original A layer. Similar results were obtained for VSi$_2$ and NbSi$_2$. Thus, the glide of $1/3<1\bar{2}10>$ dislocations on (0001) basal planes in VSi$_2$, NbSi$_2$ and TaSi$_2$ is found to occur in a conventional manner by a single shearing operation. On the other hand, the stacking sequence of the fault region in Mo(Si,Al)$_2$ (Fig. 4(d)) can not be achieved only by a single shearing operation but a sequence of shears occurring on two adjacent basal planes is required to achieve the observed stacking sequence. The stacking sequence of Fig. 4(d) is produced by shearing an A layer sitting on a C layer and all successive upper layers by $-\mathbf{b}_2$ and synchronously shearing a B layer sitting on the A layers by $-\mathbf{b}_3$. As a result, the stacking sequence of the C11$_b$-type, which is based on the AB (BC, in this particular case) stacking sequence, is locally formed in five layers adjacent to the fault. Similar results were obtained for CrSi$_2$. Thus, the glide of $1/3<1\bar{2}10>$ dislocations on (0001) basal planes in CrSi$_2$ and Mo(Si,Al)$_2$ is found to occur through a synchroshear mechanism.

DISCUSSION

The temperature dependence of CRSS for basal slip

The CRSS for basal slip normalized to the shear modulus corresponding to the $<1\bar{2}10>$ slip direction on (0001) slip planes is plotted in Fig. 5 as a function temperature normalized to the melting temperatures. The shear modulus used in plotting Fig. 5 were calculated from single-crystal elastic constants experimentally

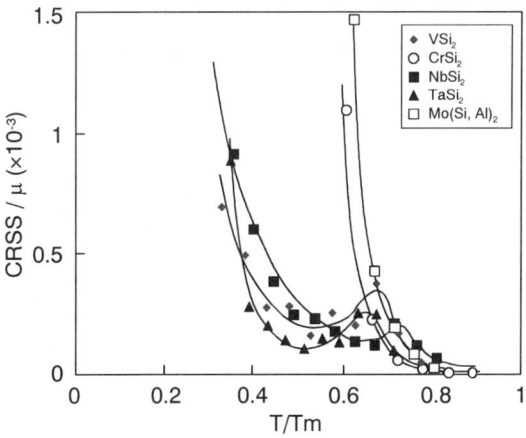

Fig. 5. The normalized plot of the CRSS-temperature curves of Fig. 2.

measured by Nakamura [11], Chu et al. [10] and Tanaka et al. [12]. In the normalized plot of Fig. 5, surprisingly, the CRSS-temperature curves for the five C40 disilicides can be classified into two groups, depending on the mechanisms for basal slip. The curves for VSi_2, $NbSi_2$ and $TaSi_2$ that deform through a conventional shear mechanism almost coincide with each other at low temperatures, exhibiting the onset temperatures for plastic flow around 0.3 T/T_m. On the other hand, the curves for $CrSi_2$ and $Mo(Si,Al)_2$ that deform through the synchroshear shear mechanism are located away from those for the former three C40 disilicides in the plot, but their curves almost coincide with each other, exhibiting the onset temperatures for plastic flow around 0.6 T/T_m. These indicate that the ease of occurrence of basal slip in C40 disilicides depends on deformation mechanism and, at the same time, that if the deformation mechanism is identical, there is no significant difference in the ease of the occurrence of basal slip.

Factors controlling the operation of the synchroshear mechanism for basal slip

The operation of the synchroshear mechanism for basal slip in C40 silicides is not a consequence of the geometrical requirements as in the case of α-Al_2O_3 [13] since the synchroshear mechanism is operative in $CrSi_2$ and $Mo(Si,Al)_2$ but not in $NbSi_2$, VSi_2 and $TaSi_2$. In the following, we discuss the possible factors controlling the operation of the synchroshear mechanism in C40 silicides. In a previous paper [9], we pointed out that the relative stability of the C40 phase with respect to the $C11_b$ phase may be one of possible factors that control the operation of the synchroshear mechanism in C40 silicides, since the stacking sequence of the $C11_b$-type is locally formed in five layers adjacent to the fault region between two-coupled synchro-partials. This may be true since Carlsson and Meschter [14], who have calculated energies of $C11_b$, C40 and C54 structures in transition-metal disilicides by *ab initio* band-structure calculations, have pointed out that the energy difference between $C11_b$ and C40 structures is very small for disilicides formed with Cr and Mo when compared to disilicides formed with V, Nb and Ta. The relative stability of C40 $Mo(Si,Al)_2$, which is presently investigated, with respect to the $C11_b$ form is considered also to be small, since it is formed by transforming the $C11_b$ structure of $MoSi_2$ by Al additions. Thus, we may conclude that as the stability of the C40 phase with respect to the $C11_b$ phase is reduced, the more easily the synchroshear mechanism operates.

As we pointed out in our previous paper [9], the directionality in atomic bonding may also be one of possible factors controlling the operation of the synchroshear mechanism since the directionality in atomic bonding is reported to decrease in the order of $CrSi_2 > VSi_2 > NbSi_2 > TaSi_2$ through the analysis of elastic constants [10,12]. Tanaka et al. [12] have further pointed out through a survey of solubility ranges of $C11_b$ and C40 phases in pseudobinary disilicide systems that the phase stability of the $C11_b$ phase with respect to the C40 phase is closely related to the directionality in atomic bonding in these disilicides; the lower is the relative stability between $C11_b$ and C40 phases, the stronger is the directionality in atomic bonding.

CONCLUSIONS

(1) C40 transition-metal disilicides are classified into two groups, depending on the onset temperature for plastic flow.

(2) The low-temperature group, which consists of VSi_2, $NbSi_2$ and $TaSi_2$, exhibits the onset temperature for plastic flow around 0.3 T/T_m (melting temperature) and deforms by a conventional shear mechanism. In contrast, the high temperature group, which consists of $CrSi_2$ and $Mo(Si,Al)_2$, exhibits the onset temperature around $0.6T/T_m$ and deforms by a synchroshear mechanism.

(3) The operation of the synchroshear mechanism is facilitated in C40 disilicides with the stronger directionality in atomic bonding and the lower phase stability with respect to the $C11_b$ phase.

ACKOWLEDGEMENTS

This work was supported by the Japan Society for Promotion of Science grant on Advantage High-temperature Intermetallics (JSPS-RFTF96R12301). The authors would like to thank Dr. Y. Yukawa and Mr. T. Shiraki, Nikko Superior Metals, Co. Ltd., and Dr. H. Shiraishi, Sumitomo Sitix Co. Ltd., for supplying high-purity Mo, Nb, Ta and Si, respectively.

REFERENCES

[1] A.K. Vasudevan and J.J. Petrovic, Mater. Sci. Eng., **A155**, 1 (1992).

[2] W.J. Boettinger, J.H. Perepezko and P.S. Frankwicz, Mater. Sci. Eng., **A155**, 33 (1992).

[3] D.M. Shah, D. Berczik, D.L. Anton and R. Hecht, Mater. Sci. Eng., **A155**, 45 (1992) .

[4] Y. Umakoshi, T. Nakashima, T. Yamane and H. Senba, in *Proc. on Int. Conf. on Intermetallic Compounds -Structure and Mechanical properties-*, edited by O. Izumi (The Japan Institute of Metals, Sendai, Japan, 1991) pp. 639-644.

[5] Y. Umakoshi, T. Nakashima, T. Nakano and E. Yanagisawa, E., in *High Temperature Silicides and Refractory Alloys*, edited by C.L. Briant, J.J. Petrovic, B.P. Bewlay, A.K. Vasudevan and H.A. Lipsitt (Mater. Res. Soc. Proc. **322** , Pittsburgh, PA, 1994) pp.9-20.

[6] H. Inui, M. Moriwaki, S. Ando and M. Yamaguchi, Mater. Sci. Eng., **A239-240**, 63 (1997).

[7] M. Moriwaki, K. Ito, H. Inui and M. Yamaguchi, Mater. Sci. Eng., **A239-240**, 69 (1997).

[8] H. Inui, M. Moriwaki, K. Ito and M.Yamaguchi, Phil. Mag. A, **77**, 375 (1998).

[9] H. Inui, M. Moriwaki and M. Yamaguchi, Intermetallics, 6, 723 (1998).

[10] F. Chu, M. Lei, S.A. Maloy, J.J. Petrovic and T.E. Mitchell, Acta Mater., **44**, 3035 (1996).

[11] M. Nakamura, Metall. Mater. Trans., **A25**, 331 (1994).

[12] K. Tanaka, H. Inui, M. Yamaguchi and M. Koiwa, Mater. Sci. Eng., **A261**, 158 (1999).

[13] M.L. Kronberg, Acta Metall., **5**, 507 (1957).

[14] A.E. Carlsson and P.J. Meschter, J. Mater. Res., **6**, 1512 (1991).

ATOMIC-SCALE DESIGN FOR ENHANCED LOW TEMPERATURE TWINNING IN ZrCr$_2$-BASED LAVES PHASES

W.-Y. KIM AND D.E. LUZZI
Department of Materials Science and Engineering, University of Pennsylvania, Philadelphia, PA 19104, USA, wykim@seas.upenn.edu, luzzi@lrsm.upenn.edu

ABSTRACT

Recently, we have designed and produced several transition metal Laves phases with low-temperature compressive ductility. These improved alloys demonstrate that manipulation of atomic-scale structure can have a drastic effect on meso-scale deformation behavior. To gain a basic understanding of the role of atomic-scale substitutions on the room temperature mechanical properties, a systematic investigation of the Laves ZrCr$_2$-based alloy system alloyed with Hf, Nb, Ta and Ti was conducted and is reported here. Extensive room temperature ductility was obtained in the Hf-alloyed ternary Laves phase alloy system. Mechanical twinning is found to be the predominant deformation mode at room temperature in this alloy system. It is emphasized that Hf substitution in the Zr sublattice of ZrCr$_2$ may play the most prominent role in changing the local electronic structure resulting in easier twinning.

INTRODUCTION

The AB$_2$ Laves phases, including the related cubic C15 (MgCu$_2$), hexagonal C14 (MgZn$_2$) and dihexagonal C36 (MgNi$_2$) crystal structures, comprise the single largest group of intermetallic compounds. Many of these compounds exhibit excellent high temperature mechanical properties[1], have superior hydrogen storage ability, or good superconducting properties[2-3]. However, the development of these alloys for technological use is hampered by room temperature brittleness due to their complex crystal structures.

The Laves phases are topologically close-packed materials with the A atoms larger than the B atoms (ideal ratio is 1.225). Therefore, off-stoichiometric compounds are expected to have constitutional vacancies (A-rich) or antisite defects (B-rich) and these defects may also affect the physical, mechanical and functional properties[4-5]. The interest in Laves phase increased after the discovery of room temperature compressive ductility by twinning in the HfV$_2$+Nb cubic Laves phase[6,7]. A subsequent model has suggested that the positive effect of Nb addition on HfV$_2$ alloy is due to the smaller Nb atom substituting on the Hf lattice sublattice[8-9].

In the present study, we adopt the ZrCr$_2$ Laves phase as a model alloy system to explore the effect of ternary substitution and relative atomic sizes on ductility. Hf, Nb, Ta and Ti are chosen as the ternary alloying elements because their atomic sizes vary widely, but lie between that of Zr and Cr. It can also be predicted based on known phase behavior, that these ternary elements will preferentially substitute on the Zr sublattice and should favor the cubic C15 Laves structure. The results will be discussed with respect to the relative importance of atomic size and electronic structure on twinning in these Laves phase-based alloys.

EXPERIMENTAL PROCEDURES

The purities of the raw materials used in the present work were 99.99wt.%Cr, 99.8wt.%Zr, 99.8wt.%Hf, 99.8wt.%Nb, 99.8wt.%Ta and 99.9wt.%Ti, respectively. Small ingots of 15mm diameter were prepared by the non -consumable arc-melting technique on a copper hearth in an argon atmosphere. The as-arc-melted ingots were annealed at 1373 K for 1 week in vacuum, followed by furnace cooling to room temperature. Microstructures were observed using optical microscopy (OM) and transmission electron microscopy (TEM). The volume fractions of constituent phases were measured using a computer-aided image analyzer. Phase analysis was performed using x-ray diffraction (XRD). Compression specimens with a dimension of 2x2x5 mm were cut by electro-discharge machining and mechanically polished using SiC paper and alumina powder. Compression tests were carried out in room-temperature air at an initial strain rate of 4.2×10^{-4} s^{-1}. To observe the deformation microstructure, TEM thin foils were prepared from the deformed specimens by twin-jet electropolishing in a solution of 5%HClO$_4$ and 95%CH$_3$OH at about 240K. TEM observations were carried out using a JEOL-4000 operated at 400 kV.

RESULTS AND DISCUSSION

Microstructures and Phase Analysis

In Figure 1, optical micrographs of the microstructures of the Hf-alloy (a and b of the figure) and the Ta-alloy (c of the Figure) are shown.

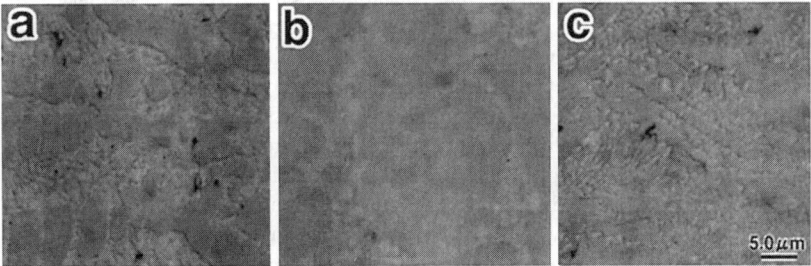

Figure 1   Optical micrographs of the microstructures of the Zr$_{10}$Cr$_{80}$Hf$_{10}$ (a), Zr$_{15}$Cr$_{75}$Hf$_{10}$ (b) and Zr$_{10}$Cr$_{80}$Ta$_{10}$ (c) alloys heat treated at 1373 K for 1 week.

The hypoeutectic microstructures are composed of a primary Laves phase and a eutectic structure of fine Laves phase with a Cr-rich solid solution. The result is consistent with the binary alloy phase diagrams (Zr-Cr, Hf-Cr, Ta-Cr), which all contain a eutectic transformation to the Cr side of the binary Laves phase[10]. With the exception of the Ti alloy system, all of the ternary systems are expected to have a eutectic reaction to the Cr side of the Laves phase field. With decreasing Cr content, the volume fraction of the Laves phase increased markedly due to the increasing volume fraction of the primary Laves phase and decreasing eutectic volume fraction.

XRD spectra for the heat-treated ternary Laves alloy systems are shown in Figure 2.

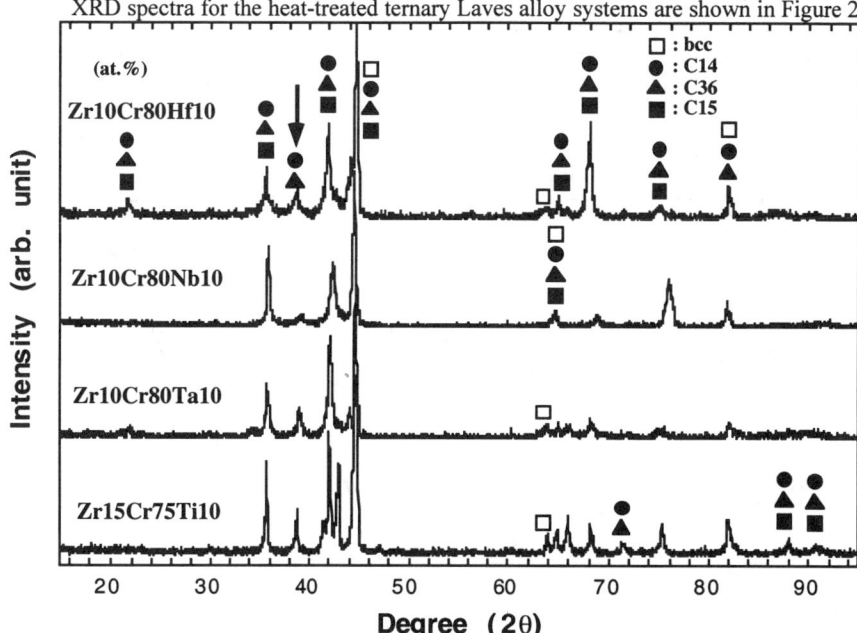

Figure 2 XRD spectra for four Laves phase alloys heat treated at 1373 K for 1week. Note that the indicated arrow corresponds to the C36 106 peak and the C14 103 peak.

The results indicate that the alloys are actually composed of the three Laves structures (C15, C14 or C36) and the Cr-rich bcc solid solution. While most diffraction peaks are common to the three Laves phases, the 106 peak of the C36 phase and the 103 peak of the C14 phase allow their presence to be uniquely determined. It is well known that AB_2 Laves phase alloys with the C15, C36 or C14 structures frequently display a phase transformation from one structure to the other with temperature change. In most Laves alloys, the C14 structure is stable at high temperatures with the C15 structure stable at low temperatures. Transformations between the different Laves phase crystal structures are sluggish and require prolonged annealing times at high temperature to complete the process. Therefore, the existence of the remnant C14 or C36 phases in the present alloys may be due to an insufficient annealing process. However, with the mechanical properties dependent upon the cubic Laves phase, the presence of the hexagonal phases will limit the ductility. Therefore, the results presented here are conservative with regard to achievable ductilities.

Mechanical Property

Compression tests were carried out on the above ternary alloy samples composed of the Laves phases (C15, C14 and C36) and the Cr-rich bcc solid solution. It has been shown that the ductility in two-phase Laves-bcc alloys is a strong function of bcc volume fraction[12]. These results were reproduced in this study using Hf-alloys, by systematically varying the two-phase alloy composition. Therefore, careful attention was paid to keep the bcc volume fraction constant among the different alloys. The bcc volume fraction of the four kinds of samples tested at room temperature was 35% with a deviation of less than 3%. The true stress-strain curves for the Hf-, Nb-, Ta- and Ti-alloys are plotted in Figure 3. Single phase Laves alloys and two phase alloys with a small bcc volume fraction were brittle with no macroscopic yielding. Extensive ductility was observed in the Hf-alloy with 35% bcc volume fraction, in contrast with the other alloys.

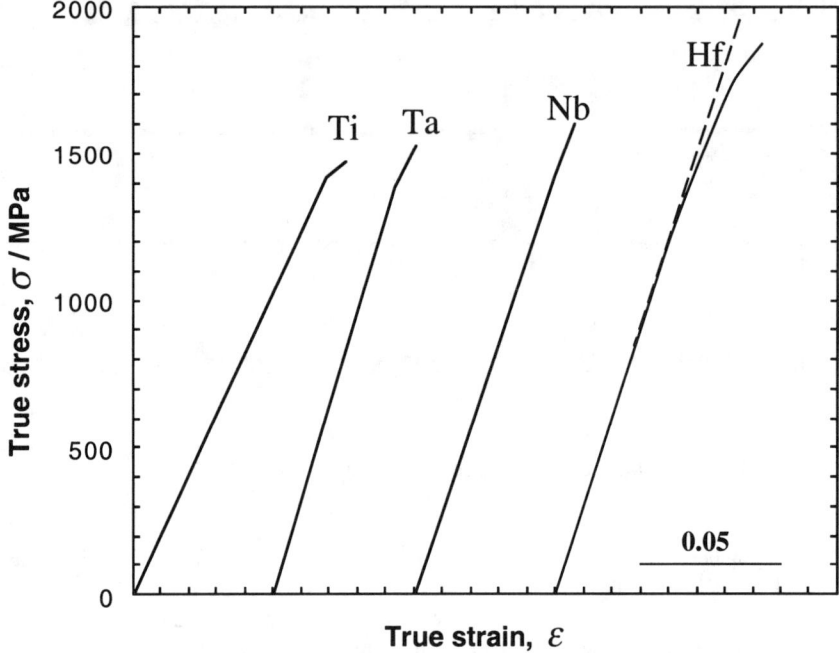

Figure 3   True stress-true strain curves for four Laves phase alloys with Ti,Ta,Nb and Hf ternary alloying tested at room temperature.

The estimated yield stress was about 1800 MPa and the plastic strain was 3.3%. Hf was found to be the most effective ternary element to enhance the room temperature ductility in the $ZrCr_2$ Laves phase. This is surprising considering that the atomic size of Hf ($1.59A^°$) is similar to that of Zr (1.61 $A^°$).

Deformation Microstructure

Figure 4 shows the deformed microstructure and corresponding SAD patterns for the $Zr_{10}Cr_{80}Hf_{10}$ alloy, which exhibited an extensive ductility of 3.3% at room temperature.

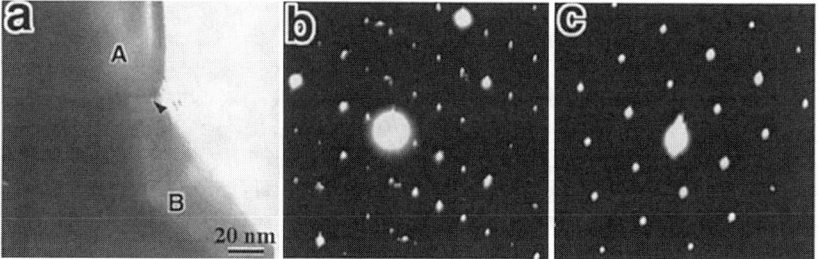

Figure 4   TEM micrograph(a) of the $Zr_{10}Cr_{80}Hf_{10}$ alloy deformed at room temperature and corresponding SAD patterns of area A(b) and B(c), respectively.

Samples were deformed in compression at room temperature to a plastic strain of 0.5% and unloaded. The electron diffraction patterns were taken from the areas marked A and B in the microstructure. The pattern of area A is from the C14/C36 crystal structure with a zone axis $[2\bar{1}\bar{1}0]_{C14}$. The diffraction pattern of area B is from the C15 $[01\bar{1}]$ zone axis. By comparing the relative orientation of the two patterns, it is clear that the two structures are twin-related with a common line of $111_{cubic}/0001_{hexagonal}$ reflections. The orientation relationship between the cubic and hexagonal structures is found to be the expected $(111)_{C15}// (0001)_{C14 (or\ C36)}$, $[01\bar{1}]_{C15}//[2\bar{1}\bar{1}0]_{C14(or\ C36)}$. The diffraction results are consistent with twinning of the cubic phase and transformation twinning of the hexagonal phases.

Factors affecting room temperature plasticity

Concerning the ternary alloying effect on the mechanical property, it has been suggested that the positive effect of Nb addition on the HfV_2 alloy is closely related to the substitution behavior of the ternary element on both the Hf and V sublattices[8-9]. In this model, when the atomic size of the ternary element is intermediate between the large A atom and the small B atom, and the number of B atoms is greater than twice the number of A atoms, proportionally more Nb atoms are on Hf sites than V sites resulting in an increase in free volume on the shear plane and easier deformation[9]. All of the ternary alloying elements and chemical compositions of the Laves alloys in the present study satisfy the criteria mentioned above, and therefore it can be expected that the shear-plane free volume, and therefore ductility, will increase with the sequence Hf<Nb<Ta<Ti. The mechanical testing results of Figure 3 directly contradict this model with Hf being the most effective to enhance room temperature ductility. This result indicates that atomic size considerations alone will be insufficient in explaining the occurrence of twinning at room temperature in the Laves phases. It is more likely that the origin of twinning lies in changes in the local electronic structure that will occur with the addition of ternary alloying elements. For example, if the local electronic structure of the $ZrCr_2$ Laves phase has more metallic character and less covalent character after ternary alloying, room temperature ductility should be enhanced.

CONCLUSIONS

The ternary alloying behavior, microstructure and room temperature deformation behavior was investigated for ternary alloys of the Zr-Cr-X (Hf, Nb, Ta and Ti) alloy systems. All alloys studied were composed of Laves phases and a Cr-rich bcc solid solution. The obtained results are summarized as follows.

1. The microstructures of the ternary alloys containing Hf, Nb and Ta are consisted of eutectic structure and primary Laves phase, suggesting that there exists a eutectic reaction in the two phase region of Laves phase and Cr-rich bcc solid solution. The Laves alloy with Ti showed a duplex microstructure consisting of the Cr-rich bcc solid solution and the Laves phase indicating the absence of a eutectic reaction.

2. Using the Zr-Cr-Hf alloy, it was found that the room temperature ductility increased with increasing bcc volume fraction. At bcc volume fractions around 35%, ductilities of over 3% were achieved while maintaining the high yield stresses (1800MPa) associated with a Laves phase matrix. This suggests that the dislocation activity within the bcc phase accommodates the Laves phase deformation.

3. Hf was the most effective alloying element to enhance the room temperature ductility via mechanical twinning despite having an atomic size similar to Zr. This result contradicts existing models in which smaller atomic sizes of the ternary alloying element should enhance twinning. These results suggest that the electronic structure of the constituent elements must be considered to explain the origin of twinning in these alloys.

ACKNOWLEDGEMENT

The authors gratefully acknowledge the support of the Department of Energy under grant # DE-FG02-97ER45641.

REFERENCES

1. Fleisher, R.L., Journal of Material Science, 22, 2281 (1987).
2. E. Olzi, F.C. Matacotta and P.Setina, J. Less-Common Met., 139, 123 (1988).
3. D. Ivey and D. Northwood, J. Less-Common Met., 115, 295 (1986)
4. P.M.Hazzledine, K.S. Kumar, D.B. Miracle and A.G. Jackson, MRS Symp. Proc., 364, 1406 (1994).
5. J.H. Zhu, L.M. Pike, C.T. Liu and P.K. Liaw, Acta mater., 47, 2003 (1998)
6. K. Inoue and K. Tachikawa, IEEE Trans. Magn.,15, 635 (1979).
7. Livingston, J. D. and Hall, E. L., J, of Mater. Res., 1990, 5 5.
8. Chu, F. and Pope, D. P., Mater. Sci. Engng., 1993, A170, 39.
9. Chu, F. and Pope, D. P., High Temperature Ordered Intermetallic Alloys VI, ed. J. Horton et al., MRS Symp. Proc., 364, 197 (1995).
10. Massalski, T.B., Murray, L.H., Bennett L.H. and Baker, H., Binary alloy phase diagram, ASM., Metals Park, OH, (1986).
11. A.K. Sinha, Prog. Mater Sci., 15 (2), 79 (1972).
12. W-Y Kim, D.E. Luzzi and D.P. Pope, submitted.

STOCHASTIC MESOSCALE MODELING OF ELASTIC-PLASTIC DEFORMATION

A. STAROSELSKY* and V.V. BULATOV **
* United Technologies Research Center, East Hartford, CT
** Lawrence Livermore National Laboratory, Livermore CA

ABSTRACT

Plastic response of a solid under stress depends on its crystallographic structure and morphology. Two of the major mechanisms of plasticity in metals are crystallographic slip and twinning. The purpose of this work is to analyze the influence of local stress distribution on slip and twin nucleation and propagation and to examine how this behavior depends on the interaction among slips, twins, and grain boundaries. We formulate a simple model in which slip and twin systems are defined at appropriate angles to each other. Plastic flow is treated as a Markovian stochastic process consisting of a series of local inelastic transformations (LITs) in the representative volume elements (RVE). The probabilities of LITs per unit time are defined in the framework of transition-state theory. By varying the types of allowed LITs and/or the scale of RVE, plastic deformation is modeled at different structural levels, from a small volume of single crystal to the aggregate response of an isotropic polycrystalline solid. An important feature of this model is that evolution of the internal stress distribution is traced explicitly throughout the simulation run. This allows us to examine conditions of slip and twinning in considerable detail. In particular, we observe that twinning occurs through a nucleation-and-growth mechanism whose rate is controlled by the size of the critical nucleus of the new phase.

INTRODUCTION

The character of the plastic response of different solids depends on their morphology and crystallographic structure as well as on loading conditions. Quantitative details of plastic flow in different solids may differ significantly, but general trends, if we consider the influence of some particular factor on plastic behavior, are quite often the same for different solids. Plasticity models based on the averaged response of continuum material have been advanced [1,2] based on the assumption that plastic strain is locally homogeneous. At the same time, it is well known that crystal plasticity is often locally heterogeneous introducing major uncertainties in the constitutive laws. In this paper we examine the effects of localized plasticity in single crystals using a 2D stochastic model incorporating, in an idealized way, slip and twinning behaviors.

The overall plastic deformation is considered to be inhomogeneous at the mesoscale associated with shear bands, slip or twinning. Following the original isotropic model developed by V. Bulatov and A. Argon [3], plastic flow is considered here to be a cumulative result of stochastic local inelastic transformations (LIT) in the representative volume elements (RVE) of the material. The superposition principle is applied to solve for heterogeneous elastic stress field building up in response to the misfit (plastic) strain produced by LITs[4]. Since the purpose of this work is to examine the influence of local stress distribution on shear bands, slip and twin nucleation and propagation, we represent slips and twins by LITs of different types oriented at special angles with respect to the material axes. We will show how twin and slip propagation processes are related to the internal stress and will attempt to elucidate critical conditions (for example, critical stress) for twin nucleation and propagation. In the following sections we present a brief description of the extension to the stochastic elastic-plastic model required to simulate slip and twinning deformation and discuss the simulation results.

Mat. Res. Soc. Symp. Proc. Vol. 578 © 2000 Materials Research Society

MODEL FORMULATION AND ALGORITHM

The main idea of this model approach is to fill the 2D plane with regular mesh elements, for example hexagonal [3]. We associate the scale of such elements with a meso-level length scale. It is assumed that each hexagonal element can spontaneously change its unconstrained shape due to one of the six possible LITs, corresponding to three elongations and three contractions along the symmetry axes of the hexagons, as shown in Fig. 1.

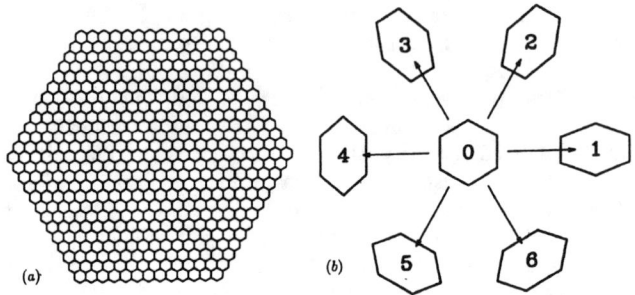

Figure 1. (a) A plane covered by the hexagonal mesh. (b) The set of inelastic transformations.

Plastic flow is treated as a stochastic (Markovian) sequence of LITs. It is assumed that the LITs are produced by the combined influence of applied stresses and thermal agitation. Thus, the probabilities of LITs developing per unit time are defined in the framework of transition-state theory [5] as follows:

$$\omega = \omega_0 \exp\left(-\frac{\Delta G(\sigma)}{kT} \right)$$

where $\Delta G(\sigma)$ is the stress dependent free enthalpy for inelastic rearrangement, kT is Boltzmann energy, ω_0 is the pre-exponential frequency factor.

The basic idea of such simulations is to account for internal misfit stresses produced by the plastic transformations occurring in the system. Each transformed element behaves like Eshelby inclusion and generates local elastic stress field, evaluated everywhere using a pre-calculated Green's function [6, 7]. Detailed spatial distribution of internal stresses is explicitly accounted for in this model and updated in all hexagonal elements after each successful LIT. In such kind of numerical simulations the number of mesh elements plays an important role in obtaining physically realistic results. The finer mesh is used the better resolution one can expect. Use of the pre-tabulated Green's tensor makes it possible to simulate plastic response of a mesh containing 10000 hexagonal elements on a desktop workstation[1]. In order to find $\Delta G(\sigma)$ we define shape increments of each element and, using Eshelby's solution as a Green function we obtain the internal stress field $\sigma_{ij}^{int\,ernal}$. The local stress state is defined as follows:

$$\sigma_{ij}^{local} = \sigma_{ij}^{external} + \sigma_{ij}^{int\,ernal}; \qquad \left\langle \sigma_{ij}^{int\,ernal} \right\rangle = 0$$

Here, angular brackets denote averaging over the entire volume of the body, and internal stress $\sigma_{ij}^{int\,ernal}$ is produced by the elastic resistance of the matrix, which has to accommodate all the previously transformed elements, if there are any.

---

[1]The use of the fast multipole methods and faster computers should allow simulations with the number of mesh elements up to millions.

Since macroscopic plastic flow is viewed as a net result of the individual microscopic plastic events (LITs), it can be modeled once the corresponding Markovian stochastic process is defined. We specify a given state of the system by a particular set of accumulated plastic strain increments in each of the lattice elements. Any given configuration is assumed to have 60000 different ways to change (6 possible LITs per each of 10000 mesh elements). The total rate of escape from the current state, which determines the overall probability rate, is the sum of all transition rates, i.e:

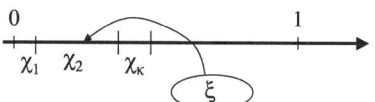

Figure 2. Selection of LITs in the Monte-Carlo simulation.

$$W = \sum_{\substack{mesh \\ elements}} \sum_{\substack{possible \\ LITs}} \omega(element, LIT).$$

The choice of the element, which undergoes a transformation at this step, and the particular type of transformation in that element is based on the contribution of each individual transition to the total rate W. The normalized probability rate for each possible transformation may be written in the form: $\chi_i = \omega_i/W$; $\sum \chi_i = 1$. Essentially, this defines a kinetic Monte Carlo algorithm that chooses type and location of the next transformation by generating a random number ξ uniformly distributed in the interval (0,1), as illustrated in Fig. 2. The probability of a transition to take place during time τ is $P(\tau) = 1 - \exp(-W\tau)$, hence the waiting time in the current configuration can be chosen as $\Delta t = -\dfrac{\ln(\varsigma_2)}{W}$, where ς_2 is another random number uniformly distributed in the interval (0,1).

The rate of inelastic response of a solid to applied stress depends sensitively on the free-energy barrier $\Delta G(\sigma)$. By playing with the function $\Delta G(\sigma)$ we may constrain or accelerate different types of transformations, and simulate different types of plastic response, including shear bands, crystallographic slip or twinning. That is essentially what we attempt to do in the following.

RESULTS AND DISCUSSION

All simulations presented in this paper start from an undeformed initial configuration that was subjected to some external stress producing a sequence of LITs. After each LIT, the local stress state in the whole meshed area was updated and used to calculate the new transition probabilities and to select the next transformation. Some of the LITs, once formed, introduce excessive distortions in the surrounding matrix and transform back (recover) almost immediately, due to a very significant increase of the probability of the reverse transformation for the same element. For as long as only such accidental LITs are produced, the material deforms nearly elastically. Then, under higher stress, LITs are produced at a higher rate forming small clusters which, after reaching certain critical size, continue to grow. Such nucleation-and-growth events are observed at all scales of the modeled deformation processes. Details and rates of these processes are controlled by the physical mechanisms of deformation, which are specified in this model by the

Figure 3. Deformation substructure in the isotropic model solid under uniaxial stress.

choice of $\Delta G(\sigma)$ for a given mechanism.

For the macroscopic length scales associated with the isotropic structure-less materials, the free-energy barrier, which an element should overcome in order to transform, is assumed to have the following linearized form [3]:

$$\Delta G = U - \Omega \cdot \sigma_{ij} \varepsilon_{ij}^T$$

Here, U is a free-energy barrier, assumed to be constant, Ω is an initial unperturbed element volume, σ_{ij} is the local stress, and ε^T_{ij} is an eigenstrain tensor, corresponding to the particular mode of transformation.

After a short transient period shear bands are observed (gray areas on Fig. 3). Plastic flow develops mostly by increasing the number of elements involved in the already formed bands. The bands grow predominantly in two nearly perpendicular directions as shown in Fig. 3. Such mode of nucleation and propagation can be explained by stress concentration at the band edges increasing the probability of transformations in the neighboring, still untransformed elements[2]. Thus, the evolution of flow structure is determined by the internal stress distribution.

Here we extend this simple model in order to incorporate some of the basic crystallographic mechanisms of plasticity and to analyze the role of local stress distribution in deformation by slip and twinning. First, we consider plastic deformation by slip only, however, the analysis of the preferred crystallographic variants will be common for both modes of deformation, slip and twinning.

Since the orientation of crystal lattice is specified in each volume element, the resolved shear stress on each slip/twin system α is $\tau^\alpha = \underline{\sigma} \cdot \mathbf{S}^\alpha$. Here, $\mathbf{S}^\alpha = \mathbf{m}^\alpha \otimes \mathbf{n}^\alpha$ is a Schmid tensor, $\mathbf{m}^\alpha = (\cos\theta, \sin\theta)^T$ is a unit vector in the slip direction, $\mathbf{n}^\alpha = (-\sin\theta, \cos\theta)^T$ is the unit normal to the glide plane, and θ is an angle between x-axis and the slip direction $\mathbf{m}^\alpha$. The orientation also determines the value of the barrier energy $U(\theta)$ for each slip system. While the barrier in the slip direction can be specified, $U(\theta) = U_0^\alpha$, it should be almost infinity in any other direction. On the other hand, because of the symmetry of our hexagonal mesh, it makes sense to have slip/twin propagation confined to one of the three symmetry directions, at angles $0^\circ, \pm 60^\circ$, and $\pm 120^\circ$ to the x-axis. For each of the three directions, we construct an angular distribution of the barrier energy so that sequential transformations form a line in the proper direction. This is accomplished by requesting that for a given angle θ,

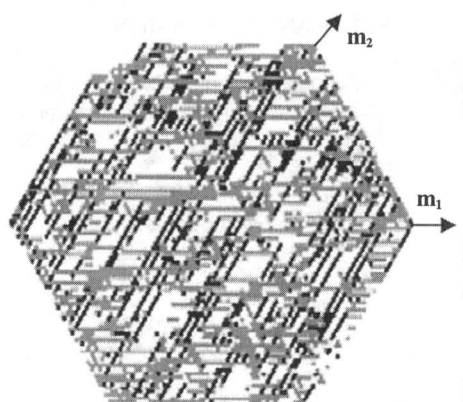

Figure 4. A flow structure due to activity of two slips systems.

$$\omega_{1,4} \sim \cos^2\left(\tfrac{3}{2}\theta\right), \quad \omega_{2,5} \sim \sin^2\left(\tfrac{3}{2}\theta\right), \quad \omega_{3,6} = 0.$$

Here, the transformations are numbered in the order of increasing angle to the slip direction, in 60° increments. At small angles this is equivalent to a parabolic distribution of $U(\theta)$ with the minimum at the slip system angle $U(\theta^{slip}) = U_0^\alpha$.

---

[2] Only the results of constant temperature simulations are reported in this work. If the effects of local heating are of interest, the resulting temperature distribution can be calculated in the entire volume. One obvious result of such process is that the mean temperature can be higher around the band tip causing further flow localization (adiabatic shear banding).

Thus, the expression for the free-energy barrier for modeling crystallographic slip is:

$$\Delta G = U(\theta) - \Omega \tau^\alpha \cdot \gamma^\alpha,$$

where τ^α is the resolved shear stress on α^{th} slip system and γ^α is a shear eigenstrain in the element along this slip system. To simulate latent hardening, we assume additionally that, if an element is already sheared, the free-energy barrier for a different mode of shear is set to a larger number. A typical flow structure after plastic deformation along two different systems is shown in Fig. 4, with the slip direction $\mathbf{m_1}$ shown as gray and a system with the slip direction $\mathbf{m_2}$ shown as black. Slip lines appear in the form of thin lines separated by untransformed elements. The lines start growing when a critical size (usually about 4-6 neighboring elements) is reached. Then the bands continue to propagate along the slip direction all the way to the intersection with another slip system or a "grain" boundary. In order to continue its growth past the obstacle, a slip line has to exceed yet another critical length in order to create sufficiently high stress concentration in the elements on the other side of the obstacle. As in clearly seen in Fig. 4, some of the shorter slip lines become locked from both ends and can not propagate further.

Next, we shift attention to the details of twinning simulations. Since in the initial stages of twinning the twinning particles should be coherent with the parent matrix, the energy of elastic accommodation is an important contribution to the thermodynamics and kinetics of twinning. This very same strain energy is also important in promoting secondary nucleation in those locations and orientations that relieve internal elastic stress. To account for such salient features of twinning the expression for $\Delta G(\sigma)$ is modified as follows. Each "twinned" element is assigned an additional energy depending on its location in the twin: (1) twin boundary interface energy γ^{twin} and/or (2) "twin edge" energy γ_0 associated with the energy of a dislocation wall at the twin tip. The second contribution is applied to the side of a mesh element that is orthogonal to the twinning direction. Therefore, the extra energy of a transformed, twinned aggregate of mesh elements is $\Gamma = \gamma^{twin} \cdot S + \gamma_0 \cdot A$, where S is the twin interface area and A is the area of "twin tip" region. Twinning transformation of a given mesh element increases its energy which is accounted for by an appropriate reduction in the barrier ΔG for the reverse transformation. On the other hand, it stimulates the neighboring elements to transform in order to decrease the area between the parent and new phases, and, subsequently, to decrease the interface energy. Consequently, free-energy barrier for the twinning transformation is calculated as follows:

$$\Delta G = U(\theta) - \Omega \tau \cdot \gamma + \Gamma,$$

where $U(\theta)$ is the barrier for transformation in the direction of a particular twin system; τ

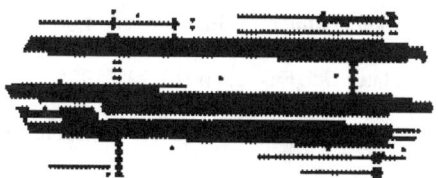

Figure 5. Simulated twin bands.

and γ are the resolved shear stress and the twinning strain, respectively; Γ is the additional "interface" energy introduced above.
As a rule, twins nucleate at a grain boundary and the resulting twinned regions have finite widths. Sometimes, two propagating twins meet forming larger twins, but more frequently the fine twin bands are observed to grow laterally (Fig. 5).

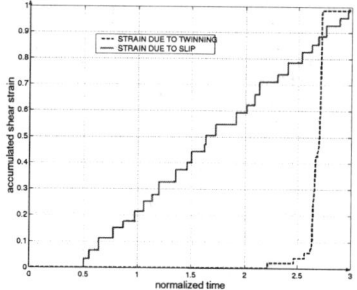

Figure 6. Accumulated plastic shear vs. time for the slip and twinning modes of deformation.

Unlike slip that takes place in either positive or negative direction, twinning can occur only in the "twinning" direction and anti-twinning is prohibited. De-twinning, i.e. the inverse transformation of an already sheared mesh element, was observed only at the early stages of deformation, prior to formation of the critical nuclei. However, once the twins grow beyond the critical size, de-twinning is no longer energetically favored. Typical results of plastic strain accumulation due to either slip (solid line) or twinning (dashed line) are presented in Fig. 6. After an initial period of mostly elastic response, plastic deformation accumulates rather quickly. Slip shear accumulation takes place in a "stop and go" fashion, each small plateau corresponding to a waiting period for the formation of the next slip. On the other hand, twinning remains dormant until much later along the deformation curve. Specifically, the higher the value of the interface energy Γ, the longer it typically takes to form the first "sustainable" twin. On the other hand, as soon as the first sustainable twin appears, further twinning propagation occurs very rapidly, nearly burst-like – the faster, the larger the interface energy Γ. In agreement with the experimental trend, twinning is less favorable when the interface energy (including the stacking fault energy) is high.

CONCLUSIONS

A self-consistent 2D model describing several generic features of crystal plasticity is presented in this paper. Plastic flow is treated as a Markovian stochastic process consisting of a series of local inelastic transformations in the representative volume elements. The probabilities of LITs per unit time are defined in the framework of transition-state theory. The approach is flexible enough to simulate different physical mechanisms of plastic deformation, such as slip, twinning and isotropic shear banding, each with its own transformation energetics. In our simulations both slip and twinning transformations follow the nucleation-and-growth kinetics. At the same time, kinetics of twinning differs significantly form that of slip. The effect is attributed to a specific contribution to the twinning energetics of the twin interface energy.

REFERENCES

1. E. W. Billington and A. Tate, *The Physics of Deformation and Flow,* McGraw-Hill Inc., 1981, 626 p.
2. M. Ohnami, *Plasticity and High Temperature Strength of Materials*, Elsevier Ltd., London, 1988, 525 p.
3. V. V. Bulatov and A. S. Argon, Modeling Simul. Mater. Sci. Eng. 2, pp.167-222, 1994
4. D. Chen and H. Nisitani, Int. J. of Plasticity, 8, pp.75-89,1992
5. T. L. Hill, *An Introduction to Statistical Thermodynamics*, Dover Publications, Inc. 1986, 508 p.
6. C. Teodosiu, *Elastic Models of Crystal Defects*, Springer, Berlin, 1982, 260 p.
7. S. L. Crouch and A. M. Starfield, *Boundary Element Methods in Solid Mechanics*, George Allen & Unwin, London, 1983, 328 p.

Dislocation Dynamics—
Experiments and Simulation

IN-SITU TEM DEFORMATION STUDIES OF DISLOCATION GENERATION AND MOTION IN HIGH-PURITY Mo SINGLE CRYSTALS

M. Jouiad*[‡], B. W. Lagow*, I. M. Robertson*, and D. H. Lassila[†]
* Frederick Seitz Materials Research Laboratory, University of Illinois at Urbana-Champaign, Urbana IL 61801, b-lagow@uiuc.edu
† Materials Science and Technology Division, Lawrence Livermore National Laboratory, Livermore CA 94550
‡ Present affiliation: ENSMA Institute, 86960 Futuroscope Cedex, France

ABSTRACT

The generation and motion of dislocations in high-purity single crystals of Mo have been observed in real time by deforming electron-transparent samples in-situ in a transmission electron microscope. At 300 K and at low levels of stress, a novel dislocation source was observed that generated a long, straight screw dislocation. The source was a dislocation tangle that existed in the annealed material. An edge dislocation emerged from the tangle, trailing behind it the screw dislocation. These screw dislocations were immobile at this stress level. At higher stresses, the same dislocation tangle generated many dislocations, but now by a pole mechanism. The nature of these tangles and the source operation mechanisms will be described.

INTRODUCTION

In bcc metals, the deformation mode is strongly temperature dependent [1, 2]. It is well known that there are two different deformation modes [2]. At low temperatures ($T < T_c$ ~400° C for Mo) and at low levels of applied stress, screw dislocations are much less mobile than edge dislocations at. Hirsch [3] attributed the low mobility of screw dislocations in this regime to non-planar dissociation of the dislocation core. This was confirmed later by molecular dynamic simulations [4] and recently it was observed for the first time by high-resolution electron microscopy [5]. At higher temperatures ($T > T_c$), all dislocations have similar mobilities and dislocation sources become active.

Previous work has focused on the low temperature properties of bcc metals, where a great deal of research has been aimed at determining the mechanisms responsible for high flow stress, the strong work-hardening rate, and the activation of anomalous slip planes [6-12]. The present study is focused on in-situ transmission electron microscopy deformation experiments in high-purity single crystal Mo at both 100 K and 300 K. These experiments permit dislocation source operation, dislocation mobility and dislocation interactions to be observed in real time and at high spatial resolution and provide one of the few ways of observing dislocation dynamics. These observations are needed as input data and as a means of validation of mechanisms for multi-scale modeling efforts. In this paper we concentrate on a qualitative description of the source mechanisms operating during room temperature deformation.

EXPERIMENTAL PROCEDURE

Conventional disc and rectangular straining stage samples were cut by using an electric discharge machine from a high-purity [100] oriented single-crystal of annealed Mo.

(The material was received in this condition from Lawrence Livermore National Laboratory.) The thickness of the cut samples was reduced to a final thickness of about 100-micrometers by mechanical grinding. The rectangular straining stage samples had dimensions of 3x10x0.1mm and were cut so that the tensile axis was near <100>. Electron transparent samples were prepared by jet polishing in a twin jet polisher. The electrolyte was a 20% sulfuric acid in methanol solution, and the polishing conditions were 25 V, 1.6 A, and –10° C. The samples were strained in a single-axis tilt straining stage, which is capable of stage displacement rates as low as a few nm per second. The electron microscopy was performed in the JEOL 4000 EX controlled environment transmission electron microscope [13], which is equipped with a Gatan TV rate camera system. The dynamic events were recorded on Super VHS videotapes for later analysis. Static images were recorded using the conventional plate system, the straining stage being sufficiently stable to permit this.

RESULTS

Although the Mo samples were given a high temperature anneal under ultra-high vacuum conditions, a significant number of dislocations were retained in the material. Examples of the different dislocation structures are shown in the bright-field images presented in Figure 1. In some regions of the sample only isolated straight dislocations existed, Figure 1(a), whereas in others dislocation tangles were observed, Figures 1(b) and 1 (c). The density of the dislocation tangles also varies: in some regions the tangles are isolated from each other, whereas in others the dislocation tangles are extensive and involve many dislocations.

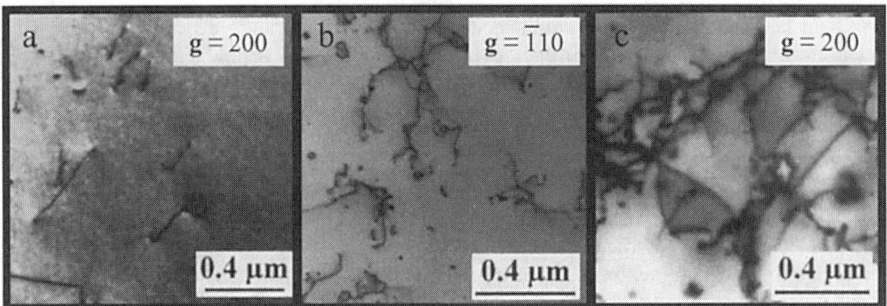

Figure 1: Initial distribution of dislocations. Images show areas with (a) isolated dislocations, (b) isolated tangles, and (c) complicated tangles.

The formation of dislocation tangles is common in materials exhibiting non-planar glide [14], but it is surprising that they exist in such densities after the annealing treatment. $g \cdot b$ analysis of the dislocation tangles shows that they primarily consist of different $b = a/2<111>$ dislocations, with short $b = a<100>$ dislocations forming at junctions in the tangle. Similar observations have been made in Nb [2], so this is not unexpected. These junction dislocations will contribute to the stability of the tangle [15], until the applied stress is sufficient to unzip the junction. During straining, dislocation tangles have been observed to impede the motion of other dislocations, as well as to act as dislocation sources.

On initial straining, and occasionally under the influence of just the electron beam, edge dislocations were observed being emitted from the dislocation tangles. An example of

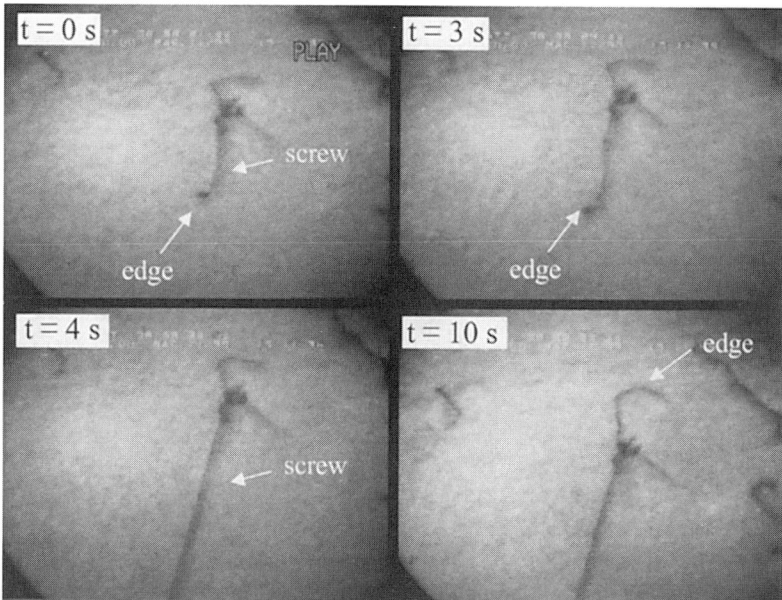

Figure 2: Evolution of a tangle with time at low stress. Edge dislocations leaving the tangle trail immobile screw dislocations behind them.

this during straining is shown in the sequence of images presented in Figure 2, which shows two edge dislocations leaving a tangle. As these edge dislocations move away, they trail behind them long, immobile screw dislocations, which remain pinned at the tangle.

As the straining is continued this source ceases to operate and screw dislocations in the matrix begin to move and other sources begin to operate. The matrix dislocations are observed to readily cross-slip between the primary and cross-slip planes. An example of this is shown in the images presented in Figure 3, which were produced by superimposing two

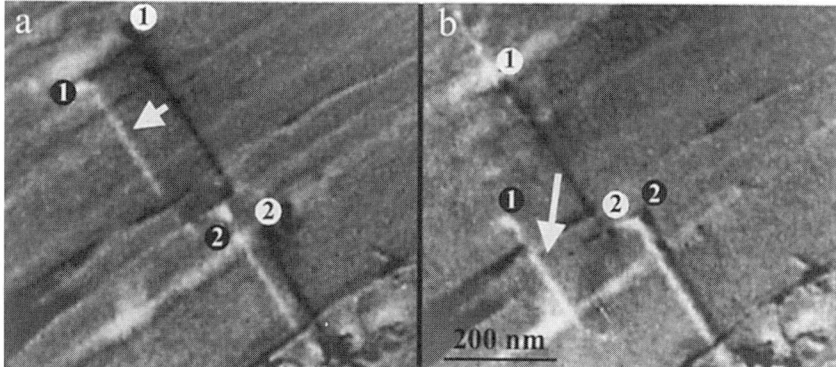

Figure 3: Cross-slip of screw dislocations as the sample is strained. Dark images indicate initial position in each image. White arrows indicate direction of motion of dislocation 1.

frames, one a positive image (black dislocations) and the other a negative image (white dislocations). In each image, the positive image shows the initial position of the dislocation and the negative image the final position. Analysis of slip traces during post-mortem observations of these samples indicates that screw dislocations switch between {110}- and {112}-type planes.

Two types of dislocation source have been observed to operate as the straining continues. In the first, loops are nucleated at forest dislocations (Figure 4). As indicated by the images, the loop expands, the edge component of the loop expanding more rapidly than the screw component, resulting in the formation of long straight screw dislocations. The process then repeats, generating more long straight screw dislocations. This mechanism has also been observed in NiAl [16].

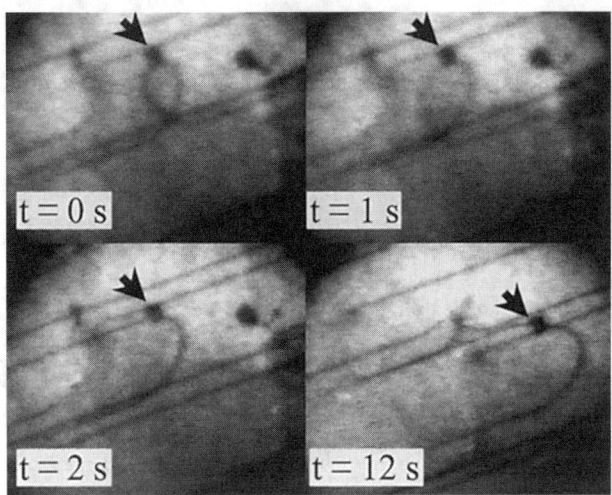

Figure 4: Operation of a looping source. Arrows mark pinning point common to all images. Edge segments break away from the loop, trailing a pair of screw dislocations.

In the second type of source, tangles begin to act as dislocation pole sources. An example is illustrated in the series of images presented in Figure 5 and is shown schematically in Figure 6. The screw dislocation is pinned at the point indicated by the arrow at t = 0s. A highly mobile edge segment is nucleated on the dislocation line, which moves in the direction indicated in the schematic Figure 6(b). The motion of this edge dislocation generates long straight screw dislocations, one of which moves as indicated in Figure 6(d) and the other remains pinned at the obstacle. The source is thus reset, and the pinned screw dislocation nucleates another edge segment, which moves in the opposite direction; the initial stage of this is shown in Figure 5 at t = 3s and schematically in Figure 6 (d). The conditions to generate this type of pole source are not yet known, although a similar source mechanism has been recently simulated [17]. It is worth mentioning that the tangle shown in Figure 2 also became a pole source, although the dislocation responsible for activating the source was neither of the original screw dislocations that were generated by the propagation of an edge dislocation.

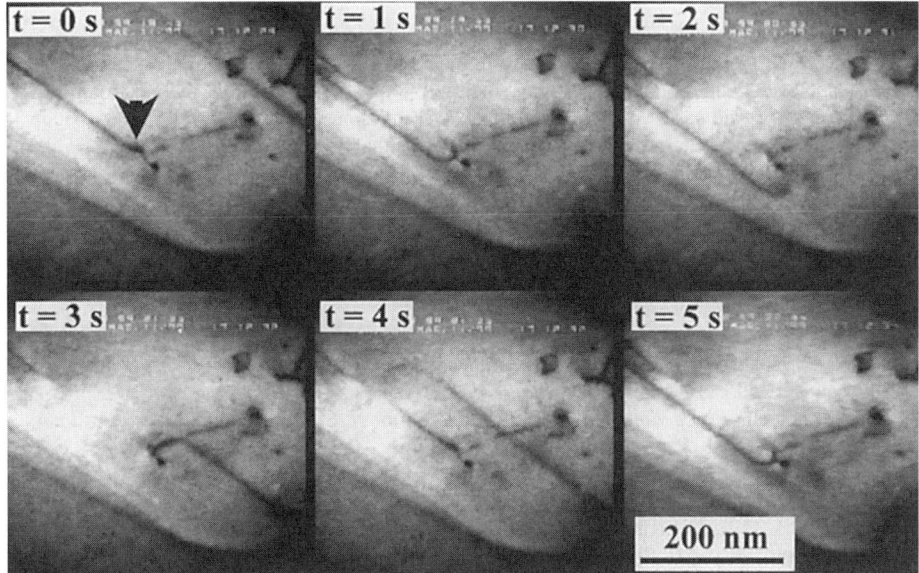

Figure 5: Operation of a dislocation pole source. Pinning point is arrowed. Pinned screw dislocation nucleates an edge segment at the pinning point *(t = 1s)*. Edge segment sweeps away from the source, trailing two screw dislocations *(t = 2s)*. The process then repeats in the opposite direction (*t* = 3-4s). This resets the source and the process repeats (*t* = 5 s).

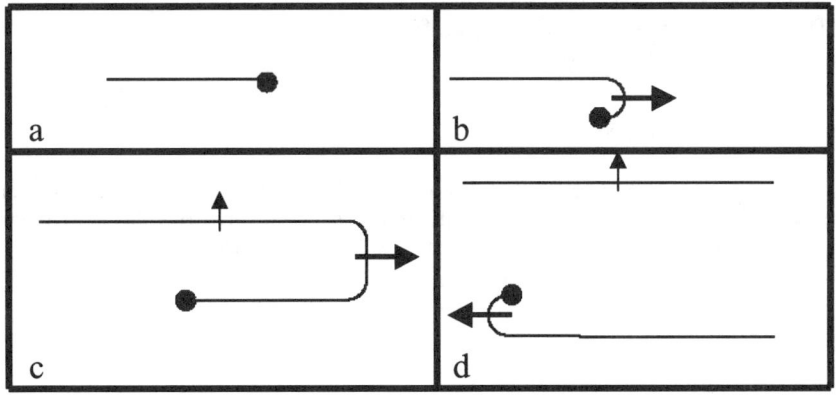

Figure 6: Schematic of pole source operation. Dislocation pinned at pinning point (circle). Relative velocities of edge (vertical) and screw (horizontal) dislocations indicated by width of arrows.

CONCLUSIONS

In-situ straining of single crystal Mo in the TEM has shown that edge dislocations are readily nucleated from preexisting tangles; the thermal stresses induced by the electron beam is sufficient in some cases to induce motion. The emission and propagation of these edge dislocations generates long straight screw dislocations that are immobile at low stress levels. Screw dislocations are mobile at higher stress levels and they are generated by the activation of looping and dislocation pole sources. The nature of these tangles needs to be understood and incorporated into models if the models are to correctly predict the behavior of materials during deformation.

REFERENCES

[1] J. W. Christian, *Met. Trans. A* **14A** 1237 (1983).
[2] F. Louchet and L. P. Kubin, *Phys. Stat. Sol. A* **56** 169 (1979).
[3] P. B. Hirsch, *Proceedings of the Fifth International Conference on Crystallography* (Cambridge University Press, 1960), p. 139.
[4] V. Vitek, *Crystal Lattice Defects* **5** 1 (1974).
[5] W. Sigle, *Phil. Mag. A* **79** 1009 (1999).
[6] L. Kaun, A. Luft, J. Richter and D. Shultze, *Phys. Stat. Sol.* **26** 485 (1968).
[7] R. Lachenmann and H. Shultz, *Scripta Met.* **4** 709 (1970).
[8] R. L. Fleisher, *Acta Met.* **15** 1513 (1967).
[9] F. Louchet and L. P. Kubin, *Acta Met.* **23** 17 (1975).
[10] H. Matsui and H. Kimura, *Scripta Met.* **7** 905 (1973).
[11] S. Takeuchi and E. Kiramoto, *Scripta Met.* **8** 785 (1974).
[12] W. Wasserbach, *Phys. Stat. Sol. A* **147** 417 (1995).
[13] T. C. Lee, D. K. Dewald, J. A. Eades, I. M. Robertson, and H. K. Birnbaum, *Rev. Sci. Instr.* **G2** 1438 (1991).
[14] D. Kulmann-Wilsdorf and N. Hansen, *Met. Trans. A* **20** 2393 (1989).
[15] Y. T. Chou, *Mater. Sci. Eng.* **10** 81 (1971).
[16] D. Caillard, *Mater. Res. Soc. Symp. Proc.* **552** 737 (1999).
[17] W. Tang, L. P. Kubin and G.R. Canova, *Acta Mater.* **46** 3221 (1998).

ACKNOWLEDGEMENTS

This work is supported by Lawrence Livermore National Laboratory. The use of the electron microscope facilities in the Center for Microanalysis in the Frederick Seitz Materials Research Laboratory at the University of Illinois is greatly appreciated.

DISLOCATION MULTIPLICATION IN THE EARLY STAGE OF DEFORMATION IN Mo SINGLE CRYSTALS

L. M. Hsiung and D. H. Lassila
Lawrence Livermore National Laboratory, Materials Science and Technology Division, L-369, P.O. Box 808, Livermore, CA 94551-9900, hsiung1@llnl.gov

ABSTRACT

Initial dislocation structure in annealed high-purity Mo single crystals and deformation substructure in a crystal subjected to 1% compression have been examined and studied using transmission electron microscopy (TEM) techniques in order to investigate dislocation multiplication mechanisms in the early stage of plastic deformation. The initial dislocation density is in a range of $10^6 \sim 10^7$ cm^{-2}, and the dislocation structure is found to contain many grown-in superjogs along dislocation lines. The dislocation density increases to a range of $10^8 \sim 10^9$ cm^{-2}, and the average jog height is also found to increase after compressing for a total strain of 1%. It is proposed that the preexisting jogged screw dislocations can act as (multiple) dislocation multiplication sources when deformed under quasi-static conditions. The jog height can increase by stress-induced jog coalescence, which takes place via the lateral migration (drift) of superjogs driven by unbalanced line-tension partials acting on link segments of unequal lengths. The coalescence of superjogs results in an increase of both link length and jog height. Applied shear stress begins to push each link segment to precede dislocation multiplication when link length and jog height are greater than critical lengths. This "dynamic" dislocation multiplication source is suggested to be crucial for the dislocation multiplication in the early stage of plastic deformation in Mo.

INTRODUCTION

The main purpose of this study is to examine, analyze the initial dislocation structure and deformation substructure of Mo single crystals in order to provide detailed physical mechanisms to facilitate multi-scale modeling and dislocation dynamics simulation [1]. For the success of simulation, it is of paramount importance to have a systematic and rigorous study on dynamic properties of dislocations including dislocation multiplication, motion, and dislocation interaction. Since the initial dislocation structures (dislocation density, dislocation configuration, free dislocation link length, kink and jog density…) can all affect dislocation dynamics during subsequent plastic deformation, the dislocation substructures in as-annealed and quasi-statically compressed Mo crystals were studied and compared. Emphasis has been placed upon the role of initial dislocation structures in dislocation multiplication and motion during early stages of plastic deformation.

EXPERIMENT

The Mo single crystals used for dislocation dynamics experiment must have low dislocation density in order to analyze the structure with TEM and to subsequently establish initial conditions for simulation. Mo single crystals were obtained from Accumet Materials Company, Ossining, NY. The interstitial impurities (ppm in weight) in Mo are O: 25; N: <10; H: <5; C: <10, respectively. Prior to compression test, the test sample was heat treated at 1500 °C for 1 h, 1200 °C for 1 h, and 1000 °C for 1 h at a vacuum of 8 x 10^{-11} Torr. Testing of single crystals involves compressing the test sample between two platen surfaces under precise conditions. To measure shear strain during compression, a 3-element rosette gage was bonded in the gage section on each side of the sample. The gages were applied with room temperature curing epoxy adhesive. A compression test was performed on a Mo test sample with a [118] compression axis. This orientation was chosen to study single-slip deformation involving the $(\bar{1}\,\bar{1}\,2)[111]$ slip system. The sample was compressed at a nominal strain rate of 10^{-3} S^{-1}. The sample was

119

compressed to a value of approximately 1% axial strain. TEM foils were sliced from the gage section of the tested piece with the foil sliced parallel to the ($\bar{1}\,\bar{1}\,2$) plane. TEM specimens were finally prepared by a standard twin-jet electropolishing technique in a solution of 75 vol.% ethanol and 25 vol.% sulfuric acid at ~25 V and -10 °C.

RESULTS

Initial dislocation structures in annealed crystals

The initial dislocation structures in as-annealed Mo examined from [011]-, [0$\bar{1}$1]- and [100]-sliced foils are illustrated in a 3D box as shown in Fig. 1. The dislocation density (ρ) is estimated to be on the order of $10^6 \sim 10^7$ cm^{-2}. Occasionally, cross-grid screw dislocations with Burgers vectors of $\pm$ ½ [111] and $\pm$ ½ [1$\bar{1}$$\bar{1}$] were observed in the [0$\bar{1}$1]-sliced foil, which were determined using the $\mathbf{g}\cdot\mathbf{b} = 0$ invisible criterion as shown in Fig. 2. As can be seen clearly that the dislocations of $\mathbf{b} = \pm$ ½ [1$\bar{1}$$\bar{1}$] become invisible when $\mathbf{g} = \bar{2}\,1\,1$ is applied for imaging, and those of $\mathbf{b} = \pm$ ½ [111] become invisible when $\mathbf{g} = 2\bar{1}\bar{1}$ is applied for imaging. These dislocation pairs have a near-screw character since their line vectors are nearly parallel to their Burgers vectors. Notice that the appearance of residual contrasts under $\mathbf{g}\cdot\mathbf{b} = 0$ conditions suggests that these dislocation lines are not pure screw in character.

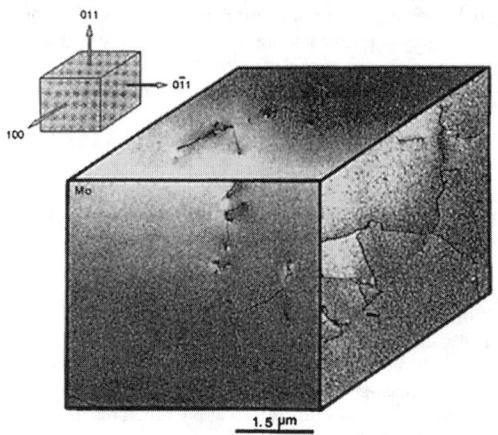

Fig. 1. Initial dislocation structures in as-annealed Mo single crystals.

According to Frank-Reed dislocation multiplication mechanism [2], dislocation can multiply by repeatedly bowing out a free segment of dislocation line lying in a slip plane, and the shear stress (τ) required bowing out a line segment (l) is given as: $\tau \approx \mu b/l$. Thus, there may exist a critical length ($l^* \approx \mu b/\tau_a$) of free segment for a given applied shear stress (τ_a). Any length of free segment l which is smaller than l^* will be permanently immobile, while length of segment greater than l^* are potentially mobile. Accordingly, an investigation of the relative density (ρ_m/ρ) of mobile dislocation in slip plane is important for studying the yield strength of crystal. An investigation made on this aspect of dislocation configuration is shown in Fig. 3. Here, an "in-plane" ½[111] dislocation segment (~5 μm in length) in the {0$\bar{1}$1} plane is shown. Notice the bend-over at two ends of the segment, which indicates that a dislocation line is not entirely lying in one crystallographic plane. However, there is an uncertainty whether the observed segment is truly a free dislocation segment without any other pinning points such as short jog segments formed on the dislocation line. In other words, there is a difficulty to define free segment length by viewing dislocation from this orientation since it is infeasible to locate the pinning points formed by jog segments along the dislocation line.

(a)

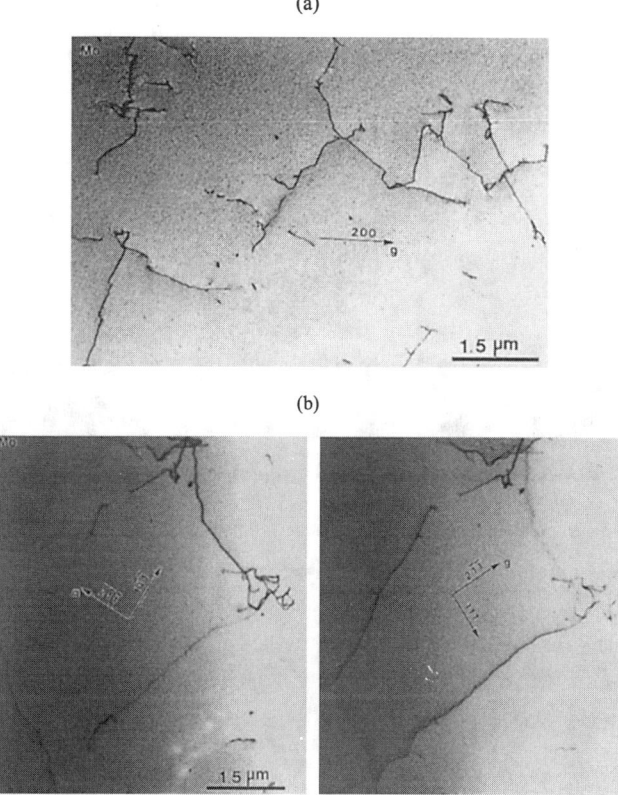

(b)

Fig. 2. Bright-field TEM images showing (a) the formation of cross-grid dislocations, and (b) a $\mathbf{g} \cdot \mathbf{b}$ analysis for the character of cross-grid dislocations; (left) $\mathbf{Z}$ (zone axis) $\approx [0\bar{1}1]$, $\mathbf{g} = \bar{2}\,\bar{1}\,1$, dislocations of $\mathbf{b} = \pm \frac{1}{2}\,[111]$ become invisible, and (right) $\mathbf{Z} \approx [0\bar{1}1]$, $\mathbf{g} = 2\bar{1}\,\bar{1}$, dislocations of $\mathbf{b} = \pm \frac{1}{2}\,[1\bar{1}\,\bar{1}]$ become invisible.

In fact, the dislocation segment length viewed from the $\{\bar{1}01\}$-sliced foils may be a measure of the line waviness along the foil normal. This can be visualized readily from a cross-sectional view of $\frac{1}{2}[\bar{1}\,\bar{1}\,1]$ screw dislocation shown in Fig. 4(a), in which the screw dislocation in the (011) plane was observed from the foil sliced parallel to the $(0\bar{1}\,1)$ plane. Here, the existence of many long superjogs (50 ~ 100 nm in height) along the $\frac{1}{2}[\bar{1}\,\bar{1}\,1]$ screw dislocation line can be seen. In addition, the dislocation line is found to skew away from the $[\bar{1}\,\bar{1}\,1]$ direction revealing that the dislocation line is also associated with many short superjogs [jog height (d) < 1 nm] or elementary jogs [jog height = interplanar spacing of $(\bar{1}\,21)$ plane = 0.135 nm]. Noted that the height of short superjogs or elementary jogs is too short to be resolvable using conventional TEM imaging techniques. This examination suggests that the short

121

dislocation segments appeared in the {$\bar{1}$01}-sliced foils is attributed to the formation of jogs along a screw dislocation line which causes it to lie across many {$\bar{1}$01} planes instead of one. Consequently, screw dislocation lines are chopped into short segments in a {$\bar{1}$01}-sliced TEM foils (~ 0.4 μm thick). Similarly, jogged ½[111] and ½[11$\bar{1}$] screw dislocations were also observed. A jogged ½[11$\bar{1}$] screw dislocation viewed from the [0$\bar{1}$1] direction is shown in Fig. 4(b). Here, many large superjogs (50 ~ 100 nm in height) can be readily seen along the dislocation line. Also notice that the lengths of each free segment linked between two superjogs are unequal.

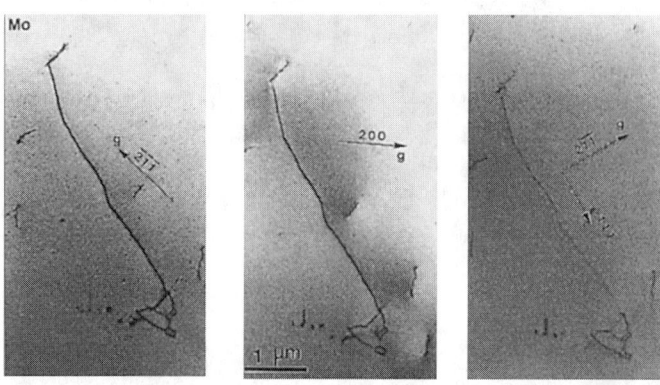

Fig. 3. A **g·b** analysis for a ± ½ [111] screw dislocation formed in (0$\bar{1}$1)-sliced Mo.

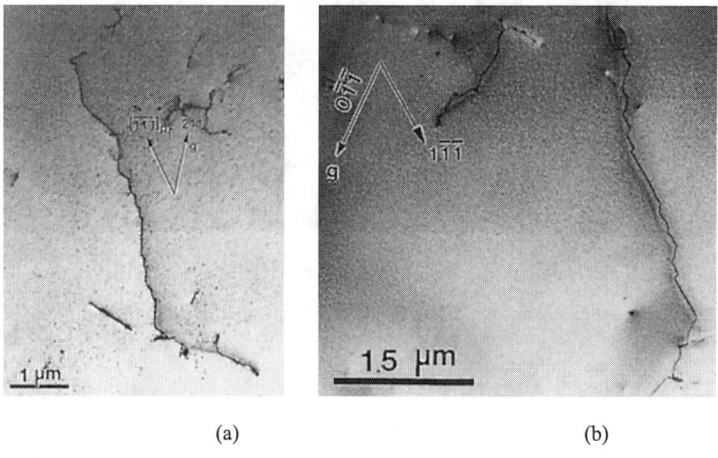

(a)   (b)

Fig 4. TEM images showing (a) a jogged ½[$\bar{1}$1$\bar{1}$] screw dislocation and (b) a jogged ½[11$\bar{1}$] screw dislocation viewed from the [0$\bar{1}$1] direction in as-annealed Mo.

Deformation substructure

Typical deformation substructures of a crystal compressed for 1 % are shown in Fig. 5. The dislocation density increases about two orders of magnitude to a range of 10^8 ~ 10^9 cm^{-2}. Although the

½[111] screw dislocations are prevalent, other secondary dislocations are also operative. Notice that many superjogs (marked by arrows) are seen along screw dislocation lines, and the average jog height, and free segment between superjogs are found to increase significantly in comparison with that observed in as-annealed samples (Fig. 4). In addition, the screw dislocation lines become much straighter and longer comparing to those in as-annealed crystals. The increase of cross-grid screw dislocations, as shown in Fig. 5(b), indicates that multiple slip systems have been activated in the early stage of plastic deformation. It is worth noting that few junction dislocations are formed at the cross points of the cross-grid dislocations during the stage.

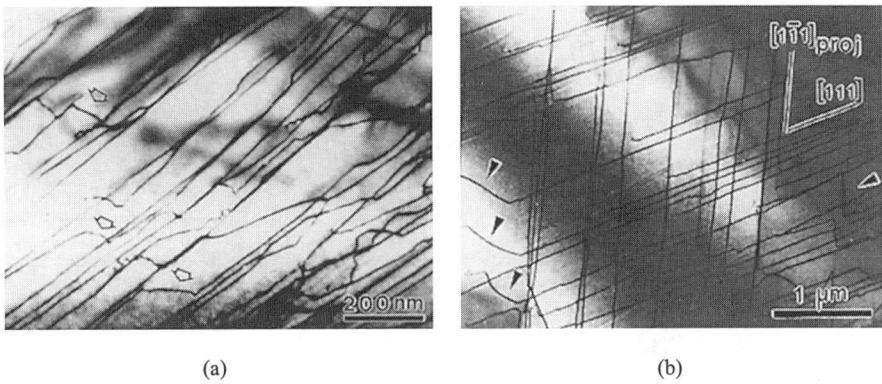

(a) (b)

Fig. 5. Typical dislocation structures observed in a crystal compressed for 1 %.

Proposed dislocation multiplication mechanisms

Based upon the observations shown in Figs. 4 and 5, dislocation multiplication during the early stage of plastic deformation in Mo single crystals can be rationalized by mechanisms proposed and illustrated in Fig. 6. Here, screw dislocation segments pinned by superjogs bow out between the superjogs under an applied shear stress (τ) to a certain curvature, yet they are immobile since the initial length (l_o) of each free segment is smaller than a critical length ($l^* \approx \mu b/\tau$) as defined earlier. Beside the force exerted on dislocation segments by the applied shear stress, each superjog is subjected to a net force parallel to the Burgers vector due to unbalanced line-tension partials acting on the segments of line between unevenly spaced jogs. Thus, each jog will move in such a direction so that the shorter segments ($\overline{CD}$ and $\overline{EF}$) become still shorter and the longer segments ($\overline{AB}$ and $\overline{GH}$) are expanded [3]. The superjogs of like-signs tend to coalescence in order to reduced line energy and resulting in the increase of jog height [4]. Consequently, the stress-induced jog coalescence renders an increase of both segment length and jog height.

Applied shear stress eventually begins to push each line segment between jogs to precede dislocation multiplication when the length of line segments ($\overline{IJ}$ and $\overline{KL}$) and height of superjog ($\overline{JK}$) are greater than critical values, i.e. $l^* \approx \mu b/\tau$ and $d_c \approx \mu b/8\pi(1-\nu)\tau$. Here, d_c can be defined by a mutual attraction force between adjacent bowing edge segments of opposite signs. That is, the originally immobile screw dislocations become multiple sources for dislocation multiplication as a result of the jog coalescence process. This "dynamic" dislocation multiplication source is suggested to be crucial for the dislocation multiplication in the early stage of plastic deformation in Mo.

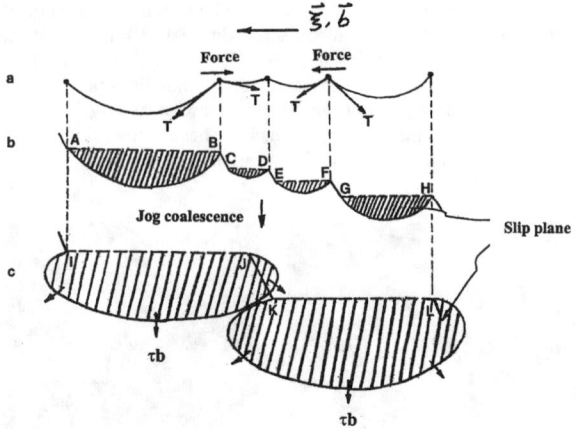

Fig. 6. Schematic illustrations of dislocation multiplication from a jogged screw dislocation. (a) A top view showing dislocation segments pinned by superjogs bowing under stress to a curvature, net forces are generated on jogs due to unbalanced line-tension partials acting on the free segments of unequal lengths. (b) A tilt view of (a) shows the initial heights of like-sign superjogs. (c) Both segment length and jog height increase as a result of stress-induced jog coalescence.

CONCLUSIONS

Screw dislocations in association with grown-in superjog segments were observed within annealed Mo single crystals. It is proposed that the jogged screw dislocations can act as a dynamic multiplication source for dislocation multiplication when deformed at quasi-static conditions as a result of stress-induced migration and coalescence of superjogs. The jog coalescence can take place via the lateral migration (drift) of jog segments driven by unbalanced line tension partials acting on line segments (between jogs) of unequal lengths. The coalescence of jog segments results in the increase of both segment length and jog height. The jog coalescence continues until the segment length and jog height are greater than critical values so that applied stress begins to push each line segment to precede multiple dislocation multiplication.

ACKNOWLEDGEMENTS

This work is performed under the auspices of the U.S. Department of Energy through contract # W-7405-Eng-48 with Lawrence Livermore National Laboratory.

REFERENCES

1. G. H. Campbell et al., *Mater. Sci. and Engrg.* **A251** (1998), p. 1.
2. F. C. Frank and W. T. Read, in "Symposium on Plastic Deformation of Crystalline Solids," Carnegie Institute of Technology, Pittsburgh, 1950, p. 44.
3. J. Weertman and P. Shahinian, *Trans. A.I.M.E.* **209** (1957), p. 1298.
4. N. Louat and C. A. Johnson, *Phil. Mag.* **7** (1962), p. 2051.

DISLOCATION FOREST INTERACTIONS:
SIMULATION AND PREDICTION

L. K. WICKHAM[1,2], K. W. SCHWARZ[2], and J. S. STÖLKEN[1]
[1]Lawrence Livermore Lab, P.O. Box 808, Livermore, CA 94551
[2]IBM T.J. Watson Research Center, P.O. Box 218, Yorktown Heights, NY 10598

ABSTRACT

Using linear elastic dislocation dynamics simulations, we show that junction formation between dislocations from various interacting slip systems can be predicted by a simple self–energy calculation. We find that this prediction is robust: dislocation curvature and external stress produce little change in the simulation results for junction formation. One key to this success appears to be a separation of timescales, where movement of the far away dislocation arms (under, for example, external stress) is typically slow compared to the process of making a junction. The self-energy calculation we describe gives a rule for dislocation encounters which should allow a considerable saving in computational effort, allowing one to impose correct interaction outcomes without calculating the interactions in detail. We also find that dislocations often come together under attraction without forming a junction. The resulting "cross–linked" state provides an additional type of connection between dislocations. We include preliminary results on the persistence of junctions and cross–linked states under stress.

INTRODUCTION

Forest interactions are essential components of work hardening. Dislocations which cross each other's path can form jogs which slow down dislocation glide, or they can become bound in structures with highly restricted motion. In particular, two dislocations can combine to form a junction, and the pinning effect of such junctions is a central component in many work hardening theories. Thus far, however, a question has remained open: just when should junction formation be expected?

Figure 1 shows junction production as modeled in the PARANOID [1] code, a 3D linear elastic dislocation simulator with capability for high spatial resolution [2]. Both the red and the blue dislocation are shown at equally spaced times. Notice that they drift together, and then curve more sharply and form a fast–zipping junction. Following this latter process in detail is extremely computationally expensive, due to the higher spatial resolution required and the small time steps necessary during such a fast event. A much more efficient strategy would be to take the model directly from early stages of dislocation interaction to the final bound structure. (This strategy should be aided by the fact that, during junction formation, the far regions of the dislocation typically move little.) Computational researchers thus have additional reason to be interested in junction prediction: they would like to use such foreknowledge to overcome a serious computational obstacle.

Mat. Res. Soc. Symp. Proc. Vol. 578 © 2000 Materials Research Society

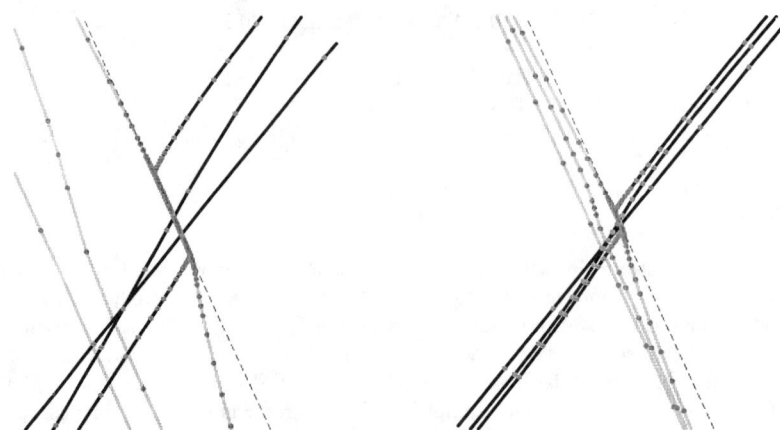

Figure 1 Three times from simulation of junction formation with high resolution. The left picture is a blowup; the second (identical) figure is viewed from a greater distance. In the first time interval, the dislocations drift and stay relatively straight. After the next (equal) time interval, the two dislocations have joined in the center along the thin dotted line, which is the line joining their two glide planes. For this final configuration, the high dot density indicates high spatial resolution.

Given the large number of geometries to be considered, and the sometimes complicated interactions of approaching dislocations, how difficult is it to build expectations of what will happen during forest interactions? We have attempted to answer this question by running a large number of dislocation pair simulations, such as that shown in figure 1. Given a starting geometry such as that in figure 2a, we find that the results are insensitive to starting dislocation separations d_1 and d_2 [3]. Thus, for a given slip system pair and set of environmental parameters, the results may be given as a function of ϕ_1 and ϕ_2, the initial angles between the dislocations and the glide plane edge.

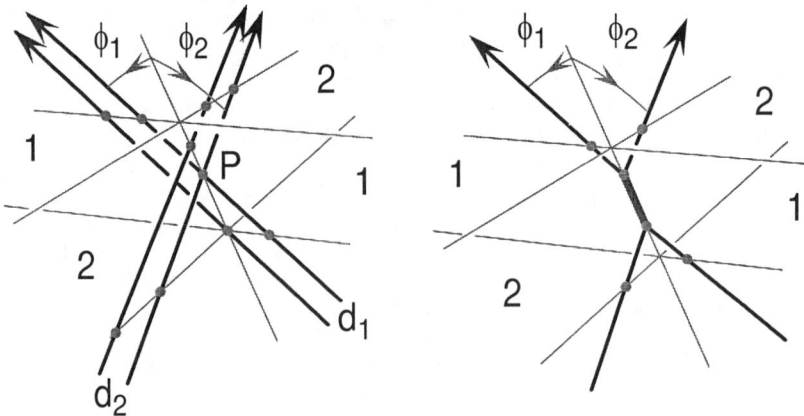

Figure 2a Starting geometry for simulations.   Figure 2b Idealized junction geometry.

RESULTS

Such results are shown in figures 3a and 3b, for two different bcc slip systems with isotropic dislocation mobility and zero external stress. Here, filled circles indicate the formation of clear junctions, and empty spaces indicate repulsive interactions. There is, however, another possible outcome: crosses indicate dislocations which attracted each other and came together in a crossed configuration, but did not zip up into a junction. Figure 4a shows one of these crossed configurations. They are qualitatively different from junctions, in that the dislocations meet at a large angle and show no signs of zipping together. Empty circles in figure 3 show borderline cases, which may have been zipping into a junction very slowly.

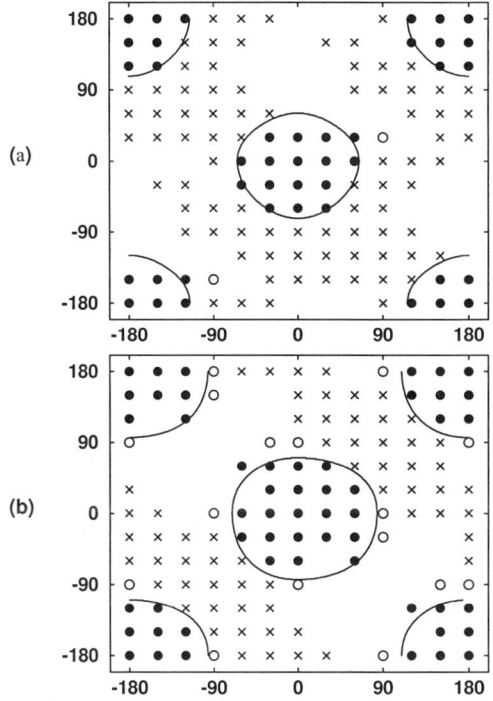

Figure 3a: results of simulations for the bcc slip system pair $(0\bar{1}1)$ $\frac{1}{2}[1\bar{1}\bar{1}]$ and $(\bar{1}01)$ $\frac{1}{2}[111]$. Figure 3b: results for the pair (110) $\frac{1}{2}[1\bar{1}\bar{1}]$ and $(\bar{1}01)$ $\frac{1}{2}[111]$. The horizontal and vertical axes show ϕ_1 and ϕ_2, the angles which the first and second dislocations make with the glide plane edge.

Before addressing the question of junction prediction, we first ask: what is the significance of forest interactions which end in a cross–link? Although we have seen little discussion of this case in the literature, such states are an inevitable result of the finding that not all attractive dislocation interactions produce a junction [4]. The importance of cross–links in work hardening must depend in part on their durability— do they hold up under stress? A final answer to this question may require experimental observation or atomistic simulation, but we can report that cross–links do seem to hold up under moderate stress in our linear elastic simulation. For example, the crossed set in figure 4a stayed joined until the resolved shear stress on each dislocation reached 23 MPa. For comparison, the junction in figure 4b stayed joined until the resolved shear stress reached 76 MPa. Both cases involved the same slip system pair, and dislocations of the same length (10000 A). (For the simulations shown in figure 4, the ends of the dislocations were pinned. All other simulations reported here had free ends.) In preliminary results, we have seen breaking stresses for cross–links which are within an order of magnitude of that for typical corresponding junctions. Thus, bound cross–links, which may only move along the glide plane edge, may play a significant role in dislocation evolution.

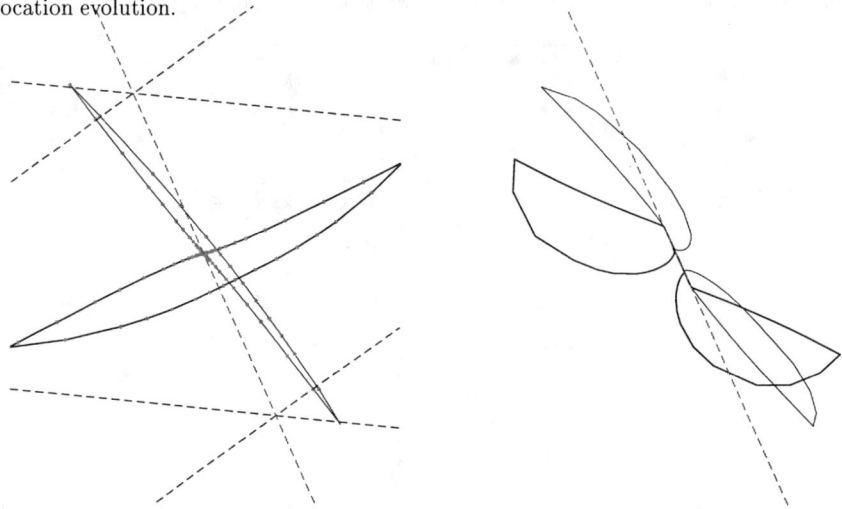

Figure 4a: cross point just above breaking stress. Figure 4b: junction just below breaking stress. In both cases, the two dislocations are shown both before and after the stress is applied. In the left figure (4a), dense dots indicate an early time, when the two dislocations were together. Here, dotted lines have been included to indicate the directions of the two glide planes.

The black contours in figure 3a and 3b do successfully show the region of space where most junction formation occurs. These contours come from a junction formation prediction based on the simple line tension model of Herring [5]. In this calculation, the junction is represented by the idealized shape of figure 2b, where the angles ϕ_1 and ϕ_2 are chosen to match the starting angles of the dislocation pair in a simulation run. Far away from the junction, the final dislocation arms still retain these original slopes.

The calculation which predicts junction formation finds the change in dislocation self-energy upon zipping of the idealized structure in figure 2b. This process lengthens the junction, shortens the arms, and tips the arms so that they make a larger angle with the glide plane edge. Junction formation is favorable when these respective energy changes add to a negative sum:

$$0 > E_J - E_1 cos(\phi_1) - E_2 cos(\phi_2) + \frac{\delta E_1}{\delta \phi_1} sin(\phi_1) + \frac{\delta E_2}{\delta \phi_2} sin(\phi_2) \qquad (1)$$

Here, E_J, E_1, and E_2 represent the self–energy per unit length of the junction, arm 1, and arm 2. We have used the linear isotropic elastic result for these energies, as appropriate for our simulation.

The success of this self–energy prediction is a little surprising, since it neglects interactions between the different dislocation segments and any curvature in the junction shape. However, we have observed that the shapes of junction arms often change very little during junction growth. Thus, the main effect of the zipping process may be to lengthen the junction and change the average arm angle, without substantially altering dislocation curvature and interaction energies near the junction.

A further check is required for the junction predictions of equation (1): do they hold up in more complicated situations? Figure 5 shows the same slip system pair as in figure 3a, under several different conditions. For the simulations which gave figure 5a, the first dislocation was a loop, and the other was a straight line. The angle ϕ_1 for the loop was taken as the original tangent to the portion of the loop which eventually interacted with the other dislocation. These results indicate that a radius of curvature of a few hundred angstroms (the typical loop size) does little to change the phase space of junction formation.

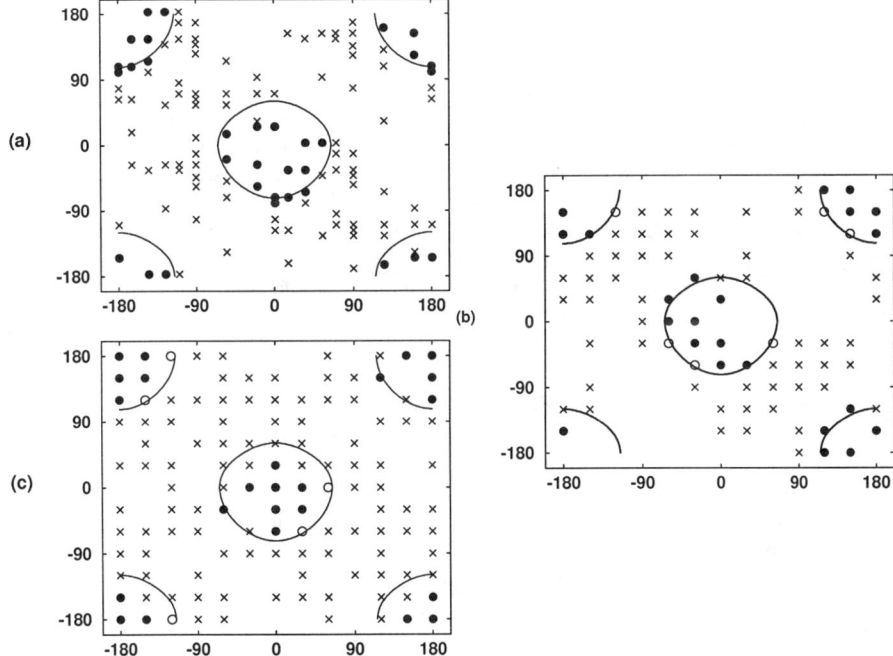

Figure 5a: results of simulations with one loop and one straight dislocation. Figure 5b: results of simulations with uniaxial stress of 140MPa. Figure 5c: results with uniaxial stress of 690 MPa. These runs used the same slip system as those of figure 3a.

For the work shown in figures 5b and 5c, an external uniaxial stress was applied along the $< 001 >$ direction. This stress was 140 MPa in the case of figure 5b, and 690 MPa in the case of figure 5c. Both positive and negative stresses were used to bring the dislocations together, and the resulting circle or cross represents the structure that was formed during the period in which the dislocations were in contact. After subsequent motion of the long dislocation arms, the cross–link or junction often pulled apart during the simulation period, as we would expect at these stress levels. In other cases, the junction or cross–link stayed during the simulation period and moved along the glide plane edge. Despite these complications, if we ask the question, "what structure will be formed during the meeting of two mobile dislocations under stress?" the answer is similar to the result for zero stress. Extremely high stresses (690 MPa) do seem to produce a shrinkage in the region of junction formation, perhaps because there is less time for junction growth before the outer arms pull the dislocation onward. More moderate (140MPa) stresses, however, appear to have little effect on the junction phase space.

CONCLUSIONS

Thus, line tension seems to dominate the process of junction formation, even in complicated situations, with dislocation curvature or external stress. Simple self–energy calculations, which neglect bending and nonlocal dislocation interactions, have had remarkable success in predicting the junction formation behavior of our linear elastic 3D dislocation simulations. Outside of the junction formation region in geometric phase space, attractive dislocations come together in the simulation to produce cross–linked states, which are capable of persisting under modest levels of external stress. Repulsive dislocations can also be forced to meet under external stress, and we have not observed any junction formation in this case. Thus, equation (1) is an accurate predictor of what should happen when two simulated dislocations are about to meet, and can be used as a powerful shortcut to bypass expensive computations of the full junction formation process. We also hope that our verification of the utility of line tension arguments will encourage their use in analytic theories about the interaction of different slip systems during work hardening.

This work was supported by Lawrence Livermore National Laboratory under contract No. W-7405-ENG-48 with the DOE.

REFERENCES

[1] PARAllel NOdal Ibm Dislocation code. In this model, the dislocation cores are able to overlap, because the singularity in dislocation interactions has been softened at separations smaller than 2 core radii. The boundary condition we typically use has each long dislocation arm connected to an infinite half-line of matched slope. The PARANOID code is described in K.W. Schwarz, J. Appl. Phys. **85**, 108 (1999), and an earlier reference on this junction work is: L. K. Wickham, K. W. Schwarz, J. S. Stölken, Phys. Rev. Lett. **83**, 4574 (1999).

[2] References on other 3D dislocation simulation efforts include: B. Devincre and L. P. Kubin, Mat. Sci. Eng. A **234-6**, 8 (1997); M. Tang, L. P. Kubin, and G. R. Canova, Acta. Mater. **46**, 3221 (1998); H. M. Zbib, M. Rhee, and J. P. Hirth, Int. J. Mech. Sci. **40**, 113 (1997); and M. Rhee, J. P. Hirth, H. M. Zbib, Acta Metall. **42** 2645 (1994).

[3] These can be varied from $\pm 20\,nm$ to $\pm 400\,nm$ without much effect.

[4] This was anticipated in G. Schoeck and R. Frydman, Phys. Stat. Sol. (b) **53**, 661 (1972).

[5] C. Herring in *The Physics of Powder Metallugy*, edited by W. E. Kingston (McGraw-Hill, New York, 1949).

FOREST HARDENING AND BOUNDARY CONDITIONS IN 2-D SIMULATIONS OF DISLOCATIONS DYNAMICS

D. GÓMEZ-GARCÍA [1], B. DEVINCRE and L.P. KUBIN
Laboratoire d'Etude des Microstructures, CNRS-ONERA, BP 72, 92322 Chatillon, France

INTRODUCTION

Dislocations are the elementary carriers of plastic flow and are ideally at the base of any physical model for plastic deformation. In the last few years, several 3-D simulations of Dislocation Dynamics (DD) have been devoted to the analysis of single crystal plasticity at small strains [1-5]. However, such DD simulations have some limitations which restrict their domain of application: (i) In many materials, there is a lack of accurate input regarding the mechanisms governed by the core properties of dislocations. (ii) The plastic strain amplitude that can be simulated is usually small (e.g. smaller than 1% in f.c.c. crystals). (iii) The boundary conditions of the simulations are generally rather crude and may introduce spurious size effects and various other artefacts. Most of the time, the questions of stress equilibrium and dislocations flux at the external surfaces are not addressed.

Some of these problems would be substantially simplified if it was possible to restrict the calculations to two dimensions, with the condition that true 3-D mechanisms are reproduced with a reasonable accuracy. This would allow treating generic problems related to dislocation patterning and strain hardening properties. Indeed, the expressions for the stress field of infinitely straight dislocations are much simpler than those for dislocation segments. In addition, efficient O(N) algorithms are available for the treatment of large densities of interacting dislocations [6]. Finally, when modeling only two active slip systems, it becomes possible to use periodic boundary conditions and, then, to equilibrate stresses and dislocations fluxes at the boundaries of the simulated area.

This is why 2-D *end-on* simulations of dislocation dynamics, which have attracted lot of interest in the past, may be still of actuality. Since the pioneering works of Lepinoux and Kubin [7] and Amodeo and Ghoniem [8], many other simulations have been developed. In the present case, use is made of a code, which was originaly designed for the study of dislocation dynamics in Ni_3Al [9]. The originality of this simulation is to incorporate, in addition to elastic interactions, a set of local rules that correct the drawbacks and artefacts of previous models. In particular, the kinetics of dislocation multiplication and the formation of sessile junctions are carefully treated. Further, the influence of the boundary conditions is considered here for the first time and tested on the stress vs. strain response in the presence of forest hardening.

NUMERICAL METHOD

Two slip systems are defined which contain infinite, parallel dislocations. The simulation plane is normal to the lines and to their Burgers vectors, b_1 and b_2. The latter are resolved into two components, an edge component lying in the simulation plane and a screw component parallel to the dislocation lines. For simplicity, only edge dislocations are considered in what follows and the Burgers vectors define the slip direction of the dislocations.

---

[1]Permanent address: Departamento de Fisica de la Materia Condensada-Instituto de Ciencia de Materiales. Universidad de Sevilla-C.S.I.C. Apartado 1065. 41080-Sevilla (SPAIN)

Mat. Res. Soc. Symp. Proc. Vol. 578 © 2000 Materials Research Society

The slip directions are taken parallel to the diagonal of the square simulation area, whose linear dimension is L.

Inertial effects being neglected, the dislocation mobility is only related to the effective stress through the velocity equation:

$$v(t) = \frac{\tau^*(t)\, b}{B},\tag{1}$$

where B is a viscous drag coefficient describing the damping by electrons and phonons ($B = 5.5\ 10^{-5}$ Pa.s in Cu at 300 K) and $\tau^*(t)$ is the effective stress. The dislocation displacements are calculated at each time step of the simulation, Δt, by means of the classical first order Euler's method.

The effective stress is made up of two contributions. (i) The resolved applied stress, τ is assumed uniform in the simulated area and identical for the two slip systems (*i.e.* the latter have same Schmid's factor). A constant strain rate is imposed through a feed-back loop on τ [10]. The simulation results are only weakly sensitive to this control provided that the imposed strain rate is smaller than $10 s^{-1}$. Thus, all the results reported below were obtained with an imposed strain rate of $1 s^{-1}$. (ii) The internal stress field, τ_{int}, is simply computed as the sum of the dislocation self-stress fields [11].

A crucial point of all *DD* simulations is that the stress field of the dislocations is long-ranged and proportional to $1/r$. As a consequence, all the dislocations should be considered when calculating the effective stress. Nevertheless, it is widly acccepted that, once dislocation self-organize, self-screening may occur and induce an upper cut-off for the pairwise elastic interactions. Such a possibility has been critically considered in the past [12]. It was found that a simple cut-off procedure always generates artefacts. Indeed, the elastic energy is abruptly decreased when the average distance between the lines becomes equal to the cut-off distance and an artificial driving force is then generated. The influence of this driving force depends on the cut-off distance (the smaller the cut-off distance, the higher the driving force) and the loading conditions (stronger effects are expected during relaxation tests under zero applied stress).

To take advantage of the self-screening effect without introducing artefacts, a new method is proposed here. It consists of using a different cut-off for each dislocation, which, in addition, is stochastically redefined at each step of the simulation. The cut-off distances used range from $L/3$ to $L/2$. This stochastic procedure adds a random background to the long distance interactions and, therefore, eliminates any preferential wavelength that could artificially induce patterning.

To treat dislocation core effects and reactions at short approach distances, simple meso-scopic rules are required. These local rules are not intended to model exactly all the details of the real 3-D properties, which would not be possible within a simple 2-D model, but to reproduce their main contribution to crystal plasticity.

Regarding direct dislocation annihilation, close dislocations pairs with opposite Burgers vector are eliminated when the dislocation spacing is smaller than $5b$. The simulation results reported below are rather unsensitive to that parameter. All the lengths involved in the simulation, including the size L, are expressed in units of b. To extract quantitative data, a value $b = 0.25 nm$ is used.

The formation of sessile junctions between attractive dislocations gliding on different slip systems is modelled using the following rules. (i) The Frank's energy criterion, which ensures that the two dislocations are attractive, must be satisfied [11]. (ii) The dislocations form a junction when their distance becomes smaller than a critical "reaction distance", r_d.

This distance was calibrated in order to yield a reaction cross-section equivalent to the three-dimensional one. As a result, $r_d = 400b$. (iii) The effective stress on the dislocations must be smaller than a critical stress for the destruction of the junctions, τ_j, whose amplitude is proportional to the square root of the dislocation density (eq. 2). Such dependency is due to the fact that the destruction of real junctions occurs by an unzipping process which depends on the length of the parent dislocations, hence on the square root of the dislocation density [13].

$$\tau_j = K_j\sqrt{\rho} \tag{2}$$

In equation 2, K_J is a fitting parameter set to 0.1 in order to to obtain a realistic value for the forest hardening. Details on this adjustment are reported in the next section.

The generation of fresh dislocations by a source mechanism cannot be ignored in a dislocation model. For this reason, the presence of a density of Frank-Read sources has to be accounted for during simulations. The position of these sources, their Burgers vector and glide plane are initially set at random and kept constant all through the numerical simulations. In addition an equivalent density of (non-multipliying) segments are present in the initial configuration. By analogy with 3-D simulations, the critical stress for source emission, τ_{FR}, is determined by the density of source segments, ρ_{source}. Hence :

$$\tau_{FR} \approx \frac{1}{2}\mu b\sqrt{\rho_{source}} \tag{3}$$

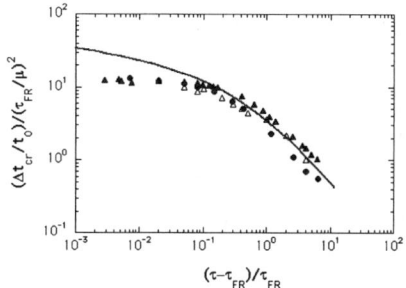

The rate of dislocation emission was estimated in the following way. The time interval (Δt_{cr}) for the emission of one dislocation loop, for given values of the applied stress and the source length, is predefined in the simulation. The numerical values were obtained by means of a 3-D simulation [2] and are shown in figure 1. A simple calculation that integrates with respect to time the bowing out of a segment under stress, neglecting the time spent beyond the critical configuration, yielded the following expression which fits well the simulated 3-D values:

Figure 1: Critical time between the emission of loops by a Frank-Read source as a function of the reduced stress and for different source lengths (l_s). The scaling terms derive from equation 4 which is represented by a full line. The data points were obtained with the help of a 3-D simulation with $l_s = 10^2$, 10^3 and $10^4 b$.

$$\Delta t_{cr} = \frac{5\mu B}{\tau_{FR}^2} \ln(\frac{\tau}{\tau - \tau_{FR}}) \tag{4}$$

Based on this calibration, two dislocations with opposite Burgers vectors are nucleated on each side of a source and at time intervals Δt_{cr} when the effective stress is larger than τ_{FR}. The distance between the emitted dislocations is set to $(\mu b)/\tau^*$, in order to minimize the elastic interactions between successive loops.

Finally, both free and periodic boundary conditions were tested. In the former case, dislocations reaching a boundary simply vanish from the simulation. In the latter, dis-

133

locations leaving from one side are instantaneously reintroduced through the opposite side. However, the height of the new slip plane is shifted by $L/(5\sqrt{2})$ with respect to the old one to avoid early annihilation and steady-state behavior in the active glide planes.

1 RESULTS

Forest hardening
We first examine whether a domain of dislocation densities can be defined where the simulation results are in good agreement with experiment. For this purpose, the flow stress level τ was computed as a function of the dislocation density. These computations were performed with a constant density of dislocations (i.e., without dislocation sources). Periodic boundary conditions were used. The results are summed up in figure 2. To allow comparison with experiment, the present results are plotted in reduced units, $\alpha = (\tau - \tau_0)/(\mu b \sqrt{\rho})$, with $\tau_0 = 1.5\ 10^{-4}\mu$. The

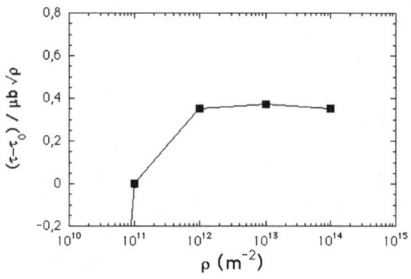

Figure 2: Variation of the flow stress in reduced units as a function of the dislocation density.

constant α is, indeed, a measurable quantity whose value is about $0.3 - 0.4$, in pure f.c.c. crystals.

From figure 2, one can see that the simulation results reproduce well the expected value of α from $\rho = 10^{12}m^{-2}$ up to $10^{14}m^{-2}$. For smaller densities, the probability to form junctions in the simulated area becomes very small and the flow stress is no longer governed by the dislocation-dislocation interactions but by the dislocation mobility. Inversely, for densities larger than $10^{14}m^{-2}$, the mean distance between dislocations becomes smaller than the "reaction distance" and all the dislocations are soon immobilized at junctions. In consequence, the initial densities of sources and dislocations were fixed to the value $(10^{12}m^{-2})$.

Influence of boundary conditions and simulation size
In a second step a comparison was performed between results obtained with the two types of boundary conditions mentioned above, for different values of the linear dimension L of the simulated area. The results of this study are reported in figure 3. To allow comparing the different simulation conditions, the flow stress values are measured either at a given plastic strain amplitude (0.2%) or at given dislocation density ($2.3\ 10^{12}m^{-2}$).

As shown in figure 3, the results obtained with periodic boundary conditions are not very sensitive to the simulation dimension. Both the flow stress and the density ($2.3\ 10^{12}m^{-2}$) at 0.2% are reasonably constant, except for very small system sizes. With the latter, the dislocation annihilation rate is artificially increased by the periodic conditions. For larger sizes ($L > 4$ microns), a quasi-linear increase of the dislocation density is observed upon straining with a constant applied strain rate.

Under a constant applied plastic strain rate, free boundary conditions (FBCs) yield flow stress amplitudes larger than periodic boundary conditions (PBCs). This is understood as follows. When an important fraction of the dislocation density is lost at the boundaries, the generation rate must increase in order to maintain the density of mobile dislocation at the level required by Orowan's law. Such an effect is clearly observed in the simulations performed with $L < 8\mu m$. In such conditions, the plastic strain is severely localized and most of the plastic activity occurs in few glide planes containing extremely active dislocation sources. As could be expected, when the simulation size is increased, the results tend to better agree with those obtained with the PBCs. In addition, it is worth noticing that the flow stresses are similar for the two types of boundary conditions when comparisons are performed at constant dislocation density (cf. figure 3). This clearly shows that the flow stress is always governed by forest hardennig $i.e.$, by dislocations junctions and dislocation-dislocation long range elastic interactions.

DISCUSSION AND CONCLUDING REMARKS

The 2-D simulation code presented here is able to model plasticity in multi-slip conditions and to reproduce a realistic forest hardening over two decades in dislocation density. Comparisons between different boundary conditions clearly illustrate the influence of the dislocation losses at the free surfaces. But even with the largest simulation size, when the simulated flow stress no longer depends on the type of boundary conditions, there are still strong differences in the dislocation microstructures (see figure 4). Nevertheless, this type of work opens the possibility of looking at dislocation patterning phenomena in a generic manner and within a simple framework.

In 3-D, the question of the compensation of the outgoing dislocation flux has been considered only recently [14]. In addition, one must be aware that the inbalance between outgoing and incoming dislocation fluxes is not the only problem posed by the boundary conditions, since traction-

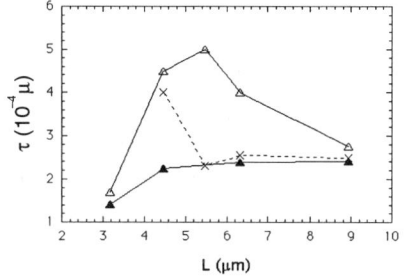

Figure 3: Variation of the yield stress on different simulated stress-strain curves. Filled and empty triangles correspond to the 0.2% plastic strain offset stress measured with periodic and free boundary conditions, respectively. The crosses represent the yield stress values measured with free boundary conditions when $\rho = 2.3\,10^{12}m^{-2}$.

free conditions must also be satisfied. These conditions, which were automatically cared of by the 2-D PBCs, is now resolved in 3-D through the coupling of DD simulations with finite element codes [3, 15].

ACKNOWLEDGMENTS

Diego Gómez Garcia wants to acknowledge the financial support awarded by the "Fundación Ramón Areces" through a postdoctoral grant during his stay in the LEM 1998-1999.

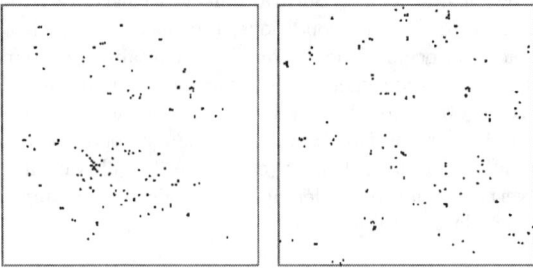

Figure 4: Simulated dislocation microstructures obtained with free boundary conditions (left) and periodic boundary conditions (right). Both computations were performed at a constant strain rate $(1s^{-1})$. Although the flow stress, the plastic strain ($\epsilon = 0.2\%$), the dislocation density and the simulation size ($L = 35000b$) are the same, the depletion of the dislocation density near the sides is clearly visible in the case of the free boundary conditions.

REFERENCES

[1] L.P. Kubin, G. Canova, M. Condat, B. Devincre, V. Pontikis, and Y. Brechet. *Solid State Phenomena*, **23-24**:455, 1992.

[2] B. Devincre and L. P. Kubin. *Modelling Simul. Mater. Sci. Eng.*, **2**:559–570, 1994.

[3] M. Fivel, M. Verdier, and G. Canova. *Mat. Sci. Engng*, **A234-236**:923–926, 1997.

[4] H. M. Zbib, M. Rhee, and J. P. Hirth. *Int. J. Mech. Sci.*, **40**:113, 1997.

[5] K.W. Schwarz. *Phys. Rev. Lett.*, **78**:4785–4788, 1997.

[6] H.Y. Wang and R. Lesar. *Phil. Mag. A*, **71**:149, 1995.

[7] J. Lepinoux and L. P. Kubin. *Scripta met.*, **21**:833, 1987.

[8] N. M. Ghoniem and R. J. Amodeo. *Phys. Rev. B*, **41**:6958, 1989.

[9] B. Devincre, P. Veyssière, and G. Saada. *Phil. Mag. A*, **79**:1609–1627, 1999.

[10] B. Devincre, P. Veyssière, L. P. Kubin, and G. Saada. *Phil. Mag. A*, **75**:1263, 1997.

[11] J.P. Hirth and J. Lothe. *Theory of dislocations*. MacGraw-Hill: New York, 1982.

[12] A.N. Gulluoglu, D.J. Srolovitz, R. Le Sar, and P.S. Lomdahl. *Scripta met.*, **23**:1347, 1989.

[13] V. Bulatov, F.F. Abraham, B. Kubin, L.P. Devincre, and S. Yip. *Nature*, **391**:669-672, 1998.

[14] A. El-Azab. *Modelling Simul. Mater. Sci. Eng.*, 1999, in press.

[15] C. Lemarchand, B. Devincre, L. P. Kubin, and J. L. Chaboche. In *MRS Symp. Proc.*, **538**:63–68, 1999.

DISCRETE DISLOCATION PLASTICITY APPROACH TO FAST MOVING DISLOCATIONS

A.Roos, E. Metselaar, J.Th.M. De Hosson, and E. van der Giessen[*], Department of Applied Physics, Materials science Centre and Netherlands Institute of Metals Research, University of Groningen, Nijenborgh 4, 9747 AG Groningen, The Netherlands, [*]Faculty of Mechanical Engineering, Delft University of Technology, Delft, The Netherlands. E-mail: hossonj@phys.rug.nl.

ABSTRACT

This paper concentrates on application of the so-called Discrete Dislocation Plasticity to high strain rate deformations. In particular the question is addressed if the DDP approach may capture the specific processes taking place at high strain rates. In particular the paper reports on tests of the validity of some approximations and provides some sample runs to show the applicability of the method. In assessing the results, one has to keep in mind two underpinning aspects: (1) the model is two-dimensional and (2) the results hold only in the regime where linear isotropic elasticity is valid. It was concluded that accelerations can not be neglected at very high strain rate deformations, both for the conventional and the relativistic case.

INTRODUCTION

In more recent years, numerical algorithms have become increasingly sophisticated and, due to the enormous increase in computing power, more accessible to solve complex questions in dislocation dynamics. However, even at the current rate of increase in number-crunching power, it is recognised that it will not be possible in the foreseeable future to simulate the deformation of a macroscopic work-piece *directly* from the motion of atoms. It is therefore necessary to split up the important processes according to the time and length scales at which they play a significant role. The processes taking place at the smaller scale then give rise to a certain effective behaviour at a larger scale. For instance, the atomic configuration around a dislocation core directly affects its scattering of lattice waves, thereby contributing to the drag force. On the other hand, when calculating the dislocation velocity due to the resolved shear stress, only this drag force is important, and not so much the precise atomic arrangement. The connection between different length scales is not always easily made. For instance, in many engineering calculations of plastic deformation, the material is considered to be a continuum. In those cases, the relation between macroscopic stress and macroscopic strain (the *constitutive relation*) is specified, always without taking into account the discrete nature of the carriers of plastic flow, the dislocations. This approach is successful for some applications, but it has the disadvantage that the material behaviour for each type of deformation has to be known in advance. Even for three-dimensional constitutive models, this is hardly ever the case.

On the other hand, some approaches exist nowadays that *do* take into account underlying microstructural processes. This method, called Discrete Dislocation Plasticity (DDP)[1], calculates the deformation of a two-dimensional computational cell by considering the long-range stresses and displacements of edge dislocations moving under influence of an externally applied deformation rate. Furthermore, the interaction between dislocations themselves and with obstacles can explicitly be put into the simulations. One area where this approach is particularly interesting lies in the regime of very fast deformation.

THEORETICAL BACKGROUND

In studies on dislocation dynamics at low to intermediate strain rates[1,2,6], the inertial forces (i.e. accelerations) are neglected. In those cases, the dislocation attains its steady-state velocity in a time that is small compared to the typical length of a time increment. In the case of fast-moving dislocations, this may no longer be the case. In this section, both points of view will be developed and compared. The *Peach-Koehler* force felt by dislocation i is a result of the combined stress-fields of all other dislocations j, the "image" stresses" and possibly of the long-range stresses of certain types of obstacles. For now, we neglect the latter. The accumulated effect of all dislocations on the Peach-Koehler force at dislocation i is obtained by summation over all i. Thus, the total Peach-Koehler force on dislocation i is

$$F^i_{\text{Peach-Köhler}} = b^i \left(\hat{\sigma}_{12} + \sum_{j \neq i} \tilde{\sigma}^j_{12} \right), \tag{1}$$

where all stress components are evaluated at $(x_1{}^i, x_2{}^i)$. It is of interest to note that, due to the velocity dependence of the resolved shear stress, the magnitude of the force $F^{i \to j}$ is not necessarily equal to $F^{j \to i}$ (only when $v^i = \pm v^j$)! In the method of discrete dislocation plasticity, the equation of motion for all dislocations is integrated in discrete time increments k of duration $\Delta t_k \equiv t_{k+1} - t_k$. Starting from a random initial distribution ($k = 0$) of dislocations in the computational cell the velocities during time increment k are calculated due to the combined forces on the dislocations at the start of the time increment. Note that a *random* configuration of dislocations is not necessarily an *equilibrium* configuration in terms of mechanical equilibrium. First, we treat the case in which the accelerations are neglected (i.e. assumed to take place instantaneously). The velocities $v_k{}^i$ are constant during the time increment. From $v_k{}^i$ and the positions $(x_1{}^i, x_2{}^i)$ at the start of the time increment k, the positions of all dislocations at time increment $k + 1$ can be calculated. Putting all the pieces together into equation (1) allows for writing an expression for the dislocation velocities $v_k{}^i$ at time increment k due to all applied forces. Initially, $v_0{}^i = 0$ for all i. For $k > 0$ and for all i, the velocities are given by the implicit relation

$$b^i \left(\hat{\sigma}_{12} + \sum_{j \neq i} \tilde{\sigma}^j_{12} \right)_k - B_{\text{TOT}}\left(v_k^i\right)v_k^i = 0, \tag{2}$$

leading to a cubic equation of the velocity with σ_{PK} written out according to equation (1). In the case of including inertial forces, the accelerations $\dot{v}_k{}^i$ are assumed to be constants during the time increment. Again using $v_0{}^i = 0$ for all i, the accelerations are given by the solution of

$$b^i \left(\hat{\sigma}_{12} + \sum_{j \neq i} \tilde{\sigma}^j_{12} \right)_k - B_{\text{TOT}}\left(v_k^i\right)v_k^i = m_e\left(v_k^i\right)\dot{v}_k^i , \tag{3}$$

(of which equation (2) is a special case). The velocities at the end of the time increment are found from

$v^i_{k+1} = v^i_k + \dot{v}^i_k \Delta t_k$ and the horizontal positions become $x^i_{1,k+1} = x^i_{1,k} + v^i_k \Delta t_k + \frac{1}{2}\dot{v}^i_k \Delta t^2_k$. When the velocity-dependence of $B(v^k_i)$ is known, an estimate can be made of the typical time needed for a dislocation to reach the velocity it would be given in the approach without acceleration.

Figures 1 compare three cases: the case of instantaneous velocity change (the solution of the cubic equation (2)), the case of constant acceleration, equation (3), and the case where the approximation of constant acceleration has been numerically integrated over extremely small time increments ($\sim 10^{-14}$ s). These particular cases are calculated for a single dislocation in an infinite medium under an effective stress σ_{applied} (so that the Peach-Koehler term in equation (3) becomes $b^i \sigma_{\text{applied}}$). Figure 1 displays the results for only Al. It can be seen that the order of magnitude of the time interval needed to reach, say, 90% of its final velocity, is of the order of a few ps for the extreme velocity change. In Cu, the moderate velocity change takes a few tens of ps. In fact, the acceleration approaches zero after 0.5 ns. From these plots it cannot be concluded *a priori* that in the simulations of Al and Cu the accelerations may be neglected, since at this point we do not know the order of magnitude of the average time increment in the actual simulations.

NUMERICAL ISSUES

In all simulations, the shear is applied at the top and bottom surfaces of the computational cell through the kinematic boundary conditions $u^{\text{applied}}_1(t) = +h\dot{\gamma}t$, $u^{\text{applied}}_2(t) = 0$ along $x_2 = h$, and $u^{\text{applied}}_1(t) = -h\dot{\gamma}t$, $u^{\text{applied}}_2(t) = 0$ along $x_2 = -h$. The total shear stress needed to sustain this shear strain rate is computed from integrating the total shear stress over the top- or bottom surface. As in any finite-element simulation, the mesh has to be small enough to capture the relevant variations of the calculated quantities. At high dislocation velocities, the fields have higher and more localized peaks than the conventional fields. The finite-element part of the simulations uses quadrilateral elements. This has the advantage of a simple linear interpolation function, but for quickly varying fields, the elements may have to become very small to follow all variations. On the other hand, since the complementary ($^\wedge$) fields correct the dislocation fields ($\sim$) *near the boundary*, only dislocations on slip planes close to the top- and bottom boundaries are expected to give a significant effect in the finite element correction. The mesh size is varied through 50×50 nm, 25×25 nm, 16.7×16.7 nm, 12.5×12.5 nm and 10×10 nm in an overall cell size of $2w \times 2h = 2 \times 2$ μm. The elastic properties of the matrix are taken to mimic Al . The calculations are carried out under two extremes of strain rates: 10^3 s^{-1} and 10^6 s^{-1}. Each case has been calculated for 5 realizations of the same microstructural parameters. To avoid the effects of sudden velocity changes, an artificially high value of the static drag parameter has been chosen. From the graphs calculated, it was concluded that apart from small fluctuations, there does not seem to be a large difference between them. To be more precise, the fluctuations in the stress-strain curves due to the variation in mesh size is of the same order of magnitude (or less) as the variation between different realizations of the initial parameters. As a rule of thumb, the number of elements in the vertical direction is taken at least equal to the number of slip planes, so that there is at least one element for each plane. This also ensures that, for horizontal slip planes, the dislocations do not come closer to the boundary than the active slip plane spacing. The size of the periodic cell should be taken large enough to sample a representative area of an actual grain, but smaller than the actual grain itself. The computational cell was varied in size through 0.5×0.5 μm, 1.0×1.0 μm, 2.0×2.0 μm and 4.0×4.0 μm. The size of the finite-element cells is kept constant at 25×25 nm. For each simulation, the total dislocation density ρ_{TOT} is 10^{14} m^{-2} and the number of slip planes is 20 μm^{-1}. Again, the mechanical properties of Al or Cu are used and in each case, several realizations are calculated at two different strain rates 10^3 and 10^6 s^{-1}. Within the variation of different realizations, a cell size of 2.0 × 2.0 μm provides sufficient accuracy for all subsequent simulations.

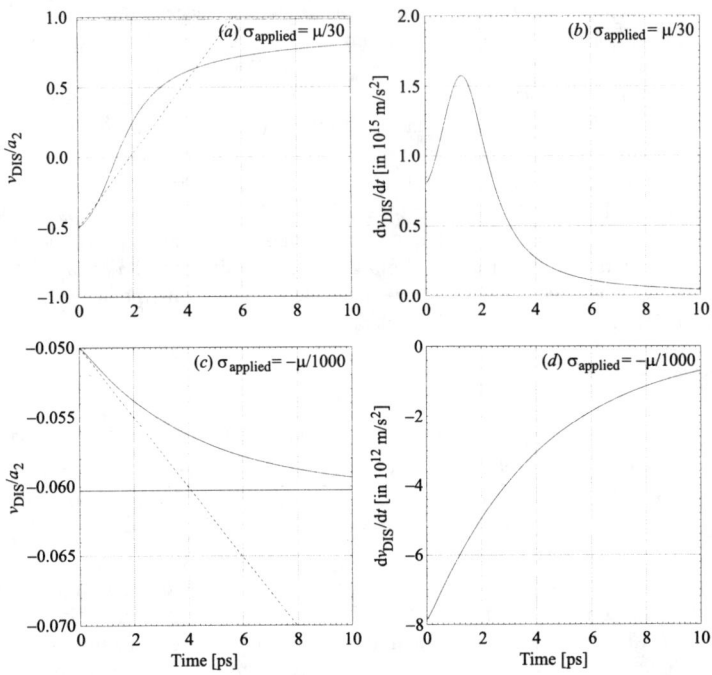

FIGURE 1 Dislocation velocity v_{DIS}/a_2 (left) and acceleration dv_{DIS}/dt (right) of an edge dislocation in Al for two different cases of applied shear stress $\sigma_{applied}$ and initial dislocation velocity. In the velocity plots, the dashed line denotes the approximation of constant acceleration (3), while the horizontal lines denote the velocity when velocity changes take place instantaneously (equation (2)). (a) and (b): $\sigma_{applied} = \mu/30$ and initial $v_{DIS} = -a_2/2$. (c) and (d): $\sigma_{applied} = -\mu/1000$ and initial $v_{DIS} = -a_2/20$.

DISCUSSION: RELATIVISTIC VS CONVENTIONAL DISLOCATION FIELDS

The question is whether the high dislocation velocities that significantly change the stress and displacement fields actually occur in the computational cell. Simulations are displayed using material parameters representing Cu, because Cu has a low static drag coefficient: at room temperature $B_{TOT}{}^{Cu}(v_{DIS} = 0, 298\ K) = 20\ \mu$Pa . In fact, to make it even easier for the dislocations to reach the high velocities, we will take the drag coefficient $B_{TOT}{}^{Cu}(v_{DIS} = 0, 100\ K) = 14\ \mu$Pa s. This temperature is at the lower temperature limit with respect to the Debye temperature ($\theta_D{}^{Cu} = 343\ K$) for which the temperature dependence of $B_{flutter}{}^0$ and $B_{wind}{}^0$ is linear. We also compared the cases with and without accelerations. In the case of accelerations, the time increment is kept very small at $t = 2 \times 10^{-14}$ s. It is important to stress that this order of magnitude is for numerical reasons only. One should not attach any physical significance to this order of magnitude, since the Debye frequency is of the order of 10^{13} s^{-1}. Any events involving *collective* motion of atoms (such as dislocation motion) cannot take place at a higher rate. The calculations without accelerations use an adaptive time step. It is determined by the next *event*, where an event is a collision between two dislocations, an annihilation, generation of a dislocation loop, and pinning or release of a dislocation at an obstacle.

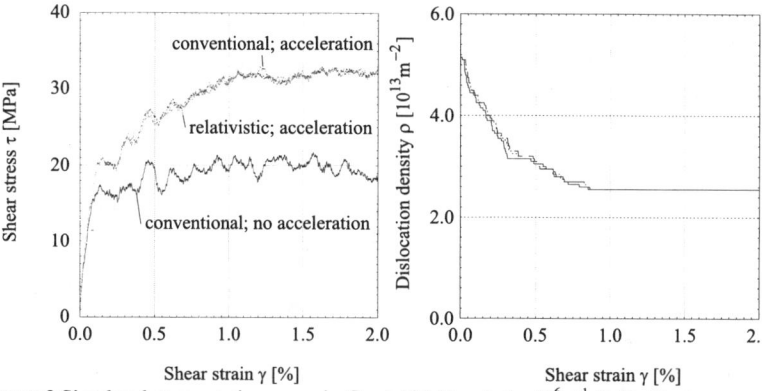

FIGURE 2 Simulated stress strain-curves in Cu at 100 K and $\dot{\gamma} = 10^6\,s^{-1}$ (left). The lower curve is the conventional case without acceleration; the light top curve denotes the relativistic case with acceleration and the remaining dark top curve the conventional case with acceleration. The picture to the right denotes the corresponding total dislocation densities: the solid line represents the conventional/no accelaration case, the dotted line the relativistic/accelaration case and the long-dashed line the conventional/accelaration case.

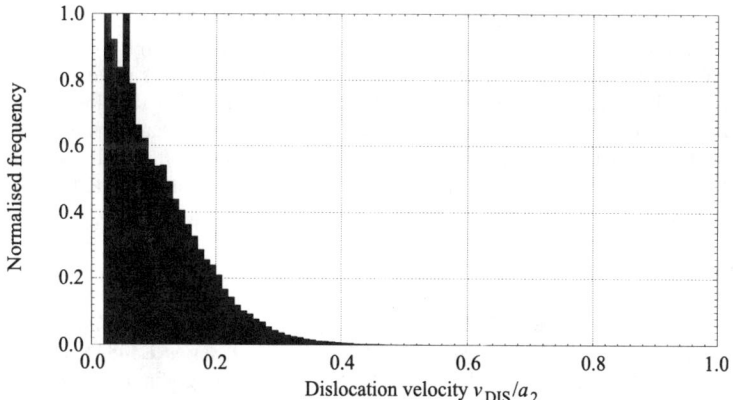

FIGURE 3: Histogram of the distributions of absolute dislocation velocities at 100 K. The black filled histogram corresponds to the conventional case without acceleration, the grey filled histogram to the relativistic case with accelerations and the transparent outlined histogram (which corresponds very closely to the grey one) corresponds to the conventional case with accelerations. The histograms are obtained by repeatedly sampling the velocity-distribution after a certain number of time increments. This plot then represents the velocities during the whole simulation. The plots are normalised to their maximum value (as a result, the relative frequencies for velocities higher than about half the shear wave velocity a_2 are not visible, although generally they are not equal to zero). For the static case, the supersonic dislocations have not been counted.

A typical example of the resulting stress-strain curves is presented in figure for 100 K. Each graph contains

(i)   a curve for the case without accelerations, conventional stress and displacement fields, and a linear stress-velocity relation (lowest curve),

(ii)   a curve for the case with constant accelerations, conventional stress and displacement fields, and a linear stress-velocity relation, and

(iii) a curve for the case with constant accelerations, relativistic stress and displacement fields, and a cubic stress-velocity relation.

The histogram of the velocity distributions of the three runs is plotted in figure 3, where each distribution has been normalized to its maximum value.

The first thing that stands out in figure 3 is the large difference between the case without, and the cases with acceleration. In the computations with the adaptive time increment, the time increments are typically of the order of a pico-second. This is smaller than the time needed for a dislocation to settle to its steady-state velocity. In order to really make sure that the difference is not due to the adaptive time stepping (instead of a fixed time increment of 2×10^{-14} s), extra simulations (not displayed) have been carried out. The only difference is that they now *also* have a fixed time increment of 2×10^{-14} s. It turns out that the stress-strain curve follows the curve with the adaptive time stepping almost exactly. The only difference then being the accelerations, it has to be concluded that they cannot be neglected at this high strain rate, both for the conventional and the relativistic case. The second point of interest is the small differences between the conventional and the relativistic case. The velocity distributions of figure 3 show that almost no dislocations move faster than a few tenths of the shear wave velocity. Actually, figure 3 shows no dislocations at all in that regime, but this is an effect of the scale. In reality, the spectrum is not zero until very close to the shear wave velocity (and beyond for the conventional case). Nevertheless, the number of dislocations that do reach the high velocities is utterly negligible with respect to the number of dislocations moving at velocities up to 20% of the shear wave velocity. A third feature of the curves of figure 3 is the fact that the flow stress attains a constant value and does not show any hardening effects. The mechanisms that give rise to hardening in the computational cell are the increase in number of forest dislocations, Taylor hardening due to the interaction of the dislocations on different slip planes, and hardening due to the formation of dislocation pile-ups at obstacles.

REFERENCES

1. H.O.KIRCHNER , L.P. KUBIN, V. PONTIKIS V. (eds.), *Computer simulation in materials science: nano/meso/macroscopic space & time scales*, ASI Series **E308**, Kluwer Dordrecht (1996) and E. VAN DER GIESSEN , A. NEEDLEMAN, Modelling Simul. Mater. Sci. Eng. **3** (1995), 689.

2. H.J.M. CLEVERINGA, E. VAN DER GIESSEN , A. NEEDLEMAN A., Acta Mat. **45** (1997), 3163.

3. J.TH.M. DE HOSSON , O. KANERT , A.W. SLEESWIJK , in *Dislocations in solids*, (ed. F.R.N. NABARRO), North Holland (1983), Vol. **6**, 441, J.Th.M. DE HOSSON , A.J. HUIS IN 'T VELD, H. TAMLER , O. KANERT , Acta Met. **32** (1984), 1205.

4. U.F. KOCKS , A.S. ARGON , M.F. ASHBY, Progress in Materials Science **19**, Pergamon (1975).

5. B. DEVINCRE , in *Computer simulation in materials science* (eds. H.O. KIRCHNER, L.P. KUBIN, V. PONTIKIS), ASI Series **E308**, Kluwer, Dordrecht (1996), 309.

6. J.WEERTMAN , J.R. WEERTMAN, in *Dislocations in solids* (ed. F.R.N. NABARRO), North Holland Publ. Corp., Vol. **3** (1980), 1.

SIMULATION OF DISLOCATION DYNAMICS IN Ni$_3$Al: A STUDY OF VELOCITY AUTOCORRELATIONS

C. K. ERDONMEZ, D. C. CHRZAN
Department of Materials Science and Mineral Engineering,
University of California, Berkeley, CA 94720

ABSTRACT

The yield strength anomaly in some L1$_2$ compounds has been linked to the thermally assisted cross slip of screw superdislocations. This work continues earlier efforts to understand the yield strength anomaly in L1$_2$ alloys using computer simulations of dislocation motion. Dislocations are modelled within isotropic elasticity theory, and simple rules are used to model the cross-slip process in the two dimensional geometry of the simulation. The velocity of a single dislocation in Ni$_3$Al is studied as a function of the applied stress. The observed velocities vary nonlinearly with the applied stress. Further, dislocations are observed to become immobile for small applied loads. At high stresses, the dislocations are observed to advance relatively unhindered by the thermally activated cross slip process. Fluctuations in the velocity of the dislocations are studied, and their autocorrelation function shows an increased correlation time near a threshold stress. This threshold stress is identified with the critical stress proposed in earlier works.

INTRODUCTION

Nickel based superalloys find use in gas turbines and jet engines where their high yield stresses at high temperatures is an asset. Ni$_3$Al precipitates of the L1$_2$ structure are used as strengthening particles in these superalloys. Single crystals of Ni$_3$Al (and other related intermetallic compounds of the L1$_2$ family) display a positive dependence of the yield stress on temperature over a significant temperature range.

The cross slip pinning model developed originally by Takeuchi and Kuramoto [1] and modified by Paidar, Pope, and Vitek (PPV) [2] serves as the starting point for most theoretical discussions of the origins of the yield stress anomaly in L1$_2$ compounds. The most widely accepted feature of this model is that a thermally activated core transformation is responsible for the yield strength anomaly. In addition, the model explains the dependence of the yield stress on temperature, sample orientation and sign of the applied stress. However, the strain-rate sensitivity and work-hardening rate cannot be justified within the PPV picture and this difficulty has prompted various theories on the details of the yielding process [2-6]. A consensus has yet to develop.

Post-mortem TEM studies of deformed Ni$_3$Al samples reveal isolated slip bands [7] and an absence of complex patterns of strongly interacting dislocations. In addition, crystals oriented for single system slip are also observed to harden. Given these observations it is possible that observed mechanical properties of single crystal Ni$_3$Al can be understood by considering the mobility of a single dislocation. Counter to this viewpoint, both Hirsch [8] and Ortiz [9] have argued that the hardening observed in these compounds stems from forest interactions. Classical dislocation theory, coupled with computer simulations can

143

address the detailed predictions of these competing ideas, and thus be used as a means of discerning between them. A logical starting point for such studies is a careful analysis of the dynamics of a single dislocation in the absence of forest obstacles.

Early simulations of dislocation dynamics [10] in $L1_2$ compounds relied on a line tension approximation and the formation of pinning points [2] to model dislocation dynamics. These simulations revealed a number of important features of the dislocation dynamics: 1) The formation of pinning points coupled through the line tension of the dislocation leads to the formation of structures mimicking Kear-Wilsdorf locks. 2) For typical stresses, dislocations advance through the lateral motion of superkinks. 3) At low stresses, the dislocations are observed to spontaneously immobilize, whereas at high stresses, the dislocations are observed to move for extended periods of time. 4) While a dislocation is mobile, its velocity may display large fluctuations.

Discrete simulations revealed the nature of the velocity fluctuations [11,12]. In the absence of thermal depinning, the dislocations move for short times before exhausting. The distribution of immobilization times and areas swept out is found to be consistent with a dynamic scaling hypothesis, and the existence of a critical stress. Dislocations moving under an applied shear stress exceeding the critical stress are observed to propagate for very long periods of time. In contrast dislocations with applied stresses below the critical stress are observed to immobilize rapidly. Introduction of a thermal depinning rate eliminated the proposed critical point, but remnants of the critical point remain [12]. In particular, it is found that the time scale characterizing the velocity-fluctuation autocorrelation function displays a sharp increase near the critical stress defined in the limit of zero thermal depinning.

It is natural to wonder if the behavior observed in discrete models will be reflected in more complete simulations of dislocation dynamics. Recently, Devincre *et al.* [11] studied simulations which include long-range interactions between segments and locking events occurring over screw segments of length $20nb$ at a time (n=1,2,3..). The microstructures observed were in qualitative accord with the results from earlier simulations and dislocations observed *post mortem* in transmission electron microscopy studies. Further, they report that the dislocations move through strain bursts very similar to those identified earlier in the simplified discrete models [11]. However, Devincre *et al.* did not attempt to characterize the statistical properties of the observed strain bursts.

In this paper, initial findings from a new computer simulation of dislocation dynamics in the $L1_2$ compounds are reported. The simulations are analyzed using techniques rooted firmly in statistical physics. Specifically, the velocity-fluctuation autocorrelation function is calculated as a function of the applied stress. The time scale characterizing the fluctuations is observed to increase as the applied stress approaches the critical stress from above. This behavior is consistent with behavior identified in the discrete models, though the exponents characterizing the transition here may differ.

MODEL

The explanation of the local glissile to sessile transformation of dislocations in Ni_3Al involves a complex dissociated core structure, multiple slip planes and anisotropic elasticity [2,14]. All of these effects can, in principle, be incorporated into computer simulations. A much simpler solution is to treat the superdislocations as a single dislocation gliding on a single slip plane and incorporate rules governing Kear-Wilsdorf lock formation and subsequent evolution [10,13]. This is the approach followed here.

The simulation program is based on a particular implementation of an algorithm for modelling 2D slip of dislocation loops and sources within isotropic elasticity theory [15]. The salient features of this approach are: 1) Dislocation lines are represented by a set of points. 2) Points are added or removed to assure that density of points remains high in regions where the velocity vector changes sharply along dislocation line. 3) For long-ranged elastic interactions between different segments, segments are assumed to be lines connecting the points. 4) Circular arcs fit between neighboring points are used to calculate the local contribution to the self-stress and to interpolate the positions of added points.

In the simulations described here, all unlocked points in a dislocation configuration are treated as possible sites for a lock nucleation event. The decomposed screw length over the circular arc fitted through a point under question is

$$L_{screw} = r \int_{\alpha}^{\beta} P_{geo}(\theta) \, d\theta \qquad (1)$$

where r is the radius of the arc and the angles α and β define the endpoints of the arc. The form of P_{geo} provided by Mills and Chrzan for straight segments was used [10], but the angle made by mixed kinks to the screw direction was set to be 60° instead of 20°.

If L_{screw} is larger than the critical nucleation length, l_c for a 'double jog' bounding an APB on a {100} plane, the locking probability is calculated as

$$P_{lock} = P_T L_{screw} \Delta t \qquad (2)$$

where P_T is a function of temperature expected to have the Arrhenius form and Δt is the timestep of the integration. If P_{lock} exceeds a random number between 0 and 1, a locking event occurs and the pinning point is replaced by two points separated by l_c (estimated to be 50 Å for Ni$_3$Al) along the screw direction. These points represent the jogs at the ends of a Kear-Wilsdorf lock.

The simulated jogs were rendered mobile to an extent determined by an input linear drag coefficient, B_{jog}, and an additional jog friction stress, σ_{jog}, the friction stress on jogs. This allows Kear-Wilsdorf locks to spread, shrink, or disappear, the last possibility leading to reconversion of jogs separated by a distance less than 49 Å to a single free point. The non-frictional force (per length) on a jog may be written

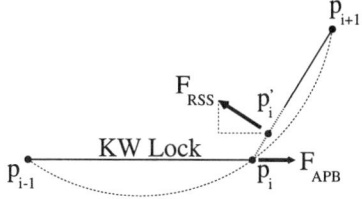

Figure 1: The geometry used in calculation of the forces on the jogs.

$$\vec{F}_{jog} = \vec{F}_{self} + \vec{F}_{APB} + \vec{F}_{RSS} \qquad (3)$$

where $\vec{F}_{self}$ is due to the stress field of the dislocation(s) acting on the jog, $\vec{F}_{APB}$ is due to the lower energy of locked structures for dislocations near the screw orientation and $\vec{F}_{RSS}$ is the Peach-Koehler force on the jog due to the applied stress. In principle, $\vec{F}_{APB}$ is dependent on superkink geometry [16], but it is approximated to be a constant 0.07 N/m.

The line direction changes abruptly near a jog. Therefore, representation of jogs by in-plane points presents difficulties for the calculation of the force terms. A further complication is that a circular arc is not a reasonable fit to the dislocation line near a jog

(Fig. 1). As a first approximation, the dislocation line may be assumed to consist of two straight segments pointing to the two neighboring points. This introduces a further difficulty: Continuum based dislocation theory predicts a destabilizing force at a sharp bend even when a cutoff procedure is employed. The problem is circumvented by calculating F_{self} not at the corner, but at a point (p_i' in Fig. 1) which is one cutoff distance away from the point. For consistency, F_{RSS} is calculated as the Peach-Koehler force due to the applied stress at point p_i'.

Following the above modifications to the basic algorithm of reference [15], the motion of Frank-Read sources and dislocation loops were simulated. Some encouraging qualitative features were noted, but these simulations were too computationally expensive due to a number of points increasing steadily with simulation time. Therefore, the computer code was modified to study single dislocations bounded by two surfaces, chosen to be 6 μm apart for the results reported. Image forces were neglected, but the dislocation line was constrained to intersect the surface at a right angle. For a particular combination of parameters, 5 runs lasting 3 μs were performed and the areal velocity from each run recorded. For the initial results reported here, the parameters governing the simulations are the dislocation drag coefficient, $B = 5 \times 10^{-2}$ Pa s, lock nucleation rate/unit length = 10^{14} s^{-1}, the jog drag is $B_{jog} = 7 \times 10^{-2}$ Pa s and $\sigma_{jog} = 20$ MPa.

RESULTS

As an initial test of the technique, the expansion of a dislocation loop under an applied stress was simulated. Similar to prior studies, the loops display an aspect ratio differing substantially from one, due to the retardation of screw segment motion through the formation of locks.

Two significant differences from simulations within a line tension approximation [10] are noted. In the current simulations, the loops no longer cease to expand under applied loads below the proposed threshold stress. Instead, the size of superkinks becomes such that the jog mobility limits the velocity of the superkinks. The jogs continue to "drift" under the applied stress, and the area of the loop increases continuously, albeit much more slowly than at higher stresses where superkink motion may be sustained for longer periods of time. Second, the self-stress of the loop causes the stress for expansion of the loop to exceed that required for motion of an isolated nearly screw-oriented dislocation. In essence, as the loop expands, the net stress felt on each of the predominantly screw segments effectively increases. If the loop is initially stressed above threshold, the stress will increase even further above threshold as the loop expands.

The structure of the simulated dislocations depends upon the choice of σ_{jog}. For larger values of σ_{jog}, the observed structures are similar to those obtained from pinning-point based simulations [10]. Large values of σ_{jog} inhibit the spreading of incipient locks. For smaller values of σ_{jog}, the locks do spread (lengths of 0.3 μm are common for the parameters employed here), though the spatial extent of the locks is considerably smaller than those observed in post-mortem transmission electron microscopy experiments. The expansion of the jogs is driven by the change in antiphase boundary energy, F_{APB}. Failure to account for this force leads to pinning "points" only.

To explore the dynamics in a quantitative and statistical manner, as well as to facilitate comparison with existing discrete models [12], the velocity fluctuations are studied. The velocity-fluctuation autocorrelation function is defined as follows. Consider a nearly screw

146

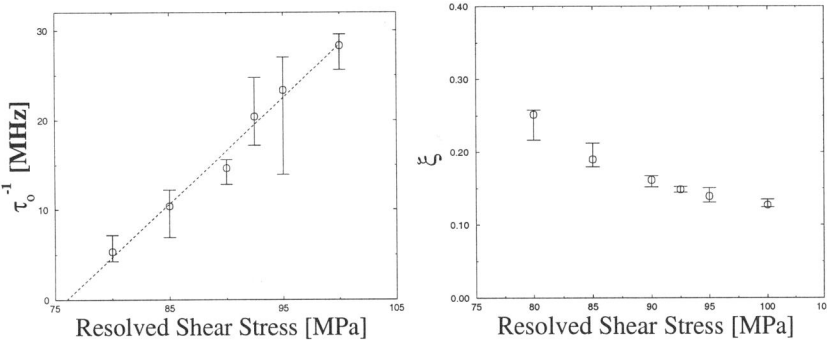

Figure 2: A plot of τ_o^{-1} and ξ as a function of applied stress. Within the accuracy of the current results, the characteristic fluctuation time depends on the applied stress as $(\sigma - \sigma_c)^{-1}$, with $\sigma_c \approx 76$ MPa, the critical stress. The value of ξ is also increasing as the apparent critical stress is approached from above.

oriented dislocation extending between two free surfaces of the solid. Define the velocity of the dislocation to be $v(t) = \frac{dA}{dt}$ with A the area swept out by the dislocation and t the time. Under constant applied stress, this velocity will have a well defined average, $\bar{v}$. The quantity $\Delta(t) = v(t) - \bar{v}$ represents the velocity fluctuations, and the velocity-fluctuation autocorrelation function, $g(\tau)$, is defined to be

$$g(\tau) = \langle \Delta(t + \tau)\Delta(t) \rangle \,, \tag{4}$$

where the $\langle \rangle$ indicate an average over time. Physically, $g(\tau)$ measures the time scale over which the dislocation "remembers" its velocity. Empirically, it is observed that $g(\tau)$ decays exponentially, with a characteristic time τ_o; τ_o is a measure of the persistence time of a velocity fluctuation.

In simple discrete models of dislocation dynamics, τ_o is observed to increase as a threshold stress is approached from above. The left hand panel of Fig. 2 displays the inverse correlation times extracted from the simulations as a function of the applied stress. The error bars represent the maximum and minimum values obtained over the five runs, and serve as an estimate of the statistical uncertainty. Two features of this plot are readily apparent. First, within the accuracy of the simulations, the inverse correlation times depend linearly on the applied stress. The implication is that

$$\tau_o \sim (\sigma - \sigma_c)^{-1} \,. \tag{5}$$

Line tension based simulations yield a value of 2.14 for the exponent in this equation [10]. The implication is that the long-ranged stresses may modify this exponent, but more data is needed.

In fact, as the critical stress is approached, obtaining good statistics becomes more difficult. One must average for times much longer than the autocorrelation time. Since this time is diverging, it is simply not possible to obtain good statistics at the critical point. This behavior is entirely analogous to critical slowing down in second order phase transitions, and has significant implications for modeling constant strain rate experiments in these materials.

Finite size also plays an important role in the dislocation dynamics [17]. In the current work, the finite length of the dislocation makes obtaining data near the threshold stress difficult: Once $\langle \Delta(t)^2 \rangle^{1/2} \sim \bar{v}$, the dislocations begin to display exhaustion of their motion [10]. The right hand panel of Fig. 2 contains a plot of $\xi = \langle \Delta(t)^2 \rangle^{1/2}/\bar{v}$. As the stress is decreased, the ratio increases toward the value one, and dislocation motion (through rapid superkink propagation) is only short lived.

CONCLUSIONS

In conclusion, a study of the velocity-fluctuation autocorrelation function of dislocations in the L1$_2$ compounds has been presented. The autocorrelation function is observed to decay exponentially, with a stress dependent correlation time, τ_o, that diverges at the critical stress, estimate to be 76 MPa for the conditions studied here. Further, the magnitude of the fluctuations is also seen to increase under the same conditions.

The authors thank Lawrence H. Friedman for useful discussions. The support of the National Science Foundation, Division of Materials Research, under grant DMR-9703427 is gratefully acknowledged.

REFERENCES

[1] S. Takeuchi and E. Kuramoto, *Acta metall.* **21**, 415 (1971).

[2] V. Paidar, D. P. Pope and V. Vitek, *Acta metall.* **32**, 435 (1984).

[3] M. J. Mills, N. Baluc and H. P. Karnthaler, *Mater. Res. Soc. Proc.* **133**, 203 (1989).

[4] P. B. Hirsch, *Progress Mater. Sci.* **36**, 63 (1992).

[5] V. Vitek and Y. Sodani, *Scripta metall. mater.* **25**, 939 (1991).

[6] F. Louchet, *J. Phys. Paris, III* **5**, 1803 (1995).

[7] E. M. Nadgorny and Y. L. Iunin, *Mat. Res. Soc. Proc.* **364**, 707 (1995).

[8] S. S. Ezz and P. B. Hirsch, *Phil. Mag. A* **72**, 383 (1995).

[9] A. M. Cuitino and M. Ortiz, *Mat. Sci. and Eng. A* **A170**, 111 (1993).

[10] M. J. Mills and D. C. Chrzan, *Acta metall.* **40**, 3051 (1992).

[11] D. C. Chrzan and M. J. Mills, *Phys. Rev. B* **50**, 30 (1994).

[12] D. C. Chrzan and M. S. Daw, *Phys. Rev. B* **55**, 798 (1997).

[13] B. Devincre, P. Veyssiere, L. P. Kubin and G. Saada, *Phil. Mag. A* **75**, 1263 (1997).

[14] M. H. Yoo, *Scripta metall.* **20**, 915 (1986).

[15] A. K. Faradjian, L. H. Friedman and D. C. Chrzan, *Modelling Simul. Mater. Sci. Eng.* **7**, 479 (1999).

[16] P. B. Hirsch, *Czechoslovak Journal of Physics* **45**, 921 (1995).

[17] D. C. Chrzan, M. D. Uchic and W. D. Nix, *Phil. Mag. A* **79**, 2397 (1999).

MODELING COLLECTIVE DISLOCATION DYNAMICS IN ICE SINGLE CRYSTALS

M.-CARMEN MIGUEL*, A. VESPIGNANI*, and S. ZAPPERI**
*The Abdus Salam International Centre for Theoretical Physics
P.O. Box 586, 34100 Trieste, Italy
**PMMH-ESPCI,10, rue Vauquelin, 75231 Paris Cedex 05, France

ABSTRACT

We propose a model to study the plasticity of ice single crystals by numerical simulations. The model includes the long-range character of the interaction among dislocations, as well as the possibility of mutual annihilation of these line defects characterized by its Burgers vector. A multiplication mechanism representing the activation of Frank-Read sources due to dislocation pinning is also introduced in the model.

With our approach we are able to probe the dislocation patterns, which result from the dislocation dynamics. Furthermore, our results exhibit features characteristic of driven dynamic critical phenomena such as scaling behavior, and avalanche dynamics. Some of these results account for the experimental findings reported for ice single crystals under creep deformation, like the power-law distributions of the acoustic emission intensity observed sistematically in experiments.

INTRODUCTION

The viscoplastic deformation of crystalline materials, such as ice single crystals, involves the motion of a large number of dislocations. Although the dynamics of an individual dislocation is a fairly well characterized phenomenon [1, 2], the collective behavior of a large number of these defects appears to be an amazingly rich but poorly understood problem. The interaction between a pair of dislocations can be attractive or repulsive, depending on the orientation of their respective Burgers vectors; it grows logarithmically with the interline distance and allows mutual annihilation of defects. Dislocations may be incorporated into a crystal in the growth process, affecting the topology of the whole lattice. Moreover, under deformation conditions, dislocations can penetrate into the material from the sample surfaces, or be generated by various mechanism, as for example, in what is usually called a Frank-Read source, which is activated after the pinning of a dislocation loop.

When a material is deformed under constant load (creep experiment), and dislocation motion is the dominant mechanism for viscoplastic deformation (other possible source is for instance crack nucleation and propagation), a constant strain-rate regime usually follows after the initial transient stage. Orowan's relation $\dot{\gamma} = \rho_m b v$ is known to prevail under such conditions, where γ is the strain of the sample, ρ_m is the density of mobile dislocations,

b is the Burgers' vector, and v is the mean velocity of the dislocations. Obviously, this is a mean-field relation which neglects temporal and spatial fluctuations of both the density and the velocity fields. As a result of their interactions, however, dislocations tend to move cooperatively giving rise to a rather complex and heterogeneous slip process. Dislocations move in groups to form slip bands. Moving dislocations can pile-up against stable dislocation configurations such as walls or boundaries, which may eventually break apart. In this process, fluctuations in the dislocation density and velocity may be comparable to or greater than the mean values, and consequently, of great importance.

The complex character of the collective dislocation dynamics reveals itself in experiments of acoustic emission (AE) [3]. Sudden local changes of inelastic strain generate AE waves. Weiss and Grasso [3] soon realised that ice single crystals are particularly well suited for the study of dislocation dynamics. The perfect transparency of this material, easily (by eye) allows to rule out the possibility of cracks being present in the material which will, otherwise, interfere with the dislocations motion. Given the amplitude threshold and the frequency range accesible to the experimental apparatus, the AE signals detected seem to correspond to the synchronous motion of several dislocations, likely to occur for example during the breakaway of a pile of these defects, or the activation of a multiplication source. The AE experiments, however, have only access to information resulting from the interplay of various magnitudes. Thus, the physical interpretation of the generated AE waves remains a major difficulty, and consequently, constitutes the main motivation of our work.

The proportionality between global AE activity and global strain-rate (Orowan's relation) has been tested experimentally. More locally, the AE intensity is thought to depend on the number of dislocations involved in a plastic instability, their length, and velocity. Various measurements of the acoustic activity recorded during a stress-constant step show that the AE signal takes place in the form of bursts spanning a wide range of amplitude values. In particular, the AE amplitude is distributed according to a power law. This behavior is a consequence of the collective motion of dislocations which spontaneously gives rise to an avalanche-like dynamics, typical of slowly driven dissipative systems. The power law distribution provides evidence of scale-free cooperative behavior, whose origin could be ascribed to nonequilibrium continuous phase transitions [4] or self-organized criticality [5].

DESCRIPTION OF THE MODEL

To characterize the plastic deformation of a material from the perspective of nonequilibrium statistical mechanics, we propose a simplified dynamic model to simulate the system in rather general conditions.

Ice single crystals deform essentially by slip on the basal plane (0001) (that we call the xy plane), i.e. the motion of dislocation lines or loops takes place by gliding on the xy planes. The possible Burgers vectors $\mathbf{b}$ are the three lattice vectors of an hexagonal lattice. For the sake of simplicity, we study a two-dimensional model representing a cross section of the crystal which is perpendicular to the basal planes and parallel to one of the lattice vectors, that is, for example, the xz plane. In this way, the dislocations constrained to move in this plane have all Burgers vectors parallel to the chosen lattice vector $\mathbf{b} = (b, 0, 0)$ and move along fixed lines parallel to the x axis. We also consider that all N dislocations are of edge type, and that, on average over many realizations, the number of dislocations with positive and negative Burgers vectors is the same.

Several simplified models containing similar basic ingredients have been proposed in the literature in the last few years [6, 7, 8, 9, 10]. A basic feature common to most models, is that dislocations interact with each other through the long-range elastic stress field they produce in the host material. An edge dislocation with Burgers vector b located at the origin gives rise to a shear stress σ^s at a point (x, z) of the form

$$\sigma^s = bD\frac{x(x^2 - z^2)}{(x^2 + z^2)^2}, \tag{1}$$

where $D = \mu/2\pi(1 - \sigma)$ is a coefficient involving the shear modulus μ and the Poisson's ration σ for the material. In our model, we further assume that the dislocation velocities are linearly proportional to the local stress. Experimental evidence supports such a relationship for low stress conditions, which is indeed the case in our model. Accordingly, the velocity of the nth dislocation, if an external shear stress σ^e is also applied, is given by

$$v_n = b_n\big(\sum_{m \neq n} \sigma^s_{nm} - \sigma^e\big). \tag{2}$$

As the number of dislocations in any real crystals exceeds by far the number of defects we can handle in a computer, one usually introduces periodic boundary conditions (PBC) to effectively extend the size of our system. Due to the long range character of the force (1), we have exactly evaluated the Ewald sums of this expression to account for the interaction of a dislocation with all the infinite periodic replicas of all the other dislocations in a finite box of dimensions $L \times L$. Contrary to what is stated in some references [8], we do not find any spureous results coming from the implementation of PBC in this fashion. Instead, we obtain artificial results when using the "nearest image" approximation and the truncation that this approximation implies.

When the distance between two dislocations is of the order of a few Burgers vectors, the high stress and strain conditions close to the dislocation core invalidate the results obtained from a linear elasticity theory (i.e. Eq .(1)). In these instances, phenomenological nonlinear reactions describe more accurately the real behavior of dislocations in a crystal. In particular in our model, we account for the *annihilation* of dislocations with opposite Burgers vectors when the distance between them is shorter than $2b$. Thus the core of one dislocation in our model has a radius of size b.

Another important feature of any computer model is the implementation of a mechanism for the *multiplication* of dislocations in the sample. It is widely believed that the Frank-Read mechanism [1] is the most relevant for a gliding process of dislocations under creep deformation. Indeed Frank-Read sources (FRS) have been observed in ice. In a FRS multiplication occurs by pinning of a dislocation segment on the basal planes due, for example, to a defect in the crystal, or to dislocation dipoles, piles, and walls. Under an applied stress, the pinned segment bows out by glide and, if the local stress concentration is less than a critical value, a metastable configuration is attained where the line tension balances the stress. Beyond this threshold value, the dislocation segment wraps around itself, creating a new dislocation loop and restoring the original configuration. Thus a sequence of loops forms continuously from the source until the local shear stress drops below the activation value. In our model, we simulate this mechanism phenomenologically: a dislocation pair is generated (i) when the fraction of immobile dislocations is high (pinning is then more likely) and (ii) the local stress is large compared to a threshold value. Rather than fixing particular values of the parameters, we use a probabilistic procedure, keeping in mind that the details of the rule should not change the collective properties of the system.

151

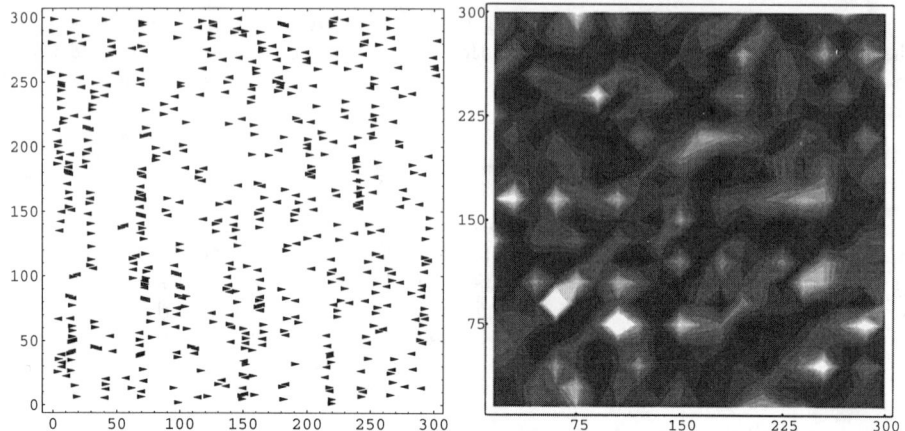

Figure 1: a) A stable arrangement of dislocations in a box of size $L = 300b$. Various structures like dislocation dipoles, piles, and dislocation walls are present. b) The corresponding elastic stress map (absolute value) in a graylevel scale. Dark color indicates a low-stress region and lighter colors areas of higher stress.

Annihilation and multiplication processes imply that the number of dislocations N in our system is not fixed in the course of time. Starting from a random configuration of dislocations, we let them relax until they find a stable arrangement (which could be an equilibrium configuration, or a long-lived metastable state). We solve numerically the N equations of motion using an *adaptive step size fifth-order Runge-Kutta method*. So far, we have considered three different box sizes $L = 100b$, $L = 200b$, and $L = 300b$, with an initial number of dislocations $N_0 = 400$, $N_0 = 800$, and $N_0 = 1500$, respectively. After the system has reached a stable arrangement, the volume fraction of dislocations $\phi = N\pi b^2/(Lb)^2$ ranges between $1 - 10\%$. Once in these conditions, we apply a small external shear stress and keep track of the various quantities describing the dynamics of the dislocations.

RESULTS AND DISCUSSION

Persistent slip bands or cell structures appearing on various length scales, are just a few examples of dislocation patterns. With our approach we are able to recover some of these structures. This is a subject which has raised a considerable amount of activity recently (see for example Refs. [11, 12] and references therein). Fig. 1a) represents a common configuration of dislocations in stable conditions. There, we can observe the formation of several dipoles, piles, as well as dislocation walls delimiting various slip bands. The corresponding elastic stress map is depicted in Fig. 1b). We plot the absolute value of the shear stress in a graylevel scale. The dark areas represent low-stress regions, and are consequently predominant in a stable configuration. The light portions mark a few regions of higher stress concentration given this particular arrangement.

Second, we have been able to keep track of several quantities which play a key role in promoting scale invariant behavior, like the root-mean-square velocity, the local stress, the average number of dislocations, etc. In Figure 2a), we represent, for example, the root-mean-square velocity V_m of a single run of the creep process. A curious feature is that after

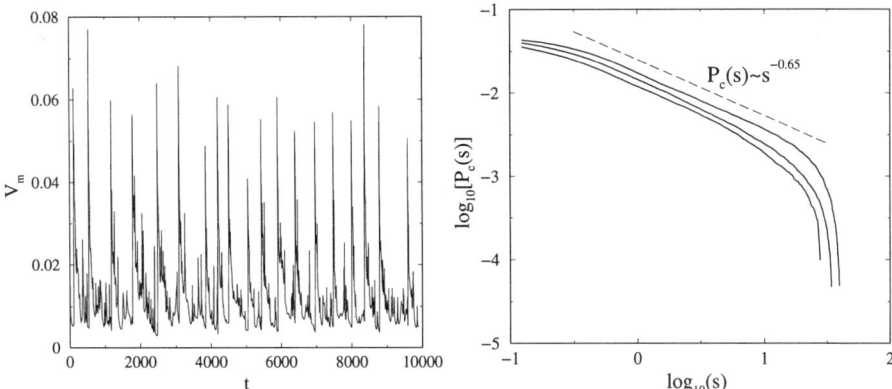

Figure 2: a) Mean velocity as a function of time in our model of collective dislocation dynamics. b) The cumulative *avalanche* size distribution for three system sizes $L = 100, 200, 300$ represented in a double logarithmic scale.

a burst, the relaxation of the velocity V_m to a still configuration is power-law like, i.e. slow and without any characteristic time.

We have defined the size s of an acoustic *avalanche* as the sum of the root-mean-square velocity of all the moving dislocations, that is $s = \sum |v_i|$. The cumulative distribution $P_c(s)$ of the avalanches obtained after averaging over several realizations is depicted in Figure 2b) for the three system sizes studied $L = 100, 200, 300$. We recover a very clear power law distribution $(P_c(s) \sim s^{-\tau})$ extending over close to two decades. The exponent $\tau \simeq 0.65$ is in reasonable agreement with experimental data [3]. The distribution cut-off for large values of s is due to the finite size of the sample. (As one would expect, the cut-off is scaling accordingly to the sytem size.) This clearly points out the presence of a very large (or infinite) characteristic size for the acoustic events. It is worth remarking that large stresses introduce a characteristic scale in the process. A more detailed study as a function of the applied stress is in progress [10].

The numerical investigation of the present model provides striking evidences for the collective critical behavior of dislocation motion under external stress. Relevant magnitudes characterizing the dynamics of dislocations show power law behavior signalling the absence of any characteristic length or time in the process. The response to an infinitesimal perturbation (slow injection of new dislocations) exhibits singular behavior in the guise of avalanches distributed over many length scales. Avalanche dynamics is the rule rather the exception in slowly driven disordered systems. Examples can be found in fracturing of wood and concrete [13], Barkhausen effect [14], and flux lines in high-T_c superconductors [15]. Under the external drive (the stress in the present case), the system jumps between metastable or pinned configurations in which the dynamics is virtually frozen. In the limit of a very slow driving (that is equivalent to a fine tuning close to the depinning critical point), the disordered energy landscape is explored quasistatically and the response function exhibits critical properties [16]. Tipically, a basic ingredient for this behavior is the presence of quenched disorder acting as the source of pinning in the system. Noticeably, the system under study does not contain any external source of disorder. Pinned states are due to the various dislocation patterns that spontaneously develop in the system. Structures such as dipoles, piles, and dislocation walls, play the role of self-generated pinning centers that create the pinning force landscape. The new scenario poses many new and interesting questions for a definitive identification and understanding of the critical nature of dislocation dynamics.

ACKNOWLEDGMENTS

We gratefully acknowledge R. Pastor-Satorras advice on the algorithm implementation. We also thank J.R. Grasso and J.Weiss for fruitful discussions regarding their experiment, and for providing us with the experimental data prior to publication.

References

[1] J.P. Hirth and J. Lothe, *Theory of Dislocations* (Krieger Publishing Company, 1992).

[2] F.R.N. Nabarro, *Theory of Crystal Dislocations* (Dover, New York, 1992).

[3] J. Weiss and J.R. Grasso, J. Phys. Chem. B **101**, 6113 (1997).

[4] *Phase Transitions and Critical Phenomena*, edited by C. Domb and M.S. Green (Academic Press, London 1972-1976), Vols. 1-17.

[5] P. Bak, C. Tang, and K. Wiesenfeld, Phys. Rev. Lett. **59**, 381 (1987); A. Vespignani and S.Zapperi, Phys. Rev. E, **57**, 6345 (1998).

[6] J. Lepinoux and L.P. Kubin, Scripta Metall. **21**, 833 (1987).

[7] R.J. Amodeo and N.M. Ghoniem, Phys. Rev. B **41**, 6958 and 6968 (1990).

[8] I. Groma and G.S. Pawley, Phil. Mag. A, **67**, 1459 (1993).

[9] R. Fournet and J.M. Salazar, Phys. Rev. B **53**, 6283 (1996).

[10] M.C. Miguel, A. Vespignani, and S. Zapperi, in preparation.

[11] P. Hahner, K. Bay, and M. Zaiser, Phys. Rev. Lett. **81**, 2470 (1998).

[12] B. Bakó and I. Groma, Phys. Rev. B **60**, 122 (1999).

[13] A. Garcimartin, A. Guarino, L. Bellon, and S. Ciliberto, Phys. Rev. Lett. **79**, 3202 (1997); A. Petri, G. Paparo, A. Vespignani, A. Alippi, and M. Costantini Phys. Rev. Lett. **73**, 3423 (1994).

[14] G. Bertotti, G. Durin, and A. Magni, J. Appl. Phys. **75**, 5490 (1994).

[15] S. Field, J. Witt, F. Nori and X. Ling. Phys. Rev. Lett. **74**, 1206 (1995).

[16] For a review on mechanisms that generate spontaneously critical behavior see R. Dickman, M. A. Muñoz, A. Vespignani and S. Zapperi, e-print cond-mat/9910454.

THE ENERGETICS OF DISLOCATION-OBSTACLE INTERACTIONS BY 3-D QUASICONTINUUM SIMULATIONS

KEDAR HARDIKAR*, R. PHILLIPS

Division of Engineering, Brown University, Providence, RI 02912
* hardikar@engin.brown.edu

ABSTRACT

The goal of this work is to study the interaction of dislocations with local obstacles to glide such as point defects, precipitates and other dislocations. The quasicontinuum method is used as the basis of this study. It is demonstrated that two types of boundary effects are of concern in the calculation of hardening parameters using finite sized simulation cells. A recently developed technique to incorporate periodic boundary conditions in the quasicontinuum method is used to eliminate surface effects which were present in earlier implementations and to simulate a dislocation of infinite extent interacting with an array of obstacles. The second type of boundary effect is due to the boundary conditions on the lateral boundaries. A method based on finite element calculations is proposed for quantifying the effect of lateral boundaries in these simulations. Preliminary results for the validation of the method are presented as well as a simulation of the interaction between a conventional edge dislocation in Al with an array of clusters of Ni atoms.

INTRODUCTION

It is well known that various interactions between dislocations and localized obstacles to glide play a crucial role in the hardening behavior of materials. Models based on the obstacle strength, line tension of the dislocation and interaction energies are used to characterize the mechanics of these interactions [1]-[4]. A precise calculation of the hardening parameters involved in modeling these reactions at a higher length scale demands an atomic scale study of the mechanisms involved. The quasicontinuum method links atomistic and continuum models through the device of finite element method which permits a reduction of the full set of atomistic degrees of freedom. This method has been successfully implemented in both 2-d and 3-d problems [5], [6], [7].

In these implementations of the method, atoms on the boundary of the simulation box are fixed at displacements corresponding to the elastic fields of the relevant defects. In a typical simulation of dislocation-obstacle interaction, the dislocation is pinned on the boundaries due to these boundary conditions and this contaminates the calculation of critical stress for the dislocation to overcome a given obstacle. In this work, periodic boundary conditions are implemented to eliminate this effect. However, as the dislocation moves from it's initial position under the action of applied stress, there is a stress on the dislocation resulting from the boundary conditions on the lateral boundaries. This effect has been of concern in atomistic simulations [8],[9]. We propose a method based on finite element calculations to account for the boundary effect in our simulations. Some preliminary calculations are presented for validation of this method. Finally, we present a simulation of the interaction between a conventional edge dislocation in Al with an array of clusters of Ni atoms.

THE QUASICONTINUUM METHOD AND PERIODIC BOUNDARY CONDITIONS

Consider a body described as a collection of a possibly huge number, N, of atoms. Once the positions of these N atoms are known in the deformed configuration, from the atomistic perspective, the total energy of the configuration can be found. In conventional lattice statics the objective is to determine the minimum energy configurations of the atoms subject to given constraints. In the quasicontinuum method, a small subset R of atoms is chosen from the total set N of atoms as "representative atoms". These are the only unconstrained degrees of freedom in the problem . A finite element mesh with nodes corresponding to the representative atoms is then defined. Now, the total energy can be determined with the constraint that the positions of most of the atoms are determined from the representative atoms by finite element interpolation. We make the assumption that the total energy may be written in the form that is additively decomposed. This presupposes the existence of well-defined site energies E_i such as in the embedded-atom method [10]. The total energy of the configuration can then be approximated as

$$E_{tot} = E_{exact}(\mathbf{x_1}, \mathbf{x_2}, \mathbf{x_3}, \ldots, \mathbf{x_N}) \approx \sum_{\alpha=1}^{R} \mathbf{n}_\alpha \mathbf{E}_\alpha \tag{1}$$

where $\mathbf{x_i}$, $i = 1, \ldots, N$, denote the positions of *all* the atoms in the deformed configurations whereas the summation runs over the *representative atoms*. Physically, the quantity n_α can be interpreted as the "number of atoms represented" by the representative atom α. In typical simulations of interest, fields are rapidly varying near the defect cores and are slowly varying in the far field. The atoms in these two regions are identified as nonlocal atoms and local atoms respectively. The energy of a non-local atom is calculated by conventional method using the positions of all the atoms in the required neighborhood. To calculate the energy of the local atom, the local deformation gradient $\mathbf{F}$ is used. To calculate the total energy of such atoms, the Bravais lattice vectors of the deformed configuration $\mathbf{b}_\alpha$ are obtained from those in the reference configuration $\mathbf{B}_\alpha$ via: $\mathbf{b}_\alpha = \mathbf{F}\mathbf{B}_\alpha$. Once the Bravais lattice vectors are specified, the energy computation becomes a standard exercise in the practice of lattice statics. Further details of the implementation can be found in [6]. The simulations are done at zero temperature.

Periodic Boundary Conditions in the Quasicontinuum Method

To implement periodicity in the quasicontinuum method it suffices to have declared the representative atoms on the two corresponding faces of the periodic cell as identical. However, this should be done in a way that is consistent with the schemes used for calculating the energies of nonlocal atoms and of the local atoms.

Fig. 1 (a) shows the representative atoms on the slip plane in a typical calculation, for a conventional edge dislocation in Al, done using the quasicontinuum method. The figure shows nonlocal atoms near the core region and local atoms away from the core. The energy of a nonlocal atom is calculated using the positions of atoms in a spherical neighborhood. The size of the neighborhood depends on the cutoff radius r_{cut} of the interatomic potentials used. Imposing periodicity in this region can therefore be achieved by explicitly manipulating the list of atomic sites declared as "neighboring sites" of a given site on the boundary. For example, in case of *atom A* in fig. 1 (a), another atom shown as *atom A1* is declared "identical" in order to impose periodicity in y-direction. The list of sites declared as neighboring sites for *atom A* would include all the atoms in the two spheres of radii r_{cut} as shown in the figure.

To compute the energy of a local atom, a representative crystallite is built using the local deformation gradient. The strain energy density is then computed using the conventional lattice statics for the deformed crystallite. The strain energy density at a particular node depends on the deformation gradients in the elements surrounding that node. Consider the local atoms at *node B* and *node B1* in fig. 1(a). These are the corresponding nodes in the direction of periodicity (y-direction) and are declared identical. Hence, the strain energy density at *node B*, depends on the deformation gradients in all the elements which have *node B* or *node B1* as one of their nodes. The corresponding triangulation in the slip plane can be seen in fig. 1(a). In this approximation scheme it is required to have identical set of representative atoms on the two corresponding faces. With the two schemes mentioned above the degrees of freedom in the energy minimization are further reduced.

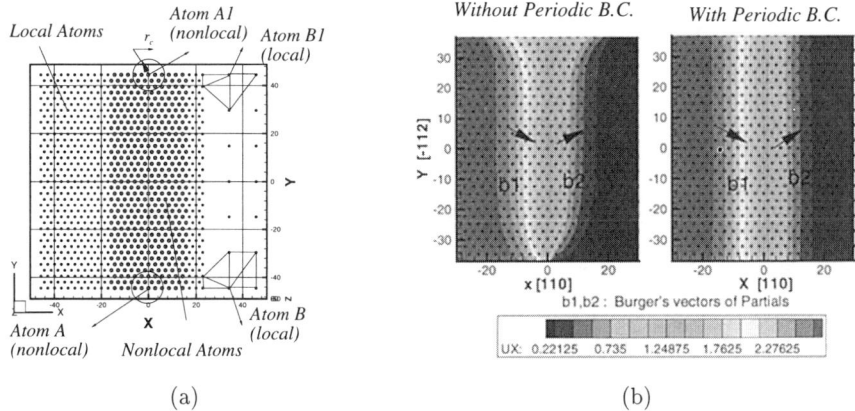

(a)   (b)

Figure 1: Periodicity in Quasicontinuum Method (a) Representative atoms on the slip plane
(b) Slip distribution without periodic b. c. and with periodic b. c.

VALIDATION

The conventional edge dislocation in fcc is the $a/2\langle110\rangle\{111\}$ dislocation. The slip distributions of a relaxed edge dislocation in Al, on the slip plane are shown in fig. 1(b). This simulation is done using 19309 representative atoms as opposed to 134458 atoms that would be required to carry out a conventional lattice-statics simulation with the same geometry. To carry out the simulation, all the nodes in the mesh are displaced in accordance with the Volterra field [2]. The equilibrium structure is then obtained as the configuration which minimizes the potential energy under given constraints on the boundaries. For the present analysis we have used the embedded atom potentials (EAM) of Ercolessi and Adams [11] developed to simulate aluminum crystals.

Fig. 1(b) shows the results as obtained using the previous implementations where all the atoms on the boundaries are held fixed except in a small region near the core (without periodic b.c.) as well as the slip distribution which results using the periodic boundary conditions along the line direction of the dislocation. The dislocation is seen to dissociate into two Shockley Partials with Burgers vectors of the type $\frac{1}{6}\langle112\rangle$. In the absence of periodicity the partial dislocations are found to have bent in the vicinity of the free surfaces. The crucial fact about this boundary condition is that the dislocation is essentially pinned at the surfaces due to the imposed boundary conditions. Also, due to the presence of free surfaces, calculated energies are not accurate for dislocation-obstacle interactions. It is clear from fig. 1(b) that the effect of free surfaces is completely eliminated by periodic boundary conditions . It should be noted here that, since the local calculation away from the dislocation core is based on continuum approximation, it can be so implemented that the surface effects are eliminated.

THE EFFECT OF LATERAL BOUNDARIES : A FINITE ELEMENT APPROACH

Consider a screw dislocation which is initially in equilibrium at the origin as shown in the fig. 2(a) and moves to a new position $(d,0)$ under the action of an applied stress, under the constraint on the boundary ∂M_e that the displacements are fixed corresponding to the initial position of the dislocation. We denote the slip surface $\overline{AO}$ by ∂M_s and slip surface $\overline{AP}$ by $\partial M_{s'}$. We denote all the linear elastic fields associated with the presence of the dislocation at O by a subscript O (e.g. $\mathbf{u_O}$) while we will use the subscript P to denote fields associated with the presence of dislocation at P. In both cases the fields correspond to an unconstrained dislocation in an infinite crystal. Let ΔE denote the difference in the elastic energy E_2 of the system when the dislocation is at P and the elastic energy E_1 when the dislocation is at O. Then following [8] we have,

$$E_1 = \frac{1}{2}\int_{\partial M_s} \mathbf{t_O} \cdot [\![\mathbf{u_O}]\!]dS + \frac{1}{2}\int_{\partial M_e} \mathbf{t_O} \cdot \mathbf{u_O}dS \qquad (2)$$

where, $\mathbf{t_O}$ is the traction vector on the relevant boundary due to stress field of the dislocation at O and $[\![.]\!]$ represents the jump in the displacement across the slip surface (i.e. Burgers vector). The total energy E_2 of the configuration with the dislocation at P (after the constrained motion) may be evaluated as follows. This configuration is taken as the superposition of the fields $\mathbf{u}_P$ due to the presence of the dislocation at P as though in an infinite crystal, and $\Delta\mathbf{u}$ which is an equilibrium displacement field with

$$\Delta\mathbf{u} = \mathbf{u}_O - \mathbf{u}_P, \forall \mathbf{x} \in \partial M_e \qquad (3)$$

and with the condition that $[\![\Delta\mathbf{u}]\!] = \mathbf{0}$ on $\partial M_{s'}$. It should be noted that $[\![\Delta\mathbf{u}]\!]$ in the interior of the region is *not* equal to $\mathbf{u_O} - \mathbf{u_P}$. The energy in the displaced configuration is

$$E_2 = \frac{1}{2}\int_{\partial M_{s'}} \mathbf{t_P} \cdot [\![\mathbf{u_P}]\!]dS + \frac{1}{2}\int_{\partial M_e} \mathbf{t_P} \cdot \mathbf{u_P}dS + \int_{\partial M_e} \mathbf{t_P} \cdot \Delta\mathbf{u}dS + \frac{1}{2}\int_{\partial M_e} \Delta\mathbf{t} \cdot \Delta\mathbf{u}dS \qquad (4)$$

where, $\mathbf{t_P}$ is the traction associated with the stress field of the dislocation at P and $\Delta\mathbf{t}$ is the traction associated with the field $\Delta\mathbf{u}$. The additional energy stored in the system due to the incompatible boundary conditions and the boundary stress are then given as

$$\Delta E = E_2 - E1 \qquad , \qquad \tau_b = -\frac{1}{b}\frac{\partial(\Delta E)}{\partial d}. \qquad (5)$$

We propose to solve the boundary value problem for the equilibrium field Δu by finite element method for the given region. Then using eq. (2), eq. (4) and eq. (5) the back stress due to the incompatible boundary conditions can be calculated numerically for any given region. The preliminary results obtained with this scheme can be found in fig. 2(b). It shows the variation of $\Delta E/\mu b^2$ with d/R for cylindrical geometry. The exact result is known for this geometry from [8]. It can be seen that the two results are consistent. We consider this result promising and this method for incorporating the effect of finite cell size in the simulations is currently under investigation for different geometries.

DISLOCATION-PRECIPITATE INTERACTIONS

Fig. 3(a) shows the interaction of a relaxed conventional edge dislocation with an array of precipitates at zero temperature. In this simulation the precipitate is introduced *by hand* and is a $13 - atom$ cluster of Ni atoms in fcc Al. We have exploited the interatomic potentials developed in [12] for Ni_3Al for these simulations. The simulation box is sheared by displacing all the atoms in accordance with the required homogeneous shear strain. The configuration is relaxed at every increment of the stress. Fig. 3(a) shows the slip distribution in one such increment. The mechanism involved could not be identified unambiguously as pure Orowan type looping or cutting of the precipitate. The leading and the trailing partial dislocations are observed to loop around the precipitate successively and subsequently the particle is sheared in the direction of the Burgers vector of the dislocation.

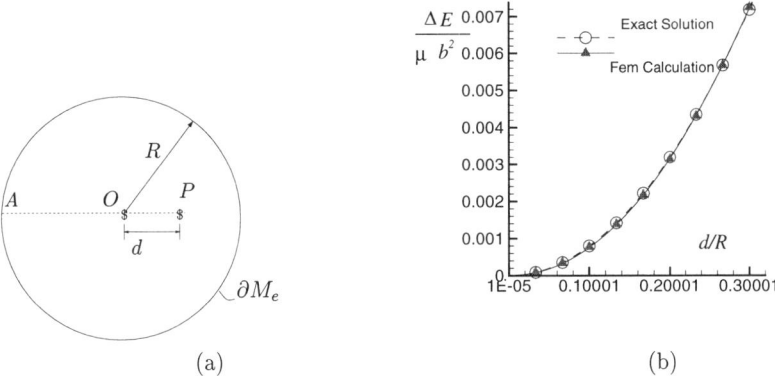

(a)   (b)

Figure 2: (a) Geometry illustrating the boundary effect (b) Preliminary results for boundary
stress by f.e.m. calculation.

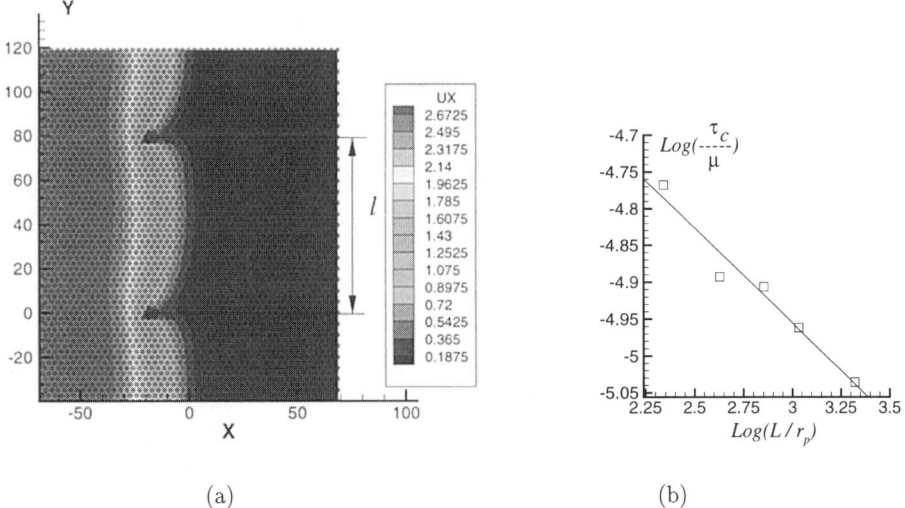

(a)   (b)

Figure 3: (a) Dislocation-Precipitate interaction : Two periodic cells (b) Critical Stress -
Order of Magnitude

Fig. 3(b) shows the preliminary results for the variation of critical stress as a function of distance (center to center) between the precipitates. For the results given here $\tau_c \sim l^{(-0.258)}$. A simple model based on the assumption of constant line tension and pinning force of a localized obstacle predicts $\tau_c \sim l^{-1}$ [2]. We believe that it is necessary to account for the effect of finite cell size accurately, in order to validate these models. This work is currently in progress.

CONCLUSION

Two sets of boundary conditions are observed to play a crucial role in the atomic-level calculations of dislocation-obstacle interactions. The effect of free surfaces present in previous implementations of the quasicontinuum method are shown to be completely eliminated by using a technique to incorporate periodic boundary conditions. The calculations of critical stress for dislocation-precipitate interactions show that a more accurate calculation of the effect of lateral boundaries is necessary. A finite element scheme is proposed to account for these finite size effects. Future work will have to exploit this scheme for arbitrary geometries, for which analytical solutions are not available.

ACKNOWLEDGMENTS

We thank M. Daw and S. Foiles for use of their DYNAMO code. This work was supported by MRSEC under the grant DMR-9632524. This support is gratefully acknowledged.

REFERENCES

1. Hull D., Bacon D. J. *Introduction to Dislocations* , 3rd ed.,(Pergamon Press, Oxford, 1984), p.235.

2. Hirth, J. P., Lothe J. *Theory of Dislocations* , 2nd ed., (Krieger Publishing Company, Malabar, Florida, 1982),pp. 681-685, p. 78.

3. A. J. Ardell, Met. Trans. A **16A**, 2131-2165 (1985).

4. M. Z. Butt and P. Feltham, J. Mater. Sci. **28**, 2557-2576 (1993).

5. E. B. Tadmor, M. Ortiz.and R. Phillips, Phil. Mag. A **73**, 1529 (1996).

6. V. B. Shenoy, R. Miller, E. B. Tadmor, D. Rodney, R. Phillips and M. Ortiz, J. Mech. Phys. Sol. **47(3)**, 611-642 (1999).

7. D. Rodney and R. Phillips, Phys. Rev. Lett. **82 (8)**, 1704-1707 (1999).

8. Shenoy, V. B. and Rob Phillips, Phil. Mag. A **76 (2)**, 367-385(1997).

9. S. Rao, C. Hernandez, J. P. Simmons, T. A. Parthasarathy and C. Woodward, Phil. Mag. A **77(1)**, 231-256 (1998).

10. M. S. Daw and M. I. Baskes, Phys. Rev. Lett. **50**, 1285 (1983).

11. F. Ercolessi and J. Adams, Europhys. Lett. **26**, 583 (1993).

12. A. F. Voter and S. P. Chen in *Characterization of Defects in Materials*, edited by R. Siegel, J. Weertman, R. Sinclair (Mat. Res. Soc. Symp. Proc. **82**, Pittsburgh PA, 1987) pp 175-180.

AMPLITUDE DEPENDENT INTERNAL FRICTION WITHIN A CONTINUUM SIMULATION

P. ALEX GREANEY, D. C. CHRZAN
Department of Materials Science and Mineral Engineering, University of California, Berkeley, CA 94720 and Materials Sciences Division, Ernest Orlando Lawrence Berkeley National Laboratory, Berkeley, CA, 94720

ABSTRACT

The mechanical losses due to the bowing of isolated Frank-Read sources under application of periodic loads is studied within a continuum simulation of dislocation dynamics. The dislocations are modelled within isotropic elasticity theory and assumed to be in the overdamped limit. Dislocation radiation effects are neglected. The mechanical losses are studied as a function of bias stress, amplitude of the periodic stress and frequency. The frequencies studied lie between 10 KHz and 1 MHz. Under high stresses applied at low frequencies, a deviation from the expected Lorentzian resonance shape is observed. The physical origins of this deviation are discussed.

INTRODUCTION

The motion of dislocations under periodic loads is known to contribute to losses measured during internal friction experiments [1]. Therefore, under ideal circumstances, internal friction experiments offer a means to explore the dynamics of dislocations. For example, in materials with a large Peierls barrier, one may deduce the dislocation kink pair formation energy [2].

However, other microscopic mechanisms, for example atomic scale diffusion, may lead to mechanical losses. Proper interpretation of internal friction experiments, therefore, requires a detailed theory-based understanding of the phenomenon in question. Identifying the proper theory, then, is a central to efforts aimed at advancing applications of mechanical spectroscopy techniques for studies of dislocation dynamics.

For describing materials with a small Peierls barrier, such as aluminium, the model most often applied is that of a vibrating string, explored originally within this context by Granato and Lücke [3]. A straight elastic string is pinned at both ends, the resonant frequencies of the string are calculated, and the losses associated with the resonances summed to yield a final expression for the total loss. Implicit in this model is the assumption that the displacement of the string is small in comparison to its length, and that the amplitudes of any residual or bias stresses are zero.

To address the large stress amplitude regime, Tyapunina and Blagoveshchenskii[4] applied a numerical approach to study large displacements of an elastic string governed by viscous dynamics. They demonstrated that the internal friction expected from a bowing Frank-Read source displays both frequency and stress amplitude dependences. However, their approach assumed a constant, isotropic line tension, and hence did not represent fully the change in dislocation self stress with the configuration of the dislocation. Further, their work did not consider the effects of a static bias stress on the expected loss.

Mat. Res. Soc. Symp. Proc. Vol. 578 © 2000 Materials Research Society

In the current work, simulations of dislocation dynamics are used to explore the mechanical losses expected from the simultaneous application of static and periodic loads to a Frank-Read source. The losses are studied as a function of periodic stress amplitude, bias stress and frequency. The results are compared with the expectations of the Granato-Lücke formalism. In particular, it is noted that under common circumstances one may observe significant deviations from the form expected based on the elastic string model.

MODEL

The dislocations are assumed to be governed by isotropic elasticity theory and embedded in an infinite and homogeneous medium. The dynamics are assumed overdamped, so that the dislocation mass may be neglected. The drag coefficient is assumed independent of velocity and isotropic with respect to the orientation of the dislocation.

A previously developed simulation code [5] incorporating these features is applied to calculate the dynamic response of the dislocations. The simulations treat the dislocations within a continuum limit, and reflect the full self-stress of the dislocations. Elastic parameters consistent with Aluminium, $i.e.$with the magnitude of the Burgers vector $b = 2.86$ Å , the shear modulus is taken to be $\mu = 26.5$ GPa, and the drag coefficient is taken as $B = 0.08$ N s m^{-2}. The dislocation is treated as if undissociated. The small scale cutoff procedure employed by Hirth and Lothe [6] is applied, and the cutoff is chosen to be $\rho = 10$ Å.

The dislocation is decomposed into segments extending between points. The points are indexed by the subscript i and the vector $\hat{\mathbf{u}}_i$ indicates the glide direction normal to the dislocation line direction at point i. The velocity of the ith point is given by

$$\frac{\partial \mathbf{u}_i}{\partial t} = \frac{(((\boldsymbol{\sigma}_i \cdot \mathbf{b}) \times \boldsymbol{\xi}_i) \cdot \hat{\mathbf{u}}_i) \, \hat{\mathbf{u}}_i}{B}. \tag{1}$$

Where $\boldsymbol{\xi}_i$ and $\boldsymbol{\sigma}_i$ are the line direction and stress at the ith point, $\mathbf{b}$ is the burgers vector. The stress $\boldsymbol{\sigma}_i$ reflects the contributions from the applied stress as well as the self-stresses of the dislocation. Radiated elastic waves are not considered.

Eqs. (1) are integrated using a fourth order Runge-Kutta integration scheme, as outlined in reference [5], with the exception that additional points are now inserted along a fitted spline, instead of inserting points along a fitted circle.

SIMULATIONS

The simulations consider dislocation configurations similar to that shown in Fig. 1. The losses associated with the application of the periodic stress are calculated as follows. The periodic and bias stresses are applied to the dislocation, and the hysteresis is recorded. The simulations are run until "steady-state" hysteresis is observed. This condition is defined practically as the point at which the area swept out over a complete cycle remains fixed from cycle to cycle (within the noise limits of the simulations). Once "steady-state" is attained, the hysteresis is averaged over eight cycles and the energy loss from the dislocation is calculated according to

$$\Delta W \propto \left\langle \int_{\tau}^{\tau + 2\pi/\omega} \sigma_a(t) d\varepsilon \right\rangle \propto \left\langle \int_{\tau}^{\tau + 2\pi/\omega} \sigma_\omega \cos(\omega t) b \frac{\partial A}{\partial t} dt \right\rangle, \tag{2}$$

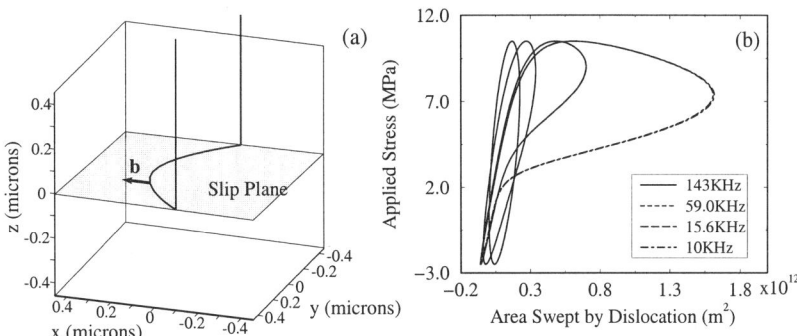

Figure 1: (a) A typical configuration of the Frank-Read Sources. Note that the arms carry on out of the diagram and the loop is closed on a parallel plane 10 μm above the slip plane. (b) The evolution of the hysteresis loop as the frequency drops through the peak loss frequency for a 1 μm edge oriented source $\sigma_{bias} = 4$ MPa and $\sigma_\omega = 6.5$ MPa.

where τ is a time in the "steady state" regime, ω is the angular frequency, and $\sigma_a(t)$ is the applied stress at time t. The time dependent strain from the motion of this dislocation, ε, is taken to be proportional to $A(t)$, the area swept out by the dislocation at time t. In short, ΔW is taken to be proportional to the area of the hysteresis loop (Fig. 1). The applied stress is given by $\sigma_a(t) = \sigma_{bias} + \sigma_\omega \cos \omega t$ with σ_{bias} a static applied stress, and σ_ω the amplitude of the periodic component of the stress. The simulations are run at a number of angular driving frequencies in the range from 10KHz to 1MHz. A peak is seen in the energy absorption as function of frequency. This absorption peak is studied for a variety of different stress conditions for both screw and edge dislocation sources of varying lengths. Typical results are depicted in Fig. 2.

DISCUSSION

The data in Fig. 2 represent the losses calculated from a 1 μm length source. For low bias stresses and high frequencies, the peak resembles the Lorentzian shape expected from a 1-dimensional linear, overdamped oscillator:

$$\Delta W = \frac{\omega D \sigma_\omega^2}{k^2 + (\omega D)^2}. \tag{3}$$

Here, D is a drag coefficient for the oscillator, and k represents the spring constant. From the form of Eq. (3), it is apparent that if one increases the value of k, the peak is shifted to higher frequencies, and the maximum loss is reduced. Increasing the drag coefficient D simply translates the entire curve to lower frequencies. Eq. (3) thus serves as a simple model for the losses observed in the more complicated dislocation problem, and the features evident in Fig. 2 are discussed in terms of the parameters k and D.

stiffness vs. length

Panel (b) of Fig. 3 displays a scaled plot of the area of a dislocation loop vs. the applied, subcritical stress. One expects this area to scale as L^2, where L is the length of the source.

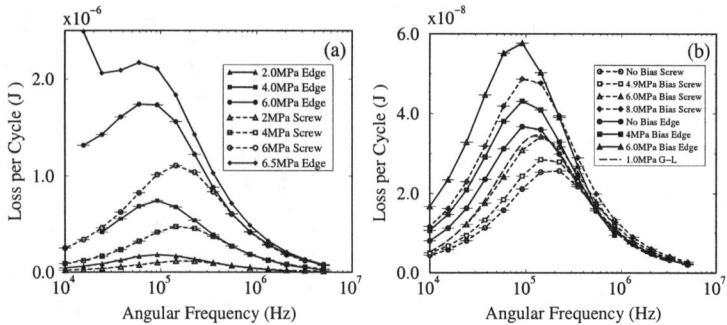

Figure 2: The mechanical loss from $1\mu m$ edge and screw sources (a) under 4 MPa bias stresses with differing σ_ω, and (b) with $\sigma_\omega = 1$ MPa for the indicated values of σ_{bias}. Panel (b) also shows the loss predicted from Granato-Lücke theory. Note that the predictions of Granato-Lücke and the current theory are similar when $\sigma_{bias} = 0$ and for low values of σ_ω.

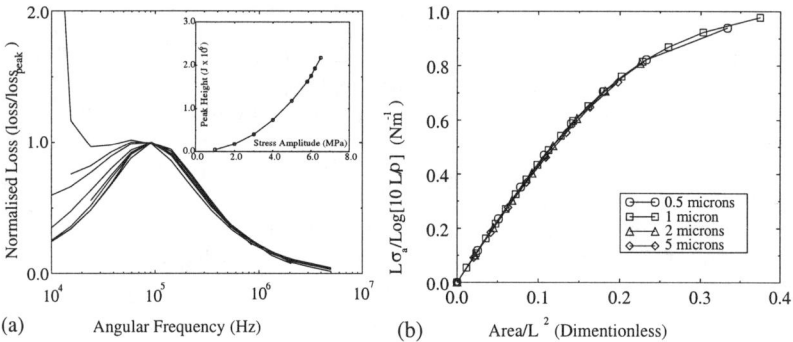

Figure 3: (a) The normalized loss for 4.0 MPa biased, $1\mu m$ edge sources subject to different driving stresses with the change in peak. Loss with stress amplitude shown inset. (b) The scaled stress-area curve for quasi-static bowing of different length edge sources.

Similarly, one expects that the critical stress to set the relevant stress scale. The critical stress scales as $(L/\log CL)^{-1}$ with C a constant. Fitting to the results of reference [5] one finds that $C \approx 1.6 \times 10^9$. The implication is that one might scale the stress axis by the factor $L/\log\frac{1.6L}{\rho}$ and the area axis by a factor of L^{-2} and produce one scaled plot of applied stress as a function of area swept out. It turns out that better data collapse is obtained for the choice $C = 10/\rho$. The plot so obtained is contained in panel (b) of Fig. 3. The data collapse is excellent, though the precise origin of the value of C which gives the collapse is not well understood.

The slope of the stress vs. area curve, then, represents the area-dependent stiffness, k. It is apparent that at higher scaled stresses, the stiffness of the loops is reduced, and eventually, near a scaled stress of 1, the stiffness becomes identically zero.

edge vs screw

Fig. 2 indicates that the losses associated with initially screw oriented segments are lower than their edge oriented counterparts. This is a consequence of the elastic strain energy difference between the screw and edge segments. The elastic strain energy of a screw dislocation is reduced relative to the edge dislocation by a factor of $1 - \nu$. Hence bowing the screw dislocation, so that part of it assumes edge character, requires more stress than bowing the edge segment to a partial screw orientation. The net result is that for the screw oriented segments, the effective value of k is larger.

stress amplitude

Fig. 2 also indicates that as the amplitude of the periodic stress is increased, the magnitude of the loss increases as well. One expects the losses to scale with the square of the periodic stress amplitude, and this behavior is observed (Fig. 3). The shift in peak frequency with increasing amplitude apparent in Fig. 2 stems from the softening of the dislocation line tension as the dislocation bows. The effective stiffness, k is thus amplitude dependent.

bias stress

For large bias stresses, the form of the loss curve can differ dramatically from the simple Lorentzian shape (Figs. 2 and 3). Specifically, the loss reveals a rapid rise at low frequencies. The physical origins of this rise are clear. Consider the situation in which the sum of σ_ω and σ_{bias} exceeds the critical stress necessary to operate the Frank-Read source. Then, as $\omega \to 0$, the dislocation behaves as an overstressed source. Earlier work has demonstrated that there is a stress dependent characteristic time to operate a Frank-Read source, and that this time increases rapidly as the critical stress is approached from above [5]. Then, the reciprocal of this characteristic time sets a characteristic frequency for losses during internal friction. Frequencies much higher than this

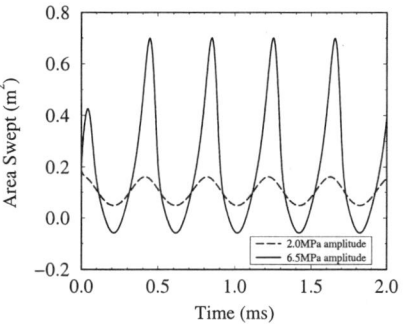

Figure 4: Area vs. time curves for $\sigma_\omega = 2.0$ MPa (dashed line) and $\sigma_\omega = 6.5$ MPa (solid line) with $\sigma_{bias} = 4.0$ MPa. The large amplitude area shows marked deviations from a simple cosine form.

characteristic frequency will yield small losses. As the characteristic frequency is approached from above, the losses will begin to increase as the dislocation is moved back and forth through the critical configuration. At even lower frequencies, the loss per cycle is not constant, as the dislocation motion is no longer periodic.

Within the model summarized in Eq. (3), the picture is as follows. For high frequencies, the displacement due to the periodic stress is small, and the stiffness of the dislocation is determined by the local slope of the stress vs. area curve shown in Fig. 3. However, as the amplitude of the periodic stress is increased, the effective stiffness of the dislocation can be driven towards the value zero, as the dislocation oscillates about the maximum shown in panel (b) of Fig. 3. This leads to the rise in losses at low frequencies. Based on this simple argument, one would expect to observe larger shifts in the peak frequency than those observed. This lack of large shift is explained as an overall decrease in the effective drag coefficient D which partially cancels the shift arising from the decrease in stiffness.

The effects of the σ_{bias} and σ_ω are readily observable in plots of area swept out vs. time for the driven dislocation. Fig. 4 displays two area vs. time curves, one for a large amplitude oscillatory stress, and a second for a small amplitude. The small amplitude curve is described well by a simple cosine. The large amplitude curve, in contrast, shows large deviations from the simple cosine form.

CONCLUSIONS

In conclusion, dislocation dynamics simulations are used to investigate the amplitude and bias stress dependence of losses due to oscillating dislocations. A simple overdamped oscillator model is used characterize the response of the dislocation to time dependent loads. The amplitude dependence of the losses scales roughly as the amplitude of the oscillatory component of the stress squared. The results are found to be in good agreement with expectations based on the theory of Granato and Lücke for the range of stresses in which that theory is likely to apply.

A striking feature of the losses is their frequency dependence. For large frequencies and small bias stresses, the frequency dependence displays a single peak. However, for larger bias stresses, and lower frequencies, the loss spectrum includes the expected peak, but also reveals a divergence at lower frequencies. This divergence arises from the fact that the effective stiffness goes to zero near the critical stress for operation of a Frank-Read source.

This work is supported by the Department of Energy, Office of Basic Energy Sciences, Division of Materials Science under contract DE-AC03-76SF00098.

REFERENCES

1. G. Fantozzi, C. Esnouf, W. Benoit and I. G. Ritchie, *Prog. Materials Sci.* **27**, 311 (1982).

2. A. Seeger, *Phil. Mag.* **1**, 651 (1956).

3. A. Granato and K. Lücke, *J. Appl. Phys.* **27**, 583 (1956).

4. N. A. Tyapunina and V. V. Blagoveshchenskii, *Phys. stat. sol.(a)* **69**, 77 (1982).

5. A. K. Faradjian, L. H. Friedman and D. C. Chrzan, *Modelling Simul. Mater. Sci. Eng.* **7**, 479 (1999).

6. J. P. Hirth and J. Lothe, *Theory of Dislocations*, Krieger Publishing Company, Malabar, Florida, second ed. (1992).

Dislocation Core Properties
and Effects

DISLOCATION STRUCTURE AND DEFORMATION BEHAVIOR OF INTERMETALLIC COMPOUNDS

M. J. Mills, G. B. Viswanathan, R. Srinivasan, M.F. Savage, R.D. Noebe[#], M. S. Daw[##]

Dept. of Materials Science and Engineering, The Ohio State University, Columbus, OH 43210
[#] NASA Glenn Research Center, Cleveland, OH 44135
[##] Motorola Corporation, Austin, TX 78721

ABSTRACT

The fine structure of dislocations in intermetallic compounds can have a profound influence on their macroscopic mechanical properties. The development of appropriate models of deformation requires consideration of dislocation core structure, possible dissociation or decomposition reactions, overall dislocation morphology and relevant dislocation interactions based on detailed transmission electron microscopy study. This empirical information may be rationalized based on both atomistic and continuum-level dislocation modeling. The specific cases of jogged 1/2<110] ordinary dislocations in γ-TiAl and a<101> superdislocations in NiAl are discussed.

INTRODUCTION

For more than two decades, research efforts have been directed toward developing an improved, fundamental understanding of the mechanical properties of intermetallic compounds. Interest in this class of materials has been driven in part by their potential for high temperature, structural applications. From the standpoint of dislocation-level modeling, these compounds also offer a remarkably rich and interesting field of study. A number of important and unique characteristics exhibited by many of these compounds, including the anomalous temperature dependence of the yield and flow strength, appear to be related directly to the intrinsic properties of dislocations. In contrast, the deformation behavior of pure metals, which one might expect to be simpler, are in many respects far more complicated than in many intermetallics. For example, the low temperature deformation of pure FCC metals is clearly dominated by extremely complex, collective interactions of large numbers of dislocations. The properties of individual dislocations are therefore of lesser importance. For intermetallics on the other hand, the study of dislocation fine structure and mobility, both from an experimental and theoretical basis, has provided significant insight into the behavior of these compounds.

A classic example is the issue of cross-slip pinning/locking of superdislocations in L1$_2$ compounds, such as Ni$_3$Al. Numerous TEM studies have shown that screw superdislocations tend not to be dissociated on the octahedral glide planes, but rather are dissociated with the APB lying on the cross-slip (010) plane [for reviews see 1-2]. HRTEM observations [3-4] and atomistic simulations (see [5] for a review) indicate however that the individual superpartials dissociate in octahedral planes. Thus, this configuration of the screw dislocation, originally hypothesized by Kear and Wilsdorf [6], is sessile to easy motion in either the octahedral or cube planes. Active debate in the literature has revolved around whether complete KW locks actually form dynamically during deformation, or whether only partial cross-slip occurs to the (010) plane [2]. Seemingly certain however is that the cross-slip process is thermally activated (thereby increasing the number of locks or pinning points as a function of temperature), and that the cross-slip process is analogous to the Escaig model of cross slip for dislocations in FCC metals [7]. The "pinning-point" model originally developed by Paidar, Pope and Vitek [8] is arguably one of the most significant successes in the application of dislocation theory as it successfully predicts the complicated orientation dependence of the yield strength of single crystals, and the tension-compression asymmetry of the yield strength. It is noted that some discrepancies between this model and experiment have been discussed recently [2].

In this paper, we will discuss several other examples in which detailed transmission electron microscopy observations have been correlated with the results from various computational approaches, including atomistic, and dislocation-level continuum modeling, to provide a better understanding of deformation in intermetallics. This discussion will be separated into two sections,

169

depending upon the nature of the dislocation fine structure: namely deformation involving: (a) nearly perfect dislocations, as exemplified by the case of ordinary dislocations in TiAl, and (b) decomposed superdislocations, as for the <101> dislocations in NiAl. In these examples, we will attempt to demonstrate the strong synergy between experiment and theory that is possible (and essential) in developing the most accurate possible deformation models.

DEFORMATION BY 1/2<110] DISLOCATIONS IN TiAl

In the tetragonal $L1_0$ structure (γ-phase) of TiAl, several dislocation types are known to be operative, including $1/2<\bar{1}12]$ and $<10\bar{1}]$ superdislocations, as well as $1/2<1\bar{1}0]$ "ordinary" dislocations [9]. Ordinary dislocations have been observed to be active following deformation at both low and high temperature, and for a range of compositions [10-12]. Based upon observed dislocation densities, ordinary dislocations are particularly important for hypo-stoichiometric compositions, which are in fact the most interesting technologically where these materials exist as two phase mixtures of γ and α_2-phases.

High resolution transmission electron microscopy (HRTEM) has revealed that $1/2<110]$ dislocations have a remarkably compact structure [13]. Shown in Figure 1 is an HRTEM image of a mixed, ordinary dislocation in Ti-52Al when viewed along a 60° orientation. The bright intensities in this image correspond to atomic column positions for these imaging conditions. A single, identifiable extra-half plane is indicated in the figure. This dislocation can potentially dissociate into two $1/2<112]$ Shockley partials, forming a complex stacking fault (CSF) in between. The CSF differs from an intrinsic stacking fault due to the composition of the nearest neighbors atoms across the fault plane. The compact nature of the dislocation indicates the CSF energy is quite large. In order to attempt a more quantitative analysis of the CSF energy, the observed displacements at the core of this dislocation have been compared directly with relaxed atomic structures obtained using three different Embedded Atom Method (EAM) functionals [13]. These functionals were fit to the bulk properties of TiAl, and were modified to produce a range of CSF energies. This analysis indicates that the CSF energy is greater than 470 mJ/m^2. This value is significantly larger than that predicted from electronic structure calculations, which are in the range of 300 mJ/m^2 using both LKKR [14] and FLAPW [15] methods. The reason for this discrepancy between experiment and theory is presently unclear.

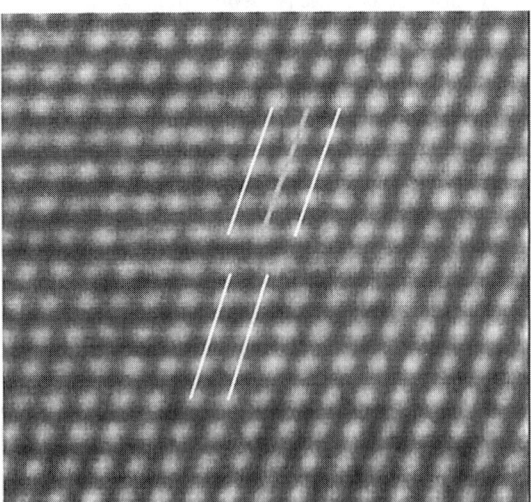

Figure 1: High resolution TEM image of $1/2<110]$ dislocation along a 60° line direction in Ti-52Al. The bright intensities correspond to atomic column positions for these imaging conditions. The compact nature of the core is indicated by the associated lattice planes.

The compact nature of the ordinary dislocation core suggests that cross slip of these dislocations should be quite easy. TEM observations of the morphology of 1/2<110] dislocations following both constant strain rate and creep conditions indicate a strong tendency for these dislocations to elongate in screw orientation, and appear to be pinned at many places along their length [11-12,16]. Figure 2 shows the typical dislocation structures in a near-gamma Ti-48Al sample tested at 768°C and 207 MPa to the minimum-creep regime. Viguier and Louchet [17] have proposed a process similar to that illustrated in Figure 3 as the origin of these pinning points. Due to large frictional forces on screw dislocations, a situation apparently supported by atomistic calculations by Simmons, et al. [18], screw dislocations must advance by forming kink-pairs on the glide plane. The high CSF energy implies that kink-pairs may be formed on more than one plane at the same time. Assuming for example that two different octahedral planes have a significant Schmid factor, then stable kink-pairs may be formed and expand on these two planes along the same dislocation line. A small jog on the screw dislocation would effectively be formed at the point of collision between the two migrating kinks. In the original Viguier and Louchet model [17], it was suggested that the dislocations which bowed between the jogs could lie on a variety of planes. Based on TEM tilting experiments, Sriram, et al [12] have recently indicated that the atomic-sized jogs may grow in height due to the collection of similar jogs at the ends of each pinned segment. Thus, the bowing segments lie on parallel glide planes, off-set by the jog height. Their observations also indicate that the jogs are relatively difficult to translate laterally based on the relative curvatures of adjacent, bowed segments, and that in addition to jog spacing distributions, there may also be a jog height distribution which is important to take into account in modeling this effect. It should also be noted that the cusped configurations along ordinary dislocations have also been attributed to extrinsic pinning due to fine-scale precipitates (e.g. oxides) [19-21], although this proposition can not explain the appearance of tall jogs.

An important ingredient of the model of Viguier and Louchet is that there also exist large frictional forces on *non-screw* dislocation orientations. If kinks/jogs travel rapidly along the dislocation, then the probability of colliding with a jog is decreased and pinning points will only rarely form, and will not limit the motion of the dislocation. Such is presumably the case in pure FCC metals with high stacking fault energies, such as aluminum. Atomistic calculations indicate that the Peierls stresses for ordinary dislocations are particularly large for the 60° orientation due to the non-planar spreading of the screw component [18].

The advance of the screw dislocation is ultimately limited by the presence of these pinning points or jogs, since they can not move conservatively in the direction of the glide force. At lower temperatures, the model of Louchet and Viguier [17] for the yield strength anomaly envisions that overall advancement of the screw dislocation takes place via the lateral glide of a macrokink along the dislocation line. The critical-sized macrokink originates at locations where the distance between pinning points is relatively large. This proposed process, called the local pinning-unzipping (LPU) mechanism, is supported by TEM observations which indicate that pinning points are aligned rectilinearly along the screw dislocations, indicating that the dislocation is initially straight after passage of a macrokink, but with time becomes cusped. Assuming that kink-pair formation is thermally activated, then the number of pinning points along the screw dislocation should naturally increase with temperature, provided that the jog/kink migration rates are relatively constant with temperature. An increase in the average linear density of pinning points with temperature in the anomalous regime has been measured by several groups [11-12].

At higher temperatures, it is possible that the jogs are dragged by the gliding screw dislocations. Viswanathan, et al [16] have recently confirmed that similar "tall" jogs are also common along 1/2<110] dislocations following creep deformation of γ-TiAl. These observations have led to the development of a dislocation model for creep which is based on the non-conservative motion of these jogs. Hirth and Lothe [22] have calculated the velocity of a screw dislocation assuming that the jogs are all of the vacancy-producing type to be:

$$v_s = (4\pi D_s / h) \{exp(\tau\Omega\lambda / hkT) - 1\} \tag{1}$$

where λ is the jog spacing, h is the jog height, Ω is the atomic volume and D_s is the self-diffusivity. In the previous jogged-screw models [23-25], it has been assumed that the jogs form as a result of the intersection between screw dislocations on different slip planes, thereby producing jogs with a height equal to b. In contrast, the process of forming *intrinsic* jogs on 1/2<110] ordinary disloca-

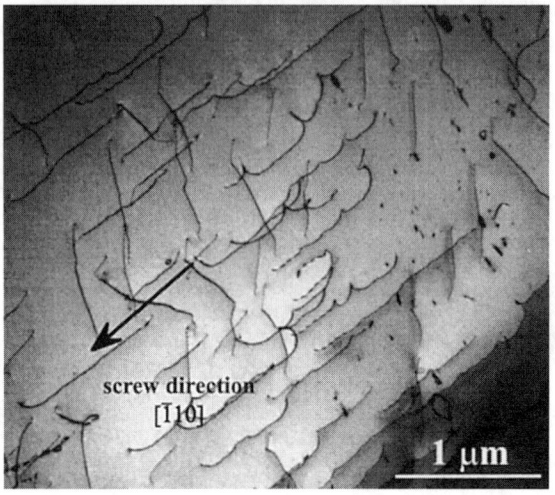

Figure 2: Bright-field TEM image of 1/2<110] dislocations in Ti-52Al. The dislocations tend to be generally aligned along the screw orientation, with frequent, local cusps. Tilting experiments indicate that the cusps are associated with tall jogs.

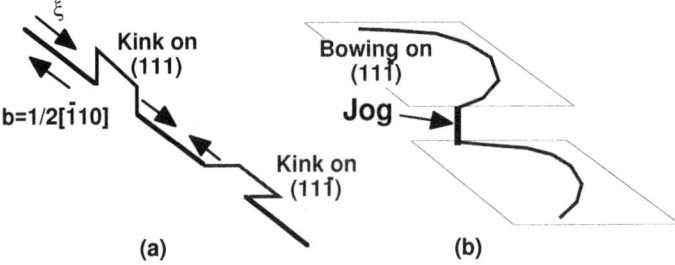

Figure 3: Illustration of intrinsic formation of non-conservative jogs along screw dislocations.

tions described above should produce a *spectrum* of jog heights. Jogs should grow in height with strain as a natural consequence of dislocation motion. An upper-bound on the value of h occurs when the jog is sufficiently tall that the oppositely-signed, near-edge segments attached to the top and bottom of the jog can by-pass each other. This maximum jog height, h_d, can be approximated using the condition for breaking of a pure edge dipole:

$$h_d = G b / \{8 \pi (1 - \nu) \tau \} \tag{2}$$

If h is less than b, then the jog no longer exists. If $h > h_d$, then the jog will no longer be dragged non-conservatively, but will instead act as a dislocation source. Thus, the range of jog heights which contribute to deformation by non-conservative motion should lie between b and h_d, and we therefore expect this "characteristic" jog height to depend on the applied stress. With the rather crude assumption that this characteristic value of h is a constant fraction of h_d (given by the term βh_d), a modified expression for the jogged-screw model has been proposed [16]:

$$\dot{\gamma} = \{\pi D / (b\beta \, h_d \, (\alpha G)^2)\} \, \tau^3 \, \{ \exp(\tau^2 \, \Omega\lambda \, /(4 \, h_d \, kT)) - 1 \} \tag{3}$$

In this expression, it is assumed that the Taylor relation $\rho_s = (\tau / \alpha \, Gb)^2$ provides a suitable description of the variation of the mobile dislocation density.

The predicted axial strain rate as a function of stress using the conventional jogged-screw model, as well as this modified expression is shown in Figure 4. For the modified model, the value of β is assumed equal to 0.5. Also shown in Figure 4 are data points taken from the work of Viswanathan [16] for Ti-48Al which was heat treated to yield a 'near-gamma' condition. The conventional jogged screw model predicts creep rates many of orders of magnitude too large relative to experiment. In contrast, the agreement between the modified model and experiment is quite good considering the approximations made in the present development. A reasonable agreement with

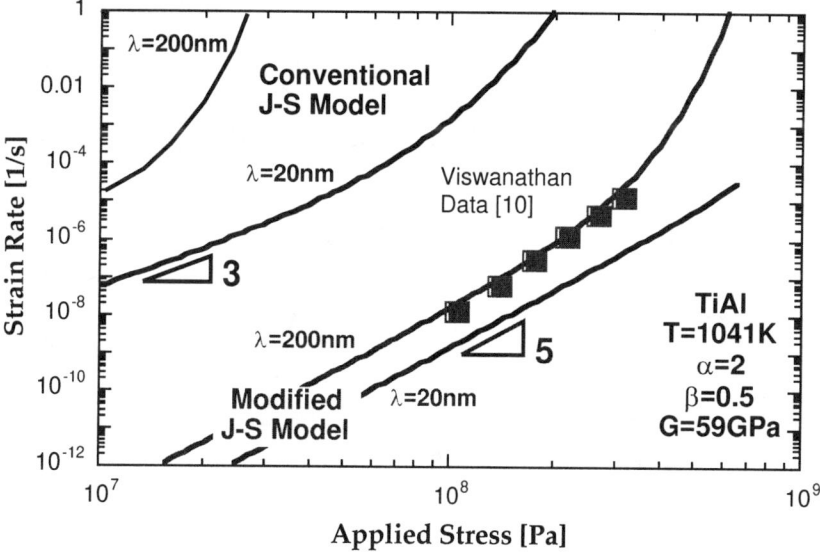

Figure 4: Predicted axial creep rate versus axial stress using the conventional (h=b) and modified jogged-screw model, with model parameters appropriate for γ-TiAl. Included for comparison are experimental data from the work of Viswanathan, et al [10].

173

the temperature-dependence for creep has also been obtained by considering only the variation of the diffusion coefficient and shear modulus. Summarizing the present assumptions for these critical parameters, they include: (a) the use of the Taylor relation for the mobile dislocation density, (b) a jog spacing which is independent of stress and temperature, and (c) characteristic jog heights which are independent of temperature. In addition, the variation of these critical parameters with strain has not been incorporated in this "steady-state" treatment.

MOBILITY AND DECOMPOSITION OF SUPERDISLOCATIONS IN NiAl

While NiAl has been the subject of extensive research as a potential replacement for Ni-based superalloys in high temperature structural applications in the aerospace industry, poor ductility and toughness at low temperatures, and a dramatic softening at higher temperatures have frustrated these efforts. The usual mode of deformation in both single crystal and polycrystalline NiAl is through motion of a<010> dislocations at all temperatures. However, on stressing single crystals of NiAl along the cube axes (also known as the "hard" orientation), <010> slip is suppressed (since there is no resolved shear stress for a<010> slip). The compressive yield strength versus temperature behavior of hard-oriented Ni-44Al is shown in Figure 5. At lower temperatures, slip occurs at very high stress values through a<111> dislocation motion [26-27], while at intermediate temperatures there is a sharp decrease in the yield strength as a function of temperature. This transition temperature will be referred to as the "knee" in the yield strength curve (T_{knee}). This coincides with a change in deformation mode from a<111> to non-a<111> slip [28-29]. There has been considerable debate in literature regarding a<010> vs. a<101> slip beyond this slip transition temperature [31]. Interestingly, a<010> slip has been reported to occur primarily in stoichiometric NiAl [28], while a<101> slip has been observed in Ni-rich NiAl [29-32]. The apparent difference in deformation mode has been rationalized elsewhere based on differences in vacancy content [31].

The deformation regime in which a<101> dislocations are operative has been the subject of investigation since their characteristics appear to be central to the dramatic softening above T_{knee}. The remarkable, characteristic morphology of the a<101> dislocations can be seen in Figure 6. The a[$\bar{1}$01] dislocations are aligned along <111> directions in the glide plane. The only line length that is not along the <111> directions is at the corners of the semi-loop-like features. A very similar dislocation microstructure has been observed by Messerschmidt et al. [33] in studies on polycrystalline NiAl doped with 0.2% Ta. Their observations of dislocations in grains with orientations

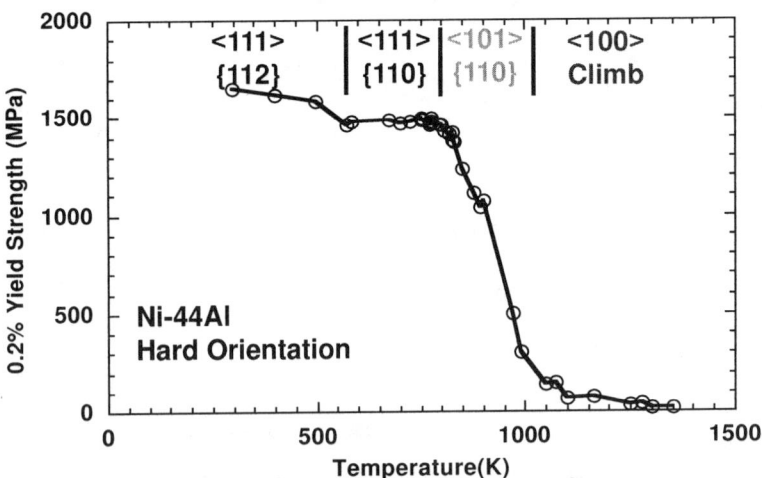

Figure 5: Yield strength as a function of temperature for Ni-44Al single crystals deformed in the hard orientation. The active deformation systems as a function of temperature are indicated.

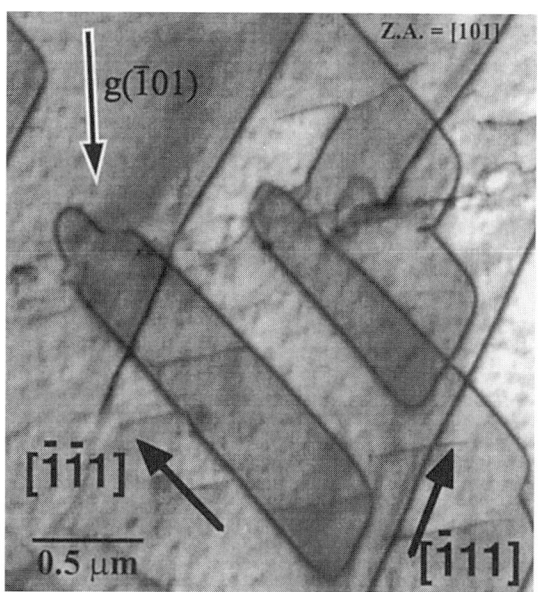

Figure 6: Morphology of a[$\bar{1}$01] dislocations on (101) glide planes in Ni-44Al after 1% deformation at 875 K. Note the remarkable alignment of these dislocations along <111> directions in the glide plane.

close to <001> relative to the tensile axis revealed rectilinear a<101> dislocations along <111> directions. At higher temperatures, or lower strain rates, a third pronounced dislocation line direction is observed along the edge orientation (a [010] direction) for the a[$\bar{1}$01] dislocations. Similar configurations have been discussed at length elsewhere [34].

This alignment is not predicted by elasticity theory. The line energy for a<101> dislocations is lowest for screw orientations, and increases monotonically to its highest value for edge orientations [34]. The <111> line direction is 35° away from screw orientation, and thus possesses an intermediate value of elastic energy. Also, both the [$\bar{1}\bar{1}$1] and [$\bar{1}$11] line directions are 35° away from screw orientation, and thus energetically equivalent. A second possibility for the preferred line directions is due to the dislocation mobility. Our observations could be rationalized if the mixed <111> or edge [010] line directions were to have a particularly large Peierls stress relative to other orientations. However, molecular dynamics calculations utilizing the EAM functionals of Ludwig and Gumbsch [35] indicate instead that there are no line directions for which the friction stress is remarkably different from the rest. In fact, the Peierls stresses are uniformly very large for the a<101>{110} system (of the order of 4 GPa). Such large values make it improbable that these dislocations can move via a simple glide process.

The rectilinear nature of the dislocation segments along both <111> and [010] line directions facilitated direct HRTEM investigation of the a<101> dislocation along these directions. Figure 7 shows a HRTEM micrograph of an a[$\bar{1}$01] dislocation along the [$\bar{1}$11] direction. The image has been Fourier filtered in order to remove some of the random background noise. The filtered image has been indexed with displacements and Burgers circuits. From the image we can observe that the Burgers circuits are consistent with displacements associated with an a[$\bar{1}$00] and a[001] dislocation along [$\bar{1}$11]. Also, the splitting of the a[$\bar{1}$01] core has occurred normal to its glide plane by about 1.5-2 nm. This observation is consistent with an earlier HRTEM observation of the local decomposition of a<101> dislocations along the [010] or edge orientation [34].

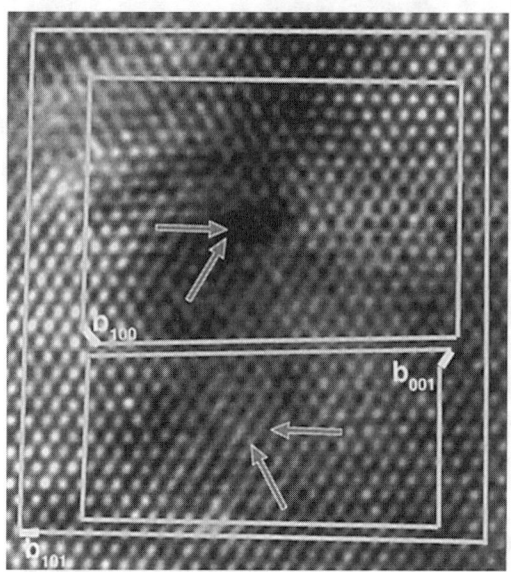

Figure 7: HRTEM image of an a[$\bar{1}$01] dislocation in Ni-44Al after 1% deformation at 875 K. The a[$\bar{1}$01] dislocation is actually decomposed into distinct a[$\bar{1}$00] and a[001] dislocations separated by about 20 Å based upon the Burgers circuit analysis shown. In this configuration, glide of the complete a[$\bar{1}$01] dislocation is impossible.

Given that the fine structure of the a<101> dislocation is actually in this decomposed state, the manner in which moves under stress is not obvious. The overall dislocation morphology (Figure 6) clearly suggests a conservation motion on the {110} glide plane. However, the dislocation's fine structure indicates that a conventional glide process is impossible since the product a[$\bar{1}$00] and a[001] Burgers vectors resulting from the decomposition are *not* contained in the glide plane. In fact, only the a[001] dislocation has any external force acting on it—a climb force which would tend to drive it off of the {110} glide plane. A representation of a possible model for the movement of this configuration is shown in Figure 8. The model incorporates the fact that as the a[001] dislocation climbs under the action of the external stress, it will cause a reduction in the local vacancy concentration, by an amount Δc, since it requires vacancies to climb. This will create an osmotic climb force on the a[$\bar{1}$00] dislocation since it generates vacancies in the process. The local vacancy concentration is increased near the a[$\bar{1}$00] dislocation by a similar amount, Δc. Thus, both a<100> dislocations are able to climb cooperatively and conservatively, with no long-range diffusion required. The a[$\bar{1}$01] velocity can be estimated if we assume that the vacancy supersaturation is determined by the action of the climb force on the a<001> dislocation. Borrowing from the equation for the climb velocity of a dislocation from Hirth and Lothe [22], we may write:

$$V_c = \frac{2\pi D_s a_o^2 \sigma}{b \, k \, T \, \ln[d/b]} \tag{4}$$

where d is the decomposition distance between the a<100> dislocations, b is the Burgers vector of the a<100> dislocation and D_s is the self-diffusion coefficient. For the 35° line orientation, there is an attractive elastic interaction which will tend to pull the two a<100> dislocations together, which

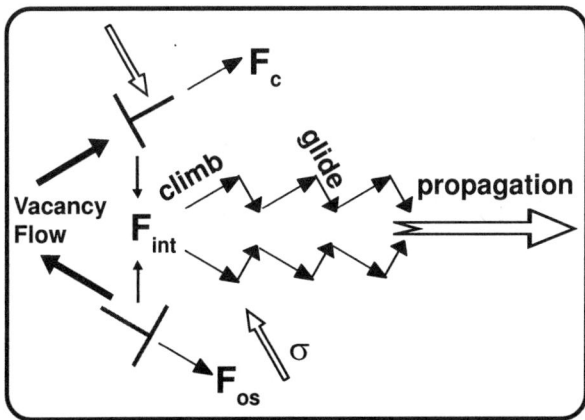

Figure 8: Proposed model for propagation of a decomposed a<101> dislocation in hard-oriented NiAl. The applied stress imposes a climb force on one of the a<001> dislocations, causing an osmotic force on the other a<100> dislocation. The climb step may proceed by the short-range diffusion of vacancies between a<100> dislocations. Elastic attractive interaction between the dislocations causes the glide step, which must occur on rational glide planes for the a<100> dislocations.

can occur by glide on two different {011} planes. The overall propagation of the configuration is therefore composed of a series of coordinated climb and glide steps. Assuming that the glide steps are much more rapid than the climb steps, the overall velocity of the a<101> dislocation is simply $\sqrt{2} V_c$. We may compare this model directly with the actual macroscopic mechanical properties. After deformation to 1% strain at 875K, the dislocation density determined from TEM observations is about 5×10^{12} /m². Given that the applied stress at this point during deformation was 1100 MPa, and using a value of D_s for Ni-42Al [36], the estimated axial strain rate is 1.5×10^{-5} /s. This predicted value agrees remarkably well with the actual imposed deformation rate of 2.5×10^{-5} /s. Thus, the proposed model would appear to provide a reasonable representation of the kinetics of a<101> motion.

This model directly addresses the two most intriguing aspects of these a<101> dislocations. First, the strong temperature dependence of the yield strength in the regime for which a<101> dislocations are active is a result of the climb step of Figure 8. Due to the temperature dependencies in Equation 4, higher deformation temperatures will lead to a more rapid overall rate of propagation.

Second, the strong preference for the <111> line directions can be understood since the glide step in the propagation process must take place on a rational glide plane for the a<100> dislocations. There are two appropriate {011} planes available for the glide step along the <111> line directions. It should be noted that the only other line direction in which a similar situation arises is [100], for which the two a<100> dislocations are pure edge and can glide on {010} planes. The prominence of this line orientation at higher temperatures and lower strain rates is therefore also consistent with the general view of propagation presented in Figure 8.

DISCUSSION AND CONCLUSIONS

In relation to the first deformation model presented above, the characteristics which appear to be necessary for the generation of jogged-screw dislocations during deformation are: (a) frictional forces on screw dislocations, (b) easy cross-slip due to high fault energies, and (c) frictional forces on mixed or edge dislocations. This model for creep may indeed be applicable to several other alloys and intermetallics. Jogged-screw configurations for a-type dislocations have been observed in the HCP α-phase of Ti-6242 alloys following high temperature creep [37]. In this case, the large frictional forces along screw orientation are presumably due to non-planar spreading of the 1/3<11~20> screw dislocation core in both prism and basal planes [38]. Strong solute interactions due to Al additions also may impede motion of non-screw dislocations. Banerjee [39] have observed similar configurations for dislocations in the orthorhombic phase of Ti-Nb-Al alloys. Finally, Xu and Arsenault [40] have found heavily jogged <100> dislocation configurations after high temperature deformation of NiAl composites.

The model described in the previous section for the propagation of a<101> dislocations is probably somewhat more specific to the case of NiAl in the hard orientation. In addition to controlling the deformation behavior above T_{knee} in off-stoichiometric NiAl, there is considerable evidence that it is the mobility of a<101> dislocations which also controls T_{knee} itself by determining the conditions for which decomposition of a<111> dislocations occurs [41-42]. It is noted that FeAl, another extensively-studied system having the B2 crystal structure, actually deforms via a<111> dislocations as the favored slip vector at lower temperatures. A transition to non-<111> activity at higher temperatures has been observed [43], although detailed TEM study is still somewhat lacking. However, the report of a<101> dislocations at higher temperatures indicates that the treatment presented here may have value to this system as well. An important deformation process which also appears to be related is that of cutting of the ordered γ' phase in a γ/γ' superalloy deformed under creep conditions. In the work of Eggeler and Dlouhy [44], a<100> dislocations which have been identified in the γ' phase are clearly linked to two independent 1/2<110> dislocations in the γ channel. While the γ' phase appears to be sheared by the glide of a<100> dislocations on {010} planes, Dlouhy and Eggeler have proposed that it is actually the propagation of these two 1/2<110> dislocations via a climb/glide process which accomplishes the shearing. It is interesting to note that the a<100> dislocations adopt a pronounced alignment along <110> line directions. As reasoned above for the case of a<101> dislocation in NiAl, this preferential alignment may be explained by the fact that it is along this line direction that both 1/2<110> dislocations in the γ' phase have {111} glide planes available for the glide step to take place.

Both of these dislocation models described incorporate climb as a critical, rate-controlling step. Thus, classical diffusion treatments are necessarily utilized. However, in both cases, crucial understanding has also been derived from atomistic calculations, even though diffusive processes are not included. Instead, the results of these calculations have been used to rationalize important assumptions made at the continuum level. Examples of such insight include the calculations of Peierls stresses as a function of line direction for both ordinary dislocations in γ-TiAl and a<101> dislocations in NiAl. In the development of these models, TEM observations have also provided essential guidance. A significant benefit of microstructure-based deformation models is that many of the critical model parameters can in principle be measured from TEM observations. However, it should be recognized that fully exploring the functional dependencies for the critical parameters requires an extensive effort for a range of deformation conditions. It is believed that modeling at both the atomistic and continuum levels will lead to improved understanding of these critical parameters, as well as their variation with strain, stress and temperature.

ACKNOWLEDGMENTS

Support for this work was provided by the National Science Foundation under DMR-9709029 and by the U.S. Department of Energy under contract no. DE-FG02-96ER45550. The authors are grateful to W. D. Nix, U. Glatzel and G. Eggeler for a number of very helpful discussions.

REFERENCES

1. Y. Q. Sun and P. M. Hazzledine, *Dislocations in Solids Volume 10*, ed. F. R. N. Nabarro and M. S. Duesbery (Elsevier Publications 1996), p. 27.
2. P. Veyssiere and G. Saada, *Dislocations in Solids Volume 10*, ed. F. R. N. Nabarro and M. S. Duesbery (Elsevier Publications 1996), p. 253.
3. M. A. Crimp, *Phil. Mag. Lett.* **60** ,45 (1989).

4. N. Baluc, H. P. Karnthaler and M. J. Mills, *Instit. Phys. Conf. Series No 93* **2**, 4633 (1988).
5. Vitek, V., Pope, D. P. and Bassani, J. L. *Dislocations in Solids Volume 10*, ed. F. R. N. Nabarro and M. S. Duesbery (Elsevier Publications 1996), p. 135.
6. B. H. Kear and H. G. F. Wilsdorf,*Trans. TSM-AIME* **224**, 382 (1962).
7. D. Escaig, *J. Phys. Paris* **29**, 225 (1968).
8. V. Paidar, D. P. Pope and V. Vitek, V, *Acta Metallurgica* **32** 435 (1984).
9. G. Hug and P. Veyssiere, *Phil. Mag. A* **57**, 499 (1988).
10. G. B. Viswanathan and V. K. Vasudevan *MRS Proceedings* **288**, 787 (1993).
11. B. Viguer, K. J. Hemker, J. Bonneville, F. Louchet and J. L. Martin, *Phil Mag.A* **71**, 1295 (1995).
12. S. Sriram, D. M. Dimiduk, P. M. Hazzledine and V. K. Vasudevan, *Phil. Mag.A* **76**, 965 (1997).
13. J. P. Simmons, M. J. Mills and S. I. Rao, *MRS Proceedings* **364**, 335 (1995).
14. C. Woodward, unpublished research.
15. M. H. Yoo and C. L. Fu, *Metall.& Mater. Trans. A* **29**, 49 (1998).
16. G. B. Viswanathan, V. K. Vasudevan and M. J. Mills, Acta mater. **47**, 1399 (1999).
17. F. Louchet and B. Viguier, *Phil. Mag. A* **71**, 1313 (1995).
18. J. P. Simmons, S. Rao, D. M. Dimiduk, *Phil Mag A* **75**, 1299 (1997).
19. B. Kad and H. L. Fraser, *Phil. Mag. A* **69**, 689 (1994).
20. D. Haussler, M. Bartsch, M. Aindow, I. P. Jones and U. Messerschmidt, *Intermetallics* **6**, 729 (1998).
21. U. Messerschmidt , M. Bartsch, D. Haussler, M. Aindow, R. Hattenhaurer and I. P. Jones, *MRS* Proceedings **364**, 47 (1995).
22. J. P. Hirth and J. Lothe, *Theory of Dislocations* (John Wiley and Sons, New York 1982)
23. N. F. Mott, *Creep and Fracture of Metals at High Temperatures*, Proc. NPL Symp., (HMSO London 1956), p. 21.
24. P. B. Hirsch and D. H. Warrington, *Phil. Mag* **6**,715 (1961).
25. C. R. Barrett and W. D. Nix, Acta Metall. **13**, 1247 (1965).
26. R.T. Pascoe and C.W.A. Newey, *Metal Sci. J.* **5**, 50 (1971).
27. Y.Q. Sun, G. Taylor, R. Darolia and P.M. Hazzledine, *MRS Proceedings* **364**, 261 (1995).
28. H.L. Fraser, M.H. Loretto and R.E. Smallman, *Phil. Mag.* **28**, 667 (1973).
29. J.T. Kim and R. Gibala, *MRS Proceedings*. **213**, 261 (1991).
30. R.D. Field, D.F. Lahrman and R. Darolia, *Acta metall. mater.* **39**, 2951 (1991).
31. M. J. Mills, R. Srinivasan, M. F. Savage, R. D. Noebe and M. S. Daw, Interstitial and Substitutional Solute Effects in Intermetalics, ed. I. Baker, R. D. Noebe and E. George (TMS Publications 1998), p. 99.
32. R. Srinivasan, M. F. Savage, R. D. Noebe and M. J. Mills, *MRS Proceedings* **552,** 745 (1998).
33. U. Messerschmidt, *Intermetallics* **6**, 729 (1998).
34. M. J. Mills, R. Srinivasan, and M. S. Daw, *Phil. Mag. A* **77**, 801 (1998).
35. M. Ludwig and P. Gumbsch, *Modeling Simul. Mater. Sci. Eng.* **3**, 533 (1995).
36. S. Shankar and L. L. Seigle, *Met. Trans.*, **9A**, 1467 (1978).
37. G. B. Viswanathan, R. W. Hayes and M. J. Mills, *Creep Behavior of Advanced Materials for the 21ˢᵗ Century*, ed. R. S. Mishra, A. K. Mukherjee and K. L. Murty (TMS Publications 1999), p. 197.
38. A. Girshick, D. G. Pettifor and V. Vitek, *Phil. Mag. A* **77**, 999 (1998)
39. A. K. Gogia and D. Banerjee, unpublished research.
40. K. Xu and R. J. Arsenault, Acta mater. **47**, 3023 (1999).
41. R. Srinivasan, M. F. Savage, M. S. Daw, R. D. Noebe, and M. J. Mills, *Scripta Mat.* **39**, 457 (1998).
42. J. S. Brown, R. Srinivasan, M. J. Mills, and M. S. Daw, *Phil. Mag. A*, submitted for publication
43. K. Yoshimi, F. Hanada and M. H. Yoo, *Acta metall. mater.* **43**, 4141 (1995).
44. G. Eggeler and A. Dlouhy, *Acta mater.* **45**, 5251 (1997).

MECHANICAL LOSS ASSOCIATED WITH STRESS ANOMALY IN Ni₃Al AND Ni₃(Al,Ta) SINGLE CRYSTALS

E. CARREÑO-MORELLI *, B.L. CHENG *, M. DEMURA **, R. SCHALLER *,
N. BALUC ***, J. BONNEVILLE ****
* Ecole Polytechnique Fédérale de Lausanne, Institut de Génie Atomique, PHB Ecublens,
CH-1015 Lausanne, Switzerland
** National Research Institute for Metals, Mechanical Properties Division, Tsukuba, Japan
*** Centre de Recherches en Physique des Plasmas, Technologie de la Fusion,
CH-1015 Lausanne, Switzerland
**** Laboratoire de Métallurgie Physique, Université de Poitiers - SP2MI - Téléport 2 BP 179
86960 Futuroscope Cedex France

ABSTRACT

The mechanical loss and shear modulus behaviors of Ni₃Al and Ni₃(Al,Ta) single crystals have been investigated in the temperature range 100 K - 1300 K. The mechanical loss spectra exhibit two temperature regimes, which are separated by a relaxation peak at nearly 950 K for a frequency of 1 Hz. This relaxation peak has been interpreted by the stress re-orientation of Al-Al elastic dipoles in the (111) octahedral plane [1, 2]. In the low temperature regime, corresponding to the anomaly domain of the flow stress, the mechanical loss of pre-deformed specimens exhibit a strong positive dependence on both the oscillation amplitude and the amount of pre-strain.

Pre-deformations, which were performed either at room temperature or at 100 K, yield a broad maximum in the mechanical loss that extends from nearly 100 K up to 550 K. This maximum is observable for only strain amplitudes larger than 10^{-4} and entirely vanishes after heating the specimens above 550 K. The increase in mechanical loss has been attributed to the bowing of the superkinks under the action of the applied stress. The gradual and irreversible decrease in damping above 300 K is interpreted in terms of pinning of the screw dislocation segments by a thermally activated process leading to the formation of Kear-Wilsdorf locks.

INTRODUCTION

The technique of mechanical spectroscopy [3] has been recently used for investigating the mechanical properties of the $Ni_{75}Al_{24}Ta_1$ intermetallic compound [4, 5]. It has been shown that the $Ni_{75}Al_{24}Ta_1$ spectrum presents over the temperature range 300 K - 1300 K a well-defined peak located at nearly 950 K for a frequency of 1 Hz [1], which is superimposed on a mechanical loss background that exponentially increases at high temperatures. This peak has been interpreted as resulting from a short-range diffusion process arising from the stress re-orientation of Al-Al elastic dipoles in the (111) octahedral plane [1], which is also indicative of the onset of local diffusion in this L1₂ structure. At higher temperatures, the mechanical loss background has been attributed to the movement of dislocations on the {001} cube planes.

Several models have been proposed for explaining the anomalous flow stress behavior of the L1₂ intermetallic alloys. It is usually considered that a cross-slip mechanism of the screw dislocations from the primary octahedral plane, where they are mobile, onto the cube plane, where they are sessile, is responsible for the flow stress anomaly (for a review see [6]). However, depending on the modeling of the dislocation path after a cross slip event, the models predict either a decrease in the dislocation mobility or in the mobile dislocation density. In these models, the cross-slip mechanism is thermally activated and enhanced with increasing temperatures, so that the positive temperature dependence of the flow stress results either from a decreasing dislocation mobility or an exhaustion of the mobile dislocation density, or eventually from both.

It is also well known that alloying may have drastic effects on the mechanical properties of Ni₃Al based intermetallic compounds [7, 8]. These differences have been interpreted by a change of the complex stacking fault (CSF) energy with composition, which has a direct influence on the cross-slip ability, and a fair correlation has been recently established between the flow stress and the CSF energy value of various Ni_3Al alloys [9].

In a first attempt of comparing mechanical loss spectra obtained with different Ni_3Al alloys, the present paper reports mechanical spectroscopy studies performed on binary $Ni_{75}Al_{25}$ and ternary $Ni_{75}Al_{24}Ta_1$ single crystals. The mechanical loss and the dynamic shear modulus behaviors have been investigated over a large temperature range that extends from 100 K up to 1300 K. For both alloys, particular attention has been focused on the temperature range of the flow stress anomaly (*i.e.* below 700 K). As in the previous studies [4, 5], virgin and pre-deformed single crystalline specimens have been used. The results obtained for the two compositions are compared and discussed in the framework of the superkink models, which have been proposed for explaining the anomalous behavior of the Ni_3Al phase.

EXPERIMENTAL

A single crystalline rod of binary Ni_3Al having the stoichiometric composition was produced at the National Research Institute for Metals, Tsukuba, Japan. It was uni-directionally grown at a rate of 25 mm/h by using a floating-zone method. The preparation and growth procedures are reported elsewhere [10, 11]. The $Ni_3(Al,Ta)$ single crystals with a nominal composition of $Ni_{75}Al_{24}Ta_1$ were provided by Prof. D. P. Pope at the University of Pennsylvania, USA. Flat rectangular specimens of sizes of 25 mm x 2 mm x 0.45 mm oriented along the <111> crystallographic direction, have been spark eroded from the rods and further carefully polished with successively finer grades of abrasive papers for removing the surface damage layers.

The mechanical loss (damping) and the shear modulus evolutions were investigated as a function of the temperature in the range 100 K - 1300 K by using an inverted torsion pendulum. This pendulum has been specially designed for allowing large oscillation amplitudes and it can be configured into two working modes: free-decay and forced vibrations. In the free-decay mode, a torsional oscillating strain is initially imposed to the specimen up to a given strain amplitude at which the applied excitation is stopped. The mechanical loss is then obtained from the waveform analysis of the free decay of the residual vibrations by the Fourier transformation method [12, 13]. The shear modulus (G) can be derived from the pendulum frequency (f) by using the relation

$$G = \frac{8\pi^2 l}{\beta w t^3} I f^2 \tag{1}$$

where l, w and t are the length, width and thickness of the specimen, respectively, I is the pendulum momentum of inertia and β is a geometrical factor [14]. In the forced-vibrations mode, the inertia masses are removed and the specimen is subjected to an alternating stress at imposed frequency. The mechanical loss corresponds to the phase lag *tan* ϕ (ϕ being the mechanical loss angle) between the stress excitation and the strain response. Here, the elastic shear modulus is measured in arbitrary units as the ratio between the amplitudes of the stress and strain signals.

Specimens were *in-situ* pre-deformed in the torsion pendulum at 300 K or at 100 K. Different pre-deformation levels were tested and the total amount of pre-deformation was usually obtained by applying to the specimen two successive torsions of opposite sign.

RESULTS AND DISCUSSION

Typical mechanical loss spectra of virgin $Ni_{75}Al_{25}$ and $Ni_{75}Al_{24}Ta_1$ single crystals are shown in Figure 1, curves a and b respectively. For both compositions, one can observe a mechanical loss peak superimposed on a background which increases with increasing temperatures. In a previous work on the $Ni_{75}Al_{24}Ta_1$ compounds [1], it has been shown that this peak is thermally activated, with an activation energy of 3.0 ± 0.1 eV, and strongly dependent on the crystallographic orientation. The peak has been ascribed to a relaxation mechanism based on the re-orientation of Al-Al elastic dipoles on the octahedral planes, indicative of the onset of local diffusion in the $L1_2$ structure. This interpretation has been used latter on by Numakura *et al.* [2], who have also shown that the characteristics of this peak are similar to the ones observed by Gadaud and Chakib [15] in Ni_3Al stoichiometric polycrystals and in Ni_3Al polycrystals of various compositions. Figure 1 shows that this mechanical loss peak is also present in $Ni_{75}Al_{25}$ stoichiometric single crystals at nearly the same temperature, but with lower relaxation strength. A point-defect relaxation peak is not expected to appear in fully ordered stoichiometric Ni_3Al. However, at high temperatures, the probability increases that zones of local disorder form, and Al enriched zones would be responsible for the damping peak. Ta impurities could act as nucleation sites for these zones and it could be the origin of a higher relaxation peak in $Ni_{75}Al_{24}Ta_1$.

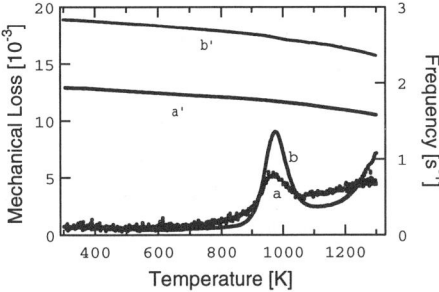 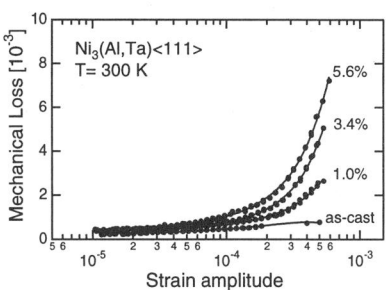

Figure 1: Mechanical loss and frequency spectra of as-cast $Ni_{75}Al_{25}$ (curves a, a') and $Ni_{75}Al_{24}Ta_1$ (curves b, b') measured at a strain amplitude of 1×10^{-5} with a heating rate of 2 K/min.

Figure 2: Strain amplitude dependence of the room temperature mechanical loss for $Ni_{75}Al_{24}Ta_1$ subjected to different amounts of pre-deformation at 300 K.

Below 800 K the mechanical loss level is rather low and almost independent of temperature. Previous experiments [4, 5] have suggested that in order to observe any dependence of the mechanical loss with the temperature in the anomaly domain of the flow stress, two conditions have to be fulfilled: specimens should be pre-deformed and the measurement oscillation amplitude must be higher than 10^{-4}. This is clearly confirmed by the results of Figure 2 showing the mechanical loss at 300 K as a function of strain amplitude for an as-cast sample and for samples which have been twist pre-deformed in-situ to 1.0%, 3.4% and 5.6%. It can be seen that:

(1) for the as-cast sample the mechanical loss increases slowly with strain amplitude,

(2) the mechanical loss is positively dependent on the pre-deformation level and strongly increases with strain amplitude when this is higher than 1×10^{-4},

(3) for as-cast and pre-deformed samples the mechanical loss level is almost constant when the strain amplitude is below about 1×10^{-4}.

The increase in mechanical loss induced by room temperature plastic pre-deformation, has been interpreted as due to an increase in the mobile dislocation density [4, 5]. Transmission electron microscopy observations [16] have evidenced that dislocations in Ni_3Al usually exhibit a stepped configuration which consists of immobile Kear-Wilsdorf locks [17] connected by segments of a mixed character, called superkinks, lying on the (111) planes. The strong increase in mechanical loss when the strain amplitude is higher than 1×10^{-4} may be explained by the existence of a critical length for the motion of these superkinks. Through a simulation of dislocation configurations at lower stress, Mills and Chrzan showed that there is a distribution of superkink lengths [18]. For small segments, the probability of finding a kink whose length lies between l and $l + dl$ is well characterized by an exponential:

$$\rho(l) \approx \exp(-l/l_0) \qquad (2)$$

where l_0 is a parameter determined by experiments. Such a distribution is important for the dynamics of dislocation motion, since the superkinks are the only potentially mobile dislocation segments. In addition, according to these authors, the distinction between mobile and immobile kinks is defined by a critical length of the form:

$$l_c = (2\mu b/\tau_a) \sin(\theta_c) \qquad (3)$$

where θ_c is the critical bowing angle for dislocation depinning, μ is the shear modulus, b is the Burgers vector of the dislocation and τ_a is the applied stress. In general, dislocation segments shorter than l_c are static and the segments longer than l_c are mobile. So the number of mobile superkinks dislocations is only a fraction of the total number of superkinks. With the increase of applied stress, the critical length l_c is reduced, so some immobile segments at low stress become mobile at high stress. Furthermore, according to the exponential form of the superkink density, a small increase of applied stress will induce a larger increase of the density of mobile segments, and consequently a large increase in the mechanical loss. In addition, another contribution to the increase of the mechanical loss at high applied stress could be the multiplication of dislocations as described by Louchet [19]. The results reported in Figure 2 clearly show that above a critical strain of about 10^{-4}, the mechanical loss dramatically increases as a function of strain amplitude. Below this critical strain most of the dislocation segments are static, the applied stress can only activate the motion of a part of mixed dislocations located on the (111) plane even if many fresh dislocations have been created by the pre-deformation.

This mechanical loss behavior is in good correlation with the results obtained by Thornton et al. [20], who showed that the anomalous increase in flow stress with temperature only appears for strain amplitudes higher than about 10^{-4}.

Mechanical loss in the low temperature domain

In order to study the dislocation mobility at low temperatures, a $Ni_{75}Al_{25}$ specimen was subjected to different amounts of pre-deformation at room temperature and then cooled-down to 100 K. Figure 3 show the evolution during the subsequent heating of the mechanical loss and vibration frequency (which accounts for the shear modulus according to relation (1)). A broad mechanical loss maximum appears between 100 K and 550 K. An increase in the pre-deformation level results in an increase in the maximum height and in a softening of the material. At about 550 K, the mechanical loss level of the as-cast state is completely recovered, independently on the pre-deformation level. After heating above this temperature, the maximum completely vanishes and is not present during cooling or successive heatings.

Figure 4 shows the mechanical loss spectra of two $Ni_{75}Al_{24}Ta_1$ specimens pre-deformed at 300 K and 100 K respectively. One can see that the damping level is higher for the specimen pre-deformed at 100 K, and that both spectra have a similar shape with their maximum located at about the same temperature. The reversible behavior exhibited by the mechanical loss and shear modulus during low temperature thermal cycling experiments (Figure 5), indicates that the mobile dislocation distribution is stable below 300 K. On the contrary, if thermal cycling is performed above 300K, a gradual irreversible decrease in the damping level is observed, which has been attributed to the locking of dislocation segments by the formation of Kear-Wilsdorf locks [4, 5].

The origin of the broad mechanical loss maximum cannot be a single relaxation mechanism. The low temperature portion of the spectrum is certainly thermally activated. In cold worked pure fcc metals, Bordoni and Hasiguti peaks [3] appear at low temperature. Bordoni relaxation arises in the overcoming of the Peierls valleys by a mechanism of double kink generation, and Hasiguti relaxations are due to the interaction between dislocations and point defects. When the concentration of impurities increases, these peaks broaden and can overlap. In hcp metals such as Mg, it has been observed that the Bordoni peak is present when the purity is 6N, but if the purity is reduced to 4N, an extremely broad maximum appears from about 40 K to 280 K [21]. The mechanical loss increase at low temperatures observed in pre-deformed Ni_3Al and $Ni_3(Al,Ta)$ would be due to the movement of dislocations, which have to overcome the Peierls valleys and point defects. Because these are thermally activated relaxation mechanisms, a shift of the low temperature part of the spectrum with the oscillation frequency is expected. This has been confirmed by measurements performed with forced oscillations at frequencies between 0.01 Hz and 1 Hz, which allows one to obtain a preliminary value for the activation energy of about 0.2 eV. Pinning of dislocations by Ta impurities could explain the lower mechanical loss level observed in $Ni_{75}Al_{24}Ta$ specimens (see Figures 3a and 4).

A striking similarity exists between the mechanical loss maximum shown in Figure 3a and the one observed by Munier et al. [22] in pre-deformed ferritic steels, particurlarly concerning the maximum shape and the high strain amplitude which is necessary to observe such a phenomenon. Nevertheless, an important distinction in the behavior of both materials should be stressed. In ferritic steels, twist deformation of the specimen (which implies long range dislocation motion and annihilation) occurs only when damping increases. The damping decrease above 300 K has been associated with a drastic reduction in the dislocation mobility due to pinning by impurities, which can migrate at higher temperature. On the other hand, in Ni_3Al intermetallic compounds, twist deformation of the specimen occurs in the full temperature range of the mechanical loss maximum. In particular, above room temperature, dislocation motion increases the probability of cross slip. The mechanical loss decrease has been attributed to a reduction in the dislocation mobility, which originates not in the interaction with extrinsic defects, but in pinning by Kear-Wilsdorf locks [4].

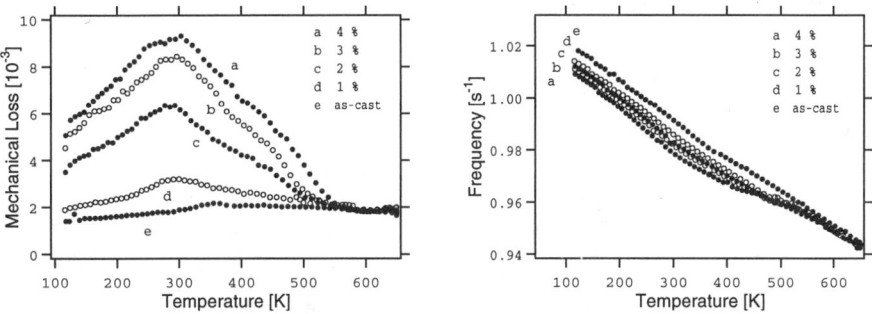

Figure 3a, b: Mechanical loss and vibration frequency of $Ni_{75}Al_{25}$ specimens subjected to different amounts of pre-deformation at 300 K. Strain amplitude: 5×10^{-4}, heating rate: 2K/min.

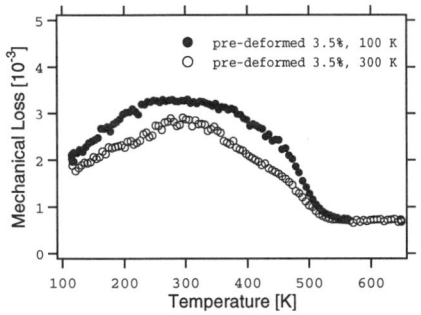

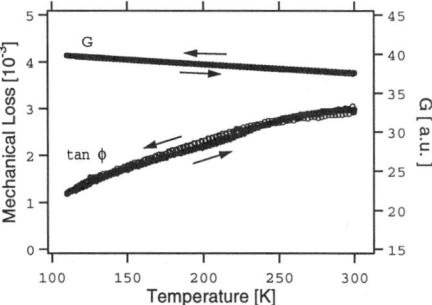

Figure 4: Mechanical loss of two $Ni_{75}Al_{24}Ta_1$ specimens pre-deformed at 300 K and 100 K. Strain amplitude: 5×10^{-4}, heating rate: 2K/min.

Figure 5: Low temperature mechanical loss and shear modulus measured during two thermal cycles for a $Ni_{75}Al_{24}Ta_1$ specimen pre-deformed 3.5% at 300 K. Forced vibrations, $f = 0.5$ Hz

SUMMARY

The mechanical loss of Ni_3Al and $Ni_3(Al,Ta)$ single crystals have been studied in the temperature range from 100 K to 1300 K. The damping spectra of both compounds show a Zener-type relaxation peak at about 950 K, the relaxation strength of which is higher in the non-stoichiometric material. It has been found that pre-deformation induces a strong increase in the room temperature mechanical loss which is certainly related to an increase in the mobile dislocation density. A broad maximum appears between 100 K and 550 K in pre-deformed specimens. At low temperatures the damping increase is due to the thermally activated movement of dislocations, which have to overcome the Peierls valleys and point defects. The mechanical loss reaches a maximum at about 300 K, followed by an irreversible decrease, which has been attributed to the pinning of dislocations by the formation of Kear-Wilsdorf locks.

ACKNOWLEDGEMENTS

The authors acknowledge the Fonds National Suisse de la Recherche Scientifique for its financial support.

REFERENCES

1. A. Mourisco, N. Baluc, J. Bonneville, and R. Schaller, Mater. Sci. Eng. A **239-240**, 281 (1997).
2. H. Numakura, N. Kurita, M. Koiwa, and P. Gadaud, Phil. Mag A **79**, 943 (1999).
3. A. S. Nowick and B. S.Berry, *Anelastic Relaxation in Crystalline Solids* (New York and London: Academic Press, 1972).
4. B. Cheng, E. Carreño-Morelli, N. Baluc, J. Bonneville, and R. Schaller, Phil. Mag A **79**, 2227 (1999).
5. B. Cheng, E. Carreño-Morelli, N. Baluc, J. Bonneville, and R. Schaller in *High-Temperature ordered Intermetallic Alloys VIII*, edited by E. P. George, M. J. Mills and M. Yamaguchi (Mater. Res. Soc. Proc. **552**, Boston, MA, 1998) pp. KK5.38.1-5.
6. P. Veyssière and G. Saada, in *Dislocations in Solids*, Vol. 10, edited by F.R.N. Nabarro and M.S. Duesbery (Elsevier, Amsterdam, 1996), p. 253.
7. P. Spätig, PhD thesis, Ecole Polytechnique Fédérale de Lausanne, Switzerland, 1995.
8. B. Lo Piccolo, PhD thesis, Ecole Polytechnique Fédérale de Lausanne, Switzerland, 1999.
9. T. Kruml, J.-L. Martin and J. Bonneville, Phil. Mag. A, *in press*.
10. M. Demura and T. Hirano, Phil. Mag. Let. **75**, 143 (1997).
11. D. Golberg, M. Demura, and T. Hirano, J. Crystal Growth **186**, 624 (1998).
12. I. Yoshida, T. Sugai, S. Tani, M. Motegi, K. Minamida and H. Hayakawa, J. Phys. E: Sci. Instrum. **14**, 1201 (1981).
13. J. Baur and A. Kulik, J. de Phys. **44**, C9-357 (1983).
14. S. Timoshenko, *Résistance des Matériaux* (Librairie Polytechnique Béranger, Paris, 1963), p. 276.
15. P. Gadaud and K. Chakib, Mater. Sci. Forum **119-121**, 397 (1993).
16. B. H. Kear and H.G.F. Wilsdorf, Trans. Metall. Soc. AIME **224**, 382 (1962).
17. M. J. Mills, N. Baluc and H. P. Karnthaler in *High-Temperature ordered Intermetallic Alloys III*, edited by C. T. Liu, A. I. Taub, N. s. Stoloff and C. C. Koch (Mater. Res. Soc. Proc. **133**, Boston, MA, 1989) pp. 203-208.
18. M. J. Mills and D. C. Chrzan, Acta Metall. Mater. **40**, 3051 (1992).
19. F. Louchet, Phil. Mag A **72**, 905 (1995).
20. P. H. Thornton, R. G. Davies and T. L. Johnston, Metall. Trans. **1**, 201 (1970).
21. S.M Seyed-Reihani, PHD Thesis, Université Claude Bernard Lyon I, France, 1981.
22. A. Munier, M. Maamouri, R. Schaller, and O. Mercier, J. Nucl. Mater. **202**, 54 (1993).

PROPERTIES OF THE DEFORMATION MICROSTRUCTURE IN Al-RICH γ-TiAl DEFORMED BY ORDINARY DISLOCATIONS

FABIENNE GRÉGORI*, PATRICK VEYSSIÈRE**
* LPMTM, Institut Galilée, Av J. B. Clément, 93430 Villetaneuse, France, fg@lpmtm.univ-paris13.fr
** LEM, CNRS-ONERA, BP 72, 92322 Châtillon cedex, France, patrickv@onera.fr

ABSTRACT

In the near vicinity of the [021] load orientation, γ-TiAl deforms via ordinary dislocations (Burgers vector $\mathbf{b}$ = 1/2<110]). As for deformation by <011] dislocations, the flow stress shows a peak at about 600°C. Results of an extensive microstructural investigation aimed at identifying the origin of this mechanical anomaly are presented. The analysis was conducted on single crystals oriented for single slip. It confirmed that ordinary dislocations tend to align themselves along the screw direction. This preferential line direction becomes gradually accentuated as the deformation temperature is raised up to the peak temperature. This effect is accompanied by a strong tendency towards forming cusps, but there is indication that the immobilisation along the screw direction takes place prior to dislocation pinning. In the vicinity of the peak temperature, screw dislocations gather in the form of bundles. No clear correlation is found between the temperature dependence of the flow stress and that of the density of pinning points. The relationship between these microstructural findings and the occurrence of a flow stress anomaly is discussed.

INTRODUCTION

γ-TiAl-based alloys are under active development for future use in automotive and aerospace applications. Amongst the many fundamental issues being addressed in the course of the current investigations, that of the mechanisms of plasticity continues to attract considerable attention in the intermetallics community. Since the microstructure of the alloy of technical interest includes two phases, i.e. $\gamma(L1_0) + \alpha_2(D0_{19})$, the general question raised is that of slip transmission through interfaces, in particular through the interfaces between the six variants of the γ phase. There exists however quite a variety of deformation microstructures for which a basic understanding of the behaviour of individual dislocations is required, and this is what, in the present and in the companion contribution [1], we are interested in.

The question, for instance, of the behaviour of ordinary dislocations remains largely open since adequate experimental conditions are difficult to achieve in order to ascertain properties that are intrinsic to the <110]{111} slip system. A good knowledge of ordinary dislocations is however important because, together with twinning, they are believed to dominate within the deformation microstructures of the two-phase alloys [2].

One difficulty in isolating characteristics of ordinary dislocations arises from the free flight distance between consecutive interfaces, which is small, thus preventing free operation of sources. Another difficulty is linked to interactions with other slip systems, such as of course <110}{111}, but also <011]{111} and twinning. This results in turn in the generation of jogs on dislocations that may partly obscure the intrinsic production of pinning events.

Properties of ordinary dislocations are nevertheless rather well documented. Analyses mostly deal, at a microscopic scale, with the characterisation of the locking abilities of this family of dislocations, and, at a macroscopic scale, with the relationship between that locking and the flow stress anomaly. Previous transmission electron microscopy studies have indeed established that ordinary dislocations behave under a remarkably unusual way in that, despite a markedly compact core, they tend to be preferentially aligned along their screw direction. On the other hand, the pinning is accompanied by a noticeable bulging of the free segments, and the corresponding curvatures are largely retained upon quenching the samples down to room temperature. These properties attest to a stable core in the screw orientation, and to a significant lattice friction, respectively. It is on these issues and in particular on the conditions of dynamical pinning, that the present paper is aimed at contributing to.

Mat. Res. Soc. Symp. Proc. Vol. 578 © 2000 Materials Research Society

EXPERIMENTS

Procedure

In order to identify properties that one can think of as being intrinsic to the <110]{111} slip system, we have employed single crystals which we have oriented so as to favour deformation under single slip. The orientation chosen is [153] which has near-maximum Schmid factor on [110](1̄11), associated with modest Schmid factor on the octahedral cross-slip plane (Table 1). Our original strategy was to investigate a possible correlation between the propensity towards cross-slip and the frequency of pinning, but that did not work because, as developed in the companion paper [1], the second orientation selected, [1 21 6], happened to activate [01̄1](1̄11) primarily.

As is imposed by the phase diagram, only Al-rich γ-TiAl can be grown under a single crystal form. With 54.3 at.% Al, the present composition is close to the smallest concentration achievable. The alloys were first homogenised at 1200°C during 100 h. They were then submitted to a pre-straining/annealing to attain several goals. The pre-straining is intended to adjust the sample compression faces to the compression rods. The subsequent heat treatment of 11 h at 900°C (under a finite load of 20 MPa) is aimed at annealing out the deformation substructure *and* at precipitating interstitial atoms (C, O, N) into large particles of H-phase. We have checked that the mean distance between H-precipitates is large enough in order not to contribute to sample strength. Because single crystals of γ-TiAl are scarce, only four deformation temperatures, i.e. room temperature, 400°C, 600°C and 800°C, could be explored. These temperatures were selected in order to include the peak of flow stress. The deformation tests were conducted at a constant strain rate of $5 \cdot 10^{-5}$ s^{-1}. The samples were all deformed to a permanent strain of 2%.

Foils suitable for TEM investigations were thinned by procedures usual to γ-TiAl, from slices cut parallel to the slip plane.

Table 1. List of Schmid factors on the various slip systems of interest for the various load orientations thought to activate <110]{111}. Also indicated are the Schmid factors of the most activated <011]{111} slip system and those for [1 21 6] used in the companion paper. The line $F_{<110]\{111\}}/F_{<011]\{111\}}$ indicates the ratio between the maximum Schmid factors of the two systems for each orientation.

slip system	[153]	[021]$^{[0]}$	[010]$^{[3]}$	[1̄32]$^{[3]}$	[1 21 6]$^{[0]}$
[110](1̄11)	0.490	0.489	0.406	0.357	0,488
[110](11̄1)	-0.061	-0.154	-0.406	-0.110	
[110](001)	0.367	0.286			
[110](11̄0)	-0.339	-0.397			
[11̄0](111)	-0.418	-0.489	-0.406	-0.467	
[11̄0](111̄)	-0.132	-0.154	-0.406	-0.009	
$F_{<110]\{111\}}/F_{<011]\{111\}}$	1.47	1.94	1.01	1.31	1.35
[101](1̄11)	0.333	0.252			0.159
[101](11̄1)					0.096
[1̄01](111)		0.252			0.124
[1̄01](11̄1)				0.356	-0.060
[011](11̄1)			-0.402		-0.314
[011](11̄1̄)			0.402		0,360
[011̄](111)			0.402		0,349
[011̄](1̄11)			0.402		0.324

Deformation tests

The temperature dependence of the flow stress is plotted in Figure 1. As anticipated it shows a peak whose position is however not ascertained within better than say ± 50°C. The critical resolved shear stress increases steadily from a temperature as low as room temperature. Up to 400°C, the level of critical resolved shear stress compares fairly well with those determined by Kawabata et al [3]. Beyond that temperature, the agreement is not as good. It is remarked that in by Inui et al's experiments [4] the critical resolved shear stresses are more than a factor of two larger, and the flow stress peak is located about 200°C above that found in Kawabata et al. and in the present studies. We have determined that those significant variations are due to differences in annealing procedures prior to deformation. Since in Al-rich γ-TiAl, the CRSS for ordinary slip is significantly larger than that for <011]{111}, it is unlikely that the [010] load orientation is capable of activating ordinary dislocations (the maximum Schmid factors on <110]{111} and <011]{111} are indeed too close; $F_{<110]\{111\}}/F_{<011]\{111\}}$ in Table 1).

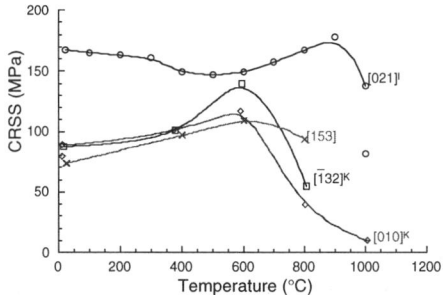

Figure 1. Temperature dependence of the critical shear stress resolved on [110]($\bar{1}$11) for the various load orientations thought or proved to activate this slip system. Note that it is in the present experiments that the anomaly is the least pronounced (K : [3]; I : [4]).

The stress-strain curves exhibit a normal behaviour with the exception of the curve at 800°C that first exhibits a yield peak, then modest work-hardening rate (μ/170, where μ is the shear modulus). Below 800°C, the work-hardening rate ranges between μ/20 and μ/50, depending on temperature and on the strain at which this rate is measured. Within the experimental uncertainties, it exhibits no anomalous temperature dependence. At 400°C there is indication of serrated flow whose amplitude, of the order of 1 MPa, is definitively distinct from experimental noise.

TEM OBSERVATIONS

Influence of temperature on deformation organisation

For the three samples deformed in the ascending region of the flow stress, we have checked that the slip system anticipated, i.e. [110](111), is actually the only activated one. Another general observation is that the microstructures are sufficiently distinct for the test temperature to be unambiguously identified from direct inspection of dislocation arrangements in foils.

The effect of temperature on deformation microstructures is as follows.

At room temperature, ordinary dislocations form tangles that are reminiscent of dislocation organisation in high stacking fault energy fcc alloys, such as aluminium. Indication of a preference for the screw orientation is encountered in places, but this is by far not representative of the microstructure at this temperature. When in the screw orientation, ordinary dislocations may be decorated by pinning points.

At 400°C, screw ordinary dislocations show a pronounced tendency towards bulging, so that the alignment with the screw orientation is true only on average. The bulging is in addition

189

highly heterogeneous as a result of the heterogeneity of the distribution of pinning points.

At 600°C, the preference for the screw orientation is even accentuated since the fraction of mixed segments (that do not pertain to bulges) is significantly less than after deformation at lower temperatures. The main microstructural signature at this temperature is the distribution of bundles of parallel screw dislocations, mostly of the same sign and sometimes all of the same sign. Within a bundle, the pinning point density is rather homogeneous while the bulging increases dramatically from one edge to the other edge of the bundle.

At 800°C, the two possible 1/2<110] Burgers vectors coexist in spite of large differences in Schmid factors (Table 1). Dislocations show frequent evidence of climb activity. The dislocation density is much less than in the microstructures of deformation at lower temperatures, and it shows a fair homogeneity, both features attesting to efficient climb-assisted recovery.

In our opinion, the phenomena observed to take place at 800°C are sufficient to explain the decrease of critical resolved shear stress, and this is why we now concentrate on the microstructures at the other deformation temperatures.

Comparison with avalaible models of the flow stress anomaly

What distinguishes 1/2<110] dislocations in γ-TiAl from dislocations in other fcc-based ordered or disordered systems is their remarkable propensity towards forming pinning points which is at the origin of one of the explanations of the flow stress anomaly. Louchet and Viguier ascribe the positive temperature dependence of the flow stress to an increased pinning frequency — itself interpreted in terms of single cross-slip on the octahedral plane [5] — and they produce an analytical model that reproduces fairly well the anomaly [5]. They however disregard two observations carried out in polycrystalline samples [6]. One is that the samples deform by a mixture of ordinary dislocations and twinning, to the extent that 75% of the grains contain microtwins at 600°C. More serious is that the pinning point density is a maximum at a temperature located 200°C *below* the temperature where the flow stress peaks.

The present experiments on single crystals deformed in single slip enable us to alleviate two difficulties. Firstly, the applied stress tensor is homogeneous which makes the measurement of densities of a cross-slip controlled mechanism less uncertain. Another difficulty, related to the mechanism of pinning, is that the contribution of intersecting slip systems to the production of jogs is eliminated. Based on a collection of a total of several thousands of pinning points at three temperatures, we have confirmed that the pinning point density does peak at a temperature located well below that of the flow stress peak [7]. It thus appears very difficult to relate in a simple manner the mechanical behaviour in γ-TiAl deformed by ordinary dislocations, to properties of pinning.

Another explanation of the flow stress anomaly has been recently proposed by Sriram et al [8]. It was proven that the pinning operates by double cross-slip and the explanation assumes that the mean free path on the cross-slip plane increases with temperature thus producing jogs of increasing heights. The pinned dislocations are regarded as still contributing to deformation, whose motion requires that the applied stress be increased in order to drag dipolar defects when the height of these increases. In this approach, the pinning point density is not of first order importance. What single crystal samples may nevertheless help elucidate is the relationship between dipolar debris and/or elongated loops and pinning points. In foils sliced parallel to the slip plane, one can indeed directly correlate these with debris. By contrast, polycrystalline samples usually provide situations where the slip plane is inclined to the foil with the consequence that supposedly trailed defects are largely eliminated in the course of the thinning procedure. The observations which we have conducted indicate that the density of debris is by far much less than that of the pinning points. It is very rare that we could observe configurations where an elongated dipole or a loop row is aligned with a given pinning point. Furthermore, we do not support the pivotal assumption of the model that the mean jog height, thus the average dipole height, actually increases with temperature. As a matter of fact, one may question why this should be the case since the mean free path in the cross-slip plane results from the contribution of two effects with opposite temperature dependencies. One is the mobility of ordinary dislocations that increases with temperature. The other is the cross-slip probability, also expected to increase with temperature, thus decreasing the free-flight time between the two

190

cross-slip events. Finally, it should be mentioned that we have direct evidence of pinned ordinary dislocations whose bulges could be set on their edge in projection, an evidence that fully support the occurrence of double cross-slip. The pinned dislocations consist of a succession of polarised bulges that assume a stair-like configuration. What we mean by 'polarised' is that the stair is either ascending or descending, a configuration which we have been unable to explain so far. From these findings, we are led to conclude that the mechanistic explanation provided by Sriram et al [8] which, as that of Louchet and Viguier, provides the correct flow stress temperature dependence, cannot be supported by sufficiently solid experimental grounds.

Additional local properties of ordinary dislocations

In regard to locking properties of ordinary dislocations, the most interesting features are to be found in the foils deformed at 400°C and 600°C. We rely on extensive microstructural maps whose size, of the order of 30 μm by 30 μm, is representative of the wholesale dislocation organisation at each temperature. In regard to the limited space available in these proceedings we cannot provide the appropriate experimental illustrations.

At 400°C, the microstructure consists of long, cusped screw dislocations that tend to gather in bundles in places. These usually contain at most 5-6 dislocations, often less, of both signs (in a bundle say 1μm wide) separated by regions, 1-3 μm wide, themselves void of screw dislocations. Rather unexpected is the observation that albeit seemingly homogeneous, one bundle or group of bundles may differ from others located 15-20 μm apart in terms of the local properties of pinning. In certain regions indeed, the majority of dislocations exhibit a pronounced bulging and the pinning point density is relatively moderate (8-12 per μm) while in other regions the bulging is modest with an increased pinning point density (in fact almost doubled, up to 20 per μm). Interestingly, within those dislocations that show a large density of small bulges, the pinning points are aligned over 8-10 μm with the [110] direction. In one instance, a dislocation that could be followed uninterruptedly over 8 μm, contained as many as 155 pinning points all rigorously aligned with the screw direction. There are even regions where neighbouring screw dislocations exhibit totally flat lines, of course within the resolution provided by the weak-beam method.

We regard the alignments of pinning points as very difficult to explain in terms of the dynamical cross-slip pinning of mobile screw segments. It is actually unclear as to why a mechanism which is thermally activated in essence, would not give rise to a scatter of kink heights. Even more surprising in the frame of cross-slip locking, is the fact that some screw segments, rectilinear over 3-4 μm thus solidly locked, are apparently void of pinning points.

Hence, after deformation at 400°C, the pinning point distribution is apparently governed by two simple rules: (i) it is evenly distributed all over a given dislocation line, (ii) it is about the same within neighbouring dislocations.

At 600°C, the situation bears strong similarities with the above, in that dislocations are still markedly pinned all over their lines. The density of mixed segments is however much less, indicating an increased difference in mobilities between mixed and screw segments. This is somewhat unexpected since the curvature of the bulges retained in post mortem samples yields values of the friction stress on mixed dislocations amounting to 100-120 MPa, that is, of the order of the critical resolved shear stress (Figure 1). Another interesting and reproducible signature of deformation at 600°C is that ordinary dislocations tend to gather in bundles comprising dislocations in majority of the same sign. This implies that dislocation organisation does not tend to minimise internal stresses. This constitutes another paradox of the deformation microstructures in γ-TiAl.

In view of the above elements, we suggest an alternative — although still incomplete — interpretation for the behaviour of ordinary dislocations. That these can be remarkably flat yet locked, indeed suggests that in γ-TiAl, screw dislocations have an intrinsically stable core in consistency with the work of Simmons et al. [9]. That the pinning point density is constant over a given line and that this density can vary dramatically from one region to the next in addition suggests that the pinning and the bulging occur as the result of an ageing process.

These conclusions are supported by two additional properties. On the one hand, there is frequent evidence that mixed segments are pinned. This property previously reported for instance by Zghal et al [10] in the course of in situ observations, implies that cross-slip is not necessarily the unique origin of dislocation pinning. On the other hand, we have found in a polycrystalline sample deformed at 400°C, that when new dislocations are created during thin foil manipulation at room temperature, these dislocations cannot be differentiated from pre-existing ones [7]. The fact that, while moved at room temperature, these show about the same configuration in terms of pinning points and bulging, as dislocations generated at 400°C, casts some doubt as to the intrinsic nature of the supposedly thermally-activated pinning mechanism.

CONCLUSION

The mechanical properties of Al-rich γ-TiAl are highly dependent on prior thermal treatments. When conducted on single crystal samples deformed in single slip, a detailed analysis of the configurations formed at various temperatures yields a significantly different view of the locking properties of ordinary dislocations. There is indeed convincing evidence that ordinary dislocations can lock themselves *before* undergoing pinning and that the latter process, which decorates the majority of dislocations present in the microstructure, takes place on dislocations once these are stopped.

Many questions remain to be clarified. An intriguing paradox is why, at variance with other systems with deep Peierls valleys such as silicon, it is only beyond a certain temperature that dislocations with 1/2<110> Burgers vectors begin to show a loss of mobility and why this mobility decreases with increasing temperature. One would indeed expect that dislocations become gradually more mobile as thermal activation helps. This raises the questions as to why the core of ordinary dislocations, known to be significantly compact, should exhibit a sessile configuration (however see [9]), and why this core has a window of existence in temperature. The general idea that we are currently exploring is that ordinary dislocations possess two core configurations. One is glissile and metastable, the other is sessile and stable. The latter is attained from the former state through a thermally-activated process. So far there is little indication from numerical simulations as to what the core configuration could be; the answer to this question should account for the fact that the propensity towards forming screw locks is rather independent of alloy composition, that is, whether the alloy is Al-rich or Ti-rich. As pinned screw dislocations do not trail as many loops or dipoles than they carry pinning points, we believe that once stopped, they are irreversibly immobile and do not contribute to deformation. By analogy with $L1_2$ alloys, the flow stress anomaly would result from two factors (i) a modest multiplication rate and (ii) an increasing exhaustion rate with increasing temperature. Of course such a macroscopic mechanical behaviour can be readily accounted for by a phenomenological analytical model that includes the appropriate rates with judiciously adjusted parameters. But, this approach would provide no added insight into the mechanisms that take place in the material under investigation and for which we definitely need a physically sound explanation.

REFERENCES

1. F. Grégori and P. Veyssière, in *Multiscale phenomena in materials — Experiments and modeling*, edited by B. Devincre, D. Lassila, R. Philips and I. Robertson (Mater. Res. Soc. Proc., Pittsburgh, PA 2000) this volume.
2. F. Appel and R. Wagner, Mater. Sci. Engng., **R22**, p. 187-268 (1998).
3. T. Kawabata, T. Kanai, and O. Izumi, Acta Metall., **33**, p. 1355-1366 (1985).
4. H. Inui, M. Matsumoro, D.-H. Wu and M. Yamaguchi, Phil. Mag. A, **75**, p. 395-423 (1997).
5. F. Louchet and B. Viguier, Phil. Mag. A, **71**, p. 1313-1333 (1995).
6. B. Viguier, K.J. Hemker, J. Bonneville, F. Louchet and J.-L. Martin, Phil. Mag. A, **71**, p. 1295-1312 (1995).
7. F. Grégori and P. Veyssière, in *Gamma Titanium Aluminides 1999*, edited by Y-W. Kim, D.M. Dimiduk and M.H. Loretto (TMS, Warrendale, PA, USA, 1999) pp. 75-82.
8. S. Sriram, D.M. Dimiduk, P.M. Hazzledine and V.K. Vasudevan, Phil. Mag. A, **76**, p. 965-993 (1997).
9. J.P. Simmons, S.I. Rao and D.M. Dimiduk, Phil. Mag. A, **75**, p. 1299-1328 (1997).
10. S. Zghal, A. Menand and A. Couret, Acta Mater., **46**, p. 5899-5905 (1998).

DEFORMATION OF Al-RICH γ-TiAl BY <011] DISLOCATIONS OVER THE DOMAIN OF FLOW STRESS ANOMALY

FABIENNE GRÉGORI*, PATRICK VEYSSIÈRE**
* LPMTM, Institut Galilée, Av. Clément, 93430 Villetaneuse, France, fg@lpmtm.univ-paris13.fr
** LEM, CNRS-ONERA, BP 72, 92322 Châtillon cedex, France, patrickv@onera.fr

ABSTRACT

The flow stress of Al-rich γ-TiAl peaks at a temperature located between about 600°C and 1000°C. We present mechanical data on samples oriented so as to deform on a <011]{111} system in single slip. These data are complemented by TEM observations in samples strained at selected temperatures (RT, 400°C, 600°C and 800°C). The present paper addresses (i) properties of dislocation organisation, (ii) the dissociation mode of <011] dislocations, (iii) locking properties of <011] dislocations and (iv) the formation of faulted dipoles.

INTRODUCTION

γ-TiAl-based alloys deform essentially by motion of dislocations with <011] Burgers vector [1]. Within the standard [001]-[010]-[$\bar{1}$10] triangle appropriate for the L1$_0$ ordered structure, there is however a narrow domain of load orientation, surrounding [021], where the alloy deforms by ordinary dislocations (**b** = 1/2<110]). At variance from these (see companion paper [2]), <011] dislocations exhibit a dissociated core which is resolved by traditional TEM methods. This core is complex and at the origin of quite a variety of configurations and reactions. Hence, investigating the microstructural cause of the flow stress anomaly under <011]{111} slip requires that deformation debris be reasonably understood in order to sort out whether or not these may take part in controlling the strain rate. The main goal of this work is to elucidate the origin of the exhaustion rate, which is expected to be potent. The present paper summarises the results of an extensive investigation on those questions [3].

EXPERIMENTS

For the sake of comparison with the [153] orientation that has but modest Schmid factor on the octahedral cross-slip plane (Table 1 in [2]), deformation was conducted along [1 21 6] so as to favour <110]{111} slip, creating a significant Schmid factor on the octahedral cross-slip plane. However, this orientation revealed a little too far from the [021] direction, thereby activating <011]{111} slip. Fortunately, only one <011] direction, [011], was operative and this occurred in the (111) plane. Four samples were deformed at RT, 400°C, 600°C and 800°C. The temperature dependence of the flow stress is plotted in Figure 1. It shows a plateau-like behaviour up to about 400°C, then it increases to peak at about 600°C. Similar to <110]{111} slip [2], the magnitude and the position of the peak of the <011]{111} slip system are highly dependent on prior thermal treatments. This is indicated by clear differences between the results of Inui et al [1], obtained on as-grown crystals, and those determined by other groups [4-5] on samples deformed following various heat treatments. That particular question is more specifically discussed in [2-3, 6].

Another analogy with <110]{111} slip is that the stress-strain curves show normal behaviour, with the exception of the curve at 800°C that exhibits a yield peak followed by a modest work-hardening rate, if any [3]. The mechanical response at 800°C lies actually beyond the scope of the present investigation since it was determined that <011]{111} is suppressed and that, as for under the [153] orientation, deformation proceeds by means of two <110]{111} slip systems with orthogonal slip directions.

Below 800°C, depending on temperature and on the strain at which this rate is measured, the work-hardening rate ranges between μ/20 and μ/50 (where μ is the shear modulus). Within the experimental uncertainties, it exhibits no clear anomalous behaviour. At variance with what occurs under <110]{111} slip [2], there is no indication of serrated flow.

193

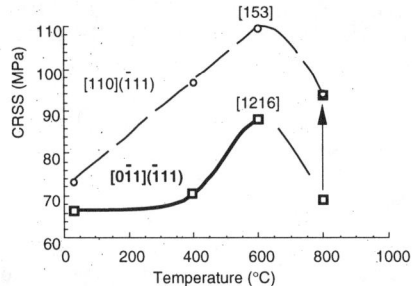

Figure 1. Temperature dependence of the critical shear stress resolved on $[0\bar{1}1](\bar{1}11)$ for the various load orientations thought or proved to activate this slip system. At 800°C, when resolved on $<110]\{111\}$ instead of $<011]\{111\}$ (arrow), the CRSS shows excellent coincidence with measurements conducted under the [153] load orientation.

TEM OBSERVATIONS

The $[0\bar{1}1](\bar{1}11)$ slip system is activated in the three samples deformed in the ascending region of the flow stress. As for $<110]\{111\}$ slip [2], the microstructures are sufficiently distinct for the test temperature to be identified from direct inspection of dislocations in thin foils.

The features which, either by their presence or by their absence, participate to the signature of the deformation temperature under $<011]\{111\}$ are :
- the coexistence of dislocations with 1/2<110] and 1/2<112] Burgers vectors,
- screw locks on <011] dislocations,
- the dissociation mode of glissile <011] dislocations,
- the trailing of faulted dipoles.

The temperature dependencies of these properties are summarised in Table 1.

Just as for <110] slip, the most noticeable characteristic is the increasing frequency of screw locks with increasing deformation temperature. It is nevertheless straightforward to discriminate between <110] and <011] slip since <011] dislocations would appear as screw segments interrupted in places by kinks, while 1/2<110] dislocation locks are markedly bulged, thus aligned only on average with the screw orientation.

As first suggested for γ-TiAl by Hug et al [7], the fact that <011] dislocations form screw locks is very likely to be at the origin of the flow stress anomaly. The conditions under which this takes place, that is, the competition between dislocation exhaustion and multiplication, should be along the same lines as in the $L1_2$ parent system (for a review see [8]).

Table 1. The principal characteristics of the deformation microstructure in the positive part of the flow dependence upon temperature.

	RT	400°C	600°C
1/2<112] & 1/2<110] dislocations	*Frequent*	*Mostly 1/2<112]*	*Both absent*
Screw locks on <011] dislocations	*Few*	*Frequent*	*Dominant*
Dissociation mode of <011] dislocations	*Asymmetrical*	*Asymmetrical Symmetrical in places*	*Increased frequency of symmetrical*
Faulted dipoles	*Profuse*	*Occasional*	*Short and seldom*

194

On the other hand, the conditions under which cross-slip locking occurs and the resulting locked configuration are issues that deserve some attention. For the sake of consistency, we first consider the split structure of mobile dislocations, i.e. in-plane dissociation. Since the so-called complex stacking fault (CSF) energy is large, the dissociation scheme of a <011] dislocation is in practice at most threefold and can be written

$$<011] \rightarrow 1/2<011] + APB + 1/6<121] + ISF + 1/6<112] \qquad (1)$$

(should the CSF energy be moderate, the left-hand side partial would be also dissociated into two Shockley partials). For reasons not clearly related to sample composition and/or deformation temperature (for a review see [9]), <011] dislocations appear either two- or threefold in their mixed parts. In the present investigation, no evidence of a threefold splitting has been observed but in the screw orientation which we shall consider separately. As far as mixed segments are concerned, two alternative reactions can take place

$$<011] \rightarrow 1/2<011] + APB + 1/2[011] \qquad (2)$$
$$<011] \rightarrow 1/6<112] + ISF + 1/6<154] \qquad (3)$$

which are referred to as symmetrical and asymmetrical, respectively. These reactions, it is worth noting, are included in reaction (1). We have shown elsewhere [9] that in order for the dissociation mode to be unambiguously determined, one has to rely on a pair of weak-beam pictures taken (i) with $\mathbf{g}$ of the {022} type and parallel to $\mathbf{b}$, and (ii) under two opposite values of the parameter $p = s_g\, \mathbf{g}.\mathbf{b}$ (i.e. $\pm s_g$ or $\pm \mathbf{g}$). Reaction (2) yields asymmetrical two-peaked images whose asymmetry is reversed upon changing the sign of p, while reaction (3) gives rise to either a single-peaked or a double-peaked image, then the two peaks show about the same intensity. We have found that as test temperature increases from RT to 600°C, <011] dislocations undergo a gradual transition from mode (3) to mode (2). No evidence other than of mode (3) was encountered after deformation at RT, whereas after deformation at the peak temperature, 600°C, a given dislocation shows evidence for both modes. We show in the following that this dissociation transition determines the temperature dependence of almost every remarkable deformation feature.

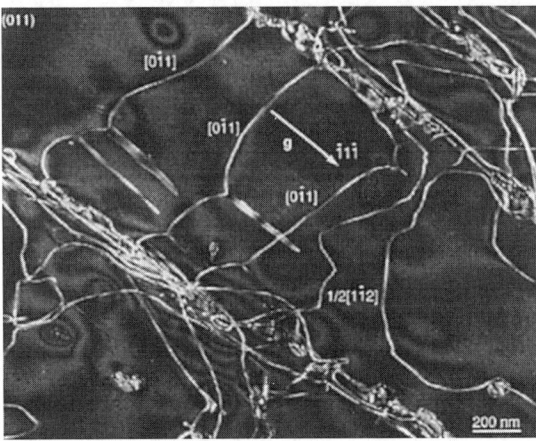

Figure 2. The microstructure after deformation at room temperature, showing the coexistence of dislocations with [011] and 1/2[112] Burgers vectors (the foil, initially sliced parallel to the (111) activated slip plane was tilted so as to set the beam along the [011] direction). Ordinary dislocations ($\mathbf{b} = 1/2[110]$) are out of contrast under this particular reflection. The rarity of screw [011] dislocations and the trailing of faulted dipoles by [011] dislocations should be noted.

By consideration of the fine structure of their contrast in relation with the remaining curvature of <011] moving dislocations, one can in addition demonstrate that in dissociation reaction (3) it is the 1/6<154] dislocation that sits in the trailing position. This is of importance for two reasons. One is the dynamical formation of screw locks by cross-slip of a 1/2<011] partial, which thus cannot occur at the leading Shockley partial. Another reason, not independent of the former is the nucleation of faulted dipoles that we address below in more detail.

Influence of temperature on the production of dislocations with b = 1/2<112] and 1/2<110]

The relative abundance of 1/2<112] and 1/2<110] dislocations decreases with increasing temperature (Table 1). The use of single crystals deformed under a single <011]{111} slip system enabled us to prove that in practice no 1/2<112] and 1/2<110] Burgers vectors other than those belonging to the same slip plane as that of the <011] dislocations, are encountered (Figure 2). In terms of Schmid factors, this is consistent with the observed Burgers vectors for ordinary dislocations. By contrast, 1/2<112] dislocations that belong to planes other than ($\bar{1}$11) are nearly twice as highly stressed as the 1/2<112] dislocations actually generated during deformation. This together with the occurrence of a number of junctions between the three dislocation families led us to conclude [9] that the presence of 1/2<112] and 1/2<110] dislocations should not proceed from multiplication at sources, but rather from the decomposition of <011] dislocations

$$<011] \rightarrow 1/2<110] + 1/2<112] \qquad (4)$$

A 1/2<110] dislocation can escape from the 1/6<154] trailing partial with the companion partial remaining unaffected as in

$$1/6<112] + ISF + 1/6<154] \rightarrow 1/6<112] + ISF + 1/3<112] + 1/2<110] \qquad (5)$$

In other words, decomposition is encouraged by reaction (3) whereas it is topologically difficult when dissociation occurs according to reaction (2). This remark is fully consistent with the nearly complete disappearance of 1/2<112] dislocations after deformation beyond 400°C.

An interesting consequence of the above interpretation is that despite the often reported profuse evidence of 1/2<112] dislocations in γ-TiAl, there might be but little overall activity, if any, of the corresponding slip system. In particular, rather than a finite multiplication rate, 1/2<112] dislocations would mark the trace of the unzipping of ordinary dislocations from <011] dislocations.

The trailing of faulted dipoles

As exemplified in Figure 2, faulted dipoles (FDs) are usually trailed by <011] dislocations, thus at variance from the frequently reported observation of FDs trailed by 1/2<112] dislocations. In fact both observations are compatible since upon escaping from a FD-trailing <011] dislocation (reaction (4)), an ordinary dislocation leaves a sessile 1/2<112] dislocation on which the pre-existing FDs should remain hanged [10].

We have shown that nucleation of FDs takes place in the course of the unzipping of a screw lock by a mobile <011] dislocation, at that particular extremity of the lock located behind the mobile dislocation [11]. A few typical examples of the relationship between the presence of a screw lock and the nucleation of a FD are provided in Figure 3. Nucleation is made possible because of the asymmetrical dissociation reaction (3), with the 1/6<154] partial in the trailing position. It is a consequence of the transition between a 3-D and a 2-D core configuration, at the screw lock and the mixed connecting segment, respectively. What occurs is that by continuity, the 1/6<154] is locally extended on two adjacent planes, with the upper part topologically influenced by the core structure of the cross-slipped 1/2<011] partial. This ensues in a core structure appropriate for the nucleation of a 1/6<1$\bar{1}$2] partial on the {111} plane adjacent to the ISF of reaction (3). As for the decomposition reaction, the disappearance of FDs with increasing test temperature can be, at least partly, correlated with the change in dissociation mode, from reaction (3) which favours the generation of a ESF-bordering zonal core, to reaction (2).

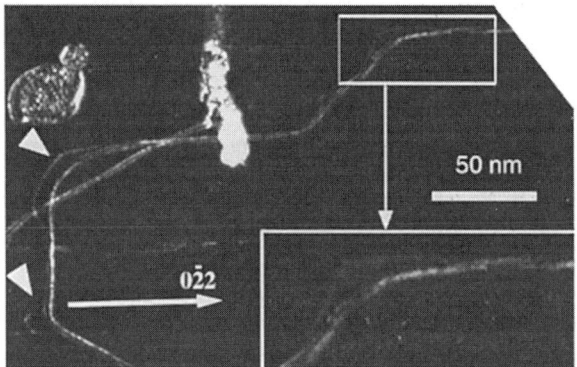

Figure 3. A dislocation with [0$\bar{1}$1] Burgers vector in γ-TiAl deformed at RT. The overall segmented aspect of the dislocation, with segments parallel the screw direction, should be noted. Indicated by arrowheads are faulted dipoles in the process of being nucleated. The detail of the boxed lock tip, enlarged in the insert, shows a threefold dissociation.

Influence of temperature on locks

An illustration of the dramatic effect of temperature on dislocation aspect is given in Figure 4. It shows that a large fraction of the dislocation density has locked in the screw orientation.

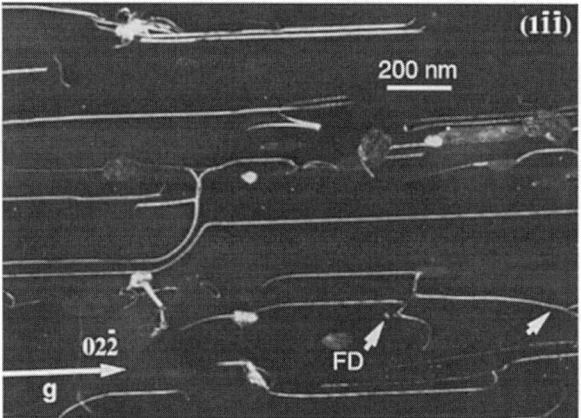

Figure 4. The microstructural organisation in γ-TiAl deformed at 600°C, is characterised by an increased density of rectilinear screw segments, themselves interconnected by mixed segments. Some short FDs are visible in places (FD arrowed).

The weak-beam method is sufficient to resolve the fine structure of screw locks. In the present Al-rich samples one finds that this fine structure is twofold at room temperature, and becomes clearly threefold at 400°C and then adopts again a twofold configuration at 600°C.

The high temperature twofold lock is compatible with the fact that dislocations tend to adopt a symmetrically dissociated configuration (reaction (2)).

The transition from twofold at RT to threefold at 400°C can be explained in two different ways both consistent with the fact that the lock is a 3D configuration (reaction (1)) that results from the cross-slip motion of a 1/2<011] partial off the slip plane. In one explanation, it is lattice friction in the cross-slip plane that would control the observed behaviour. Far too large at room temperature, lattice friction prevents the 3D configuration from reaching equilibrium and the APB strip remains very narrow. As temperature is increased, lattice friction is overcome by thermal agitation and the cross-slipped distance increases, and gives rise to an APB ribbon that can be resolved under weak-beam contrast. In favour of this explanation is the observation of two different dissociated configurations on the same screw lock (not shown here), one twofold and the other clearly threefold. An alternative explanation could be that the APB energy and maybe the ISF energy are temperature dependent. Although consistent with the transition in dissociation mode, from reaction (3) to reaction (2), this explanation is difficult to conceive since order in γ-TiAl is believed to be rather stable up to the melting temperature. The possible influence of the Ti_3Al_5 short-range ordered phase may be worth considering. This phase, whose formation is exclusive to Al-rich γ-TiAl, takes place in the 500-600°C range [12]. Further work on the temperature dependence of the fine structure of <011] dislocations is clearly needed.

CONCLUSION

When conducted on single crystal samples deformed in single slip, a detailed analysis of the configurations formed at various temperatures yields a significantly different view of the locking properties of <011] dislocations.

After deformation at RT, <011] dislocations are clearly dissociated asymmetrically into two partials with different Burgers vectors. This dissociation mode encourages the formation of debris such as faulted dipoles and 1/2<112] dislocations. It is not incompatible with the formation of screw locks, it is not really favourable either since the locking must occur dynamically on the trailing partial. After deformation at 600°C, dissociation is more frequently symmetrical, into two partials with collinear Burgers vectors. Just as in L1$_2$ alloys, this mode encourages the locking in the screw direction. In Figure 1 the RT-400°C range corresponds to the domain of predominance of the asymmetrical dissociation mode which, as suggested by the plateau-like behaviour of the flow stress, is rather inefficient with regard to locking. Beyond 400°C, the symmetrical mode is sufficiently active in order to promote extensive dislocation immobilisation. The physical origin of the dissociation transition is essentially unknown and the possible role of the short-range ordered Ti_3Al_5 phase should be put into perspective.

In regard to the anomalous temperature dependence of the flow stress, there is in principle no difficulty since what one looks for in this case is a locking mechanism. It looks like the screw locks are, just like in L1$_2$ alloys, sufficiently stable to make dislocation exhaustion increasingly potent relative to multiplication. Then, in order to reproduce the macroscopic behaviour, it is only a matter of writing a rate equation with appropriately adjusted parameters.

REFERENCES

1. H. Inui, M. Matsumoro, D.-H. Wu and M. Yamaguchi, Phil. Mag. A, **75**, p. 395-423 (1997).
2. F. Grégori and P. Veyssière, in *Multiscale phenomena in materials — Experiments and modeling*, edited by B. Devincre, D. Lassila, R. Philips and I. Robertson (Mater. Res. Soc. Proc., Pittsburgh, PA 2000) this volume.
3. F. Grégori, PhD thesis, Université Paris VI, 1999.
4. T. Kawabata, T. Kanai, and O. Izumi, Acta Metall., **33**, p. 1355-1366 (1985).
5. S. Jiao, N. Bird, P.B. Hirsch and G. Taylor, Phil. Mag. A, **78**, p. 777-802 (1998).
6. F. Grégori and P. Veyssière, in *Gamma Titanium Aluminides 1999*, edited by Y-W. Kim, D.M. Dimiduk and M.H. Loretto (TMS, Warrendale, PA, USA, 1999) pp. 75-82.
7. G. Hug, A. Loiseau and P. Veyssière, Phil. Mag. A, **57**, p. 499-523 (1988).
8. P. Veyssière and G. Saada, in *Dislocations in Solids*, edited by F.R.N. Nabarro and M.S. Duesbery (Elsevier, Amsterdam, The Netherlands, 1996), pp. 253-441.
9. F. Grégori and P. Veyssière, Phil. Mag. A, submitted 2000a.
10. B. Viguier and K.H. Hemker, Phil. Mag. A, **73**, p. 575-599 (1996).
11. F. Grégori and P. Veyssière, Phil. Mag. A, submitted 2000b.
12. T. Nakano, K. Matsumoto, T. Seno, K. Oma and Y. Umakoshi, Phil. Mag. A, **74**, p. 251-268 (1996).

STUDY OF THE MECHANICAL BEHAVIOR OF BCC TRANSITION METALS USING BOND-ORDER POTENTIALS

M. MROVEC[1], V. VITEK[1], D. NGUYEN-MANH[2], D. G. PETTIFOR[2], L. G. WANG[3] and M. SOB[3]

[1]Department of Materials Science and Engineering, University of Pennsylvania, Philadelphia, PA 19104-6272, U.S.A.
[2]Department of Materials, University of Oxford, Parks Road, Oxford OX1 3PH, U.K.
[3]Institute of Physics of Materials, Academy of Sciences of the Czech Republic, Zizkova 22, Brno, Czech Republic

ABSTRACT

Deformation properties of body-centered-cubic transition metals are controlled by the core structure of screw dislocations and their studies involve extensive computer simulations. In this paper we present the recently constructed bond-order potentials (BOP) that are based on the real-space parametrized tight-binding method. In order to examine the applicability of the potentials we have evaluated the energy differences of alternative structures, investigated several transformation paths leading to large distortions and calculated phonon dispersions. Using these potentials we have calculated γ-surfaces that relate to the dislocation core structures and discuss then the importance of directional bonding in studies of dislocations in transition metals.

INTRODUCTION

The low temperature and high strain-rate plastic deformation of body-centered-cubic (bcc) transition metals is dominated by the structure of the cores of 1/2<111> screw dislocations [1-3]. Consequently, theoretical investigation of mechanical properties of these materials habitually involves extensive atomistic simulations [2-4]. The essential precursor of such calculations is an appropriate description of atomic interactions. The fact that it is the level of the filling of the d-band that controls the stability of the bcc lattice relative to alternate structures (see e. g. [5]) suggests that the angular character of bonding originating from d-bonds should be properly accounted for. Nevertheless, the bulk of atomistic studies of dislocations and other extended defects in bcc metals have been made using central-force potentials; in the early studies pair-potentials [6, 7] and more recently many-body potentials of the embedded atom type [4, 8]. The first calculations that include the non-central character of bonding in transition metals are those employing potentials derived from first-principles generalized pseudopotential theory (MGPT potentials) [9-11]. Although they confirmed the general characteristics of the screw dislocation cores found in central-force studies, they also indicate that structural features specific to different bcc metals may not be appropriately captured by central force schemes.

While a full analysis of the effect of non-central interactions on dislocation cores has not yet been made, recent theoretical and experimental investigation of the structure of the $\Sigma 5$ (310) symmetrical tilt grain boundary in Nb and Mo clearly demonstrates the inability of central forces to differentiate between these two transition metals. Specifically, the electron microscopic studies reveal that this grain boundary is symmetric in Nb [12] while in Mo the symmetry is broken due to the relative displacement of the grains parallel to the tilt axis [13]. This difference is not found in studies employing central force potentials [12, 14] but it is revealed in calculations in which the covalent character of bonding has been included [12-14], comprising calculations employing the bond-order potentials discussed in this paper.

All aspects of interatomic bonding are, of course, fully accounted for in LDA based ab initio calculations. However, these are severely limited by the number of independent atoms that can be considered and, moreover, studies of extended defects, such as dislocations, are greatly restricted if periodic boundary conditions need to be applied (see, e. g. [15, 16]). Hence, real-space methods, albeit approximate, are strongly preferred. The above mentioned MGPT potentials of Moriarty [9, 17] are one such method. Another approach in which the covalent character of bonding is included explicitly is the tight-binding method. This has often been used in reciprocal space but recently it has been reformulated in terms of Bond-Order Potentials (BOP) [18-22] into a

Mat. Res. Soc. Symp. Proc. Vol. 578 © 2000 Materials Research Society

real-space method that scales linearly with the size of system since the diagonalization of the Hamiltonian matrix is avoided by direct calculation of the density matrix and/or bond order.

We have constructed BOP for four bcc transition metals, niobium, tantalum, molybdenum and tungsten; in this paper we concentrate on molybdenum and tantalum. First we summarize briefly the main aspects of BOPs and their construction. We then present the testing of these potentials by calculating energies of alternative structures, phonon dispersion curves and energies of distorted structures obtained on several transformation paths. Finally, we have calculated the γ-surfaces that are a precursor to future studies of dislocation cores, and discuss here the similarities and differences with those calculated using central-force potentials.

BOND-ORDER POTENTIALS

In this scheme the cohesive energy is composed of three parts: E_{pair}, E_{bond} and E_{env}. E_{pair} represents the electrostatic interaction between the atoms and the overlap repulsion between the valence d electrons. It is principally repulsive and described by a pair-potential. E_{env} is the environmentally dependent part of the energy which characterizes the s, p ion core repulsion [23]. It is represented by a repulsive central-force many body potential which was proposed to have the screened Yukawa form [24, 22]. The non-central character of the atomic interactions is hidden in the bond energy, E_{bond}, .

Within BOP approach $E_{bond} = \sum_{I, J(i \neq j)} H_{I,J} \Theta_{J,I}$ where $\Theta_{J,I}$ and $H_{I,J}$ are the bond order and Hamiltonian matrix elements, respectively. The index $I \equiv [i, L_i]$, where i numbers the atomic sites in the system studied, $L_i = (\ell, m)_i$ where ℓ denotes the quantum orbital moment and m the quantum magnetic moment for spherical symmetry. The bond-order matrix is determined for a given Hamiltonian in terms of the derivative of certain diagonal elements of the Green's function [18, 19] that are evaluated using the recursion method of continued fractions [25] and Lanczos algorithm [26]. In order to damp down the long-range Friedel oscillations that are present in metallic materials, a fictitious finite temperature of the electrons is introduced in order to achieve rapid real-space convergence of the bond-order potentials [27, 28]. The relatively complex procedures involved in evaluation of E_{bond} are all part of the suit of computer codes available as the Order N (OXON) package.

For $i \neq j$ $H_{I,J}$, the bond integrals, are determined within the orthogonal two-center approximation and for the transition elements only d-orbitals ($\ell = 2$) are used in the tight-binding scheme. The validity of this approximation was tested by comparing the electronic densities of states calculated in this approximation with those evaluated using the ab-initio all electron tight-binding LMTO method [29]. In order to determine the functional form of the bond integrals, $H_{I,J}$ were evaluated as functions of the separation of atoms i and j using the first-principles TB-LMTO which allows to establish a direct link with the parametrized tight-binding method [29, 30]; variation of the interatomic distance was achieved by evaluating the integrals for several volumes per atom. The bond integrals are then represented analytically by the functional form suggested in [31]

$$dd\alpha(R_{ij}) = dd\alpha^0 \left(\frac{R_0}{R_{ij}} \right)^{na} exp\left\{ na\left[\left(\frac{R_0}{R_c} \right)^{nc} - \left(\frac{R_{ij}}{R_c} \right)^{nc} \right] \right\} \qquad (1)$$

where α denotes σ, π or δ (quantum magnetic moment in cylindrical symmetry) and R_{ij} is the separation of atoms i and j; the bond integrals are cut-off between the second and third neighbors such that for $R_1 < R < R_{cut}$ they are represented by polynomial splines smoothly approaching zero at R_{cut}. $dd\alpha^0$, R_0, R_c, na and nc, together with the number of d electrons, are adjustable parameters used to fit the ab initio calculated bond integrals in the vicinity of the first nearest neighbors. (Note that a different form was used in [22] for $dd\alpha$). Using this form $dd\sigma$ is also reproduced well at second nearest neighbors but $dd\pi$ and $dd\delta$ are at this separation reproduced less satisfactorily. Other parameters entering the bond part are the number of the recursion levels in the continued fractions used to evaluate the Green's functions and the effective electronic temperature. Details of these parameters will be presented elsewhere [32].

200

The on-site elements, which do not enter directly into the expression for E_{bond} but are needed when evaluating the bond order, are adjusted self-consistently to maintain local charge neutrality with respect to each atom. This condition which reflects the perfect screening properties of metallic materials [5], is achieved efficiently within the BOP scheme as described in [20, 21].

It should be noted that E_{bond} is constructed based solely on ab-initio calculated data with no empirical input. As a next step in the construction of the BOP the E_{env} is fitted to reproduce the Cauchy pressure, $c_{12} - c_{44}$, and, finally, the pair potential is fitted so as to reproduce the lattice parameter, remaining elastic moduli and cohesive energy. Thus the fitting procedure is sequential with empirical input concentrated in the repulsive part of the cohesive energy. Details of both E_{env} and E_{pair} will again be presented elsewhere [32].

However, atomistic simulations require not only evaluation of the energy but also its derivatives with respect to the positions of atoms, i. e. forces acting on individual atoms. The calculation of the derivatives of the pair and environmental parts of the energy is straight forward. The derivatives of the bond part can be found using the Hellmann-Feynman theorem provided the bond order is computed sufficiently accurately [28]; evaluation of these derivatives is also part of the Order N (OXON) package.

TESTING OF THE POTENTIALS

The first requirement for the validity of the constructed potentials is that the bcc structure is stable relative to possible alternative structures. This has been tested by calculating the energies for the fcc, hcp, A15 and simple cubic (sc) structures with the same density as the bcc lattice using two ab-initio methods (FP-LAPW and FP-LMTO) and BOP. The results shown in Table 1 demonstrate that not only is the bcc structure most stable but BOP also reproduces closely the energies of the alternative structures.

	Molybdenum				Tantalum			
	FCC	HCP	A15	SC	FCC	HCP	A15	SC
BOP	429	460	249	1379	276	279	168	1363
FP-LAPW	409	463	218	1344	276	346	112	1248
FP-LMTO	419	460	251	1347	272	316	222	1384

	Tungsten				Niobium			
	FCC	HCP	A15	SC	FCC	HCP	A15	SC
BOP	450	481	325	1694	332	321	200	1155
FP-LAPW	459	545	163	1732	336	354	164	1102
FP-LMTO	491	504	268	1695	354	360	219	1194

Table 1
Energy differences (in meV/atom) between alternative structures and the bcc structure for the four transition metals studied.

The next test of the potentials is provided by comparison of calculated phonon dispersion curves with experimental data [33, 34]. The phonon spectra were computed for all four metals for three high symmetry directions using the method of frozen phonons. For Mo and Ta the results are shown in Fig. 1. Except for the zone boundary phonons, that display higher frequencies than the experimental values, BOP reproduces well the shapes of the dispersion curves. However, for Mo (and also W) the T2 mode phonon at the zone edge in the [1$\bar{1}$0] direction is too soft. This problem is common in semi-empirical schemes and it may be related to the fact that at the second nearest neighbors $dd\pi$ and $dd\delta$ are not represented precisely enough by eq. (1). Indeed, as shown in [35], these bond integrals can only be reproduced with sufficient precision if an environmental dependence is introduced.

Since the potentials are intended for atomistic modeling of extended defects they have to be applicable when the atomic environment is considerably different from that in the ideal lattice. While this can never be fully tested, an assessment can be made by investigating the highly distorted structures encountered along certain transformation paths. Three distinct transformation paths, tetragonal, trigonal and hexagonal, described in detail in Ref. [36], have been investigated. The energy as a function of a parameter characterizing these paths was calculated using BOP and by the ab-initio FLAPW method [37]. Results of these calculations are not shown here but some of them can be found in the earlier publication [22] and will be presented in detail elsewhere [32]. In general, BOP reproduces the ab-initio data very closely and is able to trace correctly even certain subtle features that originate from the directional character of bonding. An example is the occurrence of two energy minima in the vicinity of the fcc structure along the trigonal path in the case of Mo and Nb; for the fcc structure the energy is at maximum [22]. In contrast such minima are not found in the case of Ta where the energy is at minimum for the fcc structure. All these features of the trigonal deformation path are correctly reproduced by BOP.

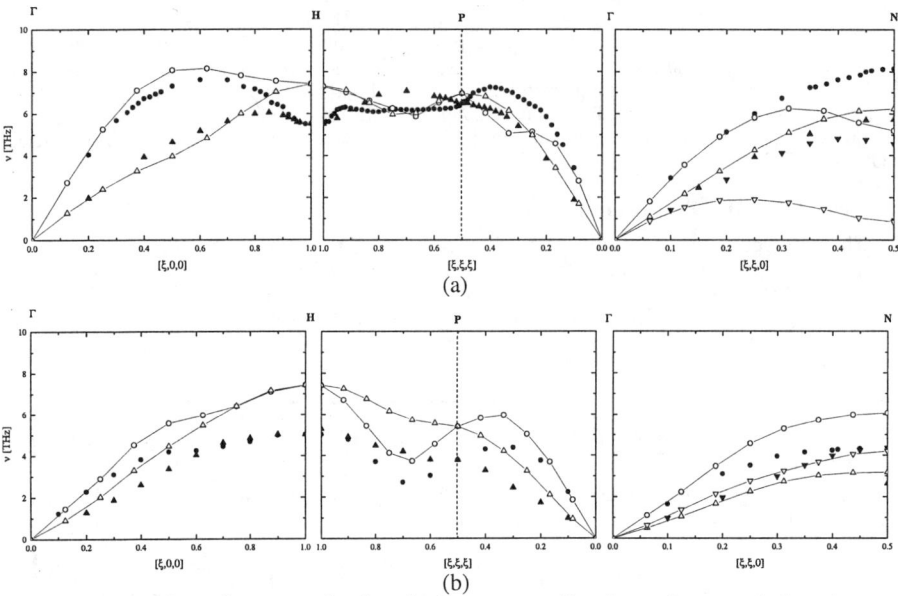

Fig. 1. Phonon dispersion curves for three high symmetry directions. Open symbols and curves represent calculated values and full symbols experimental data [33, 34]. (a) Mo; (b) Ta.

{110} γ-SURFACES FOR MOLYBDENUM AND TANTALUM

It has been shown in a number of studies that γ-surfaces, which can be calculated relatively easily, are an effective tool for the qualitative analysis of dislocation cores [2-4, 38, 39]. For example, Duesbery and Vitek [4] discussed the differences between the screw dislocation cores of metals such as Mo and Ta on the basis of the shapes of γ-surfaces for {110} planes into which the cores spread. However, these calculations were performed using many body potentials of the Finnis-Sinclair type [40] and thus no directional bonding was involved.

In this study we have calculated {110} γ-surfaces using the constructed BOPs and the results for Mo and Ta are shown in Fig. 2 in terms of maps of equi-energy contours. The principal characteristics of these γ-surfaces, no local minima corresponding to metastable stacking faults and symmetries related to the crystal structure, are the same as in all previous calculations [2, 3]. However, the shapes of the two γ-surfaces, in particular in the important region corresponding to the displacements along the [111] direction shorter than |1/6[111]|, the maximum spread of the Burgers vector into one {1̄10} plane, are distinctly different. In the case of Ta the energy contours are almost circular while in the case of Mo they deviate very significantly from the circles. The significance of this difference is that in Ta the deviation of the displacement vector of the same length away from [111] direction does not lead to any energy gain while on Mo such deviation towards [1̄10] direction corresponds to lowering of the energy. This implies that in Mo it will be energetically favorable for the screw dislocation spread into three {110} planes to possess significant edge components while these components will be much smaller in Ta. A similar, but much less distinct difference was found when using central force potentials [4], and the present calculations suggest that this difference between Mo and Ta will be increased by directional bonding. Since the edge components in the core of screw dislocations are directly related to how important is the effect of non-glide components on the deformation behavior [4, 41] and thus break down of the Schmid law, the directional bonding appears to enhance the difference between Mo and Ta in this important aspect of the plastic deformation.

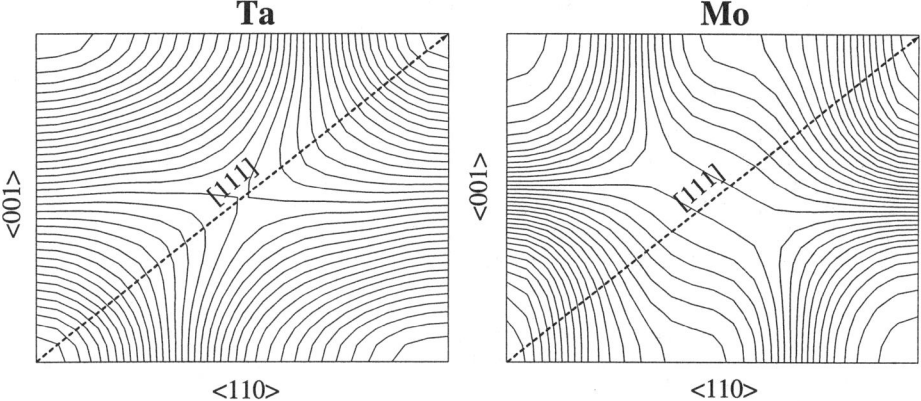

Fig. 2. Contour plots of {110} γ-surface for tantalum and molybdenum.

CONCLUSIONS

The bond-order potentials based on two-center orthogonal tight-binding approximation have been constructed for bcc transition metals Mo, W, Nb and Ta. These potentials reproduce a number of ab-initio calculated properties of these materials, namely relative energies of alternate structures, energy variation along transformation path leading to very significant deviations away from the ideal bcc lattice. They also lead to a satisfactory agreement between calculated and observed phonon dispersion curves. However, for Mo and also W the calculated T2 mode phonon at the zone edge in the [110] direction is much softer than experimentally observed one. This problem is most likely related to the fact that ddπ and ddδ bond integrals are not well represented by the dependence on the separation of the corresponding atoms only but their environmental dependence needs to be taken into account [35].

The γ-surfaces calculated for Mo and Ta suggest that directional bonding enhances the difference in the edge component in the cores of screw dislocation in these two materials and thus the difference in their response to non-glide stresses during plastic deformation.

ACKNOWLEDGMENTS

This research was supported in part by the Advanced Strategic Computing Initiative of the U.S. Department of Energy through LLNL, grant no. B501663 (MM and VV) and by the NSF - International Programs Grant no. INT-96-05232 and the Grant Agency of the Academy of Sciences of the Czech Republic, project no. A1010817 (LGW, MS). Screened LMTO calculations were performed in Oxford on the HP EXEMPLER computer that was funded by Higher Education Funding Council of England and by Hewlett-Packard.

REFERENCES

1 J. W. Christian, *Metall. Trans. A* **14**, 1237 (1983).
2. M. S. Duesbery, *Dislocations in Solids*, edited by F. R. N. Nabarro (Amsterdam, North Holland), Vol. 8, p. 67 (1989).
3. V. Vitek, *Prog. Mater. Sci.* **36**, 1 (1992).
4. M. S. Duesbery and V. Vitek, *Acta Mater.* **46**, 1481 (1998).

5. D. G. Pettifor, *Bonding and Structure of Molecules and Solids*, (Oxford University Press, Oxford, 1995).
6. V. Vitek, R. C. Perrin and D. K. Bowen, *Philos. Mag. A* **21**, 1049 (1970).
7. Z. S. Basinski, M. S. Duesbery and R. Taylor, *Can. J. Phys.* **49**, 2160 (1971).
8. K. Ito and V. Vitek, *Multiscale Modelling of Materials*, edited by V. Bulatov, T. de la Rubia, R. Phillips, E. Kaxiras and N. Ghoniem (Pittsburgh, MRS), Vol. 538, p. 87 (1999).
9. J. A. Moriarty, *Phys. Rev. B* **49**, 12431 (1994).
10. W. Xu and J. A. Moriarty, *Phys. Rev. B* **54**, 6941 (1996).
11. W. Xu and J. A. Moriarty, *Comp. Mat. Sci.* **9**, 348 (1998).
12. G. H. Campbell, S. M. Foiles, P. Gumbsch, M. Rühle and W. E. King, *Phys. Rev. Lett.* **70**, 449 (1993).
13. G. H. Campbell, J. Belak and J. A. Moriarty, *Acta Mater.* **47**, 3977 (1999).
14. T. Ochs, C. Elsässer, M. Mrovec, V. Vitek, J. Belak and J. A. Moriarty, *Philos. Mag. A,* to be published (2000).
15. R. W. Nunes, J. Bennetto and D. Vanderbilt, *Phys. Rev. B* **58**, 12563 (1998).
16. A. Valladares, J. A. White and A. P. Sutton, *Phys. Rev. Lett.* **81**, 4903 (1998).
17. J. A. Moriarty and M. Widom, *Phys. Rev. B* **56**, 7905 (1997).
18. M. Aoki, *Phys. Rev. Lett.* **71**, 3842 (1993).
19. M. Aoki and D. G. Pettifor, 1993, *Physics of Transition Metals*, edited by P. M. Oppeneer and J. Kübler (Singapore, World Scientific), p. 299 (1993).
20. A. P. Horsfield, A. M. Bratkovsky, M. Fearn, D. G. Pettifor and M. Aoki, *Phys. Rev. B* **53**, 1656, 12694 (1996).
21. D. R. Bowler, M. Aoki, C. M. Goringe, A. P. Horsfield and D. G. Pettifor, *Model. Sim. Mat. Sci. Eng.* **5**, 199 (1997).
22. M. Mrovec, V. Vitek, D. Nguyen-Manh, D. G. Pettifor, L. G. Wang and M. Sob, *Multiscale Modelling of Materials*, edited by V. Bulatov, T. de la Rubia, R. Phillips, E. Kaxiras and N. Ghoniem (Pittsburgh, MRS), Vol. 538, p. 529 (1999).
23. D. G. Pettifor, *J. Phys. F: Metal Phys* **8**, 219 (1978).
24. D. Nguyen-Manh, D. G. Pettifor, S. Znam and V. Vitek, *Tight-Binding Approach to Computational Materials Science*, edited by P. E. A. Turchi, A. Gonis and L. Colombo (Pittsburgh, MRS), Vol. 491, p. 353 (1998).
25. R. Haydock, *Solid State Physics*, edited by H. Ehrenreich and D. Turnbull (New York, Academic Press), Vol. 35, p. 216 (1980).
26. C. Lanczos, *J. Res. Natl. Bur. Stand.* **45**, 225 (1950).
27. A. P. Horsfield and A. M. Bratkovsky, *Phys. Rev. B* **53**, 15381 (1996).
28. A. Girshick, A. M. Bratkovsky, D. G. Pettifor and V. Vitek, *Philos. Mag. A* **77**, 981 (1998).
29. O. K. Andersen, O. Jepsen and D. Glötzel, *Highlights of Condensed Matter Theory*, edited by F. Bassani, F. Fumi and M. P. Tosi (Amsterdam, North Holland), p. 59 (1985).
30. O. K. Andersen, O. Jepsen and G. Krier, *Lectures on Methods of Electronic Structure Calculations*, edited by V. E. E. Kumar (Singapore, World Scientific), p. 63 (1994).
31. L. Goodwin, A. J. Skinner and D. G. Pettifor, *Europhys. Lett.* **9**, 701 (1989).
32. M. Mrovec, V. Vitek, D. Nguyen-manh, D. G. Pettifor, L. G. Wang and M. Sob, to be published (2000).
33. B. M. Powell, P. Martel and A. D. B. Woods, *Can. J. Phys.* **55**, 1601 (1977).
34. A. D. B. Woods, *Phys. Rev.* **136,** A781 (1964).
35. D. Nguyen-Manh, D. G. Pettifor and V. Vitek, to be published (2000).
36. V. Paidar, L. G. Wang, M. Sob and V. Vitek, *Model. Sim. Mat. Sci. Eng.* **7**, 369 (1999).
37. P. Blaha, K. Schwartz, P. Sorantin and S. B. Trickey, *Comp. Phys. Comm.* **59**, 399 (1990).
38. V. V. Bulatov and E. Kaxiras, *Phys. Rev. Lett.* **78**, 4221 (1997).
39. O. N. Mryasov, Y. N. Gornostyrev and A. J. Freeman, *Phys. Rev. B* **58**, 11927 (1998).
40. G. J. Ackland and R. Thetford, *Philos. Mag. A* **56**, 15 (1987).
41. M. S. Duesbery, *Proc. Roy. Soc. London A* **392**, 145, 175 (1984).

TRANSITION TEMPERATURES IN PLASTIC YIELDING AND FRACTURE
OF SEMICONDUCTORS

P. PIROUZ[1], L. P. KUBIN[2], J. L. DEMENET[3], M. H. HONG[1], A. V. SAMANT[1]

[1]Department of Materials Science and Engineering, Case Western Reserve University, Cleveland, OH, 44106-7204, U.S.A..

[2] LEM, CNRS-ONERA, B.P. 72, Av. de la Division Leclerc, 92322 Châtillon Cedex, France.

[3] LMP, CNRS, SP2MI, Bd 3, Teleport 2, BP 179, 86960 Futuroscope Cedex, France

ABSTRACT

Recent experiments on deformation of semiconductors show an abrupt change in the variation of the critical resolved shear stress, τ_y, with temperature, T. This implies a change in the deformation mechanism at a critical temperature T_c. In the cases examined so far in our laboratory and elsewhere, this critical temperature appears to coincide approximately with the brittle-ductile transition temperature, T_{BDT}. In this paper, the deformation experiments performed on the wide bandgap semiconductor, 4H-SiC, over a range of temperatures and strain rates are described together with the characterization of induced dislocations below and above T_c by transmission electron microscopy. Based on these results, and those of Suzuki and coworkers on other compound semiconductors, some understanding of the different mechanisms operating at low and high temperatures in tetrahedrally coordinated materials has been gained, and a new model for their brittle-ductile transition has been proposed.

INTRODUCTION

Recently, it has been possible to perform deformation experiments on a few (six) different compound semiconductors extending the deformation temperature to regions in which they are brittle. Specifically, Suzuki and coworkers have deformed InP, InSb, GaAs, and GaP using compression tests under hydrostatic pressure in order to prevent fracture of the deformation samples before they plastically yield [1-4]. Some low-temperature tests on GaAs under roughly the same conditions were actually performed some years ago by Rabier and coworkers [5,6]. In our deformation tests, Samant [7] investigated the plastic behavior of two wide bandgap semiconductors, 4H-SiC and 6H-SiC, by compression over a wide range of temperatures and strain rates. It should be mentioned that because single crystals of these two semiconductors were not available until recently, prior to the work of Samant, there were only a few reports of deformation tests on single crystal 6H-SiC and practically none on 4H-SiC. Some significant works that existed were reports by Fujita *et al.* [8] on compression experiments on Acheson-grown crystals over the temperature range 1300-1600°C, and creep tests on Cree-grown 6H-SiC by Corman [9]. In Samant's experiments, the rather high initial density of dislocations in the samples ($\sim 10^3$-10^4 cm^{-2}), careful alignment of the sample in the deformation jig, together with the use of very low strain rates, made it possible to deform the materials at temperatures hundreds of degrees below their usual range of BDT. The results of these experiments have been reported in Refs. [7,10,11]. More recently, Demenet repeated Samant's experiments on 4H-SiC (grown at Cree Research, Inc.) obtaining much more data in the temperature range 900-1360°C [12].

EXPERIMENTAL PROCEDURES

The details of the experimental procedure for deforming the SiC crystals are given elsewhere (see Ref. [7]). Briefly, a single crystal ingot of 4H-SiC grown at the Cree Research, Inc. was oriented for single glide using X-ray diffraction. From the ingot, 2.2x2.2x4.7 mm^3 samples were cut and carefully polished on all faces. The compression tests were made over the temperature range 900-1360°C at a strain rate of $3.6 \times 10^{-5}\ s^{-1}$. Following the mechanical tests, thin foils parallel to the (0001) slip plane were prepared from the samples deformed at temperatures above and below the transition temperature $T_c \approx 1100°C$. In addition to strain contrast experiments to characterize the Burgers vector of the dislocations, the large-angle convergent-beam electron diffraction (LACBED) technique was used to identify the nature of dislocation cores [13].

RESULTS

Some of Demenet's results are shown in Fig. 1 in the form of a plot of $\ln(\tau_Y)$ vs. $1/T$.

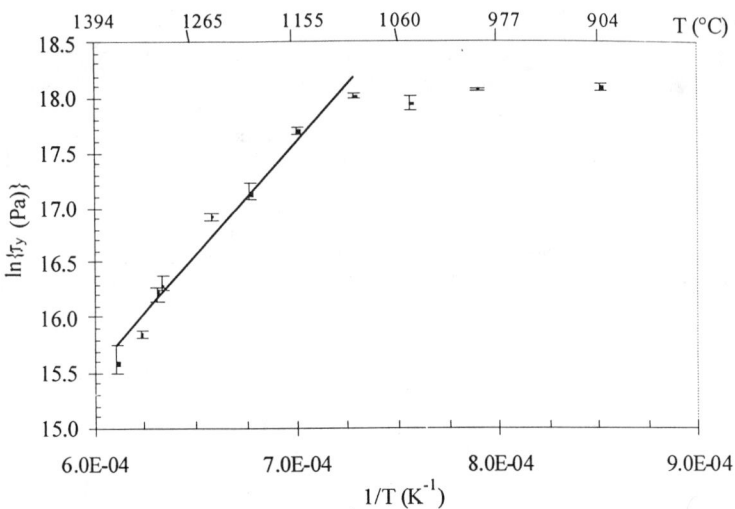

Fig. 1. Plot of $\ln(\tau_Y)$ vs. $1/T$ for 4H-SiC deformed at a strain rate of $3.6 \times 10^{-5}\ s^{-1}$.

As in previous experiments of Samant [7,10,11] on 4H- and 6H-SiC, there is a sharp transition at $T_c \approx 1100°C$. It should be noted that while macroscopic plastic deformation of both 4H- and 6H-SiC polytypes is relatively easy at $T > 1100°C$, it becomes very limited at $T < 1100°C$ and, when the samples have not failed catastrophically, deformation is very often accompanied by appearance of extensive microcracking. These observations indicate that the BDT temperature of both polytypes is ~1100°C, consistent with experiments of Maeda and coworkers who concluded that the BDT temperature of 6H-SiC is above 800-1000°C [8,14]. Following the compression experiments [12], TEM of the deformed samples revealed that a very low density of dislocations was generated by deformation at $T < T_c$ and these were predominantly single leading partials all with the same Burgers vector $\mathbf{b}_l = 1/3[1\bar{1}00]$. An example is shown in Figs. 2(a-c). The cores of five single partial dislocations produced by deformation at temperatures below T_c showed them to be silicon in every case. On the other hand, the samples deformed at $T > T_c$ contained a high density of total dislocations all dissociated, i.e. they were in the form of leading/trailing pairs bounding a ribbon of stacking fault. The dislocations were predominantly screw type with a Burgers vector of $\mathbf{b} = 1/3[\bar{2}110]$; an example of these dislocations is shown in Fig. 2(d).

DISCUSSION

Similar to Fig. 1, plots of the critical yield stress versus temperature for all the six compound semiconductors mentioned previously show an abrupt transition at a critical temperature T_c [2,4,10]. It is possible that this transition exists in all the tetrahedrally coordinated materials and signifies a change of the deformation mode in the material. Also, intriguingly, in every case the transition temperature in the $\tau_Y(T)$ plot appears to coincide approximately with the brittle-to-ductile transition temperature, T_{BDT}, of that material, taking due account of the strain rate. In the following, we present a model to describe this transition in the yield stress and try to relate it to the BDT [15].

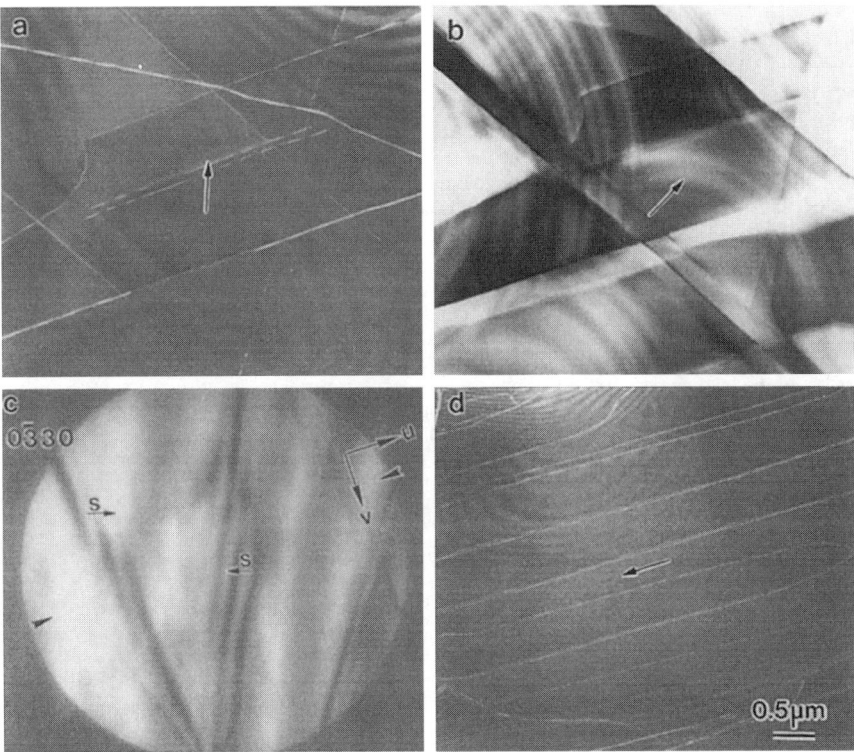

Fig. 2. (a) BF micrograph of dislocations in 4H-SiC deformed at 900°C showing single leading partial dislocations; (b) same region as (a) showing overlapping stacking faults; (c) a LACBED pattern of a single leading partial in (a). (d) BF micrograph of dislocations in 4H-SiC deformed at 1300°C showing dissociated (leading/trailing) total dislocations.

Dislocations in tetrahedrally-coordinated materials glide on the {111} slip plane in cubic crystals and (0001) slip plane in hexagonal materials and have Burgers vectors, $\mathbf{b} = 1/2\langle 1\bar{1}0 \rangle$ in the former and $\mathbf{b} = 1/3\langle 11\bar{2}0 \rangle$ in the latter. Because of the high Peierls potential in such crystals, the dislocations lie in the Peierls valleys that are parallel to $\langle 1\bar{1}0 \rangle$ in cubic crystals and $\langle 11\bar{2}0 \rangle$ in hexagonal crystals. As a result an ideal dislocation loop has a hexagonal shape with the segments parallel to the Peierls valleys, i.e. all the segments are 60° or screw dislocations. In all the tetrahedrally coordinated materials that have been studied to date, the dislocations are dissociated according to the following reactions:

$$\frac{1}{2}\langle 1\bar{1}0\rangle \rightarrow \frac{1}{6}\langle 1\bar{2}1\rangle + \frac{1}{6}\langle 2\bar{1}\bar{1}\rangle$$

$$\frac{1}{3}\langle 2\bar{1}\bar{1}0\rangle \rightarrow \frac{1}{3}\langle 1\bar{1}00\rangle + \frac{1}{3}\langle 10\bar{1}0\rangle$$

i.e. the partials are either 90° or 30° dislocations. The separation of the partials is of course determined by the stacking fault energy, γ, of the material. The latter ranges from ~2.5 mJ/m^2 in SiC to ~280 mJ/m^2 in diamond [16-18]. In addition, in the case of compound semiconductors, the core of all the dislocations (dissociated or non-dissociated) consists of only one atom species. Thus, an idealized dissociated dislocation loop, of say hexagonal SiC, will have the configuration shown in Fig. 3; depending on expansion or contraction, the inner and outer loops correspond to the leading and trailing partials, respectively. Notice that a 60° dislocation dissociates into a 90° and a 30° partial where both of the partials have the *same* core (either both Si(g) or both C(g)). On the other hand, a screw dislocation dissociates into two 30° partials that have *different* core structures (e.g. one is Si(g) while the other is C(g)). Also note that half the segments in both the inner and outer loops have a silicon core while the segments in the other half have a carbon core. We shall later argue that the core nature of the different partials in such dislocation loops plays a significant role in the mode of deformation that occurs in tetrahedrally-coordinated materials.

As can be seen in Figs. 2, our TEM results show that the microstructure of the SiC samples that were deformed below T_c consisted predominantly of only Si(g) leading partials while the microstructure of samples deformed above T_c consisted predominantly of dissociated screw dislocations (leading/trailing pairs). Let us now consider the production of dislocations during deformation of such a tetrahedrally–coordinated crystal. In general, the easiest way to produce dislocations in a crystal that contains appropriate sources is by the Frank-Read mechanism. However, if, as argued in this paper, only partial Frank-Read sources are activated (by the operation of the leading partial dislocations only) at low temperatures, then such a source will not be a multiplication site for dislocations but can operate only once. Subsequently, nucleation of dislocations will occur in the bulk or from the sample surfaces.

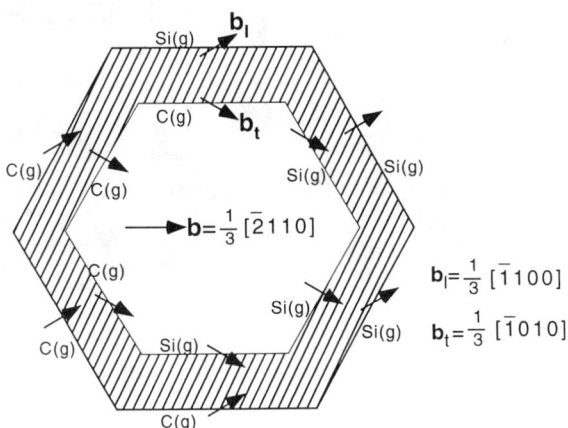

Fig. 3. A dissociated hexagonal loop on the (0001) slip plane in SiC.

In the energetics of dislocation nucleation, thermal activation is often thought to play a negligible role. The reason for this is thought to be that the available thermal energy, kT, at a temperature T is too small compared to the large energy barrier, ΔH_n, required for dislocation nucleation. In terms of the reaction rate theory, thermally activated processes are statistical phenomena whereby atoms that vibrate with a frequency v_0 ($\sim 10^{13}$ s^{-1}) have a Boltzmann probability $exp(-\Delta H_n /kT)$ of overcoming the energy barrier. Accordingly, the frequency, v, with which such an activation will occur is given by $v = v_o \, exp(-\Delta H_n /kT)$, i.e. the mean time, t, during which a successful activation occurs is $t = 1/v = (1/v_o) exp(\Delta H_n /kT)$. For dislocation nucleation, the energy barrier, ΔH_n, is thought to be so large ($\geq 100kT$) that the time t for a successful operation becomes unreasonably large (e.g., $t \approx 2.7 \times 10^{30} s$ for an activation enthalpy of $100kT$). However, when a shear stress, τ, is applied to the crystal, the energy barrier to overcome decreases by an amount $\alpha \tau V^*$, where V^* is the activation volume $\left[= -(\partial H / \partial \tau)_T \right]$ and $\alpha \tau$ is the stress concentration at a heterogeneity (e.g. scratch at the surface), to become $\sim \left(\Delta H_n - \alpha \tau V^* \right)$. Thus, the probability of overcoming it becomes $exp \left[-\left(\Delta H_n - \alpha \tau V^* \right) /kT \right]$ and the frequency of successful dislocation nucleation events will be:

$$v = v_o \, exp \left[-\left(\Delta H_n - \alpha \tau V^* \right) /kT \right] \qquad \qquad ...(1)$$

Based on the experimental results, we shall assume that, in the above equation, the leading and trailing partials have different activation enthalpies for dislocation nucleation, ΔH_n^l and ΔH_n^t, respectively, and $\Delta H_n^l < \Delta H_n^t$. In addition, the resolved shear stress on the leading and trailing partial dislocations will be different in Eq. (1) because they have different Burgers vectors.

Since our experiments were performed on hexagonal SiC, we shall use this material as an example, but the arguments are essentially general and can be applied to other semiconductors. Fig. 4(a) shows schematically the orientation of the parallelepiped sample used in our compression experiments. The sample is oriented for single glide, i.e. an orientation in which the $(0001)\langle \bar{2}110 \rangle$ system is primarily activated. The dislocations most probably nucleate heterogeneously as half-loops from the crystal surface; an example of one (denoted as A) on a particular (0001) slip plane is shown in Fig. 4(a). For comparison, the nucleation of a dislocation (full-)loop (denoted as A') on the same (0001) slip plane within the sample is also shown. In fact, nucleation of four types of loops on a (0001) slip plane may be envisaged as illustrated in Fig. 4(b). In cases A and A', only the leading partial has nucleated while in cases B and B', nucleation of the leading partial has been followed by that of the trailing partial.

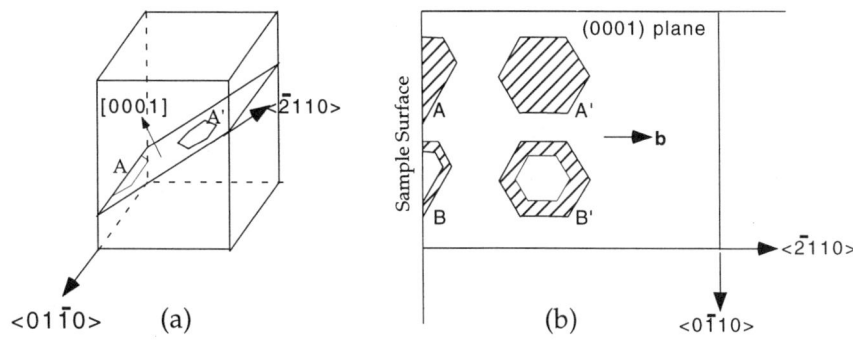

Fig. 4. (a) The geometry of a deformation sample oriented for single glide; (b) four possible ways of dislocation nucleation on the slip plane.

The formation of a dislocation loop presumably takes place by, first, the nucleation of the leading partial followed by the nucleation of the trailing partial. Assuming that intrinsic faults are favored over extrinsic ones, then the order of nucleation of the two partials can be determined from the Thomson tetrahedron [19]. It is important to note that once the leading partial forms, it creates a stacking fault behind it that changes the state of the lattice for nucleation of the trailing partial. The stacking fault of course adds an extra stress term, γ/b, to the external applied stress that should make it simpler for the trailing partial to nucleate. As we said, it will be assumed that ΔH_n^l and ΔH_n^t are different and that $\Delta H_n^l < \Delta H_n^t$. In fact, as shown in Ref. [20], there is some independent evidence for this hypothesis. There is, however, even more evidence for a related property of the leading and trailing partials in tetrahedrally coordinated crystals: the difference in their mobilities (for the case of Si and Ge, see, e.g., [21,22]). It appears that this difference increases as the temperature decreases. Of particular importance is the mobility difference between the two 30° partials of a dissociated screw dislocation where it is found that, even in elemental semiconductors, Si and Ge, the leading 30° partial has a higher mobility than the trailing 30° partial. This implies that the activation enthalpies for glide of partial dislocations, ΔH_g^l and ΔH_g^t, are also different and that $\Delta H_g^l < \Delta H_g^t$. The reasons for the differences in mobility and nucleation of the two partials are not really clear, but they probably have a similar origin. One obvious difference is the presence of a stacking fault after the formation of the leading partial that provides a different local environment for nucleation of the trailing partial. However, it is unlikely that the nucleation of a dislocation in a faulted region of the crystal will be significantly different from nucleation in a perfect crystal. Moreover, as was mentioned before, the presence of a stacking fault provides an extra stress term that should actually help the nucleation of the trailing partial. Another possibility is that the formation (and glide) of the leading partial may result in the creation of point defects in its wake (i.e. in the faulted region) which would have a much more significant effect on the local environment in which the trailing partial is nucleated. These point defects need not of course be extrinsic but can be intrinsic ones such as jogs or anti-phase defects (APDs) [23,24]. It should be mentioned that the interaction of point defects with dislocations and its effect on dislocation mobility goes back to the early works of Celli *et al.* [25] or Rybin and Orlov [26] on materials with a high Peierls barrier. Also, in the late seventies and eighties, the Soviet scientists performed many experiments on Si and Ge showing that the region of a crystal swept by a moving dislocation is significantly different from a virgin material, thus affecting the motion of subsequent dislocations (see, e.g., [27]). Some important evidence comes from Kisielowski's electron spin resonance (ESR) experiments that revealed the generation of point defects by screw dislocations in silicon [28]. In fact, in his classic review of dislocations in semiconductors, Alexander emphasizes that a consideration of the dislocation/point defect interactions is essential to an understanding of the physics of dislocations in these materials [29]. It is also important to realize that the point defects usually annihilate (or anneal out) at higher temperatures and their interaction with dislocations becomes less effective.

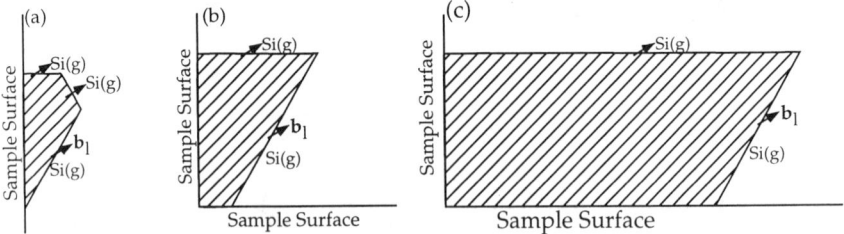

Fig. 5. Nucleation and expansion of a leading partial dislocation half-loop from the surface of the crystal at $T<T_c$.

In the case of compound semiconductors, where the atomistic core of the leading and trailing partials may be different, the core nature is another factor that must be taken into account in considering dislocation nucleation. Thus, in the case of SiC, the partial with the silicon core may have a smaller activation energy than the partial with the carbon core. As was mentioned before, half of a partial loop (with any Burgers vector) has a silicon core while the other half has a carbon core (Fig. 3). We suggest that, in the case of SiC, the Si(g) half loop of the leading partial (whose Burgers vector is determined from the Thomson tetrahedron) is preferentially nucleated as compared to a completely C(g) or a mixed Si(g)/C(g) half-loop (see Fig. 5(a)).

Assuming that at $T<T_c$, only the leading Si(g) partial (Burgers vector $\mathbf{b_l}$) is nucleated, then the sequence of events is illustrated in Fig. 5. After its nucleation from the sample surface (Fig. 5(a)), the half-loop expands and, because of the different mobilities of the segments, the semi-hexagonal shape of the half-loop is distorted. Assuming that the 90° Si(g) partial has a higher mobility than the two adjoining 30° partials, the configuration of the half loop at a later stage becomes something like that shown in Fig. 5(b). In Fig. 5(c), the 90° partial has moved out of the specimen and the lower 30° segment - that has been assumed to be more mobile than the upper 30° segment - is in the process of moving out. The final outcome will be a preponderance of 30° Si(g) dislocations left on different (and parallel) (0001) planes. Note that only a few partials can exist on each particular (0001) plane (Fig. 4(a)) because each partial drags a stacking fault behind it and the initial source (AD in Fig. 5(a)) is de-activated as a dislocation source.

Now consider the nucleation of a full half-loop from the sample surface (B in Fig. 4(b)) at $T>T_c$. The atomistic detail of the half-loop B in Fig. 4(b) is shown in Fig. 6 where the core of all the half-loop segments are shown reconstructed.

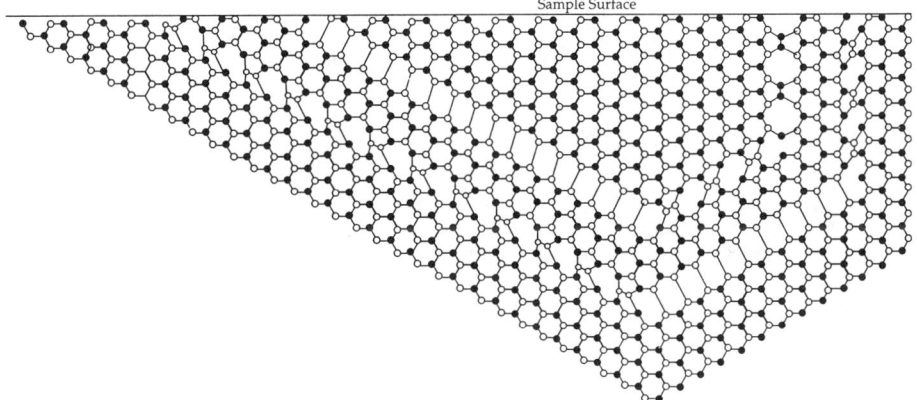

Fig. 6. Schematic illustration of the atomistic detail of a dissociated (leading/trailing) dislocation half-loop nucleated from the surface of the crystal at $T>T_c$.

The expansion of the perfect dislocation half-loop on the (0001) slip plane is shown in Fig. 7. Here, a dissociated dislocation half loop, ABCD, nucleates from the surface (7(a)) and expands (7(b)) until the mobile segments run out of the sample (Fig. 7(c)) leaving only the slow dissociated screw dislocation on the (0001) slip planes. In this figure, additional assumptions have been made that mobility of the upper 90°Si(g)/30°Si(g) segment is higher than that of the lower 30°Si(g)/90°Si(g) segment which in turn is higher than the topmost 30°Si(g)/30°C(g) screw segment. The final configuration will be a preponderance of 30°Si(g)/30°C(g) screw dislocations parallel to the $[\bar{2}110]$ direction in the deformed crystal. Clearly, a better knowledge of the partial mobilities is essential to draw clearer conclusions.

The next problem to consider is the possible relation between the transition in the $\tau_Y(T)$ plot and the BDT temperature. When a load is applied to a crystal, it is initially deformed elastically whereby elastic energy is stored in the crystal. This stored elastic energy is released if the load is taken off to let the crystal relax. If, however, loading is continued, at a rate $\dot{\tau}$, and the stored energy continuously increases, there are two main competing ways for the crystal to relax and decrease its overall energy. One way is by brittle fracture of the crystal whereby the energy is consumed in creating two new surfaces as a crack forms and propagates.

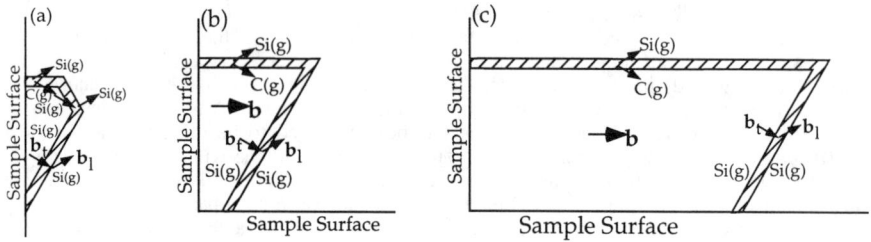

Fig. 7. Nucleation and expansion of a dissociated (leading/trailing) dislocation half-loop from the surface of the crystal at $T > T_c$.

The second way for the release of the stored elastic energy is to move pre-existing dislocations or nucleate and move fresh (as well as pre-existing) dislocations. The occurrence of brittle fracture versus plastic yielding of a crystal is determined by a competition between these two ways of releasing the stored elastic energy. In the case of crack propagation, the Griffith criterion can be written in the form:

$$\frac{\left(1 - v^2\right)K_{eff}^2}{E} \geq 2\gamma_s \qquad \ldots(2)$$

where v is the Poisson's ratio, E is the elastic modulus, γ_s is the surface energy and K_{eff} is the effective stress intensity factor given by [30]:

$$K_{eff} = K_{app} - K_d \qquad \ldots(3)$$

with:

$$K_{app} = \sqrt{\pi a}\sigma_{app} \qquad \ldots(4)$$

σ_{app} being the applied tensile stress and a the crack half-length. K_d in Eq. (3) is shielding of the crack tip introduced by the compressive stress of the dislocations around it and results in a reduction of the applied tensile stress on the crack tip.

In the above conventional picture, brittleness versus ductility of a loaded crystal is considered as a competition between shearing of the crystal by dislocation motion and rupture of the bonds at the tip of a microcrack within the crystal. The general view is that the stress needed for plastic yielding (i.e., the yield stress τ_Y) is strongly temperature dependent and decreases rapidly with increasing temperature while the (tensile) stress required for fracturing the bonds, σ_F (and the corresponding shear stress τ_F), has a much weaker temperature dependence. As a result, $\tau_F(T)$ and $\tau_Y(T)$ intersect at a critical temperature T_{BDT}, usually known as the brittle-to-ductile transition (BDT) temperature where $\tau_F < \tau_Y$ at $T < T_{BDT}$ (the brittle regime) and $\tau_Y < \tau_F$ at $T > T_{BDT}$ (the ductile regime). In the model described in this paper (as illustrated in Fig. 8) the BDT, at least in tetrahedrally-coordinated materials, is determined by a competition between nucleation and propagation of leading partial dislocations versus the nucleation and propagation of perfect (total) dislocations (i.e. that of the trailing partial). As shown by Eq. (1), such a nucleation and propagation of dislocations (partial or perfect) is temperature and stress dependent. Thus, the rupture of atomic bonds enters the picture only indirectly. It is argued that, in a constant strain rate experiment, if the temperature and stress are not sufficient to nucleate the trailing partial (and thus the perfect) dislocations, then the increasing stress will eventually reach a sufficient value (σ_F) to rupture the bonds at the crack tip leading to its propagation. In such a regime, leading partial dislocations may well nucleate at the crack tip (as well as in other favorable sites), but if, at any particular site, the leading partial is not followed by nucleation of the trailing partial, then the source will shut off and stop operating. The consequence is that dislocation nucleation will soon stop and plastic deformation is limited to the strain produced by the few partial dislocations generated before the sources stopped operating. Meanwhile, the work done on the crystal by the external load continues to increase the elastic strain energy in the crystal until the critical strain energy release rate is reached and the crystal gives in and fractures. In this way, the details of bond rupture at the crack tip or temperature dependence of σ_F (or τ_F) are irrelevant to the problem of brittle-to-ductile transition.

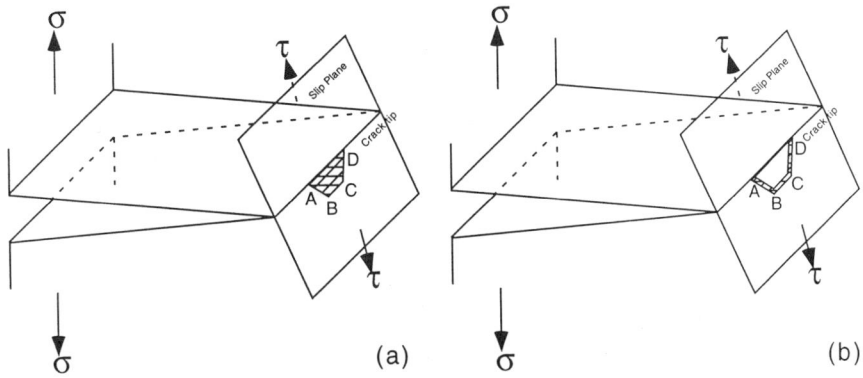

Fig. 8. Schematic illustration of dislocation nucleation at a crack tip under mode I loading: (a) at $T<T_c$, (b) at $T>T_c$.

What is of primary importance is the temperature dependence of τ_n^l and τ_n^t, where τ_n^l and τ_n^t are the minimum shear stresses required to nucleate the leading and the trailing (i.e., the total) dislocation, respectively. The BDT temperature is given by the intersection of $\tau_n^l(T)$ and $\tau_n^t(T)$ plots. Now, when an increasing K (related to the shear stress, τ, by $K=\beta^{-1}\tau$, where β is a geometrical factor) is applied at a rate of $\dot{K}$, the question to be asked is whether a successful (dislocation) nucleation event will occur during the time period in which K increases from its initial value K_o to the stress intensity factor K_{Ic} at which the crystal fails (see, Fig. 9). Note that as K (i.e., τ) increases, the effective activation enthalpy for dislocation nucleation decreases. Thus, the BDT is determined by a competition between the first occurrence of two events: either (1) K reaches K_{Ic} first and catastrophic failure occurs, or (2) $\left(\Delta H_n^t - \alpha\beta KV^*\right)$ decreases sufficiently to result in the nucleation of the (trailing partial) dislocation (at a stress intensity factor $K=K_Y$). The time taken for K to reach K_{Ic} is $(K_{Ic}-K_o)/\dot{K}$ (Fig. 9). Thus, the condition for BDT is that:

$$t \leq \frac{\left(K_{Ic}-K_o\right)}{\dot{K}}$$

At T_{BDT}, the stress intensity factor reaches the value $K=K_Y$ to nucleate dislocations and thus $t = (K_Y - K_o)/\dot{K}$ giving:

$$\frac{1}{v_o} exp\left(\frac{\Delta H_n^t - \alpha \tau_Y V^*}{k T_{BDT}}\right) = \frac{(K_Y - K_o)}{\dot{K}}$$

i.e.:
$$\dot{K} = v_o(K_Y - K_o) exp\left(-\frac{\Delta H_n^t - \delta K_Y}{k T_{BDT}}\right) \qquad \text{...(5)}$$

where $\delta = \alpha \beta V^*$ is a constant. Thus, a plot of $ln(\dot{K})$ versus $1/T_{BDT}$ should be a straight line with a slope $\left[\left(\Delta H_n^t - \delta K_Y\right)/k\right]$, and an intercept $ln[v_o(K_Y - K_o)]$. Since experiments show that a plot of $ln(\dot{K})$ versus $1/T_{BDT}$ has a slope equal to that of (total) dislocation glide (see, e.g., [31]), then Eq. (5) would imply that the activation enthalpy for trailing partial nucleation, $\Delta H_n^{eff} = \left(\Delta H_n^t - \delta K_Y\right)$, is approximately the same as that for dislocation glide.

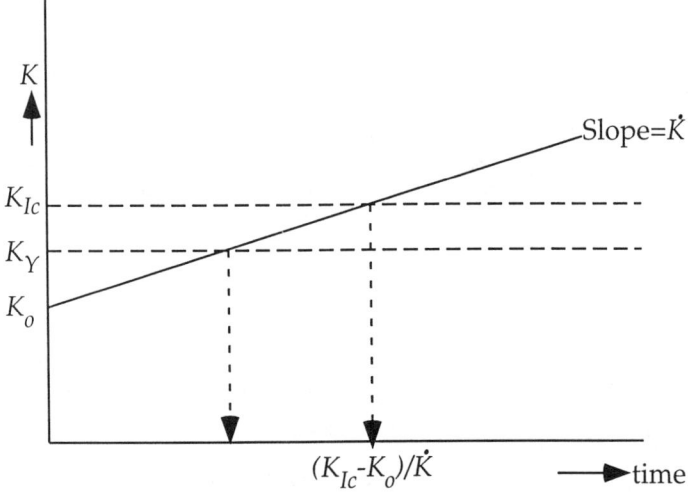

Fig. 9. Schematic plot of K versus time in a constant $\dot{K}$ experiment.

A number of experiments have shown that when high quality (very low dislocation density) semiconducting crystals are deformed under hydrostatic pressure, there is a range of temperatures, $T_{BDT} > T > T_{ll}$, over which the crystal is extremely brittle [3,4]. This is so much that if the crystals have not already fractured during the deformation test, then they usually shatter into small pieces during decompression of the sample cell. An example is GaAs that was found to be extremely brittle between 300 to 400 K even when the confining pressure was increased up to 1.2 GPa; consequently, it has been very difficult to prepare thin foils and perform TEM experiments on the samples deformed in this range [3]. However, interestingly, below this range, i.e. at temperatures less than T_{ll}, it again becomes possible to deform the crystal, albeit it at very high stresses and under hydrostatic pressure. Thus, Suzuki *et al.* [3] could deform GaAs at 300 K under a confining pressure of 400 MPa! TEM of samples deformed at these very low temperatures show a microstructure consisting of predominantly *perfect* screw dislocations. At first sight, this appears to be contrary to the argument put forward in this paper that only leading partial dislocations may be nucleated at $T<T_{BDT}$. However, according to Eq. (1), the effective activation enthalpy for dislocation nucleation, $\Delta H_n^{eff} = \Delta H_n - \alpha\tau V^*$, is lowered by the applied resolved shear stress τ. Thus, it may be that at the very high stresses (over 0.5 GPa) applied at temperatures below T_{II}, the effective activation energy, ΔH_n^{eff}, is sufficiently lowered to make nucleation of the trailing partial dislocation possible, resulting in leading/trailing pairs of partials, i.e. perfect dislocation half-loops. This is also supported by theoretical calculations of Brochard et al. [32] who have shown that the nucleation of dislocations is stress-dependent, and at a high enough applied stresses, both partials can nucleate from a surface source. The screw nature of the dislocations is probably due to the much higher velocity of non-screw segments of a half-loop that, under the very high stresses, rapidly move out of the crystal and leave the much slower screw segments in the sample (as in Fig. 7).

CONCLUSION

New deformation experiments on 4H-SiC confirm the existence of a transition in the variations of the critical yield stress with temperature at approximately 1100°C at a low strain rate of 3.6×10^{-5} s^{-1}. This transition temperature is close to the BDT of 4H-SiC. TEM experiments have verified that the low density of dislocations produced in the samples deformed below T_c are predominantly single leading partials with a silicon core, while a high density of Si(g)/C(g) dissociated screw dislocations are produced in the samples deformed above T_c. A model has been presented that associates the transition temperature T_c with a change in the mode of deformation. In this model, only leading partial dislocations are nucleated at $T< T_c$ that glide on the slip plane but contribute to a very limited extent to the straining of the crystal, specially since the sources that produced them become inoperative after they emit only one partial dislocation. On the other hand, at $T>T_c$, full dislocations can be nucleated repeatedly from sources in the crystal and their glide produces large strains and causes macroscopic yielding of the crystal. Additionally, the model relates the transition temperature T_c in the yielding of the crystal to the temperature at which the brittle-to-ductile transition in the fracture of the crystal occurs.

ACKNOWLEDGEMENTS

This work was supported by grant number FG02-93ER45496 from the Basic Energy Sciences Division of the Department of Energy. Collaboration between CWRU and LMP was initiated by the NATO grant number CRG940295. Thanks are due to Dr. Calvin Carter, Jr. of Cree Research, Inc. for providing a single crystal ingot of 4H-SiC. The authors would like to thank Drs. Christian Kisielowski and Boris Farber for useful discussions.

REFERENCES
[1] T. Suzuki, T. Nishisako, T. Taru and T. Yasutomi, Plastic deformation of InP at temperatures between 77 and 500 K, Phil. Mag. Lett. 77(4), 173-180, (1998).
[2] T. Suzuki, T. Yasutomi, T. Tokuoka and I. Yonenaga, Plasticity of III-V Compounds at Low Temperatures, phys. stat. sol. (a) 171, 47-52, (1999).
[3] T. Suzuki, T. Yasutomi, T. Tokuoka and I. Yonenaga, Plastic deformation of GaAs at low temperatures, Phil. Mag. A 79(11), 2637-2654, (1999).
[4] K. Edagawa, H. Koizumi, Y. Kamimura and T. Suzuki, Temperature Dependence of the Flow Stress of III-V Compounds. Submitted to Phil. Mag. A (1999).
[5] P. Boivin, J. Rabier and H. Garem, Plastic deformation of GaAs single crystals as a function of electronic doping. I: Medium temperatures (150-650°C), Phil. Mag. A 61(4), 619-645, (1990).
[6] P. Boivin, J. Rabier and H. Garem, Plastic deformation of GaAs single crystals as a function of electronic doping. II: Low temperatures (20-300°C), Phil. Mag. A 61(4), 647-672, (1990).

[7] A. V. Samant, *Effect of Test Temperature and Strain-Rate on the Critical Resolved Shear Stress of Monocrystalline Alpha-SiC*, Ph.D. thesis, Case Western Reserve University, (1999).

[8] S. Fujita, K. Maeda and S. Hyodo, *Dislocation glide motion in 6H SiC single crystals subjected to high-temperature deformation*, Phil. Mag. A **55**(2), 203-215, (1987).

[9] G. S. Corman, *Creep of 6H α-Silicon Carbide Single Crystals*, J. Am. Ceram. Soc. **75**(12), 3421-3424, (1992).

[10] A. V. Samant and P. Pirouz, *Activation parameters for dislocation glide in α-SiC*, Int. J. Refractory Metals and Hard Materials **16**(4-6), 277-289, (1998).

[11] A. V. Samant, W. L. Zhou and P. Pirouz, *Effect of Temperature and Strain Rate on the Yield Stress of Monocrystalline 6H-SiC*, phys. stat. sol. (a) **166**, 155-169, (1998).

[12] J. L. Demenet, M. H. Hong and P. Pirouz, *Deformation tests on 4H-SiC single crystals between 900°C and 1360°C and the microstructure of the deformed samples*, in "ICSCRM '99", ed. R. Devaty and G. Rohrer, (Trans Tech Publications Ltd., Switzerland: 1999). In press.

[13] X. J. Ning and P. Pirouz, *A large angle convergent beam electron diffraction study of the core nature of dislocations in 3C-SiC*, J. Mater. Res. **11**(4), 884-894, (1996).

[14] K. Maeda and S. Fujita, *Microscopic Mechanism of Brittle-to-Ductile Transition in Covalent Ceramics*, in "Lattice Defects in Ceramics", ed. by S. Takeuchi and T. Suzuki, *pp.* 25-31, (Jap. J. Appl. Phys. - Series 2, Tokyo, 1989).

[15] P. Pirouz, A. V. Samant, M. H. Hong, A. Moulin and L. P. Kubin, *On temperature-dependence of deformation mechanism and the brittle-ductile transition in semiconductors*, J. Mater. Res. **14**(7), 2783-2793, (1999).

[16] H. Gottschalk, G. Patzer and H. Alexander, *Stacking Fault Energy and Ionicity of Cubic III-V Compounds*, phys. stat. sol. (a) **45**, 207-217, (1978).

[17] S. Takeuchi, K. Suzuki, K. Maeda and H. Iwanaga, *Stacking-Fault Energy of II-VI Compounds*, Phil. Mag. A **50**(2), 171-178, (1984).

[18] S. Takeuchi and K. Suzuki, *Stacking Fault Energies of Tetrahedrally Coordinated Crystals*, phys. stat. sol. (a) **171**, 99-103, (1999).

[19] J. P. Hirth and J. Lothe, *Theory of Dislocations*, (McGraw-Hill, New York, 1968).

[20] X. J. Ning, T. Perez and P. Pirouz, *Indentation-induced dislocations and microtwins in GaSb and GaAs*, Phil. Mag. A **72**(4), 837-859, (1995).

[21] K. Wessel and H. Alexander, *On the mobility of partial dislocations in silicon*, Phil. Mag. **35**(6), 1523-1536, (1977).

[22] H. Alexander, H. Eppenstein, H. Gottschalk and S. Wendler, *TEM of dislocations under high stress in germanium and doped silicon*, Journal of Microscopy **118**(1), 13-21, (1980).

[23] P. B. Hirsch, *The structure and electrical properties of dislocations in semiconductors*, J. Microscopy **118**(1), 3-12, (1980).

[24] M. Heggie and R. Jones, *Solitons and the electrical and mobility properties of dislocations in silicon*, Phil. Mag. B **48**(4), 365-377, (1983).

[25] V. Celli, M. Kabler, T. Ninomiya and R. Thomson, *Theory of Dislocation Mobility in Semiconductors*, Phys. Rev. **131**(1), 58-72, (1963).

[26] V. V. Rybin and A. N. Orlov, *Dislocation Mobility in Crystals with a High Peierls Barrier*, Sov. Phys. - Solid State **11**(12), 2635-3024, (1970).

[27] I. E. Bondarenko, V. G. Eremenko, B. Y. Farber, V. I. Nikitenko and E. B. Yakimov, *On the Real Structure of Monocrystalline Silicon near Dislocation Slip Planes*, phys. stat. sol. (a) **68**, 53-60, (1981).

[28] C. Kisielowski-Kemmerich, *Vacancies and Their Complexes in the Core of Screw Dislocations: Models Which Account for ESR Investigations of Deformed Silicon*, phys. stat. sol. (b) **161**, 111 -132, (1990).

[29] H. Alexander, *Dislocations in Covalent Crystals*, in "Dislocations in Solids", ed. by F. R. N. Nabarro, pp. 114-234, (Elsevier Science Publishers B. BV, Amsterdam, 1986).

[30] R. Thomson, *Physics of Fracture*, in "Solid State Physics", ed. by H. Ehrenreich and D. Turnbull, *pp.* 1-129, (Academic Press, New York, 1986).

[31] P. B. Hirsch and S. G. Roberts, *The brittle-ductile transition in silicon*, Phil. Mag. A **64**(1), 55-80, (1991).

[32] S. Brochard, J. Rabier and J. Grilhé, *Nucleation of partial dislocations from a surface-step in semiconductors: a first approach of the mobility effect*, Eur. Phys. J. - AP **2**, 99-105, (1998).

216

ATOMIC-SCALE MODELING OF THE ANNIHILATION OF JOGGED SCREW DISLOCATION DIPOLES

T. VEGGE[1,2], O. B. PEDERSEN[2], T. LEFFERS[2], K. W. JACOBSEN[1]

[1]Center for Atomic-scale Materials Physics and Department of Physics, Technical University of Denmark, DK-2800 Kgs. Lyngby, Denmark
[2]Materials Research Department, Risø National Laboratory, DK-4000 Roskilde, Denmark.
*Email: vegge@fysik.dtu.dk

ABSTRACT

Using atomistic simulations we investigate the annihilation of screw dislocation dipoles in Cu. In particular we determine the influence of jogs on the annihilation barrier for screw dislocation dipoles. The simulations involve energy minimizations, molecular dynamics, and the Nudged Elastic Band method. We find that jogs on screw dislocations substantially reduce the annihilation barrier, hence leading to an increase in the minimum stable dipole height.

INTRODUCTION

The two determining factors in the dislocation density of deformed metals are the dislocation multiplication and annihilation. At ambient temperatures, where climb processes are limited, dislocation annihilation is expected to occur via cross slip of screw dislocations. Based on atomistic simulations the minimum stable dipole height for defect-free dislocations has previously been estimated to be of the order 1 nm for copper at low temperatures [1, 2]. The experimentally determined minimum stable dipole height found in neutron-irradiated copper single crystals at T=293 K is significantly higher: $y_s = 50$ nm[3]; however, atomistic simulations have recently shown that the presence of jogs can reduce the barrier for single screw dislocation cross slip significantly[4].

In this paper we investigate the effect of jogs on the minimum stable dipole height in Cu at low temperature. We use the *Nudged Elastic Band* (NEB) method[5, 6] to determine a minimum energy path between an initial state consisting of two screw dislocations with jogs and the final state where the two dislocations have annihilated. For more details on the approach see Rasmussen *et al.*[1]. The interatomic interactions are calculated with an *Effective Medium Theory* (EMT) many-body potential [7, 8], which is computationally fast and describes the elastic properties of Cu quite well, see Table 1.

In order to visualize the dislocations, we determine the local crystalline environment of the individual atoms via the *Common Neighbor Analysis* (CNA)[12, 13]. In the CNA the bonds between an atom and its nearest neighbors are used to determine the crystalline structure. Here we only separate the atoms into three different classes: fcc (bulk atoms), hcp (stacking faults: coloured light grey) and "other", i. e. atoms associated with the dislocation cores (dark grey).

Table 1: Comparison between reference values of elastic constants and intrinsic stacking fault energy and values calculated with EMT. The values for the elastic constants and bulk modulus, B, are in GPa. The unit for the stacking-fault energy γ is mJ/m^2. The shear modulus, μ, calculated with EMT, is a Voigt average.

Element	C_{11}	C_{44}	B	μ	γ	ν	
Cu (ref.)	176.2[1]	81.8[1]	137[1]	55[2]	45[2], 56[3]	0.32[2]	[1]Ref. [9], [2]Ref. [10], [3]Ref. [11].
Cu (EMT)	173	90.6	135	66	31	0.29	

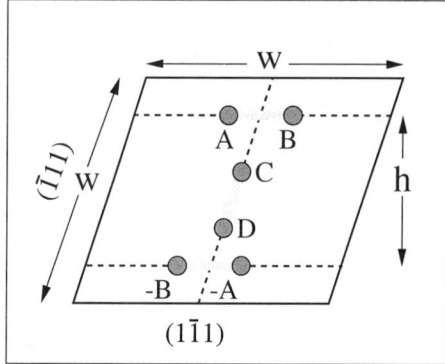

Figure 1: Sketch of the geometry of the system with possible dislocation configurations. The side lengths perpendicular to the dislocations are denoted w, and the height of the initial dipole h. The two dislocations are parallel to the [110] direction perpendicular to the paper. The dashed lines denote possible glide planes, $(1\bar{1}1)$ (horizontal) and $(\bar{1}11)$ (inclined). The dislocations are dissociated into two Shockley partials (dark grey) bounding a ribbon of intrinsic stacking faults (light grey). In the $(1\bar{1}1)$ glide plane the partials are: $A = \frac{a}{6}[21\bar{1}]$ and $B = \frac{a}{6}[121]$. In the $(\bar{1}11)$ glide plane the partials are $C = \frac{a}{6}[12\bar{1}]$ and $D = \frac{a}{6}[211]$.

SIMULATION SETUP

The simulational system consists of two screw dislocations with Burgers vectors $\mathbf{b} = \pm\frac{a}{2}[110]$ placed in a parallelepiped computational cell with periodic boundary conditions in all directions. The (non-orthogonal) sides of the cell are $w = 66\{111\}$inter-planar distances (14.5 nm), terminating along the closed packed $(\bar{1}11)$ and $(1\bar{1}1)$planes. The dislocations are placed in different $(1\bar{1}1)$planes, separated by $h \in [4, 6, 10, 15]\{111\}$planes respectively, see Figure 1. The cell extends $l_{\text{disl.}} = 44.5 \, \mathbf{b}$ (11.3 nm) in the [110] direction, and the system contains a total of 193 842 atoms. We have previously shown that this cell size is sufficient to describe the three-dimensional nature of the annihilation process for jog-free dipole heights $h \leq 15\{111\}$planes (3.3 nm)[1].

We introduce jogs on the dislocations by controlling the periodic boundary conditions of the system, i.e. by not replicating the system perfectly along the dislocation line, but displacing the next cell by a vector equal to the line vector of the jog. In this way we create a periodic array of jogs on either dislocation with opposite Burgers vectors.

We set up the jogs with a line vector $l_{\text{jog}} = \frac{a}{4}[1\bar{1}2]$, using a fractional Burgers vector cell length $l_{\text{disl.}} = 44.5 \, \mathbf{b}$, allowing the transitions: $\frac{a}{4}[1\bar{1}2] \rightarrow \frac{a}{2}[0\bar{1}1] + \frac{a}{4}[110]$ and $\frac{a}{4}[1\bar{1}2] \rightarrow \frac{a}{2}[101] - \frac{a}{4}[110]$. The jog is thus sessile in the $(1\bar{1}1)$ primary slip plane and glissile in the $(\bar{1}11)$ cross slip plane, i.e. it is a kink on the cross slipping dislocation. This choice is to a large extent dictated by the use of periodic boundary conditions, and by the fact that the simulations do not include diffusion of vacancies or interstitials. It is therefore important that the jog is glissile in the cross slip plane. Recent studies of cross slip of individual screw dislocations with different elementary jogs indicate that the reduction in the cross slip activation energy is roughly independent of the line vector of the elementary jogs[4].

It should be noted that with the described setup with periodic boundary conditions, the energy barrier for annihilation can in principle never exceed the barrier for jog-motion along the dislocation lines. If a jog moves along the dislocation line a distance, which is equal to the length of the simulation cell in that direction, it will result in a shift of the whole dislocation line by l_{jog} either closer to or away from the other dislocation. If the jogs were free to move they could therefore just continue to move along the dislocations in opposite directions until the two dislocations meet and annihilate spontaneously. In fact the simulations show that below a certain dislocation separation the barrier for jog mobility disappears and the jogs move in this way. However, at an even smaller (critical) separation the driving force for direct annihilation via cross slip becomes so large that the annihilation proceeds through this mechanism.

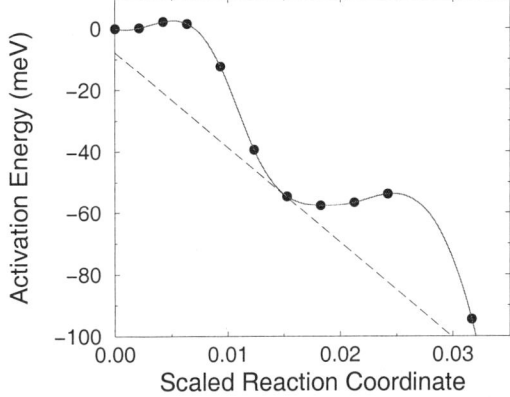

Figure 2: The annihilation energy as a function of the scaled reaction coordinate for the $h = 15\{111\}$planes dipole. The asymmetrical barrier is associated with the jog-motion along the dislocation line, effectively reducing the dipole height. The barrier for this initial motion is $E_{\text{act}} = 4$ meV. The barrier can be viewed as arising from a roughly sinusoidal "intrinsic" jog potential with an amplitude of about 15meV and a linear reduction in total energy due to the attraction from the other dislocation (dashed line).

DIPOLE ANNIHILATION FOR JOG-FREE DISLOCATIONS

Before we turn to the simulation results we recapitulate the mechanism for dipole annihilation of perfect screw dislocation dipoles[1, 15]. In that case the annihilation is initiated as the partial dislocations A $= \frac{a}{6}[21\overline{1}]$ and B $= \frac{a}{6}[121]$ start to recombine over part of the dislocation segment, and the perfect screw segment immediately redissociates on the cross slip plane into partials C $= \frac{a}{6}[12\overline{1}]$ and D $= \frac{a}{6}[211]$. The leading partial D is then attracted towards the partial -A forming a sessile stair-rod dislocation:

$$\frac{a}{6}[211]_{(\overline{1}11)} + \frac{a}{6}[\overline{21}1]_{(1\overline{1}1)} \rightarrow \frac{a}{3}[001]\text{ s.r.} \tag{1}$$

The annihilation completes as follows:

$$\frac{a}{3}[001]\text{ s.r.} + \frac{a}{6}[\overline{121}]_{(1\overline{1}1)} \rightarrow \frac{a}{6}[\overline{121}]_{(\overline{1}11)}$$

$$\frac{a}{6}[\overline{121}]_{(\overline{1}11)} + \frac{a}{6}[12\overline{1}]_{(\overline{1}11)} \rightarrow 0 \ . \tag{2}$$

SIMULATION PROCEDURE AND RESULTS

We investigate jogged screw dislocation dipoles with four different dipole heights $h \in [4, 6, 10, 15]\{111\}$planes. The first step is to determine which of the dipoles are stable under an energy minimization. When the dipoles are introduced in the computational cell they do not represent stable minimum configurations for several reasons: The atomic positions in the dislocation cores and in particular in the vicinity of the jog have not been optimized. Furthermore, the initial splitting distance between the partials is not necessarily correct.

The dipolar configurations are relaxed using the MD-min algorithm[14], where a Verlet Molecular Dynamics time step is performed and the atomic momentum set equal to zero if its dot product with the forces acting on the atom is negative or zero.

We find that the dipoles with heights $h \in [4, 6, 10]\{111\}$ planes annihilate spontaneously, while the one with $h = 15\{111\}$planes relaxes to a stable minimum-energy configuration. Varying the initial conditions (for example the initial splitting width between the partials) we find that this behavior does not depend sensitively on the details of the initial setup.

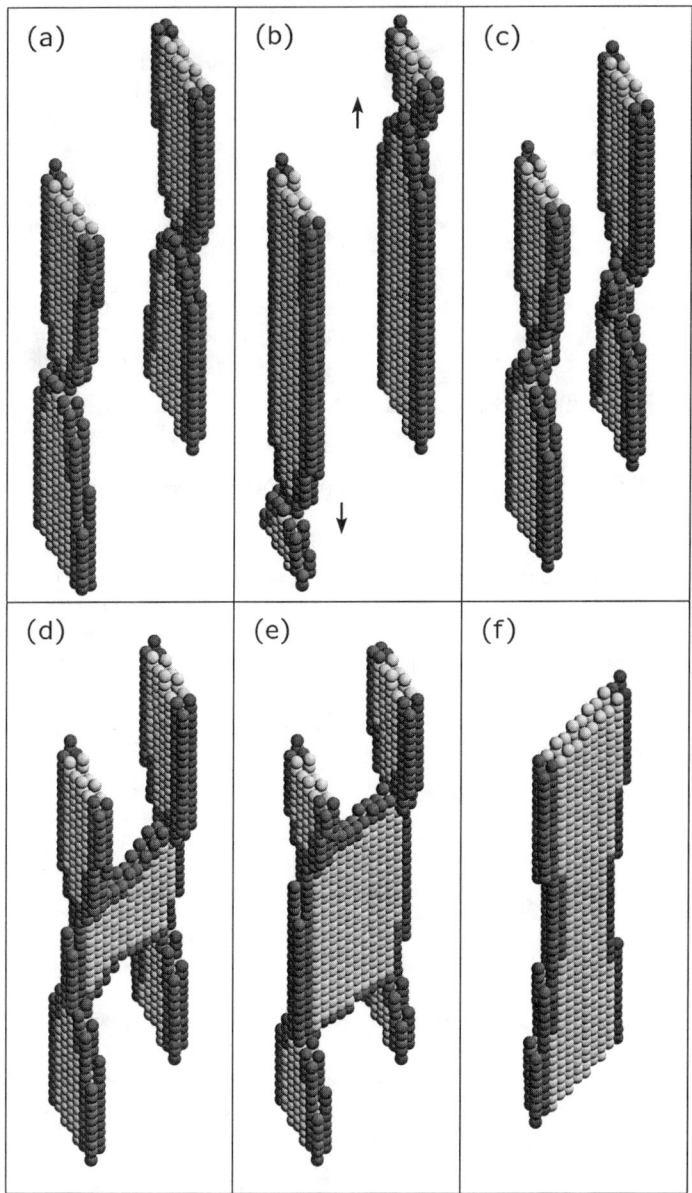

Figure 3: The annihilation of the $h = 15(1\bar{1}1)$ planes dipole. (a) The initial configuration. (b) The last meta-stable dipole configuration, as the dipole height is reduced by the motion of the jogs. (c) When $h' = 11(1\bar{1}1)$ planes, the partials A $= \frac{a}{6}[21\bar{1}]$ and B $= \frac{a}{6}[121]$ start to recombine around the jog, and redissociate with partial D $= \frac{a}{6}[211]$ moving into the cross slip plane. (d) The partials -A and -B then start redissociating, and partial -D initiate annihilation with D. (e) The continued jog-motion moves the partial dislocations into the cross slip plane. (f) The remaining partial dislocations C $= \frac{a}{6}[12\bar{1}]$ and -C complete the annihilation.

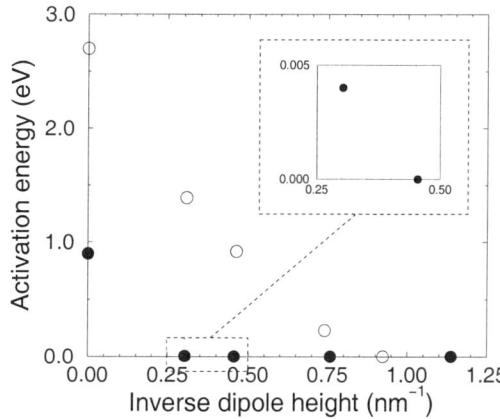

Figure 4: The activation energy as a function of the inverse dipole height ($\bullet$). The annihilation occurs spontaneously for $h^{-1} \geq 0.4$ nm^{-1}. The point for the infinitely high dipole is not from annihilation simulations, but the activation energy for cross slip of an individual screw dislocation ($l_{jog} = \frac{a}{4}[1\bar{1}2]$)[4]. The open circles ($\circ$) represent values for the annihilation of jog-free dipoles[1].

The spontaneous annihilation of the dipoles with heights $h \in [4, 6, 10]\{111\}$planes occurs via cross slip of both dislocations. First, the partials A $= \frac{a}{6}[21\bar{1}]$ and B $= \frac{a}{6}[121]$ start recombining around the jog and subsequently redissociates with partial D $= \frac{a}{6}[211]$ moving into the cross slip plane, see Figure 3 frame (c). The partials -A and -B then start redissociating, and partial -D initiate annihilation with D (frame (d)). The continued jog-motion moves the partial dislocations into the cross slip plane (frame (e)). Finally, the remaining partial dislocations C $= \frac{a}{6}[12\bar{1}]$ and -C complete the annihilation (frame (f)).

The system with $h = 15\{111\}$planes is analyzed further using NEB simulations. The annihilation is found to take place by first reducing the dipole height via displacement of the jogs with respect to each other on their respective dislocations, see Figure 3 frames (a) - (c). The asymmetric barrier associated with the motion of the jogs has been determined using the NEB technique, see Figure 2. With the initial separation we find a small activation energy of only $E_{act} = 4$ meV. This barrier can be viewed as arising from an "intrinsic" periodically varying potential for jog motion with an amplitude of about 15 meV in combination with a slowly varying attraction from the other dislocation. As the jog on one of the dislocation moves along the dislocation line the two dislocations get closer to each other and the barrier for jog motion decreases. At a dislocation height of about $h = 14\{111\}$planes this barrier vanishes completely (Figure 3 frame (b)) and the annihilation proceeds spontaneously. This last part of the process can be followed by energy minimization. First the jogs move along the dislocation lines reducing the dipole height even further until a separation of $h' = 11\{111\}$ planes is reached (see Figure 3 frame (c)), after which the annihilation via cross slip occurs spontaneously (frames (c) - (f)).

The effect of jogs on the activation energy for dipole annihilation is illustrated in Figure 4. In the limit of an infinitely high jogged dipole, we take the previously calculated activation energy for cross slip of a single screw dislocation with an identical jog (line vector: $l_{jog} = \frac{a}{4}[1\bar{1}2]$) of $E_{cs\text{-}jog} = 0.9$ eV[4] as an upper bound on the annihilation barrier, when jog migration along the dislocation is excluded.

CONCLUSION

We have shown that the presence of jogs in screw dislocation dipoles significantly affect the dipole stability. The minimum stable dipole height at zero Kelvin is increased from 1 nm for jog-free dipoles to about 2.4 nm, when jogs are present. The latter height corresponds to a dislocation separation of about 11{111}planes.

ACKNOWLEDGMENTS

We would like to thank Torben Rasmussen for helpful discussions. Center for Atomic-scale Materials Physics is sponsored by the Danish National Research Foundation. The present work was partially financed by The Danish Research Councils through Grant No. 9501775, and was done as a collaboration between CAMP and the Engineering Science Centre for Structural Characterization and Modelling of Materials Risø.

REFERENCES

[1] T. Rasmussen, T. Vegge, T. Leffers, O. B. Pedersen, and K. W. Jacobsen, Phil. Mag. A, (in press).

[2] M. S. Duesberry, Model. Simul. Mater. Sci. Eng., **35**, 35 (1998).

[3] U. Essmann and H. Mughrabi, Phil. Mag. A, **40**(6), 731 (1979).

[4] T. Vegge et al. (to be published).

[5] G. Mills, H. Jónsson and G. Schenter, Surface Science, **324**, 305 (1995).

[6] H. Jónsson, G. Mills and K. W. Jacobsen, Classical and Quantum Dynamics in Condensed Phase Simulations, edited by B. J. Berne, G. Ciccotti and D. F. Coker, World Scientific (1998).

[7] K.W. Jacobsen, J.K. Nørskov, and M.J. Puska, Phys. Rev. B, **35**,7423, (1987).

[8] K.W. Jacobsen, P. Stoltze, and J.K. Nørskov, Surf. Sci., **366**, 394, (1996), and references therein.

[9] C. Kittel, *Introduction to Solid State Physics, 4th. ed.*, Wiley, (1971).

[10] J.P. Hirth and J. Lothe, *Theory of Dislocations 2nd. ed.*, Krieger publ. comp., (1992).

[11] N. M. Rosengaard and H. L. Skriver, Phys. Rev. B **47**, 12865 (1993).

[12] H. Jónsson and H. C. Andersen, Phys. Rev. Lett. **60**, 2295 (1988).

[13] A. S. Clarke and H. Jónsson, Phys. Rev. E **47**, 3975 (1993).

[14] P. Stoltze, *Simulation Methods in Atomic-scale Materials Physics* (Polyteknisk forlag, 1997).

[15] T. Rasmussen, Mat. Res. Soc. Symp. Proc. Vol. **538**, 51 (1999).

[16] T. Rasmussen, K. W. Jacobsen, T. Leffers and O. B. Pedersen, Phys. Rev. B,**56**(6), 2977 (1997).

[17] T. Rasmussen, K. W. Jacobsen, T. Leffers, O. B. Pedersen, S. G. Srinivasan and H. Jónsson, Phys. Rev. Lett., **79**(19), 3676 (1997).

STRUCTURE OF A DISSOCIATED EDGE DISLOCATION IN COPPER

L. F. Perondi†, P. Szelestey and K. Kaski
Helsinki University of Technology, Laboratory of Computational Engineering
P.O.Box 9400, FIN-02015 HUT, FINLAND
†Permanent address: Instituto Nacional de Pesquisas Espaciais - INPE, SP, Brazil

ABSTRACT

The structure of a dissociated edge dislocation in copper is investigated. Attention is given to the structure of the Shockley partials and the equilibrium size of the fault ribbon. The studies are carried out through Molecular Dynamics simulations. The atomic interactions have been modelled through an Embedded Atom Model (EAM) potential, the implementation of which has been specially designed for this study. Our main results show that the equilibrium distance between partials is very sensitive to the type of boundary conditions imposed on the simulated system.

INTRODUCTION

The subjects of dislocation structure and dynamics have received considerable attention over the past years[1]. Dislocations play a fundamental role in a variety of phenomena, such as plastic deformation of materials [2], fast diffusion [3], melting [4], crystal growth [5], crack propagation[6], to name a few.

In the theory of plasticity, the mobility of dislocations is considered as the main factor in determining whether materials will display a ductile or brittle behaviour. Several factors influence the mobility of a dislocation. The interactions with point defects, either native or external, as well as with extended defects, such as boundaries, and with other dislocations constitute main examples. It may be said that a dislocation also interacts with the lattice that holds it. Geometrical constraints, dependent on lattice symmetry, allow conservative movement only along given planes at particular directions while the Peierls stress, a result of the discreteness of the lattice, gives rise to a periodic barrier for dislocation motion. The successful modelling and description of these *interactions* depend, ultimately, on the accurate description of the structure of the dislocation, including the splitting into partials and the formation of a stacking fault ribbon, a task that is beyond the reach of continuum theories such as linear elasticity theory.

Molecular dynamics (MD) simulations with many-body model potentials have found widespread application in such situations, *i. e.*, in the investigation of microscopic phenomena at scales where the discreteness of the lattice and non-linearities in the relation force-displacement play a significant role. The recent literature provides several examples of application of MD simulations to the study of the core of partial dislocations in fcc lattices. Examples include Refs. [7–10].

In materials with an fcc lattice, such as copper, edge dislocations glide along $\{111\}$ planes in a $< 110 >$ direction. The dislocation line lies along a $< 112 >$ direction. As is well known, energy balance favours the splitting of such dislocations into two Shockley partials, which may lie several lattice spacings apart depending on the stacking fault energy (SFE) of the material. The equilibrium distance is defined by the balance of the repulsion force between partials with the force arising from the stacking fault energy of the fault

223

ribbon that forms between the partials in the gliding plane. In a computer simulation, the equilibrium distance between the partials is an outcome of the simulation itself, and depends on the model potential, the geometry of the simulated sample and the boundary conditions. Systems with cylindrical [7] and cubic [8, 9] geometry, with either free or fixed boundary conditions, have been considered in the recent literature.

In this paper we investigate the structure of a split edge dislocation in copper. We consider a system with cubic geometry and show that the distance between partials is very sensitive to the imposed boundary conditions. The study is carried out through MD simulations. Inter-particle interactions are modelled with a recently proposed embedded atom model potential (EAM), by Chantasiriwan and Milstein (CM) [11], the derivation of which takes into account second and third order elastic constants. This potential in its original version for copper, however, gives the wrong sign for the SFE. Here, we also present a modified version, which reproduces the correct sign of the SFE, without spoiling qualities of the original version.

In the next section we describe in detail the potential used in the present study. Details about the geometry and size of the simulated system as well as a description of the boundary conditions are given in Section 3. This section is also dedicated to a discussion of our results. Section 4 contains the summary and conclusions.

MODEL POTENTIAL

The embedded atom model potential scheme proposed in Ref. [11] may be summarized as follows. The total cohesive energy of the material is written as

$$E_{EOS} = \sum_i F \left(\sum_{j \neq i} \rho(r_{ij}) \right) + 1/2 \sum_i \sum_{j \neq i} \phi(r_{ij}), \qquad (1)$$

where r_{ij} is the distance between particles i and j, F is the embedding energy, $\rho(r_{ij})$ is the electronic density at site i due to the particle at site j and $\phi(r_{ij})$ is the Coulomb interaction between particles i and j. The embedding energy together with the electron density are obtained as follows. First, the following functional forms are assumed for ρ and ϕ

$$\rho(r) \;=\; A_0 \frac{1 + b_1 cos(\alpha r) + b_2 sin(\alpha r)}{r^\beta} \qquad (2)$$

$$\phi(r) \;=\; A(r - r_m)^4 (1 + \sum_{i=1}^{6} d_i r^i), \qquad (3)$$

where A_0, A, b_1, b_2, α, β, the set of coefficients $\{d_i\}$ and r_m are fitting parameters. The parameter r_m is considered as the cut-off radius for both the density and the pair-potential. Note that the pair-potential displays a smooth cut-off by construction. Second, a modified Rose's equation is used for E_{EOS}. It has been derived by fitting the theoretical *pressure* × *volume* relation ($P = -dE_{EOS}/dV$) to the experimental data. The result given by CM is expressed as

$$E_{EOS} = \begin{cases} -E_{coh}(1 + b/\lambda + k(b/\lambda)^3) \exp(-b/\lambda) & a/a_0 \leq 0.95, a/a_0 \geq 1.05 \\ -E_{coh} + w_1 b^2 + w_2 b^3 + \sum_{i=1}^{4} \gamma_i b^{(i+4)} & 0.95 < a/a_0 \leq 1 \\ -E_{coh} + w_1 b^2 + w_2 b^3 + \sum_{i=1}^{4} \eta_i b^{(i+4)} & 1 < a/a_0 \leq 1.05, \end{cases} \qquad (4)$$

where $b = (a/a_0 - 1)$, a_0 is the lattice parameter of the reference lattice, the sets $\{\gamma_i\}$ and $\{\eta_i\}$ are fitting parameters given in Table 1[1] and $\lambda = \sqrt{(E_{coh}/9V_0B)}$, in which E_{coh}, B and V_0 are the cohesive energy per particle, the bulk modulus and the volume per particle in the reference lattice, respectively. k is also a fitting parameter and its value is given in Table 1. The constants w_1 and w_2 are related to second and third order elastic constants through the relations

$$
\begin{aligned}
w_1 &= 3V_0(C_{11} + 2C_{12})/2 \\
w_2 &= V_0(C_{111} + 6C_{112} + 2C_{123}) + w_1.
\end{aligned}
\tag{5}
$$

Finally, the set of parameters b_1, b_2, A and $\{d_i\}$ are determined through the fitting of a set of equations which relate ρ and ϕ with expressions involving the second and third order elastic constants and the unrelaxed formation energy of a vacancy. The equations involving b_1 and b_2 are independent of the equations involving A and the set $\{d_i\}$. The values for the other parameters (A_0, α, β and r_m) are chosen in an empirical way, by inspection of the phonon dispersion curves. Their values are listed on the right-hand side of Table 1. The last step then follows the usual procedure, in which through dilation of a perfect lattice one relates F to ρ by making use of Eq. 1. Table 2 displays the experimental data for copper used in the construction of the potential while Table 3 displays, on the second column, the original values of the density and pair-potential parameters.

γ_1	$-17\,259.61681$	η_1	$18\,726.46078$	k	0.06
γ_2	$-541\,086.9574$	η_2	$-541\,086.9574$	$\alpha\,(\text{Å}^{-1})$	7
γ_3	$-7\,195\,460.939$	η_3	$7\,198\,976.218$	β	10
γ_4	$-35\,988\,498.49$	η_4	$-35\,987\,737.82$	$A_0\,(C\text{Å}^7)$	1.0

Table 1: Parameters γ, η, k, α, β and A_0 of Eqs. 2 and 4 as given in Ref. [11]. Values of γ and η are in eV. See footnote.

$a_0(\text{Å})$	3.62	$C_{111}(Mbar)$	-12.71
$E_v(eV)$	0.67	$C_{112}(Mbar)$	-8.14
$E_{coh}(eV)$	0.67	$C_{123}(Mbar)$	-0.50
$C_{11}(Mbar)$	1.08	$C_{144}(Mbar)$	-0.03
$C_{12}(Mbar)$	0.62	$C_{166}(Mbar)$	-7.80
$C_{44}(Mbar)$	0.62	$C_{456}(Mbar)$	-0.95

Table 2: Experimental data for copper as given in Ref. [11].

The parameters given in Ref. [11] lead to a negative value for the stacking fault energy in copper. By increasing the cut-off radius (r_m) of the potential and repeating the fittings that involve the parameters of ϕ, we have derived a new set of values that satisfy the previous

---

[1]According to our computations, there seems to exist a problem with units in Ref. [11] and these values have to be divided by 1.60219 before being inserted in Eq. 4. The same holds for the constants w_1 and w_2 and the parameter A.

equations while giving the correct sign for the stacking fault energy. Table 3 displays the old and new values for the parameters and a comparison of computed physical quantities obtained from the two versions of the potential. As shown in the table, the new parametrization of the potential gives approximately the same values for the formation energy of point defects while giving an SFE value with the correct sign. These are preliminary conclusions – to certify the accuracy of the description given by the proposed potential further computations, such as of the phonon dispersion relations, are still necessary.

	Ref. [11]	Present work
$r_m(\mathring{A})$	4.50	5.40
$A(eV/\mathring{A}^4)$	25.04620427	0.1436702762
$d_1(\mathring{A}^{-1})$	-2.098826585	2.4855880870
$d_2(\mathring{A}^{-2})$	1.840347959	-5.3507629375
$d_3(\mathring{A}^{-3})$	-0.8604116108	3.6641531830
$d_4(\mathring{A}^{-4})$	0.225611227	-1.2040526784
$d_5(\mathring{A}^{-5})$	-0.03140081975	0.1952439417
$d_6(\mathring{A}^{-6})$	0.001808853498	-0.0126038106
SFE (erg/cm^2)	-5.05	17.05
$E_f^v\ (eV)$	1.30	1.28
$E_f^i\ (eV)$	3.20	3.32

Table 3: List of the parameters of the pair-potential ϕ as given in Ref. [11] and in the modified version developed in the present work. The lower part of the table shows a comparison of properties computed in the two versions. E_f^v and E_f^i stand for the relaxed formation energy of a vacancy and a dumbbell interstitial, respectively.

RESULTS AND DISCUSSION

A system in the form of a parallelepiped with faces in the directions $[\bar{1}10]$ (x), $[\bar{1}\bar{1}2]$ (y) and $[111]$ (z) has been used in the simulations. Two halves of $(\bar{1}10)$ planes, four planes apart, were initially removed and the atoms between them were given a displacement corresponding to the Burgers vector of a partial dislocation, thus introducing a stacking fault at the gliding plane in this region. The resulting Shockley partials have their dislocation lines in the y-direction and exhibit Burgers vectors $\frac{1}{6}[\bar{2}11]$ (δB) and $\frac{1}{6}[12\bar{1}]$ (Aδ). In all simulations we use periodic boundary conditions along the y-direction. We have investigated the effect of different boundary conditions along the x- and z-directions. Unless otherwise specified, the system dimensions are $60a_0/2^{1/2}$ (120 $(\bar{1}10)$ planes) in the x-direction, $3\ 6^{1/2}a_0/2$ (18 $(\bar{2}\bar{2}1)$ planes) in the y-direction and $23a_0/3^{1/2}$ (23 (111) planes) in the z-direction. After the introduction of the partials, there are 8214 atoms in the system. We next discuss our results.

For free boundaries in the gliding direction, the partials end up at the external borders, even for systems with x-dimension of around 140 unit cells. This value is considerably larger than the value of approximately 50 unit cells predicted from elasticity theory for a system with the present SFE energy and when consideration is given to the image forces due to the free boundaries. The migration of the partials is accompanied by a large bending

Figure 1: Perspective view of the free boundary system showing bending in the z-direction.

of the system along the z-direction. Fig. 1 shows a typical configuration at intermediate times, *i. e.*, when the partials are in the middle of their way to the borders. Fig. 2 shows a close view of the atomic structure near the core of a partial. This structure changes with the amount of bending of the system. The large bending makes it difficult to justify any simple computation of the equilibrium distance based on linear elasticity theory.

The partials always migrate to the boundary also in the situation when the lower border in the z-direction is held fixed, for systems with x-dimension of 60 unit cells. The time it takes for the partials to reach the borders is relatively smaller than in the free boundary case, probably because due to the boundary constraint there is no bending.

With fixed borders in the x-direction and free boundaries in the z-direction, it is observed that the partials always equilibrate at a position far from the fixed boundaries. This configuration has been observed in previous simulations, as shown in Refs. [8, 9]. The equilibrium distance between partials as a function of the system size, obtained in our simulations, were (87,120), (120,160) and (199,240), where the numbers in each pair represent the equilibrium distance and system size, respectively, in units of number of x-planes. It is seen that the equilibrium distance exhibits a noticeable dependence on the x-dimension of the system.

CONCLUSIONS

The results given in this note indicate that care must be exercised when interpreting the results of simulations of split dislocations in samples with cubic geometry. In particular, the observed equilibrium distance between partials was shown to depend markedly on the boundary conditions imposed on the system. With free boundaries in the direction of dislocation motion we always observed migration of the partials to the borders, while if the referred borders are held fixed we observe a dependence of the equilibrium distance between partials on system size.

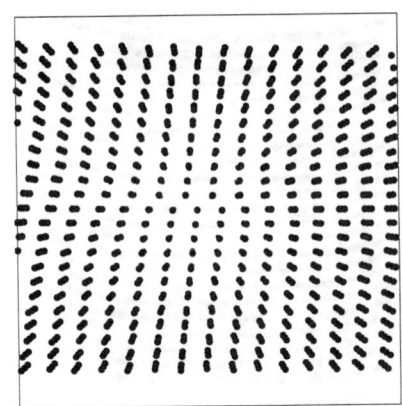

Figure 2: Perspective view of the core structure of a partial dislocation in the free boundary system.

REFERENCES

[1] T. Suzuki, S. Takeuchi and H. Yoshinaga, *Dislocation Dynamics and Plasticity*, Springer-Verlag, Berlin, 1989; J. P. Hirth and J. Lothe. *Theory of dislocations*, John Wiley & Sons, New York, 1982;D. Hull and D. J. Bacon, *Introduction to dislocations*, Butterworth-Heinemann, Oxford, 1984;

[2] D. Kuhlmham-Wilsdorf, Phil. Mag. A, **79**, 955 (1999).

[3] A. R. Allnatt and A. B. Lidiard, *Atomic Transport in Solids*, Cambridge University Press, Cambridge, 1993.

[4] See for instance H. Kleinert, *Gauge Fields in Condensed Matter*, World Scientific, Singapore, 1989.

[5] E. A. Fitzgerald, Mat. Sci. Reps. **7**, 87 (1991).

[6] B. Lawn, *Fracture of Brittle Solids*, Cambridge University Press, Cambridge, 1993.

[7] R. G. Hoagland, A. F. Voter and S. M. Foiles, Scrip. Materialia **39**, 589 (1998).

[8] J. Huang, M. Meyer and V. Pontikis, Phys. Rev. Letts. **63**, 628 (1989); J. Huang, M. Meyer and V. Pontikis, Phys. Rev. B **42**, 5495 (1990).

[9] J. von Boehm and R. Nieminen, Phys. Rev. B **53**, 8956 (1996).

[10] M. Doyama, in *Mesoscopic Dynamics of Fracture*, ed. by H. Kitagawa, T. Aihara and Y. Kawazoe (Springer-Verlag, Berlin, 1998), pp. 49.

[11] S. Chantasiriwan and F. Milstein, Phys. Rev. B **58**, 5996 (1998); S. Chantasiriwan and F. Milstein, Phys. Rev. B **53**, 14080 (1996).

ATOMISTIC SIMULATION OF TRANSONIC DISLOCATIONS

JONATHAN A. ZIMMERMAN*, FARID F. ABRAHAM** and HUAJIAN GAO*
*Division of Mechanics and Computation, Stanford University, Stanford, CA 94305-4040
**IBM Almaden Research Center, San Jose, CA 95120

ABSTRACT

Recent work has been done on the analysis of elastic stress singularities, such as cracks and dislocations, which propagate at supersonic speeds. Gumbsch and Gao have performed atomistic simulations in which dislocations are created and travel at transonic velocities (speeds which are greater than the material's shear wave speeds but less than the longitudinal wave speed) close to the theoretical value corresponding to the radiation-free state for glide motion. Gao et al. have derived expressions for this radiation-free velocity in both isotropic and anisotropic media. We have performed molecular dynamics simulations showing dislocation nucleation at crystal surface ledges. Dislocations were nucleated at either sub- or transonic velocities, depending upon the ambient temperature, and accelerated to transonic speeds. This paper shows the velocity profiles for the emitted dislocations and compares velocities observed with a theoretical minimum-radiation speed derived by a Stroh-type anisotropic elasticity analysis performed by Barnett and Zimmerman. Our findings are particularly exciting considering these simulations were not specifically engineered for the purpose of creating transonic defects, but show agreement with theory nonetheless.

INTRODUCTION

The motion of strength-limiting material defects is of substantial interest to materials scientists and engineers. Defects such as cracks and dislocations are atomic in nature, yet create long-distance stress fields which are able to be modeled by linear elasticity. Eshelby [1] derived the stress and displacement fields for a dislocation in an isotropic material moving at a uniform velocity. He noted that these expressions became unbounded when the dislocation traveled at the material's shear wave speed. Eshelby also showed a mathematically-possible solution, which did not require external work, occurred for a glide dislocation moving at a speed $\sqrt{2}\,c_{shear}$. Supersonic motion of a dislocation generally requires external work be provided because the dislocation radiates energy in the form of shock waves as it propagates through a material.

Weertman [2] later developed stress expressions for both singular dislocations and dislocations with a Burgers vector distribution of finite width. He derived expressions for the work required for supersonic motion, which reduced to zero for the case mentioned by Eshelby. Stroh [3] derived fields for dislocations moving at a uniform subsonic speed in an anisotropic elastic material and suggested that the functional form of these solutions altered from standard logarithmic functions for subsonic motion, to heaviside functions for supersonic motion in order to represent the line discontinuity of the shock front. Gao et al. [4] have shown that for anisotropic solids, it may be possible for a dislocation to exhibit "radiation-free" transonic (a speed between the smallest and largest limiting speeds of the material) motion depending upon the elastic moduli and Burgers vector of the dislocation. Barnett and Zimmerman [5] have developed full solutions for the stress and displacement fields of a dislocation with any Burgers vector moving in an anisotropic material at any velocity. These solutions are used to form an expression for the work necessary to compensate

Mat. Res. Soc. Symp. Proc. Vol. 578 © 2000 Materials Research Society

for the energy radiated. They have shown it is possible to determine a "minimum-work" velocity, the speed within the transonic regime at which the least amount of energy is radiated. When this "minimum-work" equals zero, the solution matches the "radiation-free" expressions of Gao *et al.*

Although supersonic motion of dislocations is not commonly observed, it has been noticed in atomistic simulation of materials. Abraham *et al.* [6] observed dislocations emitted transonically from a crack-tip in a Lennard-Jones solid. Gumbsch and Gao [7] performed molecular dynamics simulations of isotropic BCC tungsten and showed dislocations created with speeds close to the theoretical "radiation-free" speed. They suggest that such speeds are physically possible for situations of large background stress, such as mechanical twinning, and for dislocation nucleation from a stress concentration. Zimmerman [8] has shown such nucleation to occur in biaxially-stretched FCC metals with surface ledges, a situation emulating a heteroepitaxial thin film that acquires surface steps during deposition.

This paper will show the results of molecular dynamics simulations of dislocations nucleated at crystal surface ledges at both 0 K and at finite temperatures. Images of dislocation emission will be shown, along with profiles of the dislocations velocities with time. Details about the velocity profiles will be discussed with regard to the energy barriers associated with elasticity theory.

THEORY

Molecular dynamics simulations were performed on a copper hexahedron modeled with the embedded atom method [9] using the potential functions developed by Voter and Chen [10, 11]. The slab was stretched in horizontal and thickness directions, containing periodic boundary conditions, while the vertical surfaces remained free. Two different crystal orientations were used; in one case the vertical surfaces were (001) planes while in the other case they were (110) planes. The (001) system contained approximately 56,000 atoms, constructed from a primitive 2-atom cell which is replicated 70 times in the horizontal direction, 40 times in the vertical direction, and 10 times in the thickness direction. The (110) system is of the same size with 50 cells horizontally by 56 cells vertically by 10 cells wide. The dimensions of the 2-atom cell are $\frac{a_0}{\sqrt{2}}$ in each $\langle 110 \rangle$ direction and a_0 in the [001] direction, prior to any stretching or relaxation, where $a_0 = 3.615$ Å. The top surfaces of both systems contained a trough, two oppositely-oriented surface ledges, which was a single monolayer deep and five cells wide.

FCC metals slip by the motion of dislocations with $\langle 110 \rangle$-type Burgers vectors on $\{111\}$ planes, which usually appear as a pair of $\langle 112 \rangle$-type partial dislocations bounding a stacking fault. In our simulations, dislocation emission occurred as $\langle 112 \rangle$ partials created at the surface ledge and left a stacking fault behind as the partial moved into the bulk. Sometimes, a trailing partial with a different $\langle 112 \rangle$ Burgers vector also formed and traveled into the bulk, leaving a whole crystal behind. Energy minimization simulations were used to determine how much biaxial stretch was necessary to emit the leading partials. This value was 3.66% for the (001) system and 6.4% for the (110) system. These levels of deformation were also used in our molecular dynamics calculations; however, restart configurations were taken from the energy minimization simulations (before dislocation formation), with velocities reset to a random distribution at a desired temperature.

RESULTS

Simulations of the (001) Geometry

The simulation for the (001) crystal is shown in Figure 1. This image is shaded according to local temperature (*i.e.* atomic kinetic energy), with the "hot" atoms shaded the lightest and the "cool" ones the darkest. The figure shows two partial dislocations being emitted,

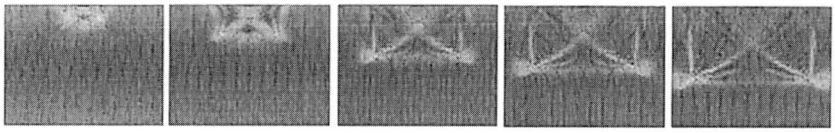

Figure 1: Snapshots of a MD simulation of copper with a (001) free surface and a surface trough. Atoms are shaded according to temperature.

one from each surface ledge, leaving stacking fault trails behind. A feature made visible by shading according to temperature is the appearance of shock waves in the material, indicating that the dislocations traveled through the material at supersonic speeds. This was confirmed by measurement of the distances the dislocation cores traveled between captured frames, taken at intervals of 1000 time steps, where the characteristic time step was $\Delta t = 0.03453 \times 10^{-14}$ sec. Figure 2 shows profile of the dislocation velocity as a function of time (represented by frame number). The average dislocation speed for this simulation was

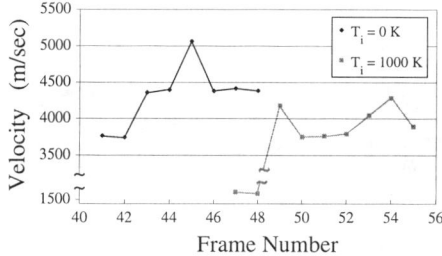

Figure 2: Velocity profiles for simulations of systems with a (001) free surface.

4314 m/sec with an uncertainty of ±290 m/sec.

Figure 3 depicts the same crystal with atomic velocities reset to an equivalent temperature of 1000 K. This figure is shaded according to local values of normal Lagrangian strain, calculated using the current and reference (*i.e.* stress-free) configurations of the system. This method of calculating continuum mechanical strain produces extremely high values for atoms bordering a stacking fault plane. Prior to dislocation emission, the system's temperature dropped to roughly half of its initial value ($\sim$ 500 K), converting the rest of the kinetic energy added to the system to potential energies. At a later time, a dislocation was emitted and traveled into the bulk. Due to the random nature of the velocity distribution, nucleation was seen to occur only at the right-hand ledge, rather than at both ledges simultaneously, as previously seen. The velocity profile for this simulation was also seen in Figure 2 and shows that nucleation occurred somewhat later than for the 0 K simulation (a difference of 2.4 ps), and the dislocation traveled at somewhat lower speeds. The dislocation at 500 K

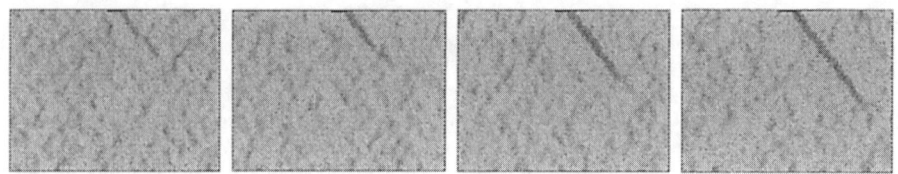

Figure 3: Snapshots of a MD simulation of same (001) system with an initial temperature set to 1000 K. Atoms are shaded according to values of normal Lagrangian strain in the horizontal direction with darkest shading denotes the largest magnitudes of strain.

initially moved subsonically ($v \approx 1594$ m/sec), then accelerated into the transonic regime ($v \approx 3958$ m/sec) once it has moved away from the free surface.

The analysis of Barnett and Zimmerman [5] was used for motion of a dislocation in a $\langle 112 \rangle$ direction on a $\{111\}$ plane in a cubic crystal. The limiting speeds for Voter and Chen EAM copper are $v_3 = 2136$ m/sec, $v_2 = 2359$ m/sec and $v_1 = 4998$ m/sec. When the Burgers vector and the velocity are along the same $\langle 112 \rangle$ directions, the case for the (001) free surface, the "minimum-work" speed is 4186 m/sec. This value is in good agreement with both the observed value of 4314 m/sec at 0 K and the value of 3958 m/sec reached in the higher temperature simulation.

Simulations of the (110) Geometry

A simulation and subsequent analysis was also performed for the (110) free surface copper crystal in which the Burgers vector of the leading partial dislocation is not along the same $\langle 112 \rangle$ direction as the direction of motion. This simulation is shown in Figure 4. The atoms

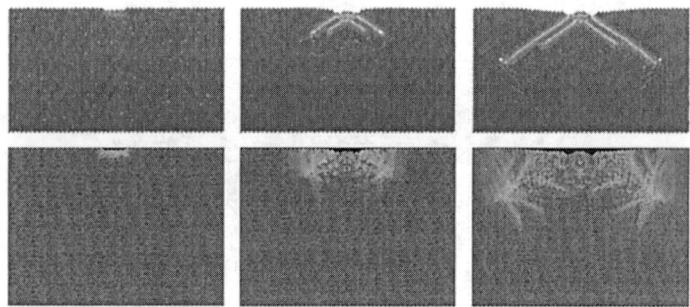

Figure 4: Snapshots of a MD simulation of copper with a (110) free surface and a surface trough. The top sequence shows atoms shaded according normal Largrangian strain in the thickness direction with lightest shading denoting the largest magnitudes of strain. The bottom sequence shows atoms shaded according to temperature.

in the top sequence in Figure 4 are shaded according to their value of normal Lagrangian strain in the thickness direction, while the same respective frames are shown in the bottom sequence shaded according to the local value of temperature.

In this simulation, two dislocations were emitted from each ledge on adjacent $\{111\}$ glide planes. The upper dislocations were full $\langle 110 \rangle$ dislocations consisting of two $\langle 112 \rangle$ partials which have both been emitted. The lower dislocations were leading $\langle 112 \rangle$ partials. The upper dislocations outraced the lower dislocations, leaving a separation distance between dislocation cores that grew with time. The temperature-shaded images in Figure 4 show sharp lines indicating supersonic motion of the defects. The velocity profile appears in Figure 5(a). The average speed of the upper dislocations was 2901 m/sec (a transonic

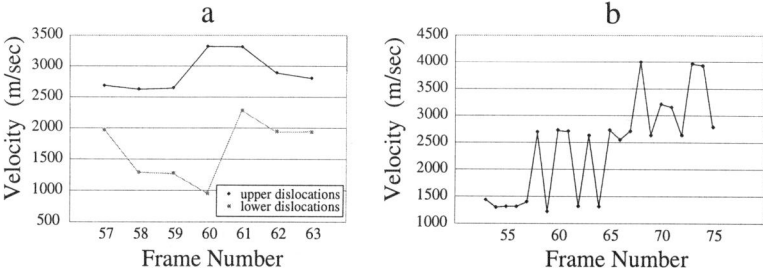

Figure 5: Velocity profile for simulations of systems with a (110) free surface: (a) 6.4% biaxial stretch with initial temperature set to 0 K (b) 5.55% biaxial stretch with initial temperature set to 300 K.

speed), while the average speed of the lower dislocations was 1668 m/sec (subsonic). The complicated contours of temperature are the result of the shocks emitted from the upper dislocations interacting with the stacking fault along the trails of the lower dislocations. The analysis of Barnett and Zimmerman was again used to determine the "minimum-work" speed for motion of a ⟨110⟩ dislocation in the inclined ⟨112⟩ direction, 4131 m/sec. While this does not agree with the simulation prediction of 2901 m/sec, a plot of required work as a function of dislocation speed shows that work is essentially constant over a range of values between the second limiting speed (2359 m/sec) and this "minimum-work" speed. Thus, a speed of 2901 m/sec can be considered a neutral equilibrium state.

It was found for the (110) system that increased temperature in the system allowed dislocation emission to occur at lower amounts of biaxial stretch. Nucleation occurred for stretches as low as 5.55%, the simulation for which is shown in Figure 6 for an initial temperature of 300 K. Here, a complete ⟨110⟩ dislocation was emitted from the right surface

Figure 6: Snapshots of a MD simulation of copper with a (110) free surface and a surface trough, biaxially stretched 5.55% with an initial temperature set to 300 K. Atoms are shaded according to values of normal Lagrangian strain in the horizontal direction with darkest shading denoting the largest magnitudes of strain.

ledge, but along the (111) plane, the opposite plane to the one expected for tension. The form of this full dislocation was again the leading partial emitted followed by the trailing partial after some amount of relaxation.

The velocity history for the dislocation emitted in Figure 6 was shown in Figure 5(b). Three distinct velocity ranges are observed: initially, the dislocation had a subsonic velocity of ~ 1325 m/sec, it later oscillated between this value and a transonic velocity of ~ 2766 m/sec, and eventually reached an even higher speed of ~ 3968 m/sec deeper into the bulk crystal. This higher value is in good agreement with the "minimum-work" speed of 4131 m/sec, and both transonic speeds fall within the range of neutral equilibrium mentioned earlier (2359 - 4131 m/sec).

CONCLUSIONS

The results shown in Figures 2 and 5 are extraordinary in light of the elasticity theory of steady-state motion of dislocations. The velocity profiles of both non-zero temperature simulations show dislocations initially moving subsonically and accelerating to transonic speeds. Elasticity solutions typically show an infinite energy barrier for accelerating a dislocation past the lowest limiting speed. These atomistic simulations do not display such a barrier. This unexpected behavior may partly be caused by the presence of the free surface which alters the elastic fields, resulting in the "minimum-work" speed to be a function of distance from the surface. However, other aspects such as the finite deformation behavior of the material and atomic nature of a dislocation core may account for this deviation from linear theory.

This paper has shown atomistic simulations of dislocations nucleated at crystal surface ledges, which travel through the crystal at transonic velocities under some conditions. The speeds achieved are in close agreement with a "minimum-work" velocity determined by anisotropic elasticity theory. These simulations also show acceleration from subsonic to transonic motion, an event contrary to linear elasticity, and show the need for more detailed analysis to predict material behavior.

ACKNOWLEDGMENTS

This research has been supported by grants and awards provided by the Department of Energy, the National Science Foundation, the Intel Foundation, the IBM Almaden Research Center and Stanford University.

REFERENCES

[1] J. Eshelby *Phys. Soc. Proc.*, vol. 62A, pp. 307–314, 1949.
[2] J. Weertman in *Mathematical Theory of Dislocations*, Joint Applied Mehchanics and Fluid Engineering Conference proceedings, pp. 178–202, A.S.M.E., 1969.
[3] A. Stroh *J. Math. Phys.*, vol. XLI, pp. 77–103, 1962.
[4] H. Gao, Y. Huang, P. Gumbsch, and A. Rosakis *J. Mech. Phys. Sol.*, vol. 47, pp. 1941–61, 1999.
[5] D. Barnett and J. Zimmerman, "Radiative solutions for supersonic dislocations in anisotropic media," 1999. in preparation.
[6] F. Abraham, D. Brodbeck, R. Rafey, and W. Rudge *Phys. Rev. Lett.*, vol. 73, pp. 272–275, 1994.
[7] P. Gumbsch and H. Gao *Science*, vol. 283, pp. 965–8, 1999.
[8] J. A. Zimmerman, *Continuum and Atomistic Modeling of Dislocation Nucleation at Crystal Surface Ledges*. PhD thesis, Stanford University, 1999.
[9] S. Foiles, M. Baskes, and M. Daw *Phys. Rev. B*, vol. 33, pp. 7983–7991, 1986.
[10] A. Voter and S. Chen in *Characterization of Defects in Materials Symposium*, vol. 82 of *Mater. Res. Soc. Proc.*, pp. 175–180, Materials Research Society, 1987.
[11] A. Voter Tech. Rep. LA-UR 93-3901, Los Alamos National Laboratories, 1993.

DIFFERENT RELAXATION MODES
OF EPITAXIAL THIN FILM ALLOYS WITH A LARGE SIZE EFFECT

J. THIBAULT*, C. DRESSLER, P. BAYLE-GUILLEMAUD

CEA-Grenoble / Département de Recherche Fondamentale sur la Matière Condensée/SP2M,
17 rue des Martyrs, 38054 GRENOBLE-FRANCE.
* Member of the CNRS.

ABSTRACT

In epitaxial layers, the mechanism of stress relaxation under consideration is in general dislocations. This paper will present experimental evidences for other modes of relaxation, which may occur in some situations especially where a strong size effect or misfit is present. In fact the stress is the driving force for intermixing, twinning, and phase transformation, which are not expected to occur in the bulk. All these experimental results have been sustained by numerical simulations that will also be presented.

INTRODUCTION

In strongly stressed epitaxial systems, the stress relaxation does not necessarily occur via the "classical" dislocation mechanism. On the contrary, the system may find transformations that correspond to metastable states in the bulk but are driven by the stress. The AuNi system grown on (001) Au is an illustrative example of this situation. Au and Ni exhibit a large size effect and the phase diagram shows a large miscibility gap with no miscibility at room temperature.

This paper will present the different relaxation modes encountered in Au/Ni multilayers, Ni thin films or $Au_{1-x}Ni_x$ alloys grown on top on a (001) Au buffer. The samples were grown by MBE at room temperature on a MgO substrate. A detail description of the growth process has been already published [1]. It will be shown that the expected dislocation mechanism occurs only for alloys with a low misfit i.e. with a low Ni content. Otherwise, intermixing, twinning or phase transformations take place. Furthermore, it will be shown that ordering occurs with annealing treatment. This study has been performed using both high-resolution electron microscopy and X-rays [2]. This paper will present the HREM results. The HREM observations were carried out on a Jeol 4000EX electron microscope (Cs = 1 mm) operating at 400 kV.

RELAXATION BY INTERFACIAL MIXING IN Au/Ni MULTILAYERS

The results presented in this paragraph were published in [3, 4]. The main results are the following ones. Despite the 13.7 % misfit defined as $(a_{Au}-a_{Ni})/a_{Au}$, a coherent growth of Ni on

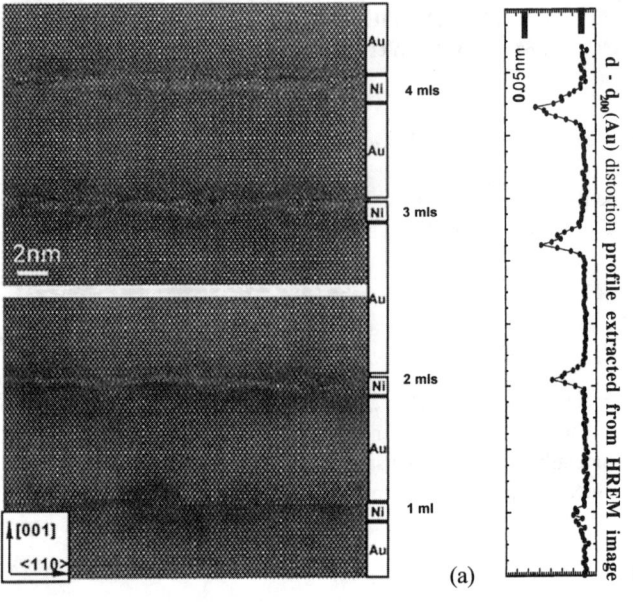

Figure 1 : HREM image (a) of a Au/Ni multilayer viewed in cross section along a <110> direction perpendicular to the [001] growing direction. It shows that a coherent growth of Ni on (001) Au takes place despite the 15% misfit up to a Ni thickness of 5 Mls (b) The distortion profile extracted from the HREM image shows that the strain is extended over a number of planes larger than the nominal one. This sustains the hypothesis that an interfacial mixing occurs during growth which partially releases the stress.

(001) Au is observed up to a Ni thickness of 5 monolayers (Mls) (fig 1a). An interfacial mixing occurs during growth that partially releases the stress. This has been evidenced by a HREM image processing which permits to extract a distortion profile (fig 1b). A molecular dynamics simulation of the growing process has also confirmed this profile [3, 5]. Under some assumptions (modified Vegard's law for the Au/Ni system and weighted combination of the elastic constants) a chemical profile has been conclusively deduced from the distortion profile and has been confirmed by line scan PEELS analysis in a VG501-STEM [6].

The main points are i) that no pure Ni plane exists and ii) that the Ni profile is larger than the number of nominally evaporated Ni layers and extends over ten to twelve planes for 3 and 4 Ni Mls respectively. Furthermore the profile is asymmetrical. The first Au/Ni interface is more abrupt than the Ni/Au interface, this latter interface being enlarged by an increasing roughness of the surface during growth of Ni on top of Au. At 5 Mls the Ni structure is close to a bcc one tetragonaly distorted. Table 1 shows the distortion and the corresponding measured strain e_{zz} along the growth direction and for comparison the elastic strain and the distortion expected from linear elasticity theory if Ni would be fully strained in tension on the Au substrate. The measurements are far lower and indicate that 9% of the strain is relaxed by the interfacial mixing which introduces a strain without stress.

	distortion $d - d_{200}$ (Au)	d_{200} (distorted Ni)	ε_{zz} (Ni)
measurement (at maximum)	- 0.04 nm	0.16 nm	-9%
elasticity	- 0.06 nm	0.1445 nm	-18%

Table 1 : measured distortion and strain in the case of 4 Ni evaporated layers and the corresponding values expected from the linear elasticity.

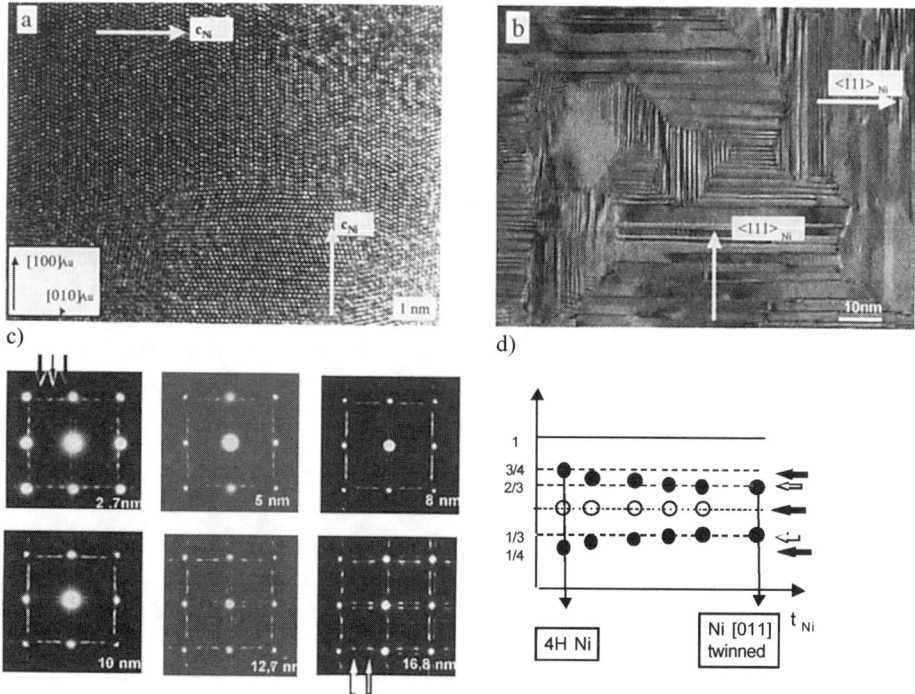

Figure 2 : Planar views of Ni thin films with different thickness grown on (001) Au.a) t_{Ni}= 4.8 nm b) t_{Ni}= 16.8 nm. c) electron diffraction patterns of planar views showing the evolution of the number and the position of the extra spots due to the Ni thin film structural changes which are shown schematically on fig 2d.

RELAXATION BY Ni PHASE TRANSFORMATION

As soon as Ni thickness is larger than 6 Mls another process occurs. Ni undergoes a phase transformation which is similar to a martensitic transformation. This has been observed both on plane views or cross section. At 4 Mls thick Ni structure is a body centred structure tetragonaly distorted, which becomes a 4H hexagonal structure. As the squared symmetry of the Au substrate is broken two variants appear whose c axis is parallel to either one or the other <001>Au in plane directions (fig. 2a). As the Ni thickness increases its structure evolves to fcc with numerous stacking fault and then to a twinned fcc (fig. 2b). At this thickness the Ni growing direction is [011]. Fig 2c and d show that the three diffraction spots characteristics from the 4H structure at t = 3 nm evolve continuously towards a situation for t > 15 nm where only two spots appear which correspond to a twinned Ni viewed along the [011] direction. 4H domains disappear at the benefit of the twinned domains, which are finally more or less periodically distributed. Twinning as a stress relaxation mechanism has been extensively studied for a long time [7]. More recently approaches linked to the specific influence of epitaxial stresses have been developped [8, 9]. In our case this can be understood as a bias for the system to minimize the dislocation density at the interface. In fact the closed packed planes interplanar distance in Ni is only 0.2% larger than d_{200}(Au), that strongly reduces the misfit in one direction. In the perpendicular direction the Ni is 6% strained in compression. Thus above 5 Mls, Ni is in compression while it is in tension below 6Mls.

The transformation has been carefully followed by numerical simulations [5]. The simulations could not predict the 4H transformation since the potential used underestimated the stacking fault energy in fcc metals, but nevertheless the 2H is predicted to be favoured. Furthermore the simulations predicted that the growth of Ni (111) thicker films (with a free surface) is favoured which is in disagreement with experimental results. This has been explained by the fact that Ni surface energy of Ni is larger than the Au one and as a consequence Ni surface is always covered by Au atoms which drastically reduces the formation energy. The point is that two Au atomic layers at the surface is found to correspond to the lowest energy. The exchange mechanism between Au and Ni atoms has been studied. The details are reported in [5].

Thus the final state of the growth process is the result of successive complex transformations unexpected from the bulk properties which are all driven by the epitaxial stress.

RELAXATION BY DISLOCATIONS IN Au_{1-x} /Ni_x ALLOYS

The only case where a dislocation mechanism was found to accommodate the stress is the one where Au_{1-x} /Ni_x alloys are grown on (001) Au and encapsulated in Au. Alloys with different compositions were grown. Fig. 3 shows an Au_{1-x}/Ni_x alloy 4 nm thick with x = 0.33. No dislocation and no other defect appears. The misfit between this alloy and Au is reduced to about -5%. For x = 0.5 the misfit is about -7%. For a 4 nm thick alloy dislocations appear (fig. 3b) indicating that the thickness is larger than the critical thickness as deduced from Matthews criterion [10]. There are mainly dissociated loops and only few are perfect dislocations. The dislocation density is very high. Since there are four (111) planes where the dissociation can take place, four types of domains appear where the dissociation is mainly on only one (111) plane. Fig. 3b shows one type of domain. The study of the further evolution of alloys with thickness indicates that micro twinning appears [11]. This reduces the dislocation and the stacking fault densities. As thickness increases, the energy due to the stacking fault increases drastically.

As a consequence, it has been observed that the system do prefer microtwinning where the twin energy is far lower than the stacking fault energy. It has to be noticed that the dislocation distribution is not homogeneous through the thin film. This has been studied in details [12]. The Burgers vector of the dislocation is either a/6<112> or a/2<011>. The misfit accommodation is spread over three zones (fig 3b). The first zone (1) is located in the alloy at the upper Au/alloy interface. It contains all the dislocations arising from the loops nucleated at the upper surface during growth. The accommodated strain calculated from the Burgers vector component parallel to the interface is about 6%. Thus the residual strain is 1%. In the two other zones 2 and 3 the dislocation density is gradually distributed. All the dislocation loops do not touch the lower alloy/Au interface. But the overall dislocation distribution in zones 2 and 3 accommodates - 6% misfit. This is consistent with the "upper" misfit or more precisely with the dislocation density found in the upper part. Nevertheless the strain gradient introduced by the

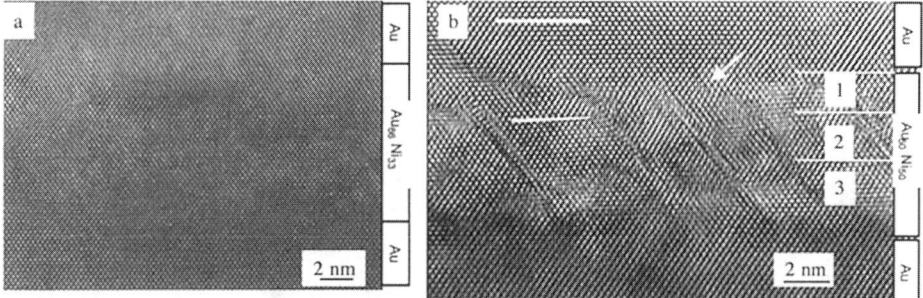

Fig. 3: Two cross-sections of Au_{1-x}/Ni_x alloy 4 nm thick, grown on (001) Au viewed along the <011> direction. For x = 0.33 (a) the alloy grows coherently, whereas in (b) x = 0.5 dislocations appear to accommodate the misfit indicating that the critical thickness is equal or larger than 4 nm for this latter composition. The arrow shows a 1/2<011> perfect dislocation. Two white lines are drawn in Au and the alloy to show the grain tilt.

non-homogeneous dislocations distribution is accompanied with grain local tilt, which arises from the Burgers vector component perpendicular to the interfaces. The resulting tilt can be seen on figure 3b: two white lines are drawn to guide the eyes. It has to be mentioned that along the alloys domains are found with tilt of opposite signs, which reduces long-range stresses.

ORDERING IN Au/Ni MULTILAYERS AND ALLOYS

The thermal stability of the previous metastable configurations was also studied [13]. An unexpected ordering of Au/Ni occurred provided the "alloy" remains residually strained. This study has been performed also by in-situ TEM and in-situ X-rays diffraction in order to cover a wide range of time and temperature [14]. The occurrence of this ordering was studied in different samples: multilayers, and thin film alloys with different thickness in order to eliminate the effects of periodicity and/or low dimensionality. In the bulk alloys, periodic modulations have been reported after quenching and aging at temperature below the miscibility gap (see the review paper [15]). The modulation occurs along the <001> axis and depends on the composition. Spinodal decomposition was suggested to account for the modulation appearance.

In fact in our samples, the very early stages of the phenomenon observed by in-situ TEM shows that the decomposition is not homogenous and that localized Guinier-Preston zones appear [16]. Furthermore the period remains constant with time and temperature: thus an ordering is more likely to occur. The period slightly depends on composition and varies from $3d_{200}$ to $4d_{200}$ with AuNi alloys content varying respectively from (0.5, 0.5) to (0.7, 0.3). It seems that alloys whose composition is less than 18% do not exhibit oscillation. The oscillations appear in a 180°C-325°C temperature range. Once established they are stable in time. If T > 350°C diffusion over large distances can take place: Ni can diffuse in adjacent Au layers and the oscillations disappear. Fig 4 a and b show the case of one Ni thin film (4 Mls thick) embedded in a multilayer as in fig.1, respectively before and after a heat treatment at 250°C for 2mn. Contrast oscillations are clearly visible, the period is about 4 d_{200}. Fig 5a and b show the decomposition of an $Au_{0.7}Ni_{0.3}$ alloy after heat treatment. The contrast oscillations are not so homogeneous. Antiphase domains are present and the period varies slightly from place to place but the mean value is found to be 4 d_{200}. The oscillation contrast becomes more homogeneous with time but is never found as strong as in fig. 4 b.

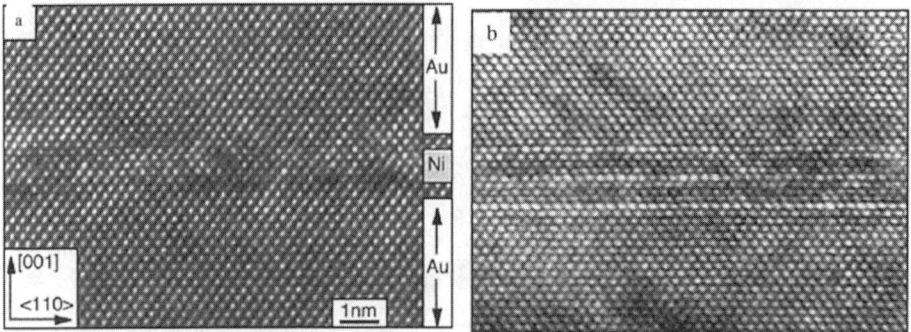

Fig. 4: Part of a 32Au/4Ni multilayer before heat treatment (a) and after 2 mn at 250°C.

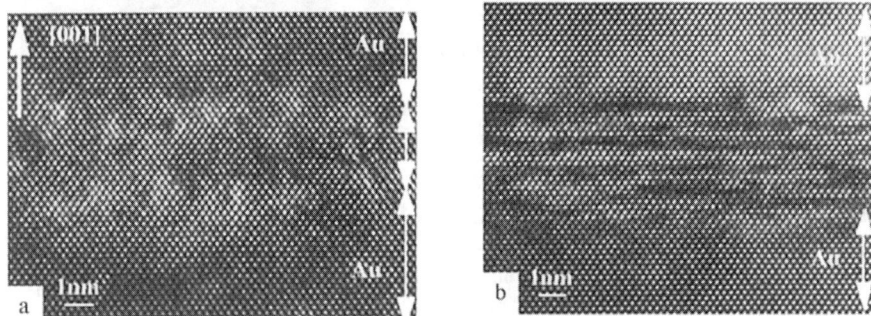

Fig 5 a and b : $Au_{0.7}Ni_{0.3}$ alloy 20 Mls thick before and after heat treatment respectively : the oscillations are still present but slightly less homogeneous than in the multilayers sample. Antiphase boundaries are present.

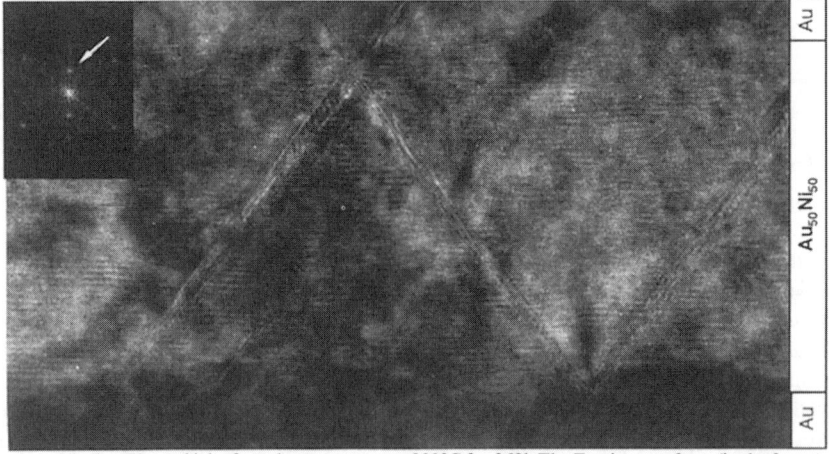

Figure 6: AuNi alloy 50 nm thick after a heat treatment at 202°C for 360h.The Fourier transform (in the frame) exhibits a spot which corresponds to a 3 d_{200} periodicity in the sample.

Fig 6 shows an $Au_{0.5}Ni_{0.5}$ alloy 50nm thick after heat treatment. As said in the previous paragraph, this $Au_{0.5}Ni_{0.5}$ alloy relaxed the epitaxial stress by micro twinning. This initial micro twinning can be still seen of figure 6. After a heat treatment at 202 °C for 360 hours, oscillations occur as shown on fig 6. The Fourier transform of the image shows that the period of the oscillation is about $3d_{200}$. The oscillation kinetics is slowed down likely by the fact that the initial alloy is partially relaxed. It has been shown that if the alloy is grown on a substrate (Pt) where the misfit is reduced to - 0.8%, no oscillation occurs [11, 17] during the experiment time.

Thus it appears that the role of the stress is the key point for the ordering to occur. Deutsch and Pasturel [18] showed that ordered phases such as Z_2 (i.e. 2Au/2Ni) could lower the energy of the system as compared to one of the disordered system. This has been confirmed independently from our experimental work by ab-initio simulations (at 0 K) published by Ozolins et al. [19]. These authors studied the stability of different alloys grown on (001)Au. They showed that the epitaxial stress is the driving force for ordering. Long period ordered phases are indeed metastable but stabilized by the stress. In fact the formation enthalpies they calculated is the lowest for 2Au/Ni and 3Au/Ni (which are the phases we observed experimentally) and are of the order of 32 meV/at. It has also to be mentioned that this energy is comparable to the stacking fault energy in Au. The Z_2 phase is less favoured (70meV/at). The main point is that the <001> direction in Ni exhibits a softening. In fact this was already pointed out by Gilles and Marty [20] and simulated by Deutsch et al. [5]. In this kind of superstructure the in-plane lattice parameter is of the order of the Au parameter, thus the misfit is strongly reduced.

CONCLUSION

All the examples presented here are illustrative of the fact that stresses are not always released by dislocations. It seems that dislocation mechanism arises in thin films when the misfit is relatively low. Furthermore, when dislocations are dissociated, this mechanism may evolve to micro twinning in cases where the stacking fault energy turns to be prohibitive when thickness increases. When the misfit is high the system find metastable configurations, which in fact reduces the dislocation density. The diffusivity and the phase diagram are changed under stress. Thus miscibility, twinning, ordering can occur which lower the energy. Furthermore as seen in the Ni phase transformation, Ni strain state changes: Ni is in tension when in small thickness (<5mls) whereas it turns to be in compression after the phase transformation for a larger thickness. Thus during growth the strain state of a film may change which certainly might have consequences at a macroscopic level.

ACKNOWLEDGEMENTS

The results presented here are only one part of a larger project which involved A. Marty, B. Gilles and G. Abadias for the MBE growth and X-rays diffraction experiments, I. Schuster for X-rays in-situ experiments and T. Deutsch and F. Lançon for simulations. The authors would like to thank their invaluable contributions.

REFERENCES

[1] B. Gilles, A. Marty, G. Patrat, J-L. Vassent, J-C. Joud, and A. Chamberod, MRS Proc., **237**, 511 (1992).

[2] C. Dressler, G. Abadias, P. Bayle-Guillemaud, A. Marty, I. Schuster, J. Thibault, Appl. Phys. Let., **72**, 2241 (1998).

[3] P. Bayle, T. Deutsch, B. Gilles, F. Lançon, A. Marty, J. Thibault, Ultramicroscopy, **56**, 94, (1994).

[4] P. Bayle, J. Thibault, Phil. Mag, **A77**, 475 (1998)

[5] T. Deutsch, P. Bayle-Guillemaud, F. Lançon, J. Thibault, Jour. of Physics: Cond. Matter **7**, 6407 (1995).

[6] P. Bayle, T.Deutsch, B. Gilles, F. Lancon, A. Marty, J. Thibault, C. Colliex, M. Tence. MRS Proc. **319**, 33 (1993).

[7] S. Pamir Alpay, A. L. Roytburg , Phys. Stat. Sol. **a 16,** 329-38 (1973).

[8] J. M. Sridhar, J.M. Rickman, D. J. Srolowitz, Acta Met., **44**, 4085 and 4097 (1996).

[9] A. L. Roytburd, J. Appl. Phys., **83**, 239-245 and 228-238 (1998).

[10] J. W. Matthews, in *Dislocations in Solids*, **vol 2**, p. 461 (1979) ed. F. R. Nabarro, North Publishing Company.

[11] M. Dynna, A. Marty, B. Gilles, and G. Patrat, Acta Met., **44**, 4417 (1996).

[12] C. Dressler, (1998), PhD thesis, Université Joseph Fourier-Grenoble-France.

[13] P. Bayle-Guillemaud, (1994), PhD thesis, Université Joseph Fourier-Grenoble-France.

[14] B. Gilles, A. Marty, G. Abadias, I. Schuster, C. Dressler, P. Bayle-Guillemaud and J. Thibault. Journal de Physique IV France, **9**, 87 (1999)

[15] J.E. Woodila and B. Averbach, Acta Met., **16**, 255 (1968).

[16] C. Dressler, P. Bayle-Guillemaud, and J. Thibault, Int. Conf. on Electron Microscopy, Cancun, **vol II**, 299 (1998).

[17] G. Abadias, (1998) PhD thesis, Institut National Polytechnique-Grenoble-France.

[18] T. Deutsch, A. Pasturel, in "Stability of Materials", "NATO ASI Sub-Series B Physics", ed. T. Gonis and P. E. A. Turchi and J. Kudrnovsky, **B355**, 381 (1996)

[19] V. Ozolins, C. Wolterton, A. Zunger, Phys. Rev., **B57**, 4816 (1998), Phys. Rev., **B57**, 6427 (1998).

[20] B. Gilles and A. Marty. MRS Proc., **333,** 621 (1994).

COMPUTER SIMULATION OF THE EFFECT OF COPPER ON DEFECT PRODUCTION AND DAMAGE EVOLUTION IN FERRITIC STEELS

J. M. PERLADO [*], J. MARIAN [*], D. LODI [*], T. DÍAZ DE LA RUBIA [**]
[*] Instituto de Fusión Nuclear (DENIM), Universidad Politécnica de Madrid, C/ José Gutiérrez Abascal, 2, 28006-Madrid, Madrid, SPAIN, jaime@denim.upm.es
[**] Chemistry and Materials Science Directorate, Lawrence Livermore National Laboratory, L–268, Livermore, CA 94550

ABSTRACT

It has long been noticed that the effect of Cu solute atoms is important for the microstructural evolution of ferritic pressure vessel steels under neutron irradiation conditions. Despite the low concentration of Cu in steel, Cu precipitates form inside the α-Fe surrounding matrix and by impeding free dislocation motion considerably contribute to the hardening of the material. It has been suggested that Cu-rich clusters and combined Cu solute atoms-defect clusters that may act as initiating structures of further precipitates nucleate during annealing of displacement cascades. In order to assess the importance of the different mechanisms taking place during collision events in the formation and later evolution of these structures, a detailed Molecular Dynamics (MD) analysis of displacement cascades in a Fe-1.3% at. Cu binary alloy has been carried out. Cascade energies ranging from 1 to 20 keV have been simulated at temperatures of 100 and 600 K using the MDCASK code, in which the Ackland-Finnis-Sinclair many-body interatomic potential has been implemented. The behaviour of metastable Cu self-interstitial atoms (SIAs) in the form of mixed Fe-Cu features is studied as well as their impact on the resulting defect structures. It is observed that above 300 K generated Cu SIAs undergo recombination with no substantial effect on the after-cascade microstructure while at 100 K Cu SIAs remain sessile and exhibit a considerable binding to interstitial and vacancy clusters. Finally, the effect that the production of vacancies via collision cascades may have on the self-diffusion of Cu solute atoms is quantitatively addressed by means of determining diffusion coefficients for Cu atoms under different microstructural conditions.

INTRODUCTION

As many nuclear power plants approach the end of their licensed operational life, one topic of current concern is the change in the mechanical properties that reactor pressure vessels (RPVs) may suffer after prolonged exposures to neutron irradiation environments. It is well known that pressure vessel steels embrittle under irradiation conditions due to the production of atomic-scale defect aggregates that hinder free dislocation motion resulting in hardening of the material. In addition to hardening caused by radiation produced clusters of defects, there is the contribution of nano-scale features such as Cu and Mn-Ni rich precipitates that form in the core of the material despite the low concentration of these elements in RPV steels [1,2]. Depending on alloy composition, Cu-rich precipitates are the dominant nanofeature in steels containing more than 0.1 at. % Cu [3] and this issue has attracted much research using a variety of techniques [4]. While there has been a number of experimental studies concerning the effect of Cu solute atoms as well as of the precipitates' structure in the microstructural evolution of irradiated steels [5,6,7], computer simulation of the precipitation kinetics has been treated recently [8,9,10,11]. This computational research effort has consisted mainly of lattice and kinetic MonteCarlo simulations of the Cu precipitation during thermal or cascade ageing and the interaction of these precipitates with cascade debris, particularly vacancies [8,11]. Results of these simulations indicate that both well-formed precipitates and solute atmospheres are formed as a consequence of highly correlated vacancy-solute transport processes and radiation-enhanced diffusion [8,12]. Moreover, there is a general belief that vacancy cluster-Cu solute atoms complexes that may act as precipitate precursors are produced during annealing of displacement collision cascades [12]. Other cascade effects of interest such as Cu SIA production and their interaction with characteristic cascade defect-structures (vacancy and SIA clusters) or more generally, the significance that solute atoms in a dilute substitutional solid solution have in the cascade evolution, are issues that occur in time scales (ps) much shorter than those related with the creation and coalescence of Cu precipitates (seconds). In this paper, a number of high-energy displacement cascades in a diluted Fe-Cu binary alloy at 600 K and 100 K has been carried out in order to investigate the process of in-cascade formation of vacancy-Cu complexes and Cu SIAs. The Cu content in the solution has been chosen to be 1.3% at. in order to enhance the statistics concerning the Cu atoms and due to the existence of several experimental works where Fe-1.34% Cu type alloys were employed [7,13].

SIMULATION MODEL

Molecular Dynamics

Although, collision events have been thoroughly analyzed with MD in pure metals [14,15,16], except for some specific papers [17], MD simulations in binary alloys have not been sufficiently studied. All the cascade simulations presented in this paper have been performed with the MDCASK code in which the new Fe-Cu potential derived by Ackland [18] from the Finnis-Sinclair many-body formulation [19] has been implemented. Details about the MDCASK code as to potential implementation, integration of the equations of motion, etc. are given elsewhere [20]. The simulations were carried out with periodic boundary conditions at constant volume. The Langevin equation of motion is applied to the atoms of the cell boundaries in order to control the temperature of the crystal.

Starting Assumptions

As in regular MD cascade simulation procedures, the solid solution was first equilibrated during several (8~10) picoseconds before the introduction into the system of the primary knock-on atom (PKA). The Cu atoms were introduced randomly into the bcc α-Fe matrix up to a proportion of 1.3 at. %. The effect that the Cu solute atoms have on the lattice parameter according to Wriedt et al. [21] has also been taken into account in our simulations, i.e. $\Delta a_o = 0.94 \times 10^{-3}$ Å per at. % Cu gives $\Delta a_o = 0.0012$ Å, which means that $a_o = 2.867$ Å rather than 2.866 Å for the bcc pure α-Fe lattice. Although this consideration is expected to have little or no effect on the results, this minute rigid dilation of the lattice may help ease down the additional strain introduced by Cu solute atoms and Cu SIAs during cascade ageing and yield more reliable results. Once the crystal was stabilized at the desired temperature, the resulting configuration was taken as starting point for all the cascades of a given energy. In order to get a reasonable statistics as to the number and type of defects produced, up to ten cascades were simulated for each PKA energy corresponding to different crystallographic directions. Both Fe and Cu PKAs were essayed for each one of the cascade energies in order to facilitate comparison between the effects of both species on atomic-scale damage production.

RESULTS

Cascade Analysis

The total number of cascades simulated for each energy can be seen in Table I. Simulations at 100 K and 600 K were performed only for 20 keV cascades since it is for high energies that variable conditions are more meaningful to microstructural evolution. The average number of Frenkel pairs produced in each cascade is also shown in Table I for cascades in the Fe-Cu solution and in pure α-Fe. It is noteworthy that the number of defects in pure Fe is consistently higher than in Fe-Cu as if the Cu atoms in the substitutional solution deadened the expansion of the cascade. A preliminary estimation of this restraining effect can be extracted calculating the volume of the cascade applying the criteria used by Gao and Bacon [22] where this cascade volume is defined as the sphere that contains at least 90% of the displaced atoms at its maximum expansion. Following this, the radius of such a sphere for a 10 keV cascade ($45a_0 \times 45a_0 \times 45a_0$ box) in the Fe-Cu alloy was found to be approximately $12.5a_0$ whereas the volume in a pure Fe block in equivalent conditions was $14.2a_0$. For 5 keV cascades ($30a_0 \times 30a_0 \times 30a_0$ box), the radius in the Fe-Cu system is $7.6a_0$ while the volume in the equivalent pure Fe block was $\sim 9a_0$. Although a satisfactory explanation for this behavior may involve other physical factors such as threshold displacement energies or linear energy transport along replacement collision sequences (RCSs), this effect seems to be somewhat an artifact of the interatomic potentials employed. The resulting RCSs are shorter in length giving rise to a lesser production of defects.

Table I: Number of cascades and defect production in the analyzed Fe-Cu alloy. Numbers in parentheses in the fourth column are data for comparison from cascades simulated in pure α-Fe at 600 K with the EAM potential [23].

PKA energy (keV)	Number of cascades		Average total number of Frenkel pairs	Average number of Cu SIAs generated
	600 K	100 K		
2	10	—	8 (10)	0
5	10	—	12 (19.5)	0
10	10	—	27 (35)	0.5
20	4	2	53 / 56 (58)	1.8 / 2.1

As reported by some authors [16,17], vacancy clustering in bcc metals is limited. In addition to this intrinsic feature, vacancies are not left to freely rearrange themselves into clusters or collapsed plates due to the effect of excited Cu solute atoms that, if sufficient annealing is allowed, form shells around the vacancy-rich core. This indeed has to do with the enhanced diffusion that Cu atoms close to the core of the cascade suffer via the vacancy mechanism. Some of the 10 keV cascades at 600 K were granted 1.15 additional ns of relaxation to check if this forced diffusion could affect the redistribution of Cu solute atoms and the vacancy clustering fraction. Results can be observed in Figure 1, where weak clouds of Cu atoms tend to form around the cascade core, as anticipated by Odette *et al.* [13]. A quantitative measure of this effect can be extracted by calculating the moment of inertia, I, of the Cu atoms in figures 1 (a) and 1 (b) ($I = \Sigma(\mathbf{r}_{Cu} - \mathbf{r}_g)^2/N_{Cu}$, where $\mathbf{r}_g$ is the center of mass of the cascade vacancies and N_{Cu} the total number of Cu atoms). The value for the initial configuration is $I \approx 232a_0^2$ whereas for the final configuration after 1.15 ns, $I \approx 228a_0^2$. With respect to the vacancy clustering fraction, the Ackland-Finnis-Sinclair potential yields a binding energy (e_b) of 0.14 eV for the 1st nearest neighbor (1nn) divacancy in pure Fe, and a value of 0.19 eV for the 2nd nearest neighbor (2nn) divacancy. Beyond this distance the vacancies are not bound at all (e_b=0) so our criterion for vacancy clustering is that they be within 2nn distances. Following this, for our Fe-Cu 20 keV cascades, we get a value of 0.25 at 600 K and 0.22 at 100 K for the vacancy clustering fraction, while, for instance, Stoller [24], for pure Fe, obtained values of 0.37 and 0.50 respectively, using the 2nn clustering criterion. Nevertheless, there is probably some potential-dependence contribution to this difference. These numbers, however, indicate that, even though vacancies are more strongly bound to each other (e_b=0.14 eV) than to Cu atoms (e_b=0.09 eV), the effect of the solute over the clustering tendency of the vacancies is to hinder it.

Cu Self-Interstitials

As anticipated before, Cu SIAs are metastable defects that appear as final links of RCSs in displacement cascades. When found isolated, Cu interstitials are observed forming mixed Fe-Cu split dumbbells (formation energy, e_f = 6.18 eV). The activation energy for the Cu atom to take the split lattice position (recombine) has been calculated to be approximately 0.16 eV, i.e. forcing the formation of a pure Fe-Fe dumbbell via the interstitialcy mechanism. This means that in the cascades at 600 K, Cu SIAs undergo recombination quickly and have no further significant effect on the crystal lattice. However, at 100 K Fe-Cu dumbbells remain sessile and their average lifetime has been extrapolated to be as long as 0.5 µs.

On the other hand, in some cascades at 600 K Cu interstitials were found in the form of <111> crowdions (e_f = 6.40 eV) forming small dislocation loops with a considerable binding energy (e_b = 1.15 eV). These Cu interstitials, however, did not seem to be any obstacle to the further propagation of the loops.

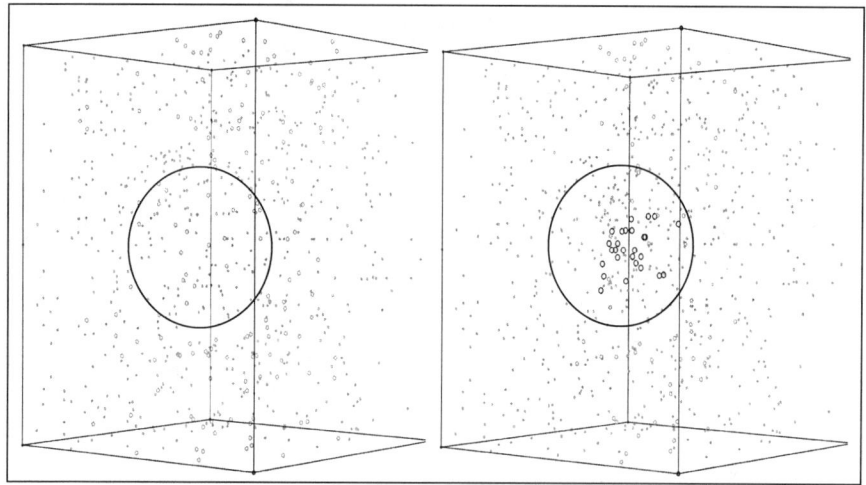

Figure 1. (a) Initial atomic configuration of the computation box. Only Cu solute atoms are shown (gray dots). (b) Final distribution of Cu atoms and vacancies (big circles) 1.15 ns after the 10 keV-PKA event. Note the increased density of solute in the encircled area (cascade core).

Enhanced Diffusion Kinetics

The combination of excess vacancies and excited Cu solute atoms hastens the process of solute clustering and gives rise to atomic configurations like that shown in Fig. 1 (b). A way to quantify this enhanced Cu solute atom-migration is calculating the self-diffusion coefficient, D, via MD simulations. In principle, a coarse estimation of D (vacancy mechanism) can be obtained through the following expression:

$$D = C \cdot \omega \cdot N_v \cdot a_o^2 \qquad (1)$$

where C is a constant that includes the nature of the 1nn jumps in the bcc lattice, ω is the jump frequency, a_o the lattice parameter and N_v is the probability to find a vacancy in a given lattice position, i.e. the percentage of vacancies in the crystal. As to ω, it is ordinarily taken as $\omega = v \cdot \exp(-E_m / KT)$, where E_m is the energy required to move an atom to the saddle point position, K the Boltzmann's constant, T the absolute temperature of the crystal and v is a typical vibration frequency of the material ($\sim 10^{13}$ Hz). We have calculated the Cu-vacancy interchange energy by a static relaxation calculation and computed a value of 0.739 eV, similar to the number employed by Soisson et al. (0.69 eV) [9], but quite different to the one obtained by Ackland et al. (0.60 eV) [19]. In a system at equilibrium conditions N_v is usually taken as $N_v = N_o \cdot \exp(-E_f / KT)$, E_f being the vacancy formation energy. Hence, in the 1000-1500 K temperature interval, N_v ranges between 2×10^{-9} and 2×10^{-6}, having taken $E_f = 1.7$ eV. However, during cascade ageing highly non-equilibrium conditions are found and the increased local strain due to the excess vacancies tends to be counterbalanced by the diffusion of energetic Cu atoms. Under these extreme conditions, it is not exaggerated to consider $N_v \approx 0.01$ in a domain around the vacancy-rich core of the cascade. We have calculated the temperature dependence of D assuming an Arrhenius behaviour:

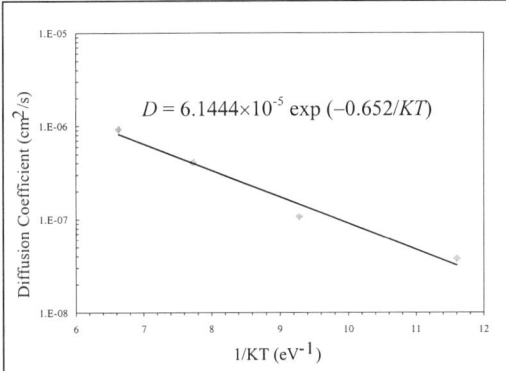

$$D = 6.1444 \times 10^{-5} \exp(-0.652/KT)$$

Figure 2. Arrhenius plot of the diffusion coefficient for Cu solute atoms in highly far from equilibrium conditions (excess cascade vacancies)

$$D = D_o \cdot \exp(-E_m / KT) \qquad (2)$$

where D has been calculated for four different temperatures ($0.55T_m$, $0.70T_m$, $0.85T_m$ and $0.95T_m$) in a Fe-1.3% at. Cu solution with 1% concentration of vacancies (figure 2). Atoms were allowed to evolve during 0.75 ns and then D_{Cu} was estimated using:

$$D_{Cu} = \frac{1}{6} \frac{d}{dt} \langle R_{Cu}^2(t) \rangle \qquad (3)$$

From (2) and (3) we obtained: $D_{Cu} = 6.144 \times 10^{-5} \cdot \exp(-0.652/KT)$ (cm$^2 \cdot$s^{-1}). This means that, for $T = 1500$ K, $D_{Cu} \sim 3.96 \times 10^{-7}$ cm$^2 \cdot$s^{-1} which is reasonably close to the value deduced from (1) at 1500 K and $N_v = 0.01$ (4.80×10^{-7} cm$^2 \cdot$s^{-1}).

CONCLUSIONS

Preliminary results of the effects of Cu solute atoms at low concentration in Fe in displacement cascades have been presented. The development of the thermal spike is restrained by the containment effect of Cu solute atoms. The formation of precursor Cu solute atom-clouds as well as of Cu-vacancy clusters has been demonstrated as a consequence of cascade enhanced diffusion kinetics. Further work in order to improve the statistics as to cascade defects and diffusion parameters determination is under progress.

ACKNOWLEDGEMENTS

This work has been performed at DENIM within the *VENUS* Research Project framework under contract P970530432 (CSN/UNESA Coordinated Research Programme) and under the auspices of the U.S. Department of Energy by LLNL under contract W7405-ENG-48.

REFERENCES

[1] G. E. Lucas, G. R. Odette, R. Maiti and J. W. Sheckherd in *Influence of Radiation on Materials Properties: 13th International Symposium, Part II*, edited by F. A. Garner, C. H. Genager and N. Igata (ASTM-STP **956**, ASTM, Philadelphia, PA, 1987) p. 379
[2] S. B. Fisher and J. T. Buswell, Int. J. Pressure Vessel Piping **27**, 91 (1987)
[3] G. R. Odette in *Microstructure of Irradiated Materials*, edited by I. M. Robertson, L. E. Rehn, S. J. Zinkle and W. J. Phytian (Mater. Res. Soc. Proc. **373**, Pittsburgh, PA, 1995) p. 137
[4] J. T. Buswell, P. J.Bischler, S. T. Fenton, A. C. Ward and W. J. Phytian, J. Nucl. Mater. **205**, 198 (1993)
[5] P. M. Rice and R. E. Stoller, J. Nucl. Mater. **244**, 219 (1997)
[6] G. M. Worrall, J. T. Buswell, C. A. English, M. G. Hetherington and G. D. Smith, J. Nucl. Mater. **148**, 107 (1987)
[7] A. C. Nicol, M. L. Jenkins and M. A. Kirk in: *Microstructural Processes in Irradiated Materials*, edited by S. J. Zinkle, G. E. Lucas, R. C. Ewing and J. S. Williams (Mater. Res. Soc. Proc. **540**, Warrendale, PA, 1998) p. 409
[8] F. Soisson, A. Barbu and G. Martin, Acta Mater. **44**, 3789 (1996)
[9] C. L. Liu, G. R. Odette, B. D. Wirth and G. E. Lucas, Mat. Sci. Eng. A **238**, 202 (1997)
[10] B. D. Wirth and G. R. Odette in: *Microstructural Processes in Irradiated Materials*, edited by S. J. Zinkle, G. E. Lucas, R. C. Ewing and J. S. Williams (Mater. Res. Soc. Proc. **540**, Warrendale, PA, 1998) p. 637
[11] C. Domain, C. S. Becquart, J. C. Van Duysen in *Microstructural Processes in Irradiated Materials*, edited by S. J. Zinkle, G. E. Lucas, R. C. Ewing and J. S. Williams (Mater. Res. Soc. Proc. **540**, Warrendale, PA, 1998) p. 643
[12] G. R. Odette and B. D. Wirth, J. Nucl. Mater. **251**, 157 (1997)
[13] T. N. Lê, A. Barbu, D. Liu and F. Maury, Scripta Metall. **26**, 771 (1992)
[14] T. Díaz de la Rubia and M. W. Guinan, Mater. Res. Forum **97-99**, 23 (1992)
[15] A. F. Calder and D. J. Bacon, J. Nucl. Mater. **207**, 25 (1993)
[16] D. J. Bacon and T. Díaz de la Rubia, J. Nucl. Mater. **216**, 275 (1994)
[17] H. F. Deng and D. J. Bacon, Phys. Rev. B **53**, 11376 (1996)
[18] G. J. Ackland, D. J. Bacon, A. F. Calder and T. Harry, Philos. Mag. A **75**, 713 (1997)
[19] M. W. Finnis and J. E. Sinclair, Phil. Mag. A **50**, 45 (1984)
[20] T. Díaz de la Rubia and M. W. Guinan, Mater. Res. Forum **174**, 151 (1990)
[21] H. A. Wriedt and L. S. Darken, Trans. Metals. Soc. AIME **218**, 30 (1960)
[22] F. Gao and D. J. Bacon, Phil. Mag. A, **71**, 65 (1995)
[23] R. A. Johnson and D. J. Oh, J. Mater. Res. **4**, 1195 (1989)
[24] R. E. Stoller in: *Microstructural Processes in Irradiated Materials*, edited by S. J. Zinkle, G. E. Lucas, R. C. Ewing and J. S. Williams (Mater. Res. Soc. Proc. **540**, Warrendale, PA, 1998) p. 679

DISLOCATION MOBILITY IN TWO-DIMENSIONAL LENNARD-JONES MATERIAL

NICHOLAS P. BAILEY*, JAMES P. SETHNA* AND CHRISTOPHER R. MYERS**

* Physics Department, Cornell University, 117 Clark Hall, Ithaca, NY 14853
** Cornell Theory Center, Cornell University, Ithaca, NY 14853

ABSTRACT

In seeking to understand at a microscopic level the response of dislocations to stress we have undertaken to study as completely as possible the simplest case: a single dislocation in a two dimensional crystal. The intention is that results from this study will be used as input parameters in larger length scale simulations involving many defects. We present atomistic simulations of defect motion in a two-dimensional material consisting of atoms interacting through a modified Lennard-Jones potential. We focus on the regime where the shear stress is smaller than its critical value, where there is a finite energy barrier for the dislocation to hop one lattice spacing. In this regime motion of the dislocation will occur as single hops through thermal activation over the barrier. Accurate knowledge of the barrier height is crucial for obtaining the rates of such processes. We have calculated the energy barrier as a function of two components of the stress tensor in a small system, and have obtained good fits to a functional form with only a few adjustable parameters.

INTRODUCTION

This paper is concerned with the motion of a single dislocation. Thus there is no dislocation-dislocation interaction; the interaction is between the dislocation and the applied stress. Furthermore we work in two dimensions, which eases the computational burden and aids visualization. Having simplified the problem to this extent, we have a chance of understanding it in detail. Once such understanding is developed, it will then make sense to proceed to more realistic, though computationally more expensive, cases (e.g. three dimensions, realistic potentials, etc.).

Our system consists of a relatively small (< 100) number of atoms in two dimensions (2D) interacting through a Lennard-Jones potential which has been truncated and made to go smoothly to zero at the cutoff distance (2.7σ; this is large enough for third neighbor interactions to be included). The system has periodic boundary conditions in the vertical direction, and rigid 'walls' on the sides. The walls are simply lines of atoms which are constrained to move as rigid bodies. In addition the atoms in each of the next-to-outermost columns are constrained to more rigidly in the x-direction, and independently in the y-direction. This system for the boundaries is due to Tomasi [5]. If a shear stress is applied to the boundary walls the dislocation will move by glide, but only if the shear stress is above a certain critical value σ_c. The applied shear stress is the resolved shear stress in this geometry. Note that the critical resolved shear stress for dislocation motion depends on the other components of stress, hence knowledge of the resolved shear stress alone is not enough to decide whether a given dislocation will move or not. At zero temperature, with $\sigma_{xy} < \sigma_c$, the dislocation cannot move, but with a finite temperature, motion still occurs as thermally activated hops over an energy barrier. This barrier corresponds to the Peierls barrier for an edge dislocation in three dimensions. Our task was to calculate this barrier as a function of all three components of the stress tensor ($\sigma_{xx}, \sigma_{xy}, \sigma_{yy}$). However so far, we have only dealt with the dependence on the first two, since we have only begun to incorporate the techniques necessary for applying a constant stress in the direction in which periodic boundary conditions are imposed. We have considered only one size of system; finite size effects are important. Extensions to all three components of stress and extrapolations to large sizes are in progress.

THEORY

Thermal Activation

We concentrate on calculating the energy barrier to hopping for shear stress less than the critical shear stress. For these values of shear, there exist so called fixed points of the dynamics. These are associated with local minima in the potential. Two nearby minima are separated by a barrier in the energy landscape (note that the saddle point of the barrier is also a fixed point, albeit an unstable one). When the energy barrier is large compared to the temperature, the transition rate between the states will have the form

$$R = \nu \exp(-\frac{E_B}{k_B T}) \tag{1}$$

where E_B is the barrier height and ν is an attempt frequency which can be calculated from the curvature of the potential landscape near the minimum and near the barrier top [7]. Because of the exponential, however, the rate is much more sensitive to E_B than it is to ν. Hence it is importance to know E_B accurately to be able to reasonably calculate such rates. For the dislocation, the rate of hopping is proportional to the velocity, and hence its mobility. The barrier height is defined as follows. For any path between the two minima we find the point along the path where the potential energy is greatest; call this value E_{max}. We then consider all paths and take the one whose E_{max} is smallest. This is the *minimum energy path*. The barrier height is $min_{paths}\{E_{max}\} - E_{intialstate}$. The location of the maximum energy along the minimum energy path corresponds to a saddle point in the potential landscape. Several methods exist for finding barrier heights. We use one which finds the whole minimum energy path, called the 'Nudged Elastic Band' method [3].

Model and Potential

We model a small piece of two-dimensional material containing a single edge dislocation atomistically, using methods of molecular dynamics. We use a classical pair potential defined as follows: Lennard-Jones (6-12, with standard parameters ϵ and σ) for $r < r_{cut1} = 2.41308788\sigma$, a quadratic in r^2 for $r_{cut1} < r < r_{cut2} = 2.7\sigma$, and zero for $r > r_{cut2}$. This potential was formulated by Chen [4]. It is continuous and smooth everywhere. Extension to other potentials and other forms of the cutoff is planned. The units for the simulation are determined by the parameters ϵ and σ in the potential, which are set to unity, hence all energies are in units of ϵ, and distances in units of σ. Units of time, stress etc. all follow from these. Often one makes a connection with physical systems by matching the parameters to those of Argon, for which Lennard-Jones is a good potential. So, $\epsilon = 119.8 K k_B$ or ~ 0.01eV, and $\sigma = 0.341$nm [2]. However since we have a 2D system, there is little useful quantitative comparison to made with experiment (for one thing, stress has different units in 2D than in 3D).

SIMULATION

We simulate an '$N \times N$' system, where N is the number of rows on the left of the dislocation; there are $N - 1$ on the right. Two extra columns are added to each side to form the boundaries. Typically $N = 7$, which corresponds to 71 atoms. Simulation runs consist of the following procedure: A set of atoms is configured as two lattices of slightly differing lattice constants, and correspondingly different numbers of rows, placed together. The atoms are relaxed by evolving the system using Langevin dynamics, after which there was a localized dislocation. Next, a shear stress is applied which is just of sufficient strength and duration to cause the dislocation to move one lattice spacing, after which the stress is reset to zero and further relaxation is done. Copies are made of the relaxed system before and after the move. The main part of the simulation consists of a loop in which stress is applied to these copies, their energy is minimized using the 'MDmin' procedure, taken from Ref. [1], and they are passed to the barrier finding routine, which uses the Nudged Elastic Band method [3]. In this method a chain of replicas of the system is created forming a line in configuration space between the local minima. Forces from the potential, and between replicas are applied, with certain corrections, and the whole chain relaxed until it lies along the minimum energy path. Once the barrier is calculated, the stress is incremented and the loop repeats, minimizing the two copies now with a different stress, and so on. Fig. 1 shows initial, final and saddlepoint configurations of the 7×7 system.

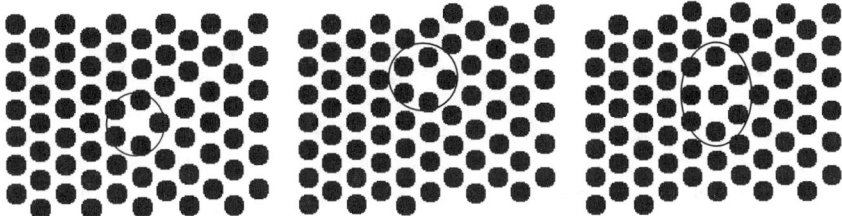

Figure 1: Initial, final and saddle point configurations. These are for zero shear stress, boundaries fixed in x-direction. Circles indicate the dislocation.

Curve-fitting: Finding the top.

   In general there will not be a replica right at the saddle point, so given the energies of the replicas it is necessary to do curve fitting to find the actual maximum energy along the path, and hence the barrier height itself. The information returned from the routine includes the configurations of the replicas and the Euclidean distance along the chain for each one, called the reaction coordinate, as well as the energies. We must subtract the work done by the external stress to find the relevant energy-quantity: at finite temperature it would be the Gibbs free energy; at zero temperature it is equal to the enthalpy, and is the quantity that is minimized in equilibrium when a constant force is applied. The set of distance-enthalpy points can be plotted in order to visualize the shape of the barrier. To find the height, a cubic is fitted to the top four points, see Fig 2; the position of the maximum can then be simply calculated. An estimate of the uncertainty can be got from considering the top five points and fitting to a quartic, and taking the difference of the two results. The difference appeared only in the fifth digit, though close to critical shear, where convergence was not as good, the relative error became large.

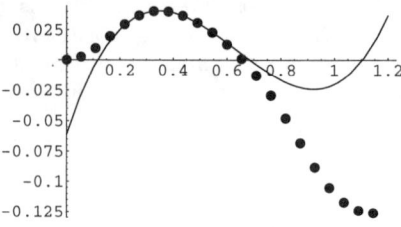

Figure 2: A single barrier profile with the fitted cubic. $\sigma_{xx} = 9.6$, $\sigma_{xy} = 0.8$

Constant pressure simulations

The first runs were done with the boundary walls free only to move in the y-direction, and fixed L_y, though it was intended that we would eventually have constant σ_{xx} and constant σ_{yy}. A material is generally under fixed stress not fixed strain, so incorporating conditions of constant stress, and hence fluctuating boundaries, is a more realistic approach. In fact we found direct evidence of the necessity for constant stress simulations when we looked at the system-size dependence of our results. The height of the barrier for $\sigma_{xy} = 0$ depended strongly on system size—it decreased by a factor of two upon going from a 7×7 system to a 13×13 one. Since we would like to believe that we can in fact get away with simulating such small systems this was a worrying fact.

Including constant σ_{xx} was straightforward: we let the boundary walls move and put a force in the x-direction on them. When the system was relaxed in the initial stage of the simulation, the final separation of the boundaries was about half a lattice constant larger than the fixed separation we had been using—the dislocation liked to take up more space than an uninterrupted column of atoms, and hence there was a significant sideways pressure in the fixed boundary simulations. When the system size was increased this pressure decreased and since it was already clear that the barrier height depended on sideways pressure, this would explain the dependence on system size in the earlier simulations. So increasing the system size at fixed σ_{xx} should have a much smaller effect on the barrier.

In fact the barrier height was an *increasing* function of the system size when σ_{xx} was held fixed. This was thought to be due to the vertical dimension of the sample being held fixed (the dimension in which periodic boundary conditions were operating), hence there was a varying effective pressure in this direction. Since there is no rigid boundary here as on the sides, a more sophisticated technique must be used, similar to Parrinello-Rahman dynamics [6]. Here, the lengths of the simulation cell in the different directions are allowed to vary dynamically. The equations of motion are suitably modified to include this extra degree of freedom. In the present case this only had to be done for the vertical direction, with the variable L_y being introduced. However there are subtleties associated with combining this technique with Nudged Elastic Band, not least the issue of defining an angle in a space which has one axis corresponding to a length and the rest to dimensionless positions.

RESULTS

The energy barrier as a function of shear stress σ_{xy} with fixed σ_{xx}, for several values of σ_{xx} is shown in Fig 4. The dots are data points from barrier calculations; the solid lines are three-parameter fits to a series expansion obtained by considering the one-dimensional

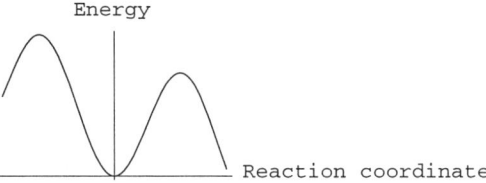

Figure 3: Schematic representation of barriers in a periodic system. The shear stress is positive, hence the system wants to move to the right. However we can calculate the barrier to hop to the left by repeating the calculation with the opposite sign of shear.

barrier problem. For stress larger than the critical value, there is no fixed point, and the defect slides with periodically varying velocity. For stress smaller than the critical value, there are two fixed points, a stable one corresponding to the local minimum, and an unstable one corresponding to the barrier top (of course there are many more really, due to the periodicity of the lattice). The appearance of two fixed points as the stress goes below σ_c (or equivalently their disappearance as stress goes above σ_c) is a *saddle-node bifurcation*. Note that we also have points for negative shear; this corresponds to hopping in the opposite direction for positive shear, see Fig. 3. For large negative shear the barrier becomes simply the energy difference between the two minima. These local minima only exist for $|\sigma_{xy}| < \sigma_c$ (beyond which the dislocation starts to slide in the appropriate direction), hence our data covers the range $-\sigma_c$ to σ_c, for several values of σ_{xx}.

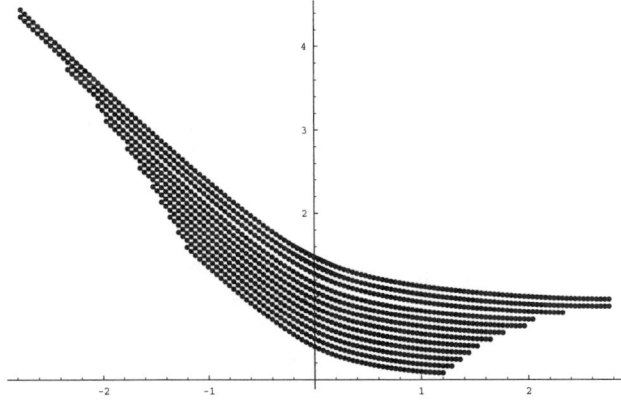

Figure 4: Energy barrier versus shear stress for $8.8 < \sigma_{xx} < 13.6$. The curves have been displaced vertically for clarity. Also included, but not readily visible are the one-dimensional fits to the data.

CONCLUSIONS

We have calculated the barrier height for dislocation hopping for range of both σ_{xx} and σ_{xy}. We have shown how this data can be parametrized to reasonable accuracy with a fit that needs only a few parameters. These parameters could be used for example in a simulation whose primitive objects are dislocations which move in a stress field. The present results, augmented by σ_{xx} dependence, could be used to calculate the motion of each dislocation given the values of the components of the stress tensor at its location (the stress field would be calculated from the elastic fields of the other dislocations plus any external sources of stress).

The next development in the simulation will be to include the dependence on the third component of the stress tensor, σ_{yy}, which corresponds to pressure on the top and bottom of the system. We have just started to be able to do runs at constant σ_{yy}, though there are subtleties in combining this with the Nudged Elastic Band method. Once we are satisfied that we can do this well, we will repeat with different inter-atomic potentials. Another aspect of simulating this system is to consider shear stress above σ_c, where the dislocation slides continuously, if not quite steadily. Preliminary studies indicate interesting behaviour, including a delocalization of the core structure at high velocity. We will add finite temperature later as well, and eventually carry over this work into three dimensions and real potentials.

ACKNOWLEDGMENTS

This project grew out of a collaboration with Jeff Tomasi, who originated our boundary condition method. It was supported by NSF grant number DMR 9873214, and was done using the Intel/NT Velocity Cluster at the Cornell Theory Center. We had helpful discussions with Tejs Vegge, Enrique Batista and Markus Rauscher.

References

[1] cond-mat/9808211 J. Schiotz, T. Vegge, F. D. Di Tolla, K. W. Jacobsen "Simulations of mechanics and structure of nanomaterials—from nanoscale to coarser scales"

[2] M. P. Allen, D. J. Tildesley, "Computer Simulation of Liquids", Oxford University Press (1987) p.21.

[3] T. Rasmussen, K. W. Jacobsen, T. Leffers, O. B. Pedersen, S. G. Srinivasan and H. Jonsson, Phys. Rev. Lett. **79**, 3676 (1997).

[4] X. Chen (private communication).

[5] J. Tomasi (private communication).

[6] M. Parrinello, A. Rahman, Phys. Rev. Lett. **45**, 1196 (1980); J. Appl. Phys. **52**, 7182 (1981); J. Chem. Phys. **76**, 2662 (1982).

[7] P. Hänggi, P. Talkner, M. Borkovec, Rev. Mod. Phys. **62**, 251 (1990).

STUDY OF ELECTRICAL PROPERTIES OF DISLOCATIONS IN ZNS USING ELECTRIC FORCE MICROSCOPY

G. F. Bai, V. F. Petrenko, and I. Baker
Thayer School of Engineering, Dartmouth College, Hanover, NH 03755

ABSTRACT

A combination of electric force microscopy (EFM) and non-contact scanning force microscopy (SFM) was used to study micro-indentation-induced dislocation bands in sphaleritic ZnS single crystals. Large local distortions in electrical potential from the dislocation bands were observed in the EFM images. For the first time, the electric charges of resting partial Zn(g) and S(g) dislocations were determined quantitatively. The results compare well with theoretical models.

INTRODUCTION

Electrically-charged dislocations play a significant role in the electrical properties of wide band gap II-VI semiconductors. Electronic states in the band gap introduced by dislocations result in a high density of electric charge at the dislocation cores that in turn results in lower charge carrier mobility and lifetime, thus impairing device performance. In applications of II-VI semiconductors, dislocations can be introduced in a number of ways. First, dislocations are produced during fabrication of integrated circuits due to the enormous elastic stresses that build up (for example, during formation of shallow isolation trenches [1]). Second, dislocations are so mobile at room temperature that they can easily multiply during the operation of optoelectronic devices due to temperature gradients and electric fields [2].

Scanning force microscopy (SFM) has been used to study various defects, such as threading dislocations, screw dislocations, and grain boundaries on semiconductor surfaces. In our previous study, both grown-in and deformation-introduced dislocations in ZnS and ZnSe were detected by SFM [3]. However, the surface topography was the only information obtained. To detect an electric charge on dislocations a mode sensitive to an electric field, for instance, electric force microscopy (EFM), has to be used. To characterize the linear density of the dislocation charge, one has to determine the density of dislocations, and simultaneously determine the local electric charges.

In this research, the electric potential of dislocations emerging on a (110) surface of ZnS crystals was measured using EFM. The topography of the surface was recorded simultaneously using a non-contact mode of SFM.

EXPERIMENTAL METHODS

ZnS specimens were cut from a monocrystal in such a way that one of the surfaces was a plane of easy cleavage (110) and one of the surfaces was close to (001); see Fig. 1(a). The electric conductivity of the ZnS was in the range of 10^{-10}-10^{-12} $\Omega^{-1} m^{-1}$. Cleaved fresh, clean, atomically smooth, and dislocation free (110) ZnS surfaces were obtained by using a razor edge in air. Micro-indentation was used to introduce new dislocations and to determine the absolute orientation of the specimen from the geometry of slip bands.

ZnS has the sphalerite structure with perfect dislocations with Burgers' vectors of $a/2<110>$ [4]. The primary slip system is well established to be $<110>\{111\}$; see Fig. 1(a).

Mat. Res. Soc. Symp. Proc. Vol. 578 © 2000 Materials Research Society

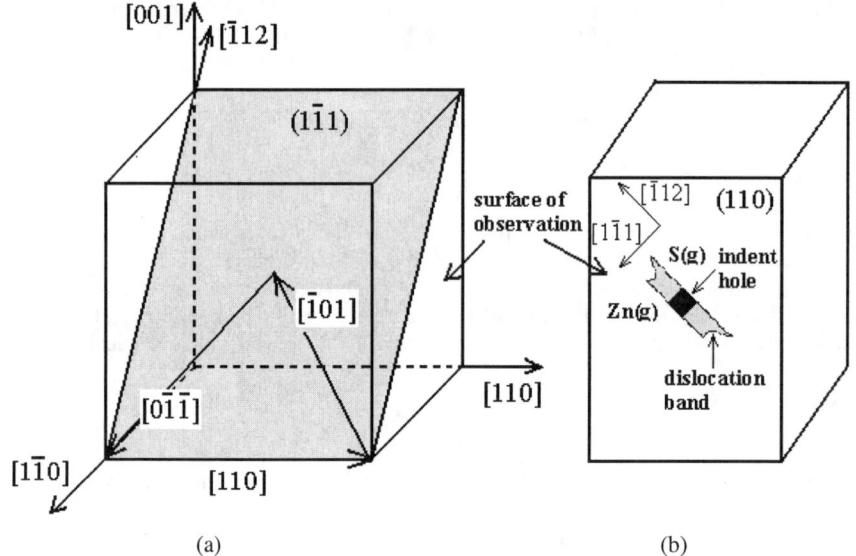

Figure 1. (a) Sample orientation and (b) dislocation bands induced by micro-indentation

Sphaleritic ZnS contains ~10% of wurtzite, which results in only one of the primary slip systems that is parallel to stacking faults being active [2].

An important feature of the sphalerite structure of ZnS is that dislocations with an edge component can be of two kinds, depending on whether the extra half-plane terminates in a row of metal or non-metal atoms [5]. Four types of edge dislocation are therefore possible [2]. It has been shown [6] that moving dislocations in ZnS are partials with Burgers' vectors of $a/6$ [112].

The induced dislocation bands were examined using a commercial SFM (an AutoProbe CP from Park Scientific Instruments). The maximum scanning field was $100 \times 100\,\mu m^2$. The typical radius of the ultralever tips was 10 nm. Scan velocities were from 25 to 33 $\mu m\,s^{-1}$ typically. Figure 2 shows EFM and corresponding non-contact SFM images of a micro-indentation. In the EFM mode, we applied an ac voltage with amplitude of 5 V and a frequency of 17 kHz between the tip and the sample. The ac frequency was far below the non-contact SFM mode operating frequency of ~80 kHz. No dc voltage was applied.

RESULTS

In II-VI crystals, the lengths of dislocations produced by micro-indentation are comparable with the width of the dislocation band [7]. This length of the dislocation is much larger than our scale of interest. Therefore, it is reasonable to make an assumption that the charge density is almost independent of the z direction. Thus, we can apply the two-dimensional Poisson's equation to the surface potential, data $V(x, y)$ acquired by EFM:

$$\frac{\partial^2 V(x, y)}{\partial x^2} + \frac{\partial^2 V(x, y)}{\partial y^2} = -\frac{\rho_e(x, y)}{\varepsilon \cdot \varepsilon_0} \tag{1}$$

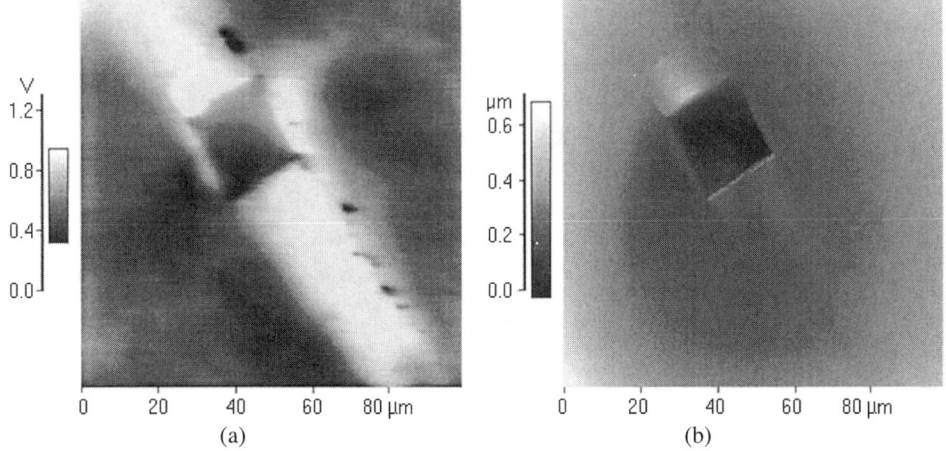

Figure 2. *(a) An EFM, and (b) corresponding topographic SFM image of a micro-indentation.*

where ε_0 is 8.82×10^{-12} $F\ m^{-1}$, for ZnS, the dielectric constant ε is around 8.6, and $\rho_e(x, y)$ is the charge density. Fig. 3(a) shows an image of $\rho_e(x, y)$ obtained from the surface potential and equation (1). In Fig. 3(a), although there is significant high frequency noise along the y direction, we can still see clearly that positive (bright) and negative (dark) charges emerge along the margins of the dislocation band. Fig. 3(b) shows an image filtered to remove this high frequency noise. Once the noise is removed, we determined that $\rho_e(x, y)$ spans a range from -30 $C\ m^{-3}$ to 20 $C\ m^{-3}$.

Since EFM data of $V(x, y)$ was acquired by scanning in the x direction, it is also reasonable to apply one-dimensional Poisson's Equation to obtain $\rho_e(x)$:

$$\frac{\partial^2 V(x)}{\partial x^2} = -\frac{\rho_e(x)}{\varepsilon \cdot \varepsilon_0} \tag{2}$$

Although some information in the y direction was neglected, the high frequency noise signal was also removed automatically and completely, as shown in Fig. 4(a). Both positive (bright) and negative (dark) charges are seen, especially along the margins of the dislocation band. By measuring the polarity of the piezoelectric effect [2], we determined the absolute orientation of each specimen and found that S(g) dislocations have a higher mobility than Zn(g), i.e., they move farther from the indentation. This is schematically shown in Fig. 1(b). From Fig. 4(b), it is also evident that the maximum of $\rho_e(x)$ is in a range of -60 $C\ m^{-3}$ to 40 $C\ m^{-3}$, which is about twice that determined from Fig. 3(b).

The dislocation density, $\rho_d(x, y)$ was calculated from the topographic data $z(x, y)$ using:

$$\frac{\partial^2 z(x, y)}{\partial x \partial y} \cdot \frac{1}{b_z} = \rho_d(x, y) \tag{3}$$

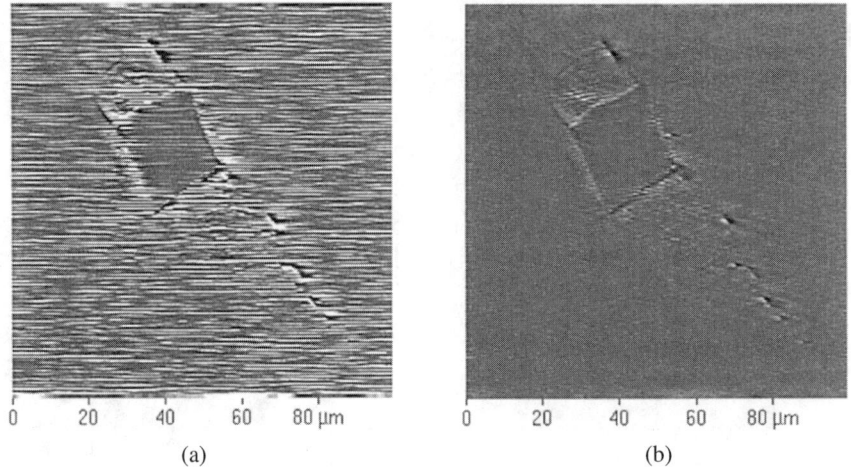

Figure 3. *Surface charges* $\rho_e(x, y)$ *around the micro-indention. Before (a) and after (b) filtering the high frequency noise.*

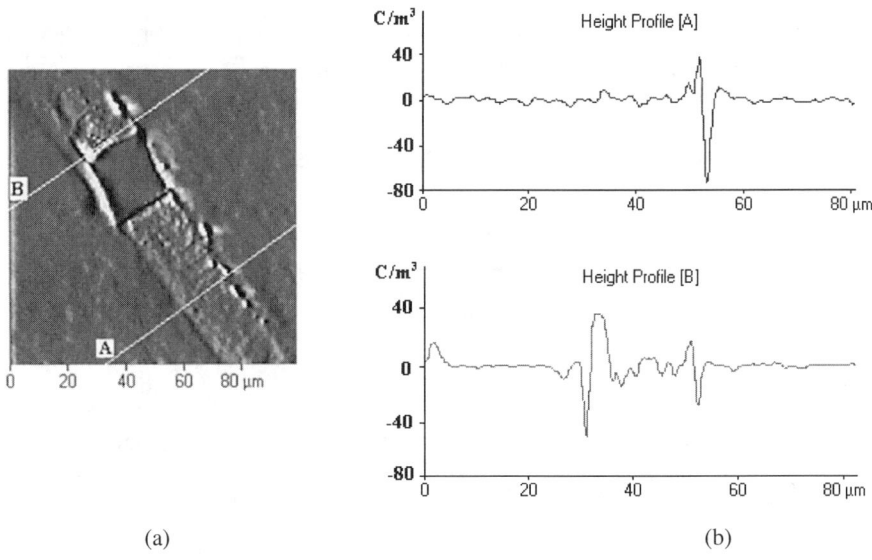

(a)   (b)

Figure 4. *(a) Charge density emerging around the micro-indentation and (b) its line profile measurements along line A and line B.*

where b_z is the component of the Burgers' vector along the z direction and $\rho_d(x, y)$ is the density of dislocations. Since perfect dislocations are never observed in ZnS [2], here we use a Burgers' vector of $a/6 <11\bar{2}>$ for isolated 30° partial dislocations. An example is shown in Fig. 5(a).

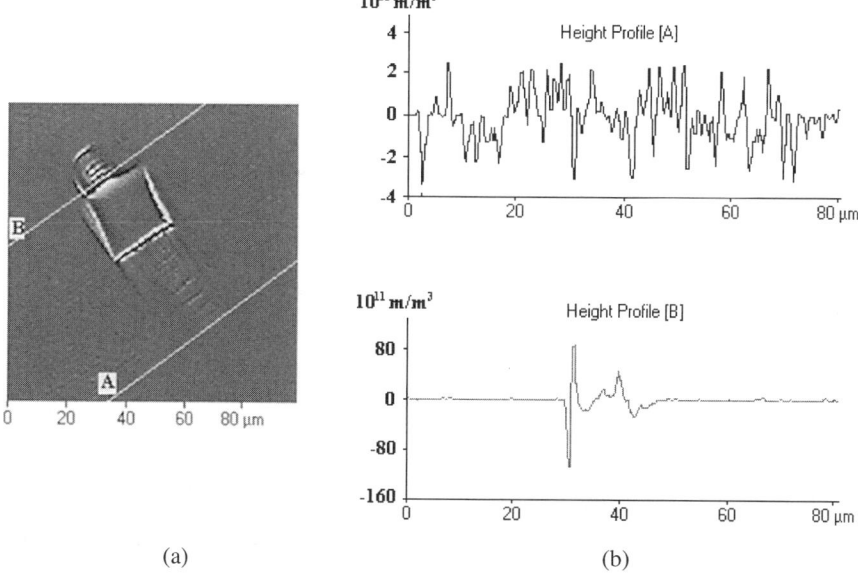

(a)   (b)

Figure 5. (a) Dislocation density around the micro-indention and (b) its line profile
measurements along the line A and B.

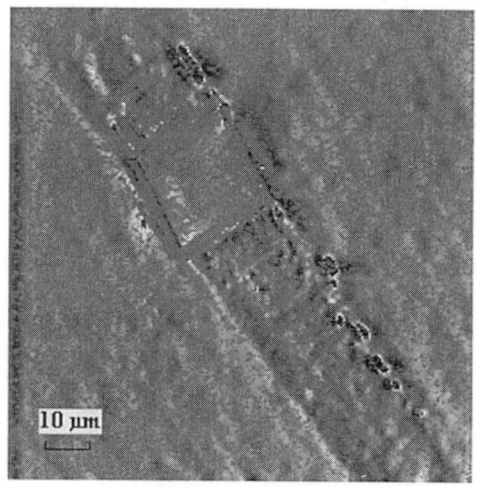

Figure 6. The line density of dislocation charge, which was determined by $\rho_e(x, y)/|\rho_d(x, y)|$

The density of dislocations $\rho_d(x, y)$ varies along the dislocation bands from 3×10^{11} to
9×10^{12} m^{-2}, as seen in Fig. 5(b). The background noise was around 1×10^{11} m^{-2}.

259

By dividing $\rho_e(x, y)$ by the absolute value of $\rho_d(x, y)$, we can estimate the line density of dislocation charges, see Fig. 6. We found that the maximum charge density is close to $-1.5 \times 10^{-10}\ C\,m^{-1}$ for S(g) and $+1.0 \times 10^{-10}\ C\,m^{-1}$ for Zn(g). This charge is comparable with the value of $-3 \times 10^{-10} C\,m^{-1}$ found by Petrenko and Whitworth [8] on moving S(g) dislocations in ZnS by using the method of dislocation currents [2]. Though the charge that we found on resting dislocations was half as much as that measured by Petrenko and Whitworth [8], it is in accordance with a theory of charge exchange between a moving dislocation and the rest of the crystal [9]. From Fig. 6, it can be clearly seen that the negative charges (dark) are often surrounded by positive charges (bright), and vice versa. The charge of Zn(g) dislocations has never been measured before because in the method of dislocation currents, motion of more mobile S(g) dislocations dominates.

CONCLUSIONS

We have developed a new method to determine the electric charge of resting dislocations. The method was successfully applied to S(g) and Zn(g) partial dislocations in ZnS. The dislocation charges value $-1.5 \times 10^{-10}\ C\,m^{-1}$ for stable S(g) dislocations was found to be comparable but half that for moving S(g) dislocations. The charges of S(g) and Zn(g) dislocations have opposite signs and different magnitudes. This is in accord with the model that combines both an ionic charge of dislocation cores and an accepter-donor effect of dangling bond in the cores [8].

ACKNOWLEDGEMENTS

This research was supported by the National Science Foundation.

REFERENCES

1. P. M. Fahey, S. R. Mader, S. R. Stiffler, R. L. Mohler, J. D. Mis, and J. A. Slinkman, IBM J. Res. Dev. **36**, 158 (1992).
2. Yu. A. Osip'yan, V. F. Petrenko, A. V. Zaretskii, and R. W. Whitworth, Adv. Phys. **35**, 115 (1986).
3. O. Nickolayev and V. F. Petrenko, J. Vac. Sci. Tech. B **12**(4), 2443 (1994).
4. D. B. Holt, J. Phys. Chem. Solids **23**, 1353 (1962).
5. P. Haasen, Acta Metall., **5**, 598 (1957).
6. A. V. Zaretskii, Yu. A. Osip'yan, V. F. Petrenko, G. K. Strukova, and I. I. Khodos, Phil. Mag. A, **48**, 279 (1983).
7. Yu. A. Osip'yan, V. F. Petrenko, and G. K. Strukova, Fiz. tvers. Tela, **15**, 1752 (1973) (Soviet Phys. solid St., **15**, 1172).
8. V. F. Petrenko and R. W. Whitworth, Phil. Mag. A, **41**, 681 (1980).
9. L. G. Kirichenko, V. F. Petrenko, and G. V. Uimin, Zh. eksp. teor. Fiz., **74**, 742 (1978) (Soviet Phys. JETP, **47**, 389).

ATOMISTIC SIMULATION AND EXPERIMENTAL INVESTIGATION OF ULTRA PRECISION CUTTING PROCESSES

R. RENTSCH
LFM, University of Bremen, Bremen, Germany

ABSTRACT

Typical applications for components and equipment with extreme quality requirements regarding surface roughness, shape accuracy and integrity of the generated surface structure can be found in optical and semiconductor industry. Ultra precision machine tools equipped with sharp, single crystalline diamond provide the necessary machining accuracy. Here the actual cutting process can take place at atomic level, which makes the acquisition of typical cutting process data difficult or impossible. However a detailed characterization and understanding of the process is vital for its effective control as well as for further tool and process development.

Therefore an approach is made that focuses on linking results from atomistic simulations with results and observations from cutting experiments. In this work the potential of molecular dynamics (MD) modeling for studying phenomena related to ultra precision cutting processes will be demonstrated. Observations and first results for machining copper will be presented.

INTRODUCTION

Since the 1970's the micro machining with single-crystalline diamond tools has developed into a sophisticated technique for the manufacture of components in a variety of advanced scientific and industrial applications. With this technique several materials are machinable, e.g. nonferrous metals like copper, brass, aluminum and electroless nickel, plastics, brittle materials like glass, ceramics, and semiconductor materials like germanium and silicon. Examples for micro - machined parts can be found in the field of optics, e.g. metal mirrors, infrared lenses, and molds for glass and plastic lenses. All these applications require sub-micrometer accuracy and a very fine surface finish of a few nanometer rms /1,2/.

Table I - Requirements for contour deviation and surface roughness of components for optical applications.

	wavelength λ	contour deviation $\lambda/10$	roughness rms $\lambda/100$
microwaves	≈ 300 µm	< 30 µm	< 3 µm
infrared light	≈ 3 µm	< 0.3 µm	< 0.03 µm
visible light	≈ 0.5 µm	< 0.05 µm	< 0.005 µm

In table I the requirements of optical components regarding form accuracy and surface roughness are summarized. They elucidate the high demands made to the manufacturing processes. In optical and semiconductor industry grinding and polishing techniques are commonly used to achieve optical surfaces. In many cases these processes could be substituted using micro machining with single crystalline diamond tools. Especially the manufacture of aspherical surfaces has been simplified due to the use of ultra precision machine tools with computer numeric control. Moreover, this technique opened the opportunity for the fabrication of microstructures, e.g. Fresnel structures and micro prisms.

Mat. Res. Soc. Symp. Proc. Vol. 578 © 2000 Materials Research Society

The extreme quality demands for high and ultra precise components, not only regarding surface roughness and shape accuracy, but also regarding the integrity of the sub-surface layer, require an effective control of the cutting process. Here the knowledge of the local stress state is of particular interest for functional surfaces, since high stresses can cause significant warping of parts and cracking of surfaces. Therefore, the detailed characterization and understanding of the process are vital for quality control, further tool and process development. The actual cutting process can take place at the atomic level. Hence atomistic modeling can provide the necessary insight into processes and allows to study local material properties and behavior in detail. Microstructural modeling employing molecular dynamics (MD) allows to directly study time-dependent processes such as surface generation, dislocation formation and stress relief.

EXPERIMENT

In order to analyze and characterize the ultra precision cutting process and to determine the influence of cutting parameters on the machining result, a project has been started combining MD simulation and cutting experiments.

First experiments were performed as inclined plunge-cut tests with continuously increasing depth of cut (inclination $0.05°$). Figure 1 shows surface micro topographies in coarse grained copper (average diameter 330 μm) obtained at 3 different places in a cutting groove. The experiments were carried-out with a sharp diamond tip (nose radius: 300 μm, edge radius < 50 nm) moving at 10 mm/min with a perpendicular standing face ($0°$ rake angle).

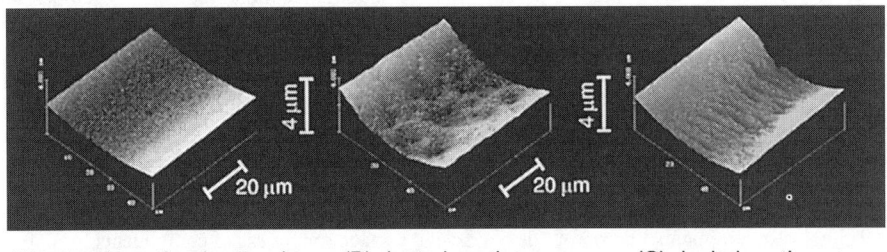

(A) smooth section (a_e<2μm)   (B) dented section   (C) rippled section

Figure 1: Micro topography at bottom of plunge-cut grooves [3]

Examining the grooves, 3 different micro-topographies were found. The micro topographic features could tentatively be divided into three distinct types: smooth sections, dented sections and rippled sections. Smooth sections are characterized by a micro roughness of less than 5nm rms (Fig.1-A). Surprisingly, smooth surfaces are always created if the depth of cut is less than 2 μm, regardless of cutting speed, tool radius, and grain orientation. Dented sections exhibit small dents approximately 10 μm in diameter, with a micro roughness larger than 50 nm rms (Fig.1-B). Rippled sections (Fig.1-C) can be associated with specific grain orientation. The micro roughness of these sections amounts to more than 100 nm rms.

For the MD simulations an orthogonal cutting process set-up with different orientations of single-crystalline copper was chosen. The basics of molecular dynamics are described in great detail elsewhere [4,5]. The employed process model, shown in Fig.2, considered about 80 000 fully dynamic work and tool atoms, and thermostat boundaries to control the process temperature. For the Cu/Cu interactions a Finnes-Sinclair type EAM potential was employed [6],

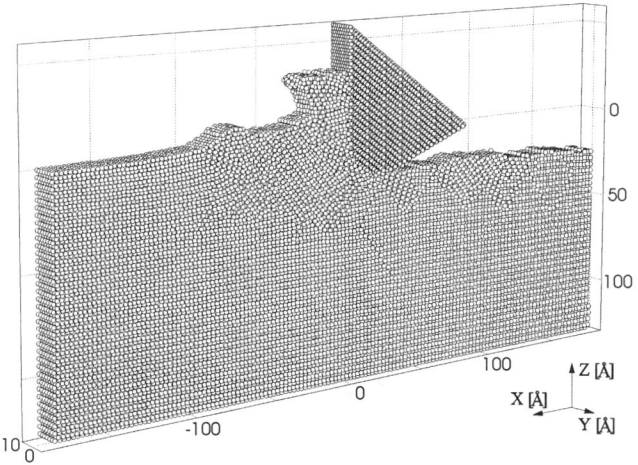

Figure 2 : Dimensions of the 3D cutting model.

because of their better description of metallic material properties. The C/C interaction in the diamond tool was approximated using a Lennard-Jones (LJ) pair potential with the bond strength of diamond. The atoms were placed on fcc lattice sites with an adjusted lattice constant in order to account for the correct atom density of diamond and, thereby, for a reasonable reaction intensity at the tool/work interface. The weak, adhesive potential for the tool/work interface was found elsewhere in the literature (see [5]). Friction conditions and wear processes at two-body interfaces, like at the tool/work contact, are of high interest in many engineering fields. One major disadvantage of using EAM, LJ or other potentials of this type is, that the type of reaction is fixed. The question as to whether a bond will form or not, beyond pure repulsion or adhesion, is already decided by choosing these potential functions. While so-called bond-angle, bond-order potentials account for the spatial dependence of the bond strength, and the thermal dependence of the reaction intensity is inherent to the MD framework (Arrhenius equation), these potentials lead to bond formation at any given level of energy. Since not all contacts and collisions lead to bond formation, like in purely exothermic reactions, the missing link is the activation barrier or energy of chemical bonding. Such a chemical potential would allow direct study of the onset and transition from friction to localized wear.

RESULTS

Structure and topography analysis

Several combinations of crystal orientation and cutting direction were analyzed. Here results of cutting on the {001} surface in <110> direction will be presented. The employed 3D model allows studying surface roughness, recovery and reconstruction directly. In figure 2, for example, it reveals dislocation loops formed during the cutting process at the generated surface. In the same way, the chip formation and other details of the cutting process can be studied. In figure 3 old bonds are represented by a simple grid of horizontal and vertical lines. Upon local deformation the grid structure distorts and shows elastic-plastic deformation. In case of a large displacement between former neighbor atoms, old bonds were not printed, in order to account for bond breaking. This convention helps to identify areas of intensive distortion, like the top layers at the generated surface, and shear bands, as seen across the chip cross-section in figure 3. There are fundamental differences in the chip formation process for the different set-up configurations. The configuration presented here led to a significantly larger chip cross-section than for the other configurations and shows a clearly scaled chip structure, which the others did not. The material in front of the chip turned for the chip generation into a more favorable orientation. Here sets of {111} planes run from the tip of the tool forward and down, and backward and down (compare figures 3 and 4). For the {111}/<121> configuration, for example, there were many fewer old bonds in the chip and at the generated surface, and less distortion. In this case the tool is positioned perpendicular to {111} slipping planes and cuts parallel to them in <121> direction. In closed-packed structures, atoms are geometrically well supported perpendicular to {111} planes. The tool can not easily generate dislocations through these planes, which causes high thrust forces. From 2D MD cutting simulations deep running dislocations were reported [5]. In the 3D model, such deep running, stable dislocations were not observed.

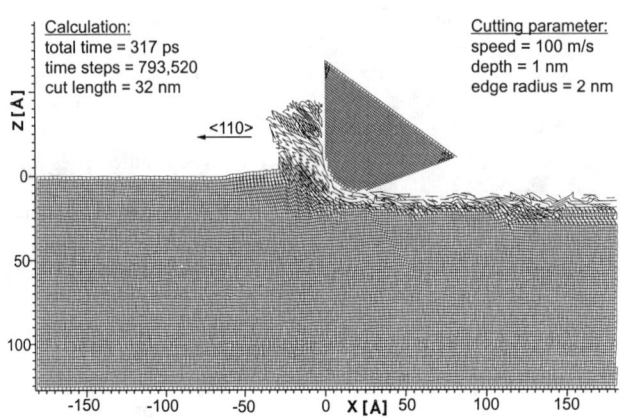

Fig. 3: Deformation graph for setup {001}/<110>, view <110>

Stress Distribution

The well-defined physical concept of MD, considering atomic interactions and classical mechanics in a thermodynamic environment, allows, at least in principle, the determination of all data of interest. Although the basis for the analysis of macroscopic, thermodynamic quantities is given by the fundamentals of MD, it was not exploited for cutting process data analysis yet. Instantaneous atomic data, of thermodynamic and mechanical quantities like temperature and stress in particular, have very limited meaning and the amount of data of large systems are difficult to handle from the point of view of efficient data analysis. In order to allow for a better comparison of MD results with results of experiments and continuum mechanics approaches, suitable sampling methods and parameter for the MD-data analysis of quantities such as stress (or hydrodynamic pressure) and temperature were tested. The quantities are expressed as local averages over space and time and allow a reasonable visualization of local material information and behavior without the need to deal with large sets of

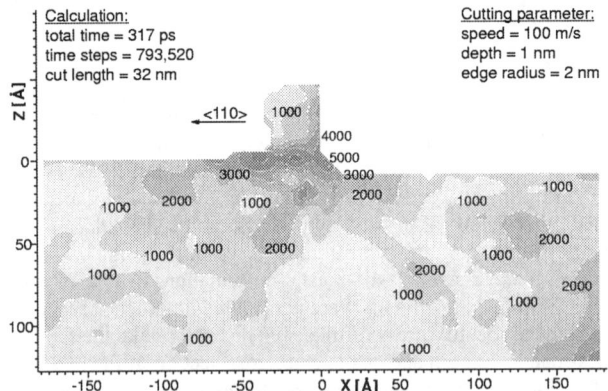

Figure 4: Maximum shear stress setup {001}/<110>, [N/mm²], view <110>

information representing each single atom site at every time step of simulation. Figure 4 shows a maximum shear stress distribution. It shows high shear stress values in a narrow band across the chip cross-section with another local maximum below. High shear stress appears at the chip/tool contact area, which is a clear hint of friction. A similar distribution was found for the hydrostatic pressure where the maximum pressure occurred under the tip of the tool, next to the area of high shear stress. Since there was no dislocation activity observed in the deeper workpiece area, the deformation here is exclusively elastic. At a ratio of depth of cut to cutting edge radius of 0.5, the tool produces a high thrust load which appears as microstructural effects in the local stress distribution as the stress needs to be resolved in the lattice by elastic-plastic deformation. In this way the {111} planes for the configuration in figure 4 (running forward and down, and backward and down relative to the cutting direction) show up as areas of slightly higher stress.

Temperature Distribution

Figure 5 shows the temperature distribution for the set-up configuration {001}/<110>. The shear zone shows a temperature of 550 K. Under the tip of the tool, where the material is sheared and deformed at high stress, most of the heat is generated. Hence, the high temperature area extends from here into the chip. The relative small chip volume and the lack of heat convection may have elevated the temperature level in the chip. On the other hand, the heat capacity of the chip serves as an important heat sink in the process as it takes away the heat from the source while the chip develops.

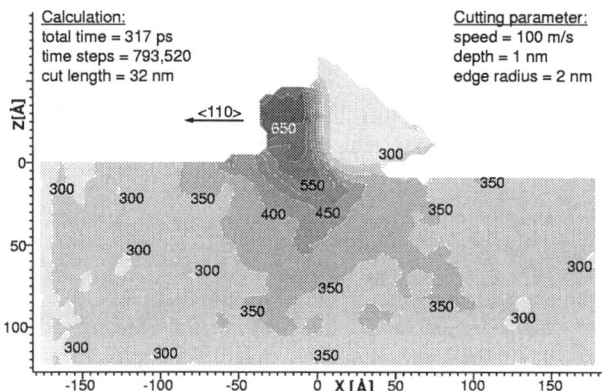

Figure 5: Temperature distribution for setup {001}/<110>, T in [K], view <110>

The temperature distribution is roughly of concentric shape with a slight shift opposite to the cutting direction. The high conductivity of diamond passes heat fast through the tool. This ability keeps the temperature in the tool low and causes a strong temperature gradient between chip and tool. The set-up configuration {111}/<121> led to a temperature level of 700 K at the bottom of the chip (maximum chip temperature of about 750 K). Here (see Fig.5), the temperature level at the chip bottom is 550-600 K (maximum 650 K). The significantly lower temperature level for set-up {001}/<110> is founded in a chip formation process that utilizes to a considerable extent twinning as shearing mechanism, which can be shown by atom structure analysis. The very short Burgers vector necessary for twinning results in a lower heat generation and, thereby, lower temperature. This is further consistent with the observation of shear bands and many old bonds in the scaled chip structure for set-up {001}/<110> (fig. 3), where as the chip in set-up {111}/<121> was formed by intensive shearing with almost no old bonds at all. The presented temperature distribution includes only thermal conductivity through phonons.

CONCLUSION

In this work first results of an ongoing effort on characterizing and modeling ultra precision cutting processes are presented. Since the actual material removal can take place at the atomic level, characteristic process data cannot easily be measured. The applied strategy takes advantage of the physically well-founded molecular dynamics method (MD), which, upon verification and evaluation by experimental results, will form the basis for an extension of cutting process modeling into the atomic range.

Beyond a critical depth of cut, cutting experiments on crystalline copper suggest specific surface generation mechanisms and led to different surface quality and cutting forces depending on the crystalline orientation. The correlated, specific crystalline orientations are not determined as yet. The MD simulations confirmed differences in the cutting process as a function of crystalline orientation (chip formation mechanism, heat generation, temperature, cutting force). The MD data were converted into continuous temperature and stress distributions and, thereby, provide improved insight into the nanometer level cutting process. The further potential of MD to support the evaluation of cutting process data lies in its micro mechanical and thermodynamical basis, that allows, at least in principle, to directly extract information about strain, strain rates and friction conditions. Such information is useful for the validation of local material properties and the development of mesoscopic and macroscopic models as well as for direct comparison with analytical and continuum mechanics models, if micro structural properties are of interest.

ACKNOWLEDGMENTS

This work was carried out during a stay at the University of Pennsylvania, Philadelphia, PA, USA, which was funded by a Feodor-Lynen research scholarship of the Alexander-von-Humboldt Foundation, Bonn, Germany, and the Laboratory for Research on the Structure of Matter (LRSM) at the University of Pennsylvania.

REFERENCES

1. IKAWA, N., DONALDSON, R.R., KOMANDURI, R., KÖNIG,W., MCKEOWN, P.A., MORIWAKI,T., STOWERS, I.F., Ultraprecision Metal Cutting - The Past, the Present and the Future, in *Annals of the CIRP*, Vol. **40** (2), (Hallwag Publ., Berne, CH, 1991), pp.587-594.

2. BRINKSMEIER,E., PREUSS,W., RIEMER,O., SCHRÖDER,M., Manufacturing and Measurement of Elliptical Mirrors for Focussing Synchrotron Radiation, *in International Progress in Precision Engineering*, edited by M. Bonis et al. (Elsevier, Compiegne, F,1995), pp. 459-462.

3. RENTSCH, R., INASAKI, I., BRINKSMEIER, E., PREUSS, W., RIEMER, O., Influence of Material Characteristics on the Micromachining Process, in *Materials Issues in Machining-III and The Physics of Machining Processes-III*, edited by D.A. Stephenson and R. Stevenson (TMS publication, Cincinnati, Ohio, USA, 1996), pp. 65-86.

4. HOOVER, W.G., *Computational Statistical Mechanics*, Studies in Modern Thermodynamics 11, (Elsevier Science Publisher, New York, 1991), pp-130-180.

5. RENTSCH, R., Process modelling by means of molecular dynamics (MD), in *Bearbeitung neuer Werkstoffe / 2nd International Conference on Machining of Advanced Materials (MAM)*, VDI-Berichte 1276 (VDI-Verlag, Düsseldorf, 1996), pp.175-195.

6. ACKLAND,G.J.; TICHY,G.; VITEK,V.; FINNIS,M.W., Phil. Mag., A. **56** (6), 735 (1987).

SURFACE RELAXATIONS OF ALUMINUM SIMULATED BY BOND ORDER POTENTIALS

S. R. NISHITANI*, S. OHGUSHI*, H. ADACHI*, M. AOKI**
* Department of Materials Science and Engineering, Kyoto University, Kyoto, 606-8501
Japan, bob@karma.mtl.kyoto-u.ac.jp
** Department of Electrical and Electronic Engineering, Gifu University, Gifu, 501-1112
Japan

ABSTRACT

An interatomic potential for aluminum was developed, which is based on empirical tight binding approximations. The model successfully reproduced the shear constants, structure energy differences, and phonon dispersion curves. This transferable potential was applied on static surface relaxations, and shows good agreements with experimental results on the oscillatory damped behavior of the multilayer relaxations and the expansion of the (111) surface.

INTRODUCTION

On materials science, the demands on understanding the mechanisms from atomistic levels have been improving interatomic potentials and now try to join the knowledge of the first principles electronic structure calculations. For realistic simulations on phase transformations or atom relaxations around the lattice defects, speedy empirical atomic potentials are still widely used on various systems. But reliable potentials are highly empirical with many fitting parameters. For describing the bondings around the lattice defects appropriately, however, the potentials should reflect the changes of the electronic structures. Thus the empirical tight binding (TB) methods have been recently focused in order to bridge the gaps between the first principles calculations and empirical interatomic potentials[1, 2]. Because sp valent metals like aluminum show a broad nearly free-electron band, which is opposite to tight binding bands of transition metals and covalent semiconductors, they have widely been thought to be hardly treated by TB framework. However, many researches successfully treat the phase stabilities of sp-valent elements by TB[3, 4, 5]. If we can apply it on the other kind of simulations, we can investigate the wide varieties of solid systems by a single and simple framework.

For realistic simulations, TB calculations have another difficulty of 'speed' comparing to widely used empirical potentials. The recursion method of TB calculations uses a mathematically efficient 'Lanczos' algorithm with the $O(N)$ relation on inequivalent atoms. Recently the authors are developing the program code 'anbop', which performs the two-center, orthogonal TB recursion method efficiently[6]. This code can be easily applied on static relaxations around the lattice defects, because the recursion routine is highly parallelized and shows the ideal inverse relation between the calculating times and the numbers of processors[7]. 'anbop' is an implementation of the concept of bond order potentials (BOP)[8, 9], whose analytic derivatives have been applied successfully on various systems from hydrocarbons[10], to covalent and metallic systems[1, 5, 11].

In this paper we investigate multilayer surface relaxations of pure aluminum using 'anbop'. Firstly we will describe the formulations and fitting procedures shortly. Then we will show the structure energy differences, monovacancy formation energy and phonon dispersion curves using the model fitted on shear constants and cohesive energy. Finally, we will show the results on surface relaxations.

Mat. Res. Soc. Symp. Proc. Vol. 578 © 2000 Materials Research Society

Table 1: TB and pair potential parameters in eqs.(3). Unit is eV except p and q.

$h_{ss\sigma}$	$h_{pp\sigma}$	$h_{sp\sigma}$	$h_{ps\sigma}$	$h_{pp\pi}$
-0.4768	1.4023	0.7617	- 0.7617	-0.0834

$E_p - E_s$	p	q	φ^0	
4.6400	6.5254	2.525	0.2828	

BOND ORDER POTENTIALS

The two-center, orthogonal TB model expresses the total energy per atom as follows[8];

$$E_{\text{atom}} = \sum_{i \neq j} \varphi\left(R_{ij}\right) + E_{\text{bond}} + \sum_{\alpha} \left(N_\alpha - N_\alpha^{\text{atom}}\right) E_\alpha \qquad (1)$$

where $i(j)$ indicates sites, and α indicates the orbitals. The first term is a simple pair interaction. The second term is the bonding contributions of the electrons, and can be given by

$$E_{\text{bond}} = \sum_{\alpha} \int_{-\infty}^{E_{\text{F}}} \left(E - E_\alpha\right) n_\alpha\left(E\right) dE \qquad (2)$$

where $n_\alpha\left(E\right)$ and E_α are the density of states and the self energy of the orbital α respectively, and E_{F} is the Fermi energy. For the calculation of the bonding energies, we use the recursion algorithm [13] and the fast analytic integrations[6]. The recursion algorithm is $O(N)$ calculation and the locality is perfect. Thus the parallel codes show ideal inverse relations between CPU times and CPU numbers[7]. The third term of eq.(1) is the promotion energy, which is associated with the change of occupancy of the atomic orbitals on forming the solid from free isolated atoms. We have assumed that each atom is locally charge neutral (LCN)[1, 7].

The fitting procedure adopted in this paper has been described in a previous paper in detail[14]. The TB model of Al is constructed by s and p orbitals with the first nearest neighbor interactions. The recursion level is four, which is equivalent to the moment level of eight. The repulsive $\varphi\left(R_{ij}\right)$ and Slater-Koster hopping parameters $h_{\alpha\beta}\left(R_{ij}\right)$[15] functions are given by

$$\begin{aligned} \varphi\left(R_{ij}\right) &= \varphi^0 \exp\left(-p\left(R_{ij}/R_0 - 1\right)\right) \\ h_{\alpha\beta}\left(R_{ij}\right) &= h_{\alpha\beta}^0 \exp\left(-q\left(R_{ij}/R_0 - 1\right)\right) \end{aligned} \qquad (3)$$

where the coefficients of the exponent functions are the values at the equilibrium distance R_0. The parameters have been fitted to the equilibrium distance, shear constants, bulk modulus, and the cohesive energy. The obtained parameters are shown in Table 1. The value of q is mainly fitted to obtain the best results of shear constants, which is not far from Harrison's suggestion of two[16]. The starting set of the Slater-Koster parameters are those proposed by Papaconstantopoulos[17]. Reducing the values of hopping integrals and the self energy difference, $E_p - E_s$, for getting the correct cohesive energy alters the shape of the density of states. A typical change is observed in the bottoms of the band from -13 eV for the original values to -9 eV for the reduced values measured from the Fermi energy. This, however, is not so much different from the value of -11.3 eV obtained by the other first principles calculation[18]. In k-space, the dispersion curves calculated with sp orthogonal bases of the first nearest neighbor interaction show queer structures on the direction between X and W. Although the interactions of further nearest neighbor interactions or d orbitals improves the k-space structures drastically, they should affect small on the total energies. We adopted only the first nearest sp orbital interactions in this research for the simplicity and speed of the models.

Table 2: Fitted physical properties and estimated values of the structure energy difference and defect energies.

	calculated	experimental
E_{atom}[eV]	-3.39	-3.39
Shear constants [GPa]		
Bulk	77	77
C_{11}	107	108
C_{12}	62	63
C_{44}	29.6	28
C'	22.4	23
Structure energy difference [eV]		
ΔE_{bcc}	0.04	0.1[20]
ΔE_{hcp}	0.05	0.06[20]
	(c/a=1.655)	
ΔE_{dia}	1.33	0.66 *
Mono vacancy formation energy [eV]		
	1.2	0.7

* The first principles calculation[12].

RESULTS

The calculated physical properties are shown in Table 2. The cohesive energy, bulk modulus and equilibrium distance were fitted. The shear constants were adjusted by the exponent of the hopping integrals, and shows good agreement with the experiments. Especially the anisotropy factor $A = C_{44}/C'$, which shows the characteristic value less than two for Al, is very good agreement with the experimental one. The correct relations are hardly obtained by the simpler empirical potentials, such as embedding atom potentials[19] or the simplest bond order potentials[11].

For checking the model, we calculate the structure energy differences of bcc, hcp and diamond lattices, and an unrelaxed mono-vacancy formation energy. Those results are also tabulated in Table 2. In structure energy differences, fcc is correctly estimated to be the most stable structure. Especially the energy difference between fcc and hcp is correctly reproduced. The diamond structure shows much different from the first principles calculations, however, because very small coordination numbers of four strengthen the bonds so much and the minimum energy locates at much smaller distance. The smallest coordination number of the surfaces concerned in this paper is seven, which is expected to be large comparing to the coordination number of four of the diamond structure. The mono-vacancy formation energy is large comparing to the experimental results, which may be reduced by the models with higher recursion levels[14]. We also check the model by the phonon dispersion curves as in Fig.1 with the experimental ones. This is performed by calculating the dynamic matrices numerically with a perfect lattice of 256 atoms. No strange behavior is observed in all direction of the errors within 20%.

Simulations on the surface relaxations using this model were performed by the energy base conjugate gradient method. First five layers of some principal surfaces were allowed to relax only in the z-direction. Note that the models are not necessary to make the slab configurations as in k-space TB calculations. The LCN needs some buffer layers, which was chosen to be large enough for the fourth recursion levels of about 20 layers. In x and

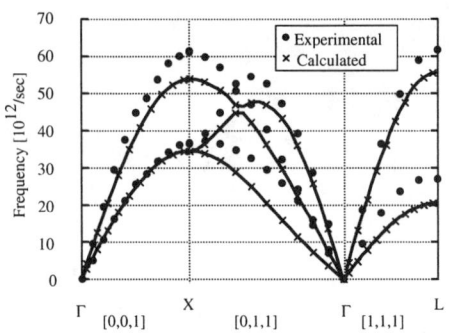

Figure 1: Phonon dispersion curves.

Table 3: Surface energies of unrelaxed and relaxed conditions.

Miller's index	Surface energy [mJ/m^2]	
	unrelaxed	relaxed
(111)	845	841
(100)	1106	1103
(110)	1211	1036
(311)	1129	1081
(331)	1026	1003

y directions, the lattices were expanded periodically.

The unrelaxed and relaxed surface energies are given in Table 3. The experimental value is 1140 mJ/m^2[21] and the orientation dependencies are very small for Al with $\gamma^{100}/\gamma^{110} = 0.98$ and $\gamma^{100}/\gamma^{111} = 1.03$ [22]. The simulated energies are scattered about $\pm20\%$ for the unrelaxed conditions, whereas it becomes $\pm10\%$ for the relaxed surfaces. Especially the (110), (311) and (331) surfaces show very large reductions, expecting large amounts of relaxations.

Surface relaxations simulated by this potential are shown in Fig.2 comparing with the experimental results[23]. The relaxed directions and amounts show very good agreement with the experimental results except (100) surface. A very small expansion in this surface was reported experimentally, whereas the recent first principles calculation suggests that the (100) surface shows a contraction[24]. The backbond strengthening, which is a straight insight of the simpler empirical interatomic potentials like EAM, can't explain the surface expansion observed in the (111) surface of Al. The oscillatory damped behavior of the multilayer relaxations of the (110), (311) and (331) surfaces are well reproduced by BOP. These behaviors have been hardly reproduced by the simplest model of EAM[25]. Thus the tight binding expressions of Al are expected to simulate the correct changes of the electronic structures near the surfaces.

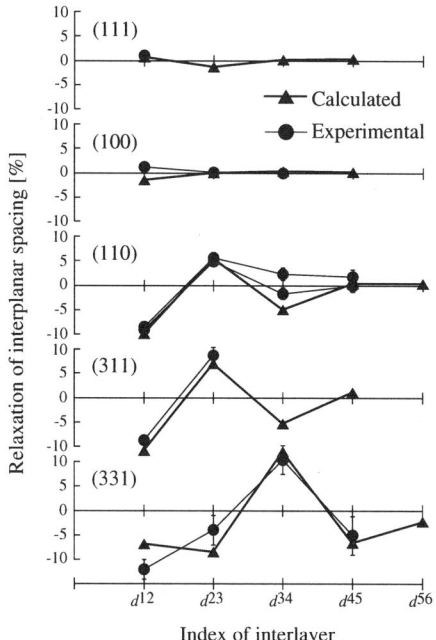

Figure 2: Calculated and experimental surface relaxations of Al. Surface indexes are indicated in parentheses. The values are expressed as a percentage of the bulk interplanar spacing, where a positive value implies an expansion and a negative value implies a contraction. The indexes in horizontal line denotes the corresponding interlayer.

CONCLUSIONS

An appropriate TB parameter set for describing equilibrium physical behaviors of Al was obtained. The model using these parameters reproduced the phonon dispersion curves and the structural energy differences. The simulations on the surface relaxation showed consistent interlayer spacings with the experimental results on some high principal surfaces. Although more minute preparations are necessary to apply these parameters on different environments, the success on the surface relaxations of Al suggests that the tight binding treatment can be applied on more complicated defect simulations of Al, such as interfaces or dislocations.

ACKNOWLEDGMENTS

The developing of 'anbop' is partially supported by the Grant-in-Aids for Scientific Research on Priority Areas, the Minister of Educations, Science, Sports and Culture, Japan, on "The Elucidation of Microscopic Mechanisms of Phase Transformations for the Microstructure Control of Materials". SRN appreciates the usage of computer facility of Kyoto University Data Processing Center.

REFERENCES

1. D. G. Pettifor, *Bonding and Structures of Molecules and Solids*, (Oxford Press, Oxford, 1995).

2. see for example, *Tight-Binding Approach to Computational Materials Science*, edited by P. E. A. Turchi, A. Gonis, L. Colombo (Mater. Res. Soc. Proc. **491**, Pittsburgh, PA, 1998).

3. G. Allan and M. Lannoo, J. Physique (Paris) **44**, 1355 (1983).

4. J. C. Cressoni and D. G. Pettifor, J Phys: Condens. Matter **3**, 495 (1991).

5. P. Alinaghian, P. Gumbsch, A. J. Skinner, D. G. Pettifor, J Phys: Condens. Matter. **5**, 5795 (1993).

6. M. Aoki, A. P. Horsfield, D. G. Pettifor, J. Phase Equilibria **18**, 614 (1997).

7. S. R. Nishitani, and M. Aoki, Trans. Mater. Res. Soc. Japan **24**, 209 (1999).

8. A. P. Sutton, M. W. Finnis, D. G. Pettifor, Y. Ohta, J. Phys. C: Solid State Phys. **21**, 35 (1988).

9. D. G. Pettifor, Phys. Rev. Lett. **63**, 2480 (1989).

10. D. G. Pettifor and I. I. Oleinik, Phys. Rev. B **59**, 8487 (1999); 8500.

11. P. Alinaghian, S. R. Nishitani, D. G. Pettifor, Philos. Mag. B **69**, 889 (1994).

12. I. J. Robertson (private communications).

13. V. Heine, Solid State Phys. **35**, 1 (1980).

14. S. R. Nishitani, J. Phase Equilibria **18**, 546 (1997).

15. J. C. Slater and G. F. Koster, Phys. Rev. **94**, 1498 (1954).

16. W. A. Harrison, *Electronic Structure and the Properties of Solids*, (Dover Publications, New York, 1989), p.48 and p.149.

17. D. A. Papaconstantopoulos, *Handbook of the band structure of elemental solids*, (Plenum press, New York, 1986), p.207.

18. V. L. Moruzzi and C. B. Sommers, *Calculated electronic properties of ordered alloys: A handbook*, (World Scientific, Singapore, 1995), p.13.

19. R. A. Johnson, Phys. Rev. B **37**, 3924 (1988).

20. K. F. Michaels, W. F. Lange III, J. R. Bradley, H. I. Aaronson, Metall. Trans. A **6**, 1843 (1975).

21. W. R. Tyson, Can. Metall. Q. **14**, 307 (1975).

22. R. S. Nelson, D. J. Mazey, R. S. Barnes, Philos. Mag. **11**, 91 (1965).

23. R. J. Rous, *Cohesion and Structure of Surface*, edited by F. R. deBoer and D. G. Pettifor, (Elsevier Science B.V., Amsterdam, 1995), p.7.

24. G. A. Benesh and D. Gebreselasie, Phys. Rev. B **54**, 5940 (1996).

25. S. R. Nishitani, S. Ohgushi, M. Aoki, and H. Adachi, Mater. Trans., JIM **40**, (1999), *in press*.

MODELING OF THE DISLOCATION FORMATION AT PORES AND INCLUSIONS UNDER THERMO-MECHANICAL SHEAR LOADS

R. RENTSCH*, V. VITEK**

* Laboratory for Precision Machining (LFM), Bremen University, Bremen, Germany, and Max-Planck-Institute for Metal Research, Stuttgart, Germany.
** Department of Material Science and Engineering, University of Pennsylvania, Philadelphia, PA 19104, USA.

ABSTRACT

Materials used in many technical applications contain a broad variety of defects. In both metals and ceramics, pores and inclusions influence significantly the elastic and plastic response. They can limit or broaden the range of applicability of the materials for high performance components and play, therefore, a very significant role in engineering applications.

This study focuses on the onset of dislocation formation and its intensity at pores and hard inclusions during the deformation process. For this purpose a molecular dynamics model of an fcc metal containing pores and/or inclusions was developed. This model material was then sheared at different strains and related development of the dislocation substructures investigated.

INTRODUCTION

Unlike the single crystalline structures often used in atomistic studies, real materials that form the basis for the design of engineering components contain a broad variety of defects. Such defects are grain boundaries, different phases, dislocations, pores, inclusions, and also alloying elements and impurities. Materials used in many technical applications possess a certain degree of porosity, in particular sintered materials, or heterogeneous structures, for example, due to the presence of carbides as hard inclusions. In both metals and ceramics, pores and inclusions influence significantly the elastic and plastic response [1]. They can limit or broaden the range of applicability of the materials for high performance components and play, therefore, a very significant role in engineering applications. In manufacturing engineering, which is concerned with the machining of materials to produce the required quality and shape of technical components, the machinability of materials, that is controlled by the ability to plastically deform, is the most important criterion. The fundamental requirement for high ductility is the ability to generate dislocations on a massive scale.

Hence this study focuses on the onset of dislocation formation and its intensity at pores and hard inclusions during the deformation process. First, some of the details of the employed molecular dynamics model and the simulation are described. In the following the results of shearing the model material containing pores and hard inclusions with different interface strength are discussed relative to the response of a defect-free single crystal.

COMPUTER EXPERIMENT

For the investigation of the influence of pores and inclusions on the dislocation behavior, the molecular dynamics (MD) program MOLDY was employed [2]. The system box, confining all particles in a 3D periodic arrangement, was constituted to follow Parrinello-Rahman Lagrangian dynamics [3], i.e. with constant number of particles, system pressure and temperature (NPT-system). The investigations were carried out for an fcc metal employing Ackland's EAM-type potential for copper [4]. The crystal orientation was chosen as <110>, <121> and <111> in xyz-directions of the box system with an initial size of 102 x 13 x 100 Å^3. The short y-axis leads to a quasi 2D structure. Hence, a pore or an inclusion has cylindrical shape in such a box, with the axis along the short, periodic box side. Therefore, the shearing of the box was restricted to the xz-plane (see Fig. 1).

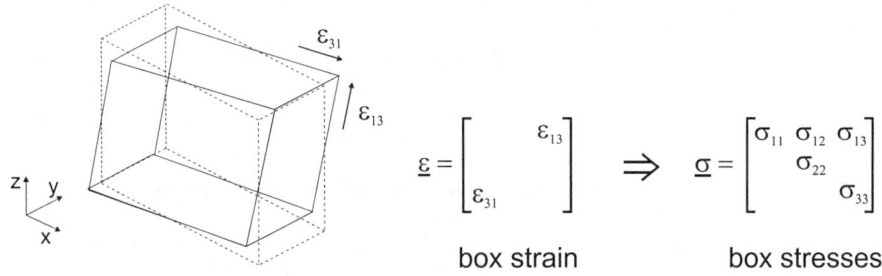

Figure 1: Shearing of the simulation box

The system response upon shearing of the box was observed in terms of the dislocation formation and stresses in the box. The results presented here, were obtained for a system at 300 K, 1 bar pressure and at a shearing speed of 50 m/s. After an initial period of relaxation of 10000 time steps (2 fs each), the shearing of the box was started and continued until dislocation formation was observed. Since the focus of this work is on the onset of dislocation formation, the shearing was stopped at this point and the system was allowed to relax thereafter.

RESULTS

First a single crystalline structure without any defect was studied for reference (see Fig.2). The dislocations observed in this system were dominantly Shockley partial dislocations. In most cases the dislocations were initiated on horizontal {111} planes (parallel to xy plane), but sometimes also on other {111} planes oblique in the box. After an initial period of relaxation of 10000 time steps, the principle stresses σ_{11} and σ_{33} are rising upon shearing, indicating a rising pressure which causes the box to expand to some degree (see Fig.2). Similarly the shear stress σ_{13} rises (here with a negative sign). In figure 2 the development of stresses for 10 % and 11% shearing of the box are

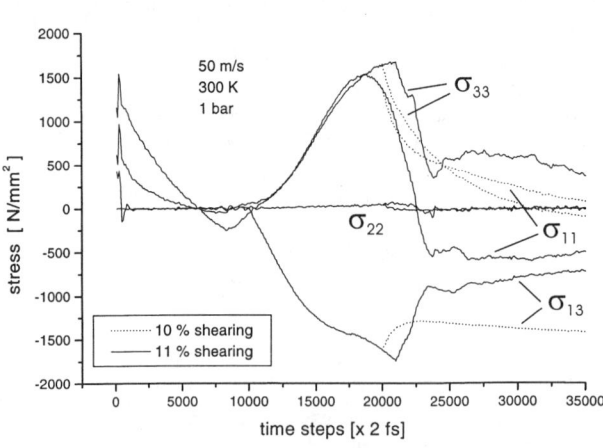

Figure 2: Box stresses when shearing single crystalline copper

shown. The difference in the development of these two sets of stresses lies in the fact, that there were dislocations generated in one case, but not in the other case. At 10 % shearing, the single crystalline structure was still intact. While the principle stresses relaxed completely by volume expansion, the shear stress remained on a high level after a short period of relaxation. At 11 % shearing, the state of stress changed due to dislocation formation, whereupon the principle stresses σ_{11} and σ_{33} dropped rapidly to a level around 500 N/mm^2 , one in a tensile (-) and the other in compressive state (+). The shear stress dropped to a level about half as big as without dislocation formation. Massive dislocation generation in single fcc crystals was observed after exceeding the maximum yield strength at lower temperatures, as here at 300 K. Rising bulk temperature led to exceeding the falling maximum yield strength already at smaller strains and generating fewer dislocations.

In real materials a perfect single crystalline structure is very scarce as most technically applied materials show structures with high defect densities. In crystalline materials dislocation densities of 10^{14} m^{-2} are common, and sintered materials show heterogeneous structures with a porosity of up to 30%. It is generally accepted, that crystal defects can act as sinks for and sources of dislocations [1]. Although this applies to both, the pores and the inclusions, their effect on the material response and mechanisms involved are significantly different. Figures 3 and 4 show simulation results of shearing fcc structures with a pore and with a hard, well bonded inclusion, respectively. At the rim of the pore in Fig.3 five partial dislocations were generated which caused a deformation of the initial, symmetrical pore shape. The dislocations were only generated at the pore, but all across its diameter. The maximum elastic strain of 10 % for the single crystal is reduced below 4 % for the structure with the pore.

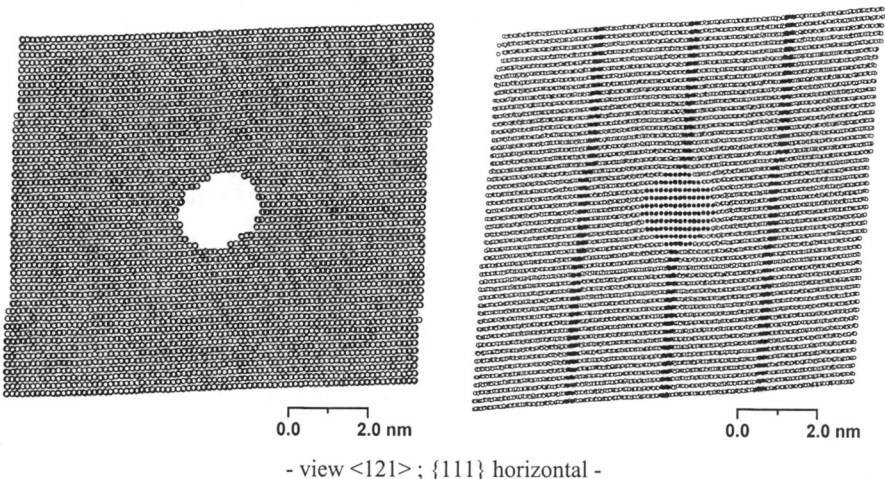

0.0   2.0 nm   0.0   2.0 nm

- view <121> ; {111} horizontal -

Figure 3: Fcc crystal with pore (4 % shearing)   Figure 4: Fcc crystal with bonded inclusion
(8 % shearing)

To exceed the maximum elastic strain and to generate first dislocations, the fcc crystal with a well bonded inclusion needed to be sheared 8 % (see Fig.4). In figure 4, three vertical layers of atoms were colored for presentation purpose to identify dislocations. It should be noted, that the inclusion was implemented as a rigid body composed of atoms which interact through atom-atom interaction with the matrix, but are fixed within the body. The term "well bonded" refers to the matrix-inclusion interface that had the strength of the matrix, which allows a distinction between the influence of the hard inclusion and the influence of the interface. Prior to shearing, there was no lattice misfit at the interface since the inclusion was built from freely relaxing copper atoms in a copper matrix. Hence, the hard inclusion was embedded in the matrix at ideal conditions. The dislocations in this structure were only generated horizontally at the top and bottom poles of the inclusion. The well bonded inclusion inhibited shearing across its width, unlike the pore.

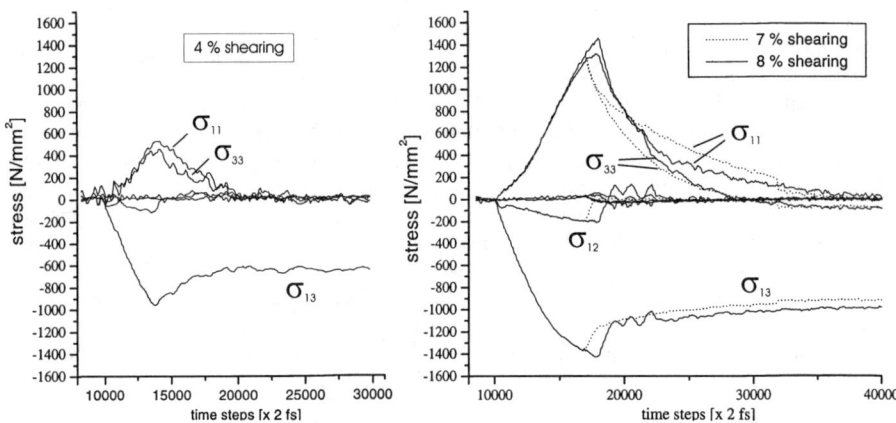

Figure 5: Course of stresses when shearing porous fcc crystal

Figure 6: Course of stresses when shearing crystal with bonded hard inclusion

Figures 5 and 6 show the development of the box stresses for a crystal with pores and inclusions respectively. In comparison to the maximum yield strength of the single crystal of $1600 - 1800$ N/mm^2, the maximum yield strength for the porous structure is reduced to 900 N/mm^2. The rather straight rise of the shear stress during the period of shearing (time step 10000 to 14000) suggests, that the dislocations were generated during the relaxation period which led to a gradual, step-like stress relief until a shear stress level of about 600 N/mm^2 was reached.

Figure 6 shows two sets of box stresses for the crystal with the bonded hard inclusion, for 7% shearing (without dislocation formation) and for 8% (with dislocations). With a maximum yield stress of 1400 N/mm^2, the crystal with the inclusion exhibits improved shear strength in comparison to the properties of a porous crystal, close to the strength of the single crystal. Although the strength and the response to shearing of the crystal with the inclusion are very similar and close to that of the single crystal, its yielding response beyond the maximum shear strength is different. In contrast to the single crystal case, where many dislocations are generated at once on exceeding the yield strength, and the stresses drop significantly (at 300 K), the crystal with the inclusion shows a rather gradual stress release during relaxation, similar to the response of a dislocation-free crystal. The dislocations were presumably formed after the shearing, during the period where the shear stress curve is going up and down (time step 19000-22000). It is interesting to notice, that the shear stress after 8% shearing remains on a high level of about 1000 N/mm^2 and is not falling to the level of $500 - 600$ N/mm^2, as seen for the single and the porous crystal. Further the shear stress at 8 % shearing is not falling below the stresses of the dislocation-free crystal at 7 % shearing, although dislocation formation is related to stress relief.

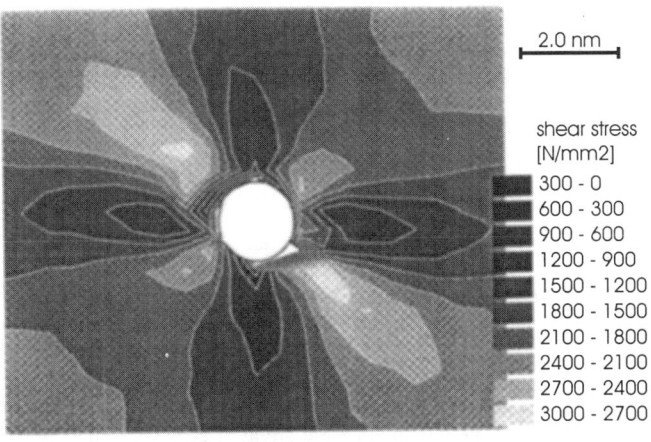

2.0 nm

shear stress
[N/mm2]
300 - 0
600 - 300
900 - 600
1200 - 900
1500 - 1200
1800 - 1500
2100 - 1800
2400 - 2100
2700 - 2400
3000 - 2700

Figure 7: Shear stresses in crystal with bonded inclusion
(at 7 % shearing)

Figure 7 shows the shear stress distribution in the relaxed simulation box after 7 % shearing (the position of the relaxed hard inclusion is whitened out). It can be seen, how the hard inclusion influences the local shear stress distribution. Instead of an equal stress distribution, the stresses concentrate around the inclusion. High shear stresses are diagonally oriented around the inclusion, with top values at its top and bottom poles. Areas of low shear stress load run vertically and horizontally, where the hard inclusion inhibits dislocation formation.

The well bonded and embedded inclusion described above represents an ideal case that allows to study the influence of the presence of hard, heterogeneous structures without the superpositional effect of interface properties. In order to study the influence of the interface in such a system, another extreme case can be considered, that of a loosely embedded, but hard inclusion. The model setup for the simulation remains the same as for the well-bonded inclusion,

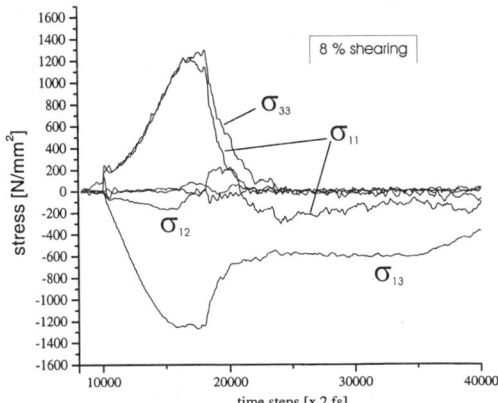

Figure 8: Development of stresses when shearing crystal with un-bonded inclusion

except for the interaction at the interface. Here, the interaction at the loosely embedded inclusion was restricted to the repulsive part of the copper matrix potential, which resulted in an un-bonded inclusion. Figure 8 shows the development of the stresses for such a model. For comparison, this model was strained to the level at which a first dislocation generated (8% shearing) in the crystal with the bonded inclusion. Because of the effectively zero-interface-strength with the un-bonded inclusion, dislocations were already initiated at about 6% shearing (time step 16000) which led to the stress fluctuation on high level until the end of shearing (time step 18000). The lack of any interface strength caused the reduction of the maximum yield stress from 1400 N/mm^2 for the full strength, bonded inclusion to 1200 N/mm^2 for the un-bonded inclusion. At the end of shearing, the simulation box relaxes again to a shear stress of 600 N/mm^2. Here, the missing bonds between matrix and inclusion atoms cause the formation of gaps or holes in areas at the inclusion surface, that are under tensile load. The interface of the un-bonded inclusion reacts like an internal surface. Hence, no energy is needed to detach the matrix from the inclusion and dislocations can be initiated at a lower shear load. Nevertheless, the un-bonded hard inclusion can still block dislocation generation and sliding towards it. In contrast to the porous crystal, the inclusion diverts compressive stresses to a wider field which enables the material to resolve higher loads before dislocations are initiated.

CONCLUSION

In this study, we focused on the onset of dislocation formation and its intensity at pores and hard inclusions during the deformation process using a molecular dynamics model. The model material, an fcc metal, was then exposed to shearing until dislocation formation was observed. For reference, a single crystal structure was analysed first. It showed the highest maximum strain and yield strength as no defects could serve as sources for dislocation formation. Such a material behavior and structure is best represented by whiskers. The dislocations observed in this system were dominantly Shockley partial dislocations. Massive dislocation generation all across the single crystal was observed after exceeding the maximum yield strength at lower temperatures. At higher bulk temperature a falling maximum yield strength was already exceeded at smaller strain and fewer dislocations were generated.

Defect structures, like pores and inclusions, can limit or broaden the range of applicability of a material for high performance components. It is generally accepted, that crystal defects can act as sinks for and sources of dislocations. Although this applies to the pore and the inclusion their effect on the material response and mechanisms involved are significantly different. The porous crystal showed the lowest maximum elastic strain and yield strength. The dislocations were only generated at the pore, but all across its diameter. Even loosely embedded inclusions, without bonding to the matrix, can significantly strengthen a material by diverting compressive stresses and blocking dislocation formation. A strong inclusion/matrix interface further improves the strength of the material. By controlling the dislocation formation intensity, the material exhibits appreciable failure properties as a tough, high-strength material. A similar reduction of the dislocation mobility by inclusions is known from particle reinforced materials and from high-carbide-content metals. However, the machinability of such a material is limited and related to intensive tool wear.

ACKNOWLEDGEMENTS

The authors like to acknowledge the generous support of this work by a Feodor-Lynen research scholarship of the Alexander-von-Humboldt foundation, Bonn, Germany, and the Laboratory for the Research on the Structure of Matter at the University of Pennsylvania, Philadelphia, USA.

REFERENCES

1. R.W. Davidge, *Mechanical Behavior of Ceramics*, Alden Press, Oxford,GB,1979, pp.11-30.

2. M. W. Finnes, Atomic Energy Authority Harwell Report No. AERE-R-13182, 1988.

3. M. Parrinello, A. Rahman, *Polymorphic transition in single crystals: A new molecular dynamics method*, J.Appl.Phys. **52**(12), Dec. 1981, p. 7182-7190.

4. G. J. Ackland, G. Tichy, V. Vitek, M. W. Finnis, *Simple N-body Potentials for the Noble Metals and Nickel*, Phil.Mag., A., Vol. **56**, No 6, pages 735 - 756, 1987.

SINGLE CRYSTAL ELASTIC MODULI OF DISORDERED CUBIC ALLOYS

CRAIG S. HARTLEY
U.S. Department of Energy, SC-131, Germantown, MD, 20874. Craig.hartley@science.doe.gov
Permanent address: Department of Mechanical Engineering, Florida Atlantic University, Boca Raton, FL 33431. Hartley@fau.edu

ABSTRACT

A review of the relationship between elastic moduli and interatomic force constants precedes the description of a method for determining the composition dependence of single crystal elastic moduli of disordered alloys having the face-centered cubic structure. The method treats the alloy as a virtual crystal, characterized by an effective pair potential between atoms. Results of calculations are presented using experimental data on Cu-rich Cu-Al alloys.

INTRODUCTION

Elastic constants of metallic materials provide a striking example of successful multiscale modeling [1]. Single crystal elastic moduli of crystalline substances can be calculated *ab initio* if the location and atomic species of each atom in the material are known and suitable interatomic potentials can be constructed. From this information, moduli can be calculated for randomly oriented [2] or textured polycrystalline material for which appropriate pole figures [3] or orientation distribution functions [4] are known. As physical models that require information on elastic behavior become more widely applied to alloys, an increasing need arises for information on their elastic moduli as functions of composition. Although considerable information exists on the elastic moduli of pure metals at room temperature, similar data on alloys is less common [5,6]. While experimental measurements constitute the most desirable means of obtaining this information, direct determinations are often impracticable. Since *ab initio* calculations are difficult to extend to disordered alloys, it is useful to explore alternative methods for predicting their elastic moduli from a limited amount of information on pure metals and selected alloys.

Early experiments on polycrystalline materials, performed primarily for the purpose of generating engineering design information, led to the conclusion that the rule of mixtures adequately described the composition dependence of elastic moduli of single-phase alloys [7,8]. Disordered terminal solid solutions in systems with limited terminal solid solubility were found to exhibit linear dependencies of elastic moduli on composition, although some curvature was observed in systems that exhibited complete solid solubility [9]. Previous attempts to calculate the composition dependence of elastic moduli of binary alloys were limited to isomorphous systems or a simple ordered structure [10,11,12]. These works either assumed central forces between nearest neighbor atoms [10,11], or neglected a component of the non-central atomic force constant matrix [12]. Using a more phenomenological approach, Hartley was successful in predicting the composition dependence of single crystal elastic moduli of binary cubic alloys in non-isomorphous alloy systems [13]. However, this method does not provide sufficient information to calculate all three independent single crystal elastic moduli as functions of composition. The following sections describe a method that permits a self-consistent calculation of single crystal elastic constants of a disordered, single phase cubic alloy from atomic force constants derived from a pair potential appropriate to the alloy.

ELASTIC MODULI AND INTERATOMIC FORCE CONSTANTS

Relationships between elastic moduli and interatomic force constants can be derived both by considering homogeneous deformations of the lattice [14,15,16]] and by comparing like terms in the secular equations for motion of an atom with those for wave propagation in an anisotropic continuum [17,18, 19]. Interatomic force constants appearing in both treatments can be

279

expressed in terms of appropriate derivatives of a pairwise interatomic potential. In a material possessing metallic bonding, the internal energy per atom can be approximated by the sum of two terms: 1) the sum of pairwise interaction energies between a limited number of neighbors, U_P, and 2) a many body term, U_V, that depends on the total volume of the crystal and accounts for the interaction of individual ions with the electron "gas" [15,16].

To determine U_P, choose any atom as the origin of a coordinate system with <100> directions as the associated coordinate axes. Expanding U_P in terms of atomic displacement about the origin [20],

$$U_P(\vec{u}) = \frac{1}{2}\{U_o + \sum_{n=1}^{s} \varphi_i'^{(n)} u_i^{(n)} + \frac{1}{2} \sum_{n=1}^{s} \varphi_{ij}''^{(n)} u_i u_j + O(|\vec{u}|^3)\} \tag{1}$$

where s is the number of atoms interacting with the atom at the origin and the primes indicate the degree of partial differentiation of the pairwise interatomic potential, φ, with respect to components of the position vector connecting the origin to each neighboring atom, evaluated at the equilibrium spacing between the atoms. Since interatomic forces are generally of short range, we consider pairwise interactions over only the first and second neighbor atoms. The term involving first derivatives corresponds to the total force exerted by neighboring atoms on the atom at the origin. To avoid imposition of the Cauchy condition on the elastic constants of the crystal it is customary to choose U_V such that the first order term in its Taylor series expansion exactly cancels this force. Higher order terms are subsumed into appropriate higher derivatives of φ.

The second derivatives form a force constant matrix such that $\varphi_{ij}''^{(n)} = f_{ij}^{(n)}$ is the force exerted on the atom at the origin in the x_i direction when an atom at the nth lattice point experiences a unit displacement in the x_j direction. Independence of the order of differentiation requires that the force constant matrix be symmetric. Neglecting terms $O(|u|^3)$ results in the harmonic approximation for the total potential energy of the crystal at absolute zero. The quasi-harmonic approximation, in which elements of $f_{ij}^{(n)}$ are regarded as temperature-dependent material parameters, is a similar form that describes the potential energy at temperatures above zero K.

In the present treatment, we assume that the magnitude of the pair potential for an atom and its first and second neighbors depends only on the position vectors connecting the center of the atom with those of the corresponding neighbors. This central potential gives rise to non-central interatomic forces characterized by two axisymmetric force constants (ASFC) for each neighbor shell: α_n, the coefficient for stretching bonds between n^{th} neighbors, and β_n, that for bending such bonds. In terms of appropriate derivatives of the pair potential:

$$\alpha_n = \left(\frac{\partial^2 \varphi}{\partial r^2}\right)_{r=r_n} ; \quad \beta_n = \left(\frac{1}{r}\left(\frac{\partial \varphi}{\partial r}\right)\right)_{r=r_n} , \tag{2}$$

where r_n represents the distance between an atom at the origin and an atom in the nth neighbor shell. Derivatives appearing in equation (1) can be expressed in terms of the ASFC:

$$\left(\frac{\partial \varphi}{\partial x_i}\right)_{r=r_n} = \ell_i \beta_n ; \quad \left(\frac{\partial^2 \varphi}{\partial x_i \partial x_j}\right)_{r=r_n} = \ell_i \ell_j (\alpha_n - \beta_n) + \delta_{ij}\beta_n . \tag{3}$$

where $\ell_i = x_i / |r|$. Applying to this model conditions that insure that the force constant matrix possesses the symmetry of the cubic crystal system reduces the number of independent force constants to two each for the first and second neighbors [21]. The relationship between the ASFC and the single crystal elastic moduli can then be obtained by comparing appropriate terms in the equations of motion of the atom at the origin [19]. The following expressions apply to the case of a face-centered cubic structure considering only first and second nearest neighbors [22]:

$$a_0 c_{11} = 2\alpha_1 + 2\beta_1 + 4\alpha_2 ,$$
$$a_0 c_{44} = \alpha_1 + 3\beta_1 + 4\beta_2 , \qquad (4)$$
$$a_0 c_{12} = \alpha_1 - 5\beta_1 - 4\beta_2 ,$$

where a_0 is the lattice parameter. It is clear from equation (4) that any linear combination of the elastic stiffnesses can be expressed in terms of an appropriate sum of ASFCs.

Since there are four independent ASFCs and only three independent elastic moduli, additional information is required to determine all of the ASFCs. In the following section, we describe an indirect approach in which ASFCs for alloy single crystals are defined in terms of three independent parameters of a composition-dependent effective potential. The resulting expressions are used in equation (4) to determine values of these parameters in terms of the elastic constants and mean nearest neighbor spacing of an fcc alloy. The composition dependence of the parameters is then estimated by a quasi-chemical approach that permits calculation of an effective potential and associated elastic constants for any single phase alloy.

EFFECTIVE PROPERTIES OF A DISORDERED ALLOY

Typical pair potentials consist of the sum of short-range repulsive terms and long-range attractive terms. Such a potential approaches positive infinity as the spacing between ions approaches zero. As the interatomic spacing increases, the potential falls to a negative value, reaching a minimum at a distance near the nearest neighbor spacing and finally approaches zero asymptotically as the interatomic spacing tends to infinity. For a condensed phase, the location of the minimum in the pair potential does not necessarily occur at a distance corresponding to a particular physical spacing of atoms.

A potential function that behaves in the manner described can be approximated near its minimum by a Taylor series expansion of powers of the distance from the minimum. Since only three independent elastic constants are available for each alloy composition and the required derivatives do not depend on constant terms, it is only possible to consider terms up to the third power of this distance. Designating by ρ_0 the interatomic spacing at which $\varphi(\mathbf{r})$ is a minimum,

$$\varphi(r) \cong \varphi_0 + \frac{\varphi_2}{2!}\left(\frac{r-\rho_0}{\rho_0}\right)^2 + \frac{\varphi_3}{3!}\left(\frac{r-\rho_0}{\rho_0}\right)^3 . \qquad (5)$$

To express the composition dependence of this potential it is necessary to expand equation (5) to form the cubic polynomial

$$\varphi(r) = \Phi_0 + \Phi_1 r + \Phi_2 r^2 + \Phi_3 r^3 . \qquad (6)$$

Coefficients in equation (6) can be expressed in terms of those in the Taylor series expansion by equating coefficients of like powers of r.

Shibuya employed a quasi-chemical approach to express the average interatomic pair potential for a disordered, binary solid solution of A and B in terms of pair potentials of the component pairs as [10]:

$$\varphi = c_A^2 \varphi_{(AA)} + (1-c_A)^2 \varphi_{(BB)} + 2c_A(1-c_A)\varphi_{(AB)} , \qquad (7)$$

where c_A is the atomic fraction of species A and the suffixes indicate the type of atomic pair to which the potential applies. Writing the potential for each component pair as a cubic polynomial and substituting the quasi-chemical approximation for the average potential into equation (6) shows that the composition dependence of each coefficient in equation (6) is given by equation (7). Like and unlike pair terms represent the corresponding coefficients in the polynomial expressions for the component pairs.

In determining the effective ASFCs from equation (3), it is necessary to evaluate the potential at appropriate interatomic distances. We assume that the first and second neighbor distances of atomic pairs depend only on the atomic species and the crystal structure, but not otherwise on the surroundings of the pair. The mean nearest neighbor spacing in the alloy can then be expressed in terms of the corresponding spacings for like and unlike pairs present in the alloy. For the binary alloy above the mean nearest neighbor distance is [13]

$$\bar{r} = c_A^2 r_{(AA)} + (1-c_A)^2 r_{(BB)} + 2c_A (1-c_A) r_{(AB)} \qquad (8)$$

where the suffixes refer to the pair species. Least squares estimates of like and unlike pair spacings are determined by fitting experimental data on the composition dependence of the lattice parameters of single-phase alloys to equation (8). The mean spacings of more distant neighbors are calculated from the geometry of the lattice. Effective force constants are calculated from equation (2) using the mean potential with derivatives evaluated at mean neighbor spacings.

COPPER-ALUMINUM

The approach described in the previous section has been used to calculate parameters of component pair potentials and nearest neighbor spacings for Cu-Al alloys using single crystal elastic constants and lattice parameters measured by Cain and Thomas [23]. Elastic constants calculated from the effective potentials so obtained are shown in Figure 1 along with the experimental data for comparison. Experimental data for elastic constants of pure Al [24] were also used in the least-squares fit.

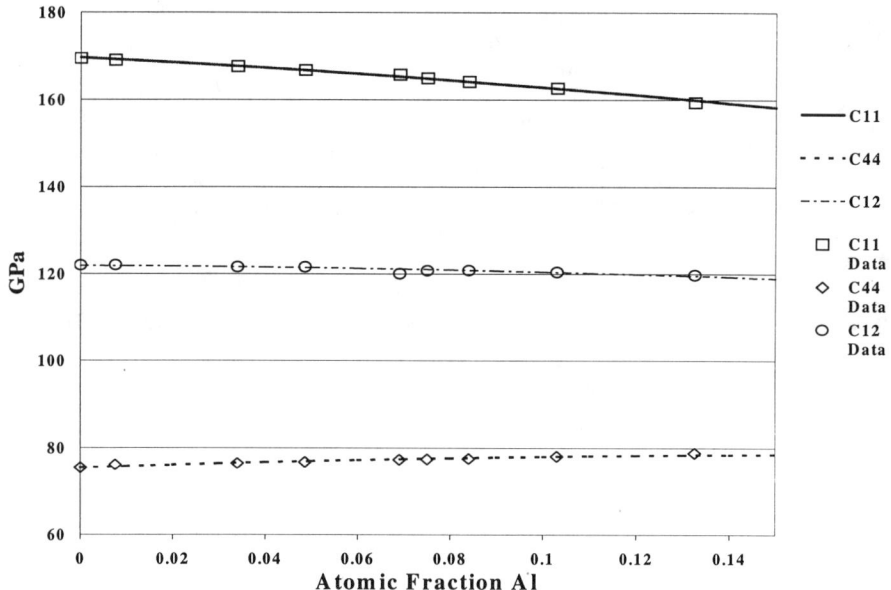

Figure 1. Elastic Constants of Cu-rich Cu-Al Alloys

Table 1 gives the results for parameters in the Taylor series expansions of the potentials and for the spacings of like and unlike pairs in the system.

Table 1. Pair Parameters for Cu-Al Alloys

Pair Species (A-B)	ρ_o (nm)	φ_2 (eV)	φ_3 (eV)	r_{AB}
Al-Al	0.3461	3.6411	-3.9393	0.2857
Cu-Cu	0.2959	5.5129	-7.6381	0.2551
Cu-Al	0.2943	6.6821	-9.5055	0.2641

Bullough and Hardy [22] employed single crystal elastic constants along with the transverse mode frequency at the center of the (100) Brillouin zone face [25] to calculate the first and second neighbor ASFCs for pure copper [26]. Table 2 compares the AFSCs determined by these investigators with those calculated from the polynomial potential obtained from Cu-Al data.

Table 2. Axisymmetric Force Constants for Copper

Reference	α_1 (N/m)	β_1 (N/m)	α_2 (N/m)	β_2 (N/m)	$M\omega_t^2$ (N/m)
This work	31.58	-4.074	1.569	1.974	77.564
B&H[22]	33.43	-2.227	-0.2770	0.128	107.00[25]
N&S [12]	33.55	-2.099	-0.4052	0	109.04

In addition, Table 2 includes the ASFCs determined directly from elastic constants using the method employed by Niu and Shimizu [12].

DISCUSSION

Locations of the minima in the polynomial potentials for each component pair all lie between first and second neighbor distances for a fcc lattice composed of the various species. This is consistent with the approximation of the potential by a Taylor series expansion about the minimum to calculate properties at the first and second neighbor positions. However, it should be noted that the composition dependence of the parameters in the Taylor series expansion is not given by the quasi-chemical approximation. It is necessary to expand the expression into the form of equation (6) before applying the quasi-chemical averaging procedure to obtain coefficients of the effective potential for an alloy.

The ASFCs of pure copper calculated from the polynomial potential are comparable in magnitude, but not always in sense, to those determined from elastic constants and neutron scattering data [22]. In all cases the nearest neighbor stretching constant, α_1, is the largest by nearly an order of magnitude. The first neighbor bending constant, β_1, is the same order of magnitude and same sign for all methods of determination, but the absolute value of β_1 calculated from the polynomial potential is higher than that determined by the other methods.

There is even less agreement among the methods on the value of α_2. Finally, the value of the second neighbor bending constant, β_2, determined by the polynomial potential is an order of magnitude larger than that obtained by direct solution using elastic constants and neutron scattering data. However, it should be noted that where significant discrepancies exist among the methods, the absolute values of the ASFCs are relatively small.

CONCLUSIONS

A self-consistent method employing a quasi-chemical approach to define an effective pair potential provides a useful means of calculating single crystal elastic constants of disordered fcc alloys over the entire composition range of a solid solution. The effective potential and its component pair potentials may also be useful for calculating elastic constants of other alloys and compounds in a given system by taking appropriate account of different crystal structures and coordination numbers. Properties that depend the absolute value of the interatomic potential are not expected to be well modeled by the polynomial potentials employed in this work, since no parameters are determined by matching to physical properties that depend on the values of particular energies.

ACKNOWLEDGEMENTS

This research was conducted during the author's appointment with the U.S. Department of Energy, Office of Basic Energy Sciences, under the Intergovernmental Personnel Act. Helpful discussions with colleagues in the Division of Materials Sciences, DOE and the Metallurgy Division, NIST, are gratefully acknowledged.

REFERENCES

1. The terms "elastic moduli" or " elastic constants" refer to either compliances or stiffnesses.

2. B. K. D. Gairola and E. Kröner, Int. J. Eng. Sci., **19**, p. 865 (1981).

3. M. A. Eisenberg,, C. S. Hartley, H.-C. Lee and C. F. Yen, J. Nucl. Mat, **88**, p. 138 (1980).

4. C. M. Sayers, J. Phys. D: Appl. Phys., **15**, p. 2157 (1982).

5. R. F. S. Hearmon, in *Landholt- Bornstein New Series, Group III,* **11,** pp. 12-21 (1979) and **18** pp. 5-12 (1984).

6. G. Simmons and H. Wang, *Single Crystal Elastic Constants and Calculated Aggregate Properties: A Handbook, 2nd. Edition* (MIT Press, Cambridge, MA, 1971).

7. P. Chevenard and A. Portevin, Compt. Rend., **181**, p. 780 (1925).

8. L. Guillet, Rev. Met., **36**, p. 497 (1936).

9. W. Köster and W. Rauscher, Zeit. f. Metallkde. **39**, p. 111 (1948).

10. Y. Shibuya, Sci. Rep. RITU, **A-3**, p. 645 (1951).

11. Y. Shibuya, Sci. Rep. RITU, **A-1**, p. 161 (1949).

12. H. Niu and M. Shimizu, J. Phys. Soc. Japan, **22**, p. 437 (1967).

13. C. S. Hartley, in *Metallic Alloys: Experimental and Theoretical Perspectives*, ed. by J. S. Faulkner and R.G. Jordan, (Kluwer Academic Press, Amsterdam, Netherlands, 1994), p. 171.

14. K. Fuchs, Proc. Roy. Soc. (London), **A153**, p. 622; **A157**, p. 444 (1936).

15. R. A. Johnson, Phys. Rev. B, **6**, p. 2094 (1972).

16. R. A. Johnson, Phys. Rev. B, **9**, p. 1304 (1974).

17. M. Born and K. Huang, *Dynamical Theory of Crystal Lattices*, Oxford University Press, Oxford, England (1956).

18. J. De Launay, in *Solid State Physics,* ed. F. Seitz and D. Turnbull (Academic Press, **2**, New York, 1956) p. 285.

19. G. L. Squires, Arkiv för Fysik, Band 25, nr 3, p. 21 (1963).

20. Summation from 1 to 3 over repeated Latin suffixes is implied unless otherwise indicated.

21. J. L. Zaretsky, *Lattice Dynamics of hcp and bcc Zirconium* (Ph.D. Dissertation, Iowa State University, 1979).

22. R. Bullough and J. R. Hardy, Phil. Mag., **17**, p. 833 (1968).

23. L. S. Cain and J.F. Thomas, Jr., Phys. Rev. B, **4**, p. 4245 (1971).

24. D. Gerlich and E.S. Fisher, J. Phys. Chem. Solids, **30**, p. 1197 (1969).

25. S. K. Sinha,, Phys. Rev., **143**, p. 422 (1966).

26. The definition of ASFC employed by B&H is twice that that employed in the present work.

A SYSTEMATIC INVESTIGATION OF STRAIN RELAXATION, SURFACE MORPHOLOGY AND DEFECTS IN TENSILE AND COMPRESSIVE InGaAs/InP LAYERS

C. FERRARI*, L. LAZZARINI*, G. SALVIATI*, M. NATALI°, M. BERTI°, D. DE SALVADOR°, A.V. DRIGO°, G. ROSSETTO^, G. TORZO^
*CNR-MASPEC Institute, Parco Area delle Scienze 37/A, I-43010 Fontanini- Parma, Italy
° INFM, Department of Physics, University of Padova, Via Marzolo 8, I-35131 Padova, Italy
^ CNR-ICTIMA Institute, Corso Stati Uniti 4, I-35127 Padova, Italy

ABSTRACT

The results of a systematic investigation by transmission electron microscopy (TEM), cathodoluminescence (CL), Rutherford backscattering (RBS), X-ray diffraction and topography and scanning force microscopy (SFM) techniques on several InGaAs/InP compressive and tensile strained layers covering the misfit range from -2.3 to 1.5×10^{-2} and grown by the metal organic vapor phase epitaxy (MOVPE) technique are reported. In compressively strained films the same dependence for the residual strain vs the film thickness as for the InGaAs/GaAs is found whereas a different strain release rate and different extended defects are found in tensile stressed InGaAs alloy. In particular in tensile stressed samples, grooves, planar defects and cracks are present in addition to the interfacial network of misfit dislocations. The correlation between the observed planar defects and the mechanisms of strain relaxation in the case of tensile strained layers is discussed.

INTRODUCTION

The strain relaxation mechanism for lattice mismatched epitaxial layers under compressive strain has been extensively studied, in particular for the InGaAs/GaAs system (Drigo et al 1989 [1], Lavoie et al. 1995 [2]). For layers under tensile strain such as InAlSb/InSb it has recently been suggested (Maigné et al. [3]) that the relaxation mechanism might be different since the residual strain is significantly larger than in InGaAs/GaAs layers with comparable misfit. However a quantitative comparison between the two systems is not necessarily appropriate because the materials exhibit different physico-chemical, mechanical and thermal properties.

The $In_xGa_{1-x}As/InP$ material system is ideal for the direct comparison of the relaxation processes involved in compressive and tensile strains. In fact, varying the In composition below or above x=0.53, the epitaxial layer shows negative or positive lattice mismatch.

EXPERIMENT

Thirty-seven InGaAs epitaxial films were grown on (001) semi-insulating InP substrates via metalorganic vapor phase epitaxy (MOVPE) in order to cover large intervals of Indium concentration (0.2 to 0.73) and film thickness (8 to 2400 nm).

TEM bright-field and dark-field images of <001> plan view and <110> cross-sections were recorded with a JEOL 2000 FX microscope at 200 kV on mechanically and chemically thinned samples finished by argon ion-milling (Gatan 600 Duo Mill).

The indium composition and strain of the layers were determined by high-resolution X-ray diffraction (Philips MRD) measuring rocking curves of symmetric (004) and asymmetric (444) reflections along the <110> in-plane directions both in grazing incidence and grazing emergence geometry. X-ray topographs were recorded by using a conventional Lang camera using the 115 asymmetric diffraction condition and the $CuK\alpha$ wavelength. RBS measurements were performed with 2MeV 4He+ beams at the Van de Graaf accelerator (LNL Legnaro) to determine the layer thickness and to cross-check the indium compositions.

The surface morphology was investigated by a scanning force Park CP microscope operated in contact-mode, using ultra-lever™ tips with nominally 10 nm tip radius.

RESULTS

Strain Release in InGaAs/InP Layers

In the compressive case (In concentration in the interval $0.61<x<0.73$) the strain relaxation along the two 110 directions is nearly symmetric. The residual strain as a function of the layer thickness follows the empirical curve previously found for MBE grown InGaAs/GaAs layers with In concentration <0.2 [1]. The main result is that strain relaxation in compressive InGaAs layers is independent of both growth conditions and Indium composition [4].

Tensile strained InGaAs layers (In concentration varying in the interval $0.2<x<0.36$) exhibit a relevant asymmetry in the strain release between [110] and [1-10] directions, the strain along the [110] direction relaxing at a layer thickness lower than for the [1-10] direction (Lazzarini et al. [5]). Such a difference increases if the misfit is decreased. Fig. 1) reports the average residual strain of InGaAs/InP tensile layers in a In composition range between 0.35 and 0.2. It is evident that the onset of strain release, which appears at the points where the strain-thickness curves change slope, is shifted toward higher layer thickness values with respect to compressive layers. Once initiated, however, the strain relaxation proceeds much faster than for compressive layers. It is evident from the figure that it is not possible to fit the experimental curves with a single curve. As shown later the types of extended defects in tensile layers are very different from compressively strained ones. This suggests that extended defects are responsible for the retarded strain release.

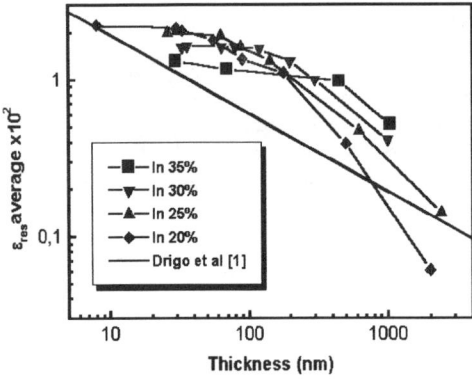

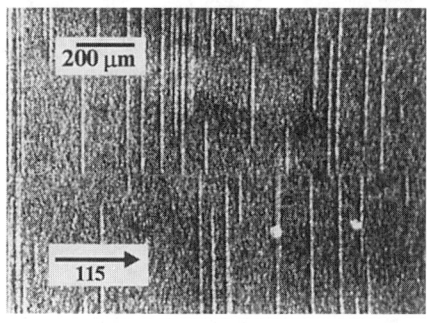

Fig. 1: average residual strain along [110] and [1-10] of InGaAs/InP tensile layers

Fig. 2 : CuKα (115) reflection topograph of a 0.2 μm thick InGaAs/InP layer close to the lattice match.

Extended Defects Characterization

In compressive InGaAs/InP relaxed layers the only extended defects present are misfit dislocations (MD), 60° mixed in character, arranged in the usual orthogonal network at the interface between the epilayer and the substrate and whose density accounts for the measured relaxation. Surface morphology is affected by the typical cross-hatch pattern observed in relaxed heterolayers.

Fig. 2 reports a detail of a X-ray topography of a tensile InGaAs/InP layer close to the lattice match showing single misfit dislocation segments a few mm long which end in correspondence of an extended defect line. The very faint contrast of the line excludes this as a simple 60° misfit dislocation line since the product **b·h** would not be zero, as required by the contrast extinction rule. In comparison with compressively strained InGaAs/GaAs layers which exhibit much longer MD lengths (Ferrari et al.[6]), the topograph of fig. 2 demonstrates the effective interaction between structural defects in tensile strained layers during the strain release process.

The analysis of the surface morphology has been performed by scanning force microscopy. For tensile strained layers one of the main morphology differences with respect to compressive samples is the development of V-shaped grooves along [1-10] and [110], which appears in correspondence with the onset of measurable strain relaxation along the respective perpendicular directions. Grooves appear at the beginning of strain release along the [1-10] direction and only a lower density of broad [110] oriented grooves appear. While grooves deepen and widen as the layer thickness increases, their density does not change significantly. On the contrary, fig. 3 shows the rapid increase of groove density with the layer misfit.

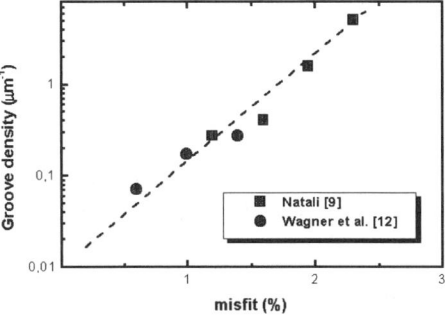

Fig. 3: groove density vs layer misfit along [110] direction

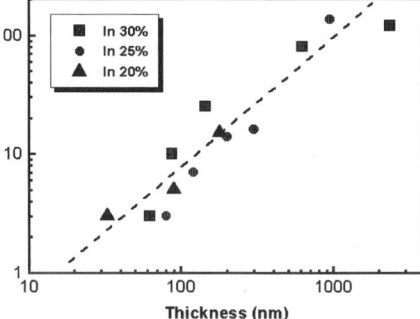

Fig. 4: average groove depth vs thickness along [110] direction measured by SFM for different composition values

The average groove depth as a function of layer thickness along the [110] direction and for different composition values is shown in fig. 4. These data show that no matter the degree of strain release the average groove depth is approximately 10% of the layer thickness. Nevertheless a smaller groove density with depths of the same order of the layer thickness are evidenced by SFM observations and TEM micrographs along the [1-10] direction (fig 5). The groove side-walls often coincide with the {111} planes but faceting occurs also on higher index {11h} planes. Okada et al. [7] demonstrate that the appearance of surface faceting in MBE grown InGaAs/InP tensile strained layers can be explained in term of the reduction of elastic energy in the layer.

Figure 6 shows a plan-view TEM image of a network of strongly interacting stacking faults and/or twins which reproduces the groove in-plane arrangement as it is observed by SFM. We thus believe that stacking fault of twins are correlated to the formation of grooves and coexist with a network of misfit dislocations contributing to the strain release. In fact only in low misfit samples where the groove density is low, the density of misfit dislocation accounts for the amount of relaxed misfit measured by XRD.

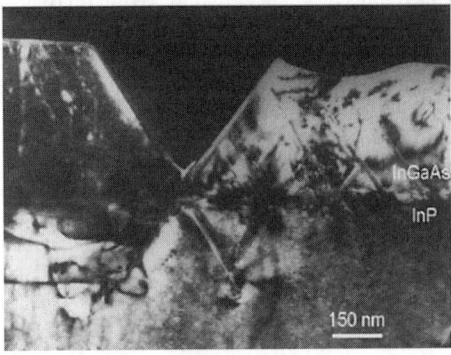
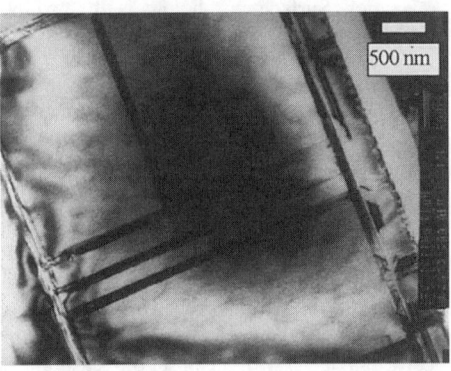

Fig. 5: TEM image of a groove running along [1-10] direction with depth equivalent to the layer thickness

Fig. 6: [110] zone axis plan-view TEM image showing stacking faults and/or twins

DISCUSSION AND CONCLUSIONS

Layers under compression present the same defects and morphology and relax following the same thickness dependence as that found for MBE grown InGaAs/GaAs showing no composition or growth technique effect. Since the larger Indium composition and the higher growth temperature of the present samples significantly increase the dislocation glide velocity [8] it is possible to conclude that the strain release rate cannot be due to kinetic limitation to the dislocation glide.

In the tensile InGaAs/InP relaxed layers, in addition to the interfacial misfit dislocation network, V-shaped grooves are present. In cross section TEM observations stacking faults and twins were observed in the groove bottom part. From the observation that misfit dislocations and grooves appear first along the [1-10] direction parallel to the first MDs we can deduce that the grooves must have a relevant role in the strain relaxation mechanism. Two possible mechanisms are suggested to evaluate such a contribution:

a) the presence of a surface faceting or grooves can reduce the total energy of the system by elastic strain release with a decrease of the local strain value

b) grooves reduce the local film thickness and act as a barrier for the glide of misfit dislocation threading segments running in the orthogonal direction.

Okada et al. [6]) calculated the decrease of the elastic energy ΔE of a faceted surface of depth 2h and spacing λ, in the case of $h \ll \lambda$. Even if we found the depth h to be almost equivalent to the groove dimension λ, from the expression of the elastic energy in a uniformly deformed film given by Mattews et al. [10] we can evaluate the elastic energy reduction $\Delta E/E$ due to the surface morphology:

$$\frac{\Delta E}{E} \approx \frac{\pi(1-\nu)^2}{4(1+\nu)} \cdot \frac{h^2}{\lambda t} \qquad (1)$$

where ν is the Poisson ratio.

Assuming the value $h/t \approx 0.1$ determined in our samples (Natali [9]), the reduction of the strain energy depends on h/λ so that the average strain, proportional to the square root of the elastic energy, is reduced by

$$\frac{\Delta\varepsilon}{\varepsilon} \approx 0.2 \cdot \sqrt{\frac{h}{\lambda}} \qquad (2)$$

due to the surface morphology. Equation 2 partially accounts for the observed discrepancy between strain release and misfit dislocation density measured by TEM in samples with larger misfit value. Nevertheless in our samples groove depths comparable to the width were observed, so that a larger contribution to the strain reduction near the grooves is expected.

According to the model of Matthews et al. [10] the local reduction of the layer thickness due to the groove decreases the driving force for the movement of threading dislocation segments during the strain release. This effect is equivalent to the blocking due to dislocation pile-up suggested by Fitzgerald et al. [11] and explains the delay in the strain release along the [1-10] direction. Furthermore V-shaped grooves are often accompanied by the presence of stacking fault that can act as a barrier for misfit dislocation propagation.

Therefore for a given misfit value the onset of strain release would occur at thickness values larger than for compressive strained film. Also the increase of groove density as a function of layer misfit as found by Wagner et al. [12] and Natali [9] is in qualitative agreement with the increase of the blocking effect in the perpendicular direction in layers with larger misfit.

In addition to grooves and stacking faults, in tensile strained samples a significant density of cracks has also been found. Nevertheless, simple calculations show that in our samples the maximum amount of misfit relaxed by cracks turns out to be a negligible quantity and close to the amount of tensile thermal strain added to the layer during the post-growth cool-down ([4], [5]). Such finding and the evidence of crack formation after the cooling indicate that crack formation cannot have any influence in the strain release mechanism.

REFERENCES

1. A.V. Drigo, A. Aydinli , A. Carnera, F. Genova, C. Rigo, C. Ferrari, P. Franzosi and G. Salviati, J. Appl. Phys. **66**, p. 3334 (1989)
2. C. Lavoie, Appl.Phys.Lett. **67** (25), p. 3744 (1995)
3. P. Maignè, D.J. Lockwood, C. Dharma-Vardana , J.B. Webb, J. Appl. Phys. **77**, 1466 (1995)
4. A.V. Drigo, M. Natali, M. Berti, D. De Salvador, G. Rossetto, G. Torzo, G. Carta, L. Lazzarini , G. Salviati in *Lattice Mismatch and and Heterovalent Thin Film Epitaxy*, edited by E. Fitzgerald, TMS proc. 1999
5. L. Lazzarini, proc. of *Microscopy of Semiconducting Materials XI*, In press 1999
6. C. Ferrari, S. Gennari, G. Salviati, M. Natali, A.V. Drigo, G. Rossetto, M. Currie, E.A. Fitzgerald in *Lattice Mismatch and and Heterovalent Thin Film Epitaxy*, edited by E. Fitzgerald, TMS proc. 1999
7. T. Okada, G.C. Wheatherly, D.W. McComb, J. Appl. Phys. **81**, p. 2185 (1997)
8. K. Sumino and I. Yonegana, proceedings of *7th Semi-insulating III-V materials Conference* ed C.J. Miner,W.Ford, and E.R. Weber(Bristol, Institute of Physics 1992), **29**
9. M. Natali in *Strain relaxation, defect formation and surface morphology in tensile and compressive InGaAs/InP layers*, PhD Thesis, University of Padova (1998)
10. J. W. Matthews and A.E. Blakeslee, J.Cryst.Growth **27**, p. 118 (1974)
11. E.A. Fitzgerald, S.B. Samavedam, Y.H. Xie, L.M. Giovane, J.Vac.Sci.Tech. A **15**, p. 1048 (1997).
12. G. Wagner, V. Gottshalk, R. Franzheld, S. Kriegel, P. Paufler, Phys.Stat.Sol. (a) **146**, p. 371 (1994)

Fracture and Crack Propagation

A DISCRETE DISLOCATION ANALYSIS OF CRACK GROWTH UNDER CYCLIC LOADING

H.H.M. CLEVERINGA[2], E. VAN DER GIESSEN[1] and A. NEEDLEMAN[2]
[1]Delft University of Technology, Koiter Institute Delft, Mekelweg 2, 2628 CD Delft, The Netherlands
[2]Brown University, Division of Engineering, Providence, RI 02912, USA

ABSTRACT

Cyclic loading of a plane strain mode I crack under small scale yielding is analyzed using discrete dislocation dynamics. The dislocations are all of edge character, and are modeled as line singularities in an elastic solid. At each stage of loading, superposition is used to represent the solution in terms of solutions for edge dislocations in a half-space and a non-singular complementary solution that enforces the boundary conditions, which is obtained from a linear elastic, finite element solution. The lattice resistance to dislocation motion, dislocation nucleation, dislocation interaction with obstacles and dislocation annihilation are incorporated into the formulation through a set of constitutive rules. An elastic relation between the opening traction and the displacement jump across a cohesive surface ahead of the initial crack tip is also specified, which permits crack initiation and crack growth to emerge naturally. It is found that crack growth can occur under cyclic loading conditions even when the peak stress intensity factor is smaller than the stress intensity required for crack growth under monotonic loading conditions.

INTRODUCTION

The nucleation and growth of cracks under cyclic loading conditions is arguably the most important mode of failure in engineering applications. Nevertheless, although much is known about fatigue fracture from both the materials science and engineering perspectives [1], a basic quantitative understanding of the mechanisms involved is limited. In practice, phenomenological relations between the amplitude of the applied stress intensity factor and the crack growth rate are used to quantify fatigue crack growth, e.g. [2]. However, outside a limited range of conditions, additional variables are needed in a phenomenological relation for it to have predictive capability, e.g. [3, 4, 5].

Quite recently, Nguyen et al. [6] have analyzed fatigue crack growth using a cohesive surface framework to characterize the separation process and conventional continuum plasticity to characterize the material behavior. It was found that when the cohesive relation was taken to be elastic and the crack subject to cyclic loading, shake down occurred in that the deformation became elastic and the crack arrested. Crack growth was found for a cohesive relation with unloading-reloading hysteresis.

Simulations of fatigue crack growth using discrete dislocation models have also been carried out, e.g. [7, 8, 9, 10, 11]. In such studies, dislocations nucleated from the crack tip are allowed to glide on specific presumed slip planes around the crack tip. Crack growth in [10, 11] is taken to be deformation controlled in that the crack is assumed to grow by emitting dislocations from the crack tip.

In this paper, we carry out an analysis of crack growth under cyclic loading using a cohesive surface framework to describe the separation process and discrete dislocation plasticity

Mat. Res. Soc. Symp. Proc. Vol. 578 © 2000 Materials Research Society

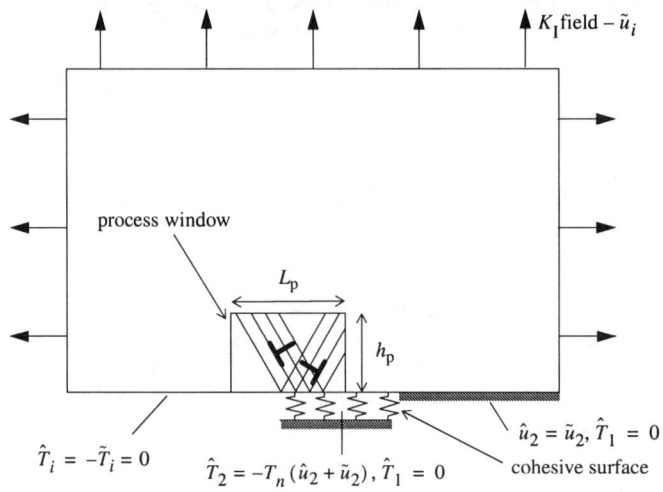

Figure 1: Mode I crack problem with the imposed boundary conditions.

to characterize the material behavior. The framework used is one recently employed [12, 13] to analyze crack growth under monotonic loading conditions. Since the fracture properties of the material are embedded in a cohesive surface constitutive relation, crack initiation and crack growth emerge as natural outcomes of the boundary value problem solution. The dislocations are modeled as line defects in an isotropic linear elastic solid. Full boundary value problem solutions are carried out for small scale yielding of a mode I crack in plane strain. The remote stress intensity factor is taken to be an oscillating function of time. Dislocation nucleation takes place from two-dimensional Frank-Read sources distributed in a process window surrounding the crack tip. There is no special dislocation nucleation from the crack tip. Random point obstacles in the process window account for precipitates and forest dislocations.

The plastic stress-strain response and the evolution of the dislocation structure along with the growth of the crack are outcomes of the boundary value problem solution. The cohesive relation is taken to be elastic and the basic issue investigated is whether the dislocation rearrangement due to the imposed cyclic loading can induce crack growth at lower levels of the applied stress intensity factor than occurs under monotonic loading conditions.

THEORY

The two-dimensional plane strain small-scale yielding problem sketched in Fig. 1 is analyzed, with dislocations restricted to a process window as shown in the figure. Symmetry about the crack plane is assumed. Remote from the crack tip, displacements corresponding to the linear elastic mode I K-field are applied, and crack initiation and growth are modeled using a cohesive surface framework, as in [14]. The boundary value problem formulation and the numerical implementation follow that in [12, 13]. The only difference is that here the applied stress intensity factor is the function of time shown in Fig. 2.

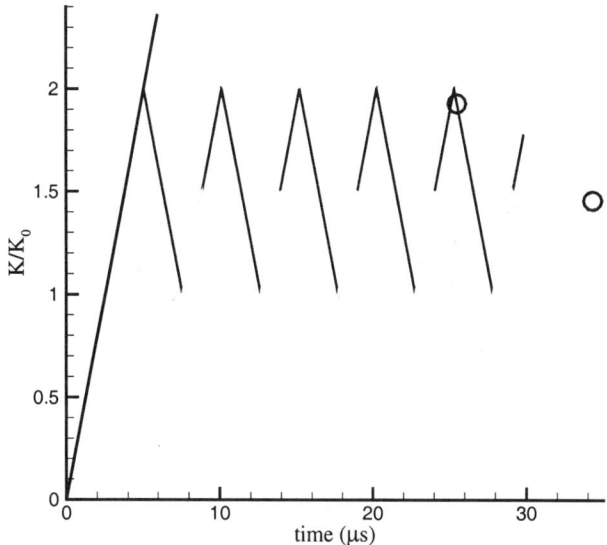

Figure 2: Time dependence of the applied stress intensity factor for the three cases analyzed. The symbols indicate the instants corresponding to Figs. 5a and 5b.

The origin of the coordinate system is at the initial crack tip, with the x_1 and x_2 directions in the crack plane and perpendicular to the crack plane, respectively. Behind the initial crack tip is a traction-free surface. Immediately ahead of the crack tip $T_1 = 0$ and $T_2 = -T_n(u_2)$, with T_n representing a traction separation law that will be specified later. Further ahead of the initial crack tip, there is no cohesive surface and the condition $\dot{u}_2 = 0$ is applied to enforce symmetry.

At each time step, an increment of the mode I stress intensity factor $\dot{K}_I \Delta t$ is prescribed. At the current instant, the stress and strain state of the body is known, so the forces on all dislocations can be calculated. Edge dislocations on three slip systems are considered; two have slip planes at $\pm 60°$ from the x_1-axis and the third has slip planes parallel to the x_1-axis. On the basis of these forces we update the dislocation structure, which involves the motion of dislocations, the generation of new dislocations, their mutual annihilation, their pinning at obstacles, and their exit into the open crack. After this, the new stress and strain state can be determined.

Superposition is used to determine the current state of the body with the new dislocation distribution, [15, 16],

$$u = \tilde{u} + \hat{u}, \quad \epsilon = \tilde{\epsilon} + \hat{\epsilon}, \quad \sigma = \tilde{\sigma} + \hat{\sigma}. \tag{1}$$

The ($\tilde{}$) fields are the superposition of the singular fields of the individual dislocations in their current configuration. It is worth noting that the formulation in [15, 16] is fully three dimensional although the implementation here is for plane strain problems with edge dislocations. The solutions that describe the stress and displacement field of an edge dislocation in a half-space are used [17], with the traction-free surface corresponding to the crack plane $x_2 = 0$. The boundary conditions are imposed through the ($\check{}$) fields, see Fig. 1.

The sum of the ($\check{}$) and the ($\hat{}$) fields in (1) gives the solution that satisfies all boundary conditions. Since the ($\hat{}$) fields are smooth in the region of interest, the boundary value problem for them can conveniently be solved using a finite element method. The size of the region analyzed is 1000μm $\times$ 500μm and a finite element mesh of 120×100 bilinear quadrilateral elements is used. The process window in Fig. 1 is specified by $L_p = 10\mu$m and $h_p = 12.5\mu$m and in it there is a fine mesh of 80×80 quadrilateral elements.

Dislocation motion is assumed to occur only by glide with no cross slip so that dislocations remain on their slip plane. The Peach-Koehler force $f^{(i)}$ acting on the ith dislocation is given by

$$f^{(i)} = n^{(i)} \cdot (\hat{\sigma} + \sum_{j \neq i} \sigma^{(j)}) \cdot b^{(i)} , \qquad (2)$$

with $n^{(i)}$ the slip plane normal, $b^{(i)}$ the Burgers vector of dislocation i and $\sigma^{(j)}$ is the stress field of dislocation j. The direction of this force is in the slip plane and normal to the dislocation line. The magnitude of the glide velocity $v^{(i)}$ of dislocation i is taken to be linearly related to the Peach-Koehler force $f^{(i)}$ through the drag relation $f^{(i)} = Bv^{(i)}$. The value for B is taken as $B = 10^{-4}$Pa s, which is a representative value for aluminum ($b = 0.25$nm).

New dislocations are generated by simulating Frank-Read sources in two dimensions by point sources on a slip plane. The sources generate a dislocation dipole when the magnitude of the Peach-Koehler force exceeds a critical value $\tau_{\text{nuc}}b$ during a period of time t_{nuc}. The source strengths are distributed randomly about a mean value $\tau_{\text{nuc}} = 50$ MPa and $t_{\text{nuc}} = 10$ ns. The distance between this dislocation pair is chosen so that their mutual attractive force is equal to $\tau_{\text{nuc}}b$. To model locking of dislocations at small precipitates or at dislocations on other slip planes we introduce obstacles as fixed points on a slip plane. A dislocation that glides against an obstacle is pinned there and is only released when its Peach-Koehler force exceeds the value $\tau_{\text{obs}}b$. The obstacle strength used here is $\tau_{\text{obs}} = 150$MPa. Annihilation of dislocations with opposite signed Burgers vector occurs when they are within a material dependent, critical annihilation distance $L_e = 6b$. Dislocations can also glide into the free surface of the open crack. When they do, they disappear from the system but leave a lattice step on the crack surface. Since slip planes are positioned symmetrically about $x_2 = 0$, when a dislocation exits the computational region across the (still closed) plane ahead of the crack, a dislocation enters the computational region on the mirror slip plane.

The constitutive relation for the cohesive surface is taken to have the universal binding form [18]

$$T_n(\Delta_n) = e\sigma_{\text{max}} \frac{\Delta_n}{\delta_n} \exp(-\frac{\Delta_n}{\delta_n}) , \qquad (3)$$

where Δ_n is the total separation of the cohesive surface, $\Delta_n = 2u_2(x_2 = 0)$, and T_n is the traction normal to the cohesive surface. As the cohesive surface separates, the magnitude of the traction increases, reaches a maximum and then approaches zero with increasing separation. The parameters used in this study are $\sigma_{\text{max}} = 0.6$ GPa and $\delta_n = 2b$ giving a work of separation, $\phi_n = e\sigma_{max}\delta_n$ of 0.815 J/m^2. The work of separation can be related to

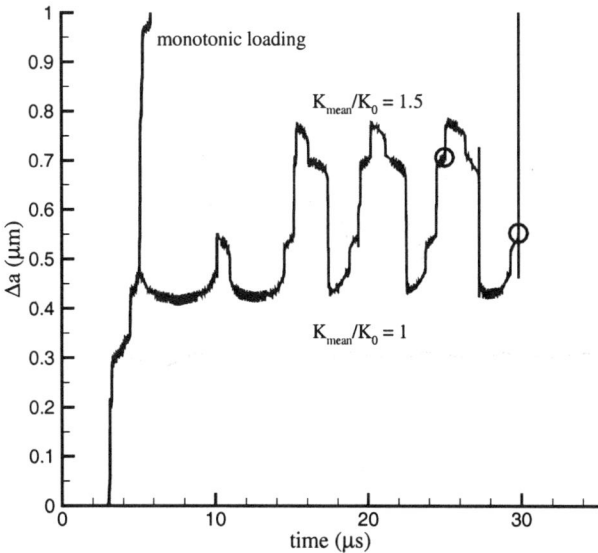

Figure 3: Time evolution of the crack growth. The symbols indicate the instants corresponding to Figs. 6a and 6b.

a reference stress intensity factor K_0 defined by

$$K_0 = \sqrt{\frac{E\phi_n}{1-\nu^2}}. \tag{4}$$

With $E = 70\,\text{MPa}$ and $\nu = 0.33$, representative for aluminum, $K_0 = 0.253\,\text{MPa}\sqrt{\text{m}}$. Crack growth in an elastic solid with the given cohesive properties takes place at $K/K_0 = 1$. The value of σ_{max} used in the calculations is about a factor of four smaller than would be appropriate for aluminum. This small value of the strength was used for numerical convenience, since the length scale over which large gradients occur is proportional to $(E/\sigma_{\text{max}})\delta$ [19].

RESULTS

The calculations are carried out for a loading rate of $\dot{K}_I = 100\text{GPa}\sqrt{\text{m}}/\text{s}$. This rather high loading rate is chosen to reduce the computer time needed for the crack growth calculations because resolving the dislocation dynamics requires a small time step of $\Delta t = 0.5\text{ns}$. The material considered has three slip systems; there are 401 slip planes oriented at $+60°$ to the x_1-axis, 401 at $-60°$ to the x_1-axis and 463 slip panes parallel to the x_1-axis. The slip plane spacing is $86b$. The material is initially dislocation-free with a random source distribution of $\rho_{\text{src}} = 78.8/\mu\text{m}^2$ and a random obstacle distribution of $\rho_{\text{obs}} = 200.1/\mu\text{m}^2$ distributed in the process region.

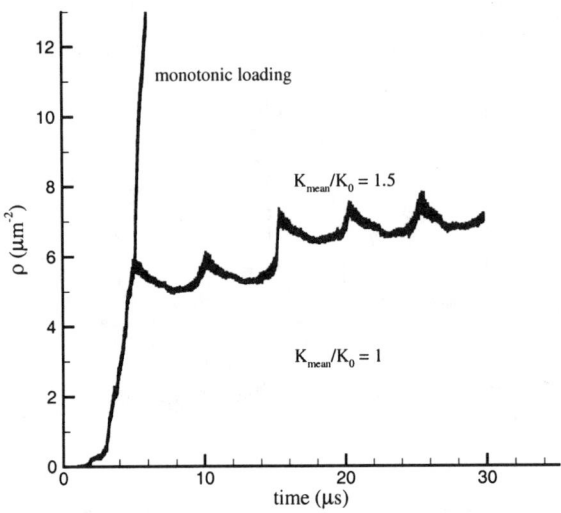

Figure 4: Time evolution for the dislocation density.

Because dislocations are free to exit the material through the crack surface, which includes newly created free surface as the crack advances, and because the cohesive surface relation (3) is reversible, our calculations pertain to circumstances where the loading is carried out in a high vacuum and no environmental effects change the properties of the newly formed surface.

Two cases are considered as shown in Fig. 2. In one case, $K_{\max}/K_0 = 2$ and $K_{\min}/K_0 = 1$, while in the other case $K_{\max}/K_0 = 1.5$ and $K_{\min} = 0.5$. Thus, in both cases $\Delta K = K_{\max} - K_{\min} = K_0$. In the first case, $K_{\mean} = (K_{\max} + K_{\min})/2 = 1.5K_0$, while in the second case $K_{\mean} = (K_{\max} + K_{\min})/2 = K_0$. Also, for comparison purposes, a calculation with K monotonically increasing was carried out. Under monotonic loading, unstable crack growth occurred at $K/K_0 = 2.4$.

Figure 3 shows the time evolution of the amount of crack growth, Δa. As in [12, 13], crack growth occurs in "spurts" under monotonic loading. The crack tip jumps from one partly stable position to another. For the cyclic loading case with $K_{\mean}/K_0 = 1$, there is some crack opening and closure during the first few fatigue cycles, but this soon levels out and shake down occurs. This is evident from the curves for dislocation density, ρ, versus time, as shown in Fig. 4. The crack growth in Fig. 3 is a cyclic loading effect, not a time effect, because a calculation, not shown here, was carried out under monotonic loading to $K/K_0 = 2$, with K subsequently held fixed. In that calculation an equilibrium state with fixed Δa was shortly reached.

The curve for $K_{\mean}/K_0 = 1$ shows a small increase in dislocation density in the first few cycles, but soon reaches a steady state. On the other hand, with $K_{\mean}/K_0 = 1.5$ crack advance occurs during multiple loading cycles. During each unloading stage Δa is recovered, but the dislocation structure at the end of a cycle is not the same as at the beginning of that cycle. Hence, the crack can grow to different lengths during different loading cycles. After the third loading peak, the crack seems to reach a stable state, but then during the sixth loading cycle the crack advances unstably. As seen in Fig. 4, the dislocation density increases during each crack advance. The jump in dislocation density is particularly large in the third loading cycle. Subsequently, a more steady increase in dislocation density occurs which is evidence for the change in dislocation structure that occurs during the later loading cycles.

The dislocation distribution that develops is shown in Fig. 5. The extent of the dislocations gives an indication of the size of the plastic zone. Also the crack profile is shown on the negative vertical axis. Figure 5a shows the dislocation distribution for $K_{mean}/K_0 = 1$ at $t = 34.3 \mu s$ which is almost at K_{max} during the seventh loading cycle. Figure 5b shows the dislocation distribution for $K_{mean}/K_0 = 1.5$ at $t = 25.5 \mu s$ which is almost at K_{max} during the fifth loading cycle. The difference in plastic zone size, as would be expected from the difference in dislocation density in Fig. 4, is clearly visible. It is important to note that all three slip systems are active, as seen especially in Fig. 5b. This indicates that dislocations on different slip systems lock each other and so form obstacles for other dislocations on these slip planes. This, in turn, strengthens the material and induces increased crack growth during the later loading cycles. The amount of crack growth can be seen in the crack profiles. In both cases, there is dislocation activity on slip planes well in front of the current crack tip, which is a consequence of the cyclic loading; earlier work on monotonic loading [12, 13] showed almost no activity on slip planes in front of the crack tip.

To gain insight into what is happening during the last loading cycle of the case with $K_{mean}/K_0 = 1.5$, when extensive crack advance takes place, Fig. 6 shows contour plots of the opening stress, σ_{22}, at the last peak load (Fig. 6a) and just before the final crack advance (Fig. 6b), as indicated by the circles in Fig. 3. The region shown is a close-up of the near-tip region in Fig. 5b and includes the dislocation structure as well as the crack profile indicating the current position of the crack tip. Figure 6b shows very high values of σ_{22} near the crack tip, close to the cohesive strength σ_{max}. The dislocation structure in Fig. 6b is somewhat different from Fig. 6a, which accounts for the additional crack advance. Figure 6b also shows more dislocation activity on slip planes in front of the crack tip.

CONCLUSIONS

Preliminary results of a plane strain analysis of a mode I crack growth under cyclic loading have been presented where plastic flow arises from the motion of large numbers of dislocations. The material has three slip systems, is initially dislocation free and dislocation nucleation occurs from Frank Read sources distributed randomly in the material. There is no special dislocation nucleation from the crack tip. The applied stress intensity factor has a prescribed cyclic time dependence. Two cases are considered. Each has the same value of the amplitude of the oscillation, $(K_{max} - K_{min}) = K_0$, where K_0 is the reference stress intensity factor in (4). For the calculation with $K_{mean}/K_0 = 1$, shake down occurs and only limited crack growth takes place whereas for the calculation with $K_{mean}/K_0 = 1.5$ extensive crack growth occurs even though K_{max} is smaller than the stress intensity factor required for extensive crack growth under monotonic loading conditions. This dichotomy is to a large extent controlled by the evolution of different dislocation structures, in particular so as to reduce the overall dislocation mobility.

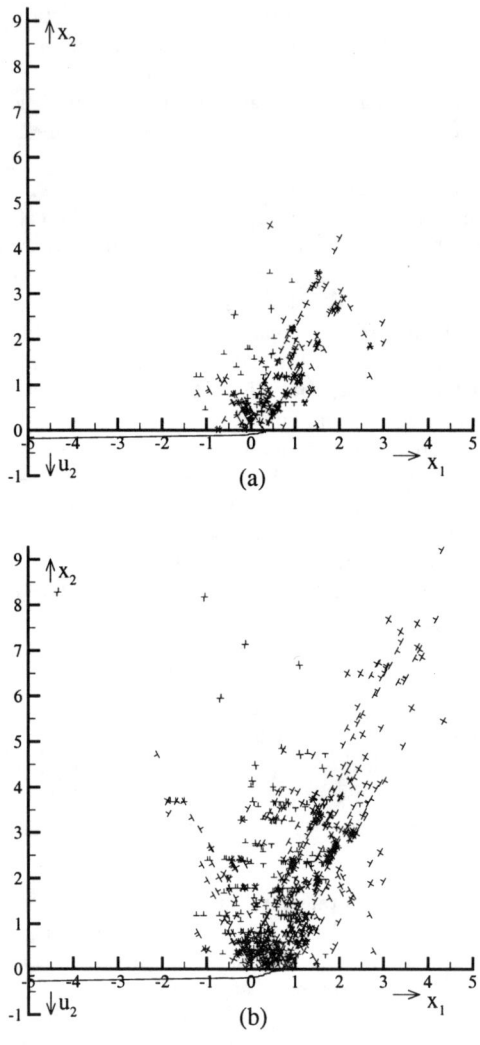

Figure 5: Dislocation distributions after a number of loading cycles, (a) for $K_{\mathrm{mean}}/K_0 = 1$ at $t = 34.3\mu s$, and (b) for $K_{\mathrm{mean}}/K_0 = 1.5$ at $t = 25.5\mu s$. All distances are in μm. The crack opening profiles (displacements magnified by a factor of 10) are plotted below the x_1-axis.

300

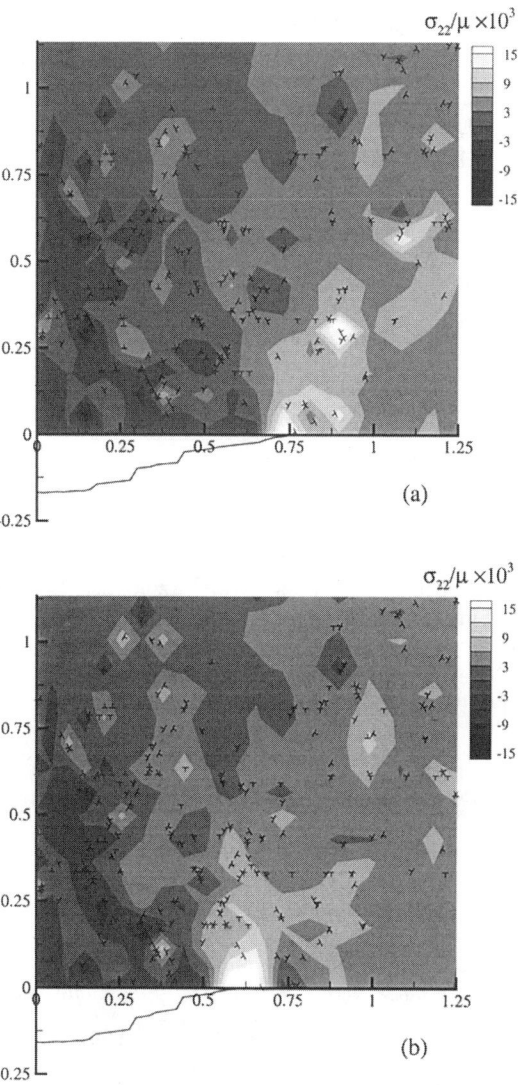

Figure 6: Distribution of the opening stress σ_{22} around the crack tip for the case with $K_{\mathrm{mean}}/K_0 = 1.5$, (a) at $t = 25.0\mu s$ and (b) at $t = 29.8\mu s$ (cf. Fig. 3). All distances are in μm. The crack opening profiles (displacements magnified by a factor of 10) are plotted below the x_1-axis.

No account has been taken of environmental effects that could change the properties of newly formed surface, although such effects can be incorporated into the framework used here. Hence, our calculations pertain to circumstances where the loading is carried out in a high vacuum and, in that sense, represent an intrinsic fatigue limit.

ACKNOWLEDGMENTS

Support from the AFOSR MURI at Brown University on *Virtual Testing and Design of Materials: A Multiscale Approach* (AFOSR Grant F49620-99-1-0272) is gratefully acknowledged.

REFERENCES

[1] S. Suresh, *Fatigue of Materials*, Cambridge University Press, Cambridge UK (1991).

[2] P.C. Paris and F. Ergogan, *J. Basic Engin.* **85**, 528–534 (1963).

[3] W. Elber, *Engin. Fract. Mech.* **11**, 573–584 (1979).

[4] J. Weertman, *Int. J. Fract. Mech.* **2**, 460–467 (1966).

[5] R.J. Donahue, H.M. Clark, P. Atanmo, R. Kumble and A.J. McEvily, *Int. J. Fract. Mech.* **8**, 209–219 (1972).

[6] O. Nguyen, E.A. Repetto, M. Ortiz and R.A. Radovitzky, *A cohesive model of fatigue crack growth*, submitted for publication.

[7] P. Neumann, *Acta Metall.*, **22** 1167–1178 (1974).

[8] K. Tanaka and T. Mura, *Acta Metall.*, **32** 1731–1740 (1984).

[9] V. Lakshmanan and J.C.M. Li, *Mat. Sci. Engin.*, **A104**, 95–104 (1988).

[10] F.O. Riemelmoser and R. Pippan, *Mat. Sci. Engin. A* **234-236**, 135–137 (1997).

[11] A.J. Wilkenson, S.G. Roberts and P.B. Hirsch, *Acta Mat.* **46**, 379–390 (1998).

[12] H.H.M. Cleveringa, E. Van der Giessen and A. Needleman, *A discrete dislocation analysis of mode I crack growth*, *J. Mech. Phys. Solids*, in press.

[13] H.H.M. Cleveringa, E. Van der Giessen and A. Needleman, *A discrete dislocation analysis of rate effects on mode I crack growth*, *Mat. Sci. Engin.*, in press.

[14] A. Needleman, *J. Mech. Phys. Solids* **38**, 289–324 (1990).

[15] E. Van der Giessen and A. Needleman, *Modeling Simul. Mater. Sci. Eng.* **3**, 689–735 (1995).

[16] V. Lubarda, J.A. Blume and A. Needleman, *Acta Metall. Mater.* **41**, 625–642 (1993).

[17] L.B. Freund, *Adv. Appl. Mech.* **30**, 1–66 (1994).

[18] J.H. Rose, J. Ferrante and J.R. Smith, *Phys. Rev. Lett.*, **47**, 675–678 (1981).

[19] J.W. Morrisey and J.R. Rice, *J. Mech. Phys. Solids*, **46**, 467–488 (1998).

X-RAY MICROBEAM DIFFRACTION MAPPING OF DIFFERENT TEXTURE SCALES IMPORTANT IN FATIGUE CRACKING OR IN LARGE DEFORMATIONS

S.R. STOCK*, Z.U. REK**
*School of Materials Sci. & Eng., Georgia Institute of Technology, Atlanta, GA 30332-0245
stuart.stock@mse.gatech.edu
**Stanford Synchrotron Radiation Laboratory, Stanford University, Palo Alto, CA

ABSTRACT

Polychromatic synchrotron x-ray microbeam diffraction allows one to record Laue patterns simultaneously from all of the grains within the column of material irradiated. This paper describes 3-D texture mapping in polycrystalline Al and Cu samples. Transmission methods for determining the depths of diffracting volumes are outlined. The focus is on how the texture scale between micro- (i.e., individual grain orientations) and macro-texture defines fatigue crack paths in Al-Li 2090 T8E41: very large volumes of near single crystal material lead to large asperity formation, and at least forty volume percent of the material consists of these volumes. Results on grain subdivision in Cu after large deformations are also briefly reviewed.

INTRODUCTION

Synchrotron microbeam x-ray diffraction can quantify texture or strains over a wide range of length scales **on the same sample**, and repeated, nondestructive interrogation of the interiors of bulk samples by these techniques will undoubtedly provide increasingly detailed data as more experimenters use the latest generation of synchrotron x-radiation sources. Data following the evolution of individual samples throughout the course of mechanical or other testing or processing presents a great opportunity for implementing multi-length-scale modeling of complex phenomena. The opportunity also presents a challenge, however, namely how to isolate and quantify behavior of one or more controlling processes buried within large volumetric data sets.

Two examples are used to illustrate how polychromatic microbeam x-ray diffraction quantifies texture at very different length scales. A brief description of data collection and apparatus precedes discussion of results obtained with microbeam x-ray diffraction. The first example examines grain subdivision accompanying large deformations in Cu, i.e., texture at the micrometer level. The second describes how a specific type of mesotexture intermediate between micro- and macro-texture leads to roughness-induced fatigue crack closure and to extremely low fatigue crack growth rates in Al-Li 2090 T8E41; this texture scale is between the scale of texture defined by individual grains and the average texture of the sample.

MICROBEAM TRANSMISSION LAUE DIFFRACTION

Microbeam transmission Laue patterns were recorded with polychromatic bending magnet radiation at Stanford Synchrotron Radiation Laboratory (SSRL) beamline 2-2 (3.0 GeV, between 20 and 100 mA beam currents). A 0.01 mm diameter pinhole collimator was used to define the beam dimensions; different sample-collimator separations defined the diameter of the sampling volume between 0.02 mm and 0.08 mm (~25 mm to produce the former for Cu samples and 550 mm to produce the latter for Al samples). Image storage plates, positioned perpendicular to the x-ray beam, recorded the diffraction patterns. Because highly strained samples produce diffracted beams with considerable asterism and the diffraction "spots" from polycrystalline samples often overlap, absorption edge filters were used to index the transmission Laue patterns. Further details appear elsewhere [1-4].

LARGE DEFORMATIONS IN COPPER

Oxygen-free high-conductivity Cu samples with an initial grain size of ~0.06 mm were tested to two effective plastic strain levels (50% compression, 100% torsion and 50% compression followed by 50% torsion), and the resulting macrotextures (111, 200 and 220 pole figures) were equivalent to those reported in the literature [5]. Translating the sample across the 0.02 mm microbeam with 0.01 mm steps allowed the grain subdivision structures to be mapped spatially and orientationally [3, 4]. Quantitative measurements showed that the angular spread of domain orientations depended on the effective plastic strain to the two-thirds power [4, 6], in agreement with a postulated relationship between geometrically necessary dislocation boundaries [7, 8]. Incorporation of the x-ray microbeam-measured dependence of grain subdivision into a crystal plasticity model improved agreement between experimental and calculated pole figures [4].

DIFFERENT LENGTH-SCALES AND FATIGUE CRACKING IN Al-Li 2090 T8E41

Structure at several length scales produces the superior fatigue crack propagation resistance of L-T oriented samples of Al-Li 2090 T8E41 (i.e., load axis along the plate's rolling direction L and crack propagation along the plate's long-transverse direction T). Under ambient conditions, fatigue crack growth rates for this near peak aged Al-Cu-Li-Zr alloy are typically an order of magnitude lower than those of other Al alloys [9, 10]. The T8E41 "temper" (solution treatment followed by a 6% stretch and 24 hr aging at 435 K) produces highly elongated, pancake-shaped grains (~ 2 mm x 0.5 mm x 0.05 mm along the L, T and S or short-transverse directions, respectively) [11]. The outer surfaces of 12.7 mm thick plates are partially recrystallized while the center of the plates are unrecrystallized, and the macrotexture differs substantially from the plate center to the outer surface [12].

In the absence of side grooves, fatigue cracks in L-T oriented compact tension samples of Al-Li 2090 T8E41 often grow with such a strong mode II component that the standard analytical expressions for stress intensity no longer apply. When samples were side-grooved, the average crack path follows the nominal crack plane defined by the notch, but the crack is hardly planar: large asperities (peaks on one crack face and valleys opposite) dominate the fracture surface, giving it significant crystallographic characteristics. Fracture occurs predominantly along slip planes under monotonic or fatigue loading [12, 15] because coherent, uniformly distributed Al-Li precipitates cause inhomogeneous slip and strain localization [13, 14]. The resulting transgranular fracture in side-grooved samples from the center of plates of Al-Li 2090 T8E41 creates crack faces which correspond geometrically to specific maxima in 111 pole figures [12, 15]. Serial sectioning of loaded samples along the L-S plane [16] and in situ x-ray microtomography of intact samples (described below) show prominent mixed mode I-III surface contact on these faces.

The specific macrotexture and strong tendency for fracture surfaces to follow {111} occur not only in the center of plates of Al-Li 2090 T8E41 but also in other Al alloys and tempers; what distinguishes 2090 is the presence of an additional level of texture between the grain-scale microtexture and the sample-average macrotexture. Haase and coworkers [1, 2, 17, 18] found a specific type of mesotexture consisting of groups of 5 - 10 adjacent grains possessing nearly identical orientations. Because the near single crystal volumes extend as much as 0.4 mm along a sample's S direction, quite large expanses of near {111} crack faces result; in the absence of such mesotexture the crack one expects the crack to cut back and forth, never deviating far from the nominal crack plane. One expects, therefore, that transitions between differently oriented near-single crystal volumes mark the edges of large asperities, and this is observed (Fig. 1, above).

The presence of side-grooves appears to be important in the formation of large asperities With near single crystal regions extending as much as 0.4 mm along the plate's S direction, it is not surprising that large mixed mode I-III surfaces result. When the mesotexture-dictated deflection forces part of the advancing crack far from the plane of the side-grooves, the side-grooves act to force the crack back to its nominal plane. Haase et al [1, 2, 17, 18] noted that fully 40-45 percent of the volume of the 2090 plate centers consisted of these near-single crystal regions, and large asperities often result because of the high likelihood of encountering an adjacent large near single crystal volume in which slip (and cracking) on the favored {111} will bring the crack back to the side-groove plane.

In the absence of side-grooves it would seem that the crack is as equally likely to continue to deflect away from its the nominal plane as it is to reverse its orientation; analysis of microbeam data from fatigue-cracked samples without side-grooves is currently underway. Large asperities are observed less often in these samples and large average deflections are seen. The results indicate that the spatial and orientational correlation of grain orientations is important, but the question of why cracks begin macroscopic deflections where they do is not yet answered. With or without side-grooves, mixed mode surfaces result and are important in fatigue crack propagation.

Roughness-induced crack closure underlies the low fatigue crack growth rates observed for Al-Li 2090 T8E41; this reduces the driving "force" for crack extension from $\Delta K = K_{max} - K_{min}$ to $\Delta K_{eff} = K_{max} - K_{cl}$, where ΔK is the applied stress intensity range, K_{max} and K_{min} are the stress intensities at the maximum and minimum points of the fatigue cycle and K_{cl} is the closure stress intensity. The rationale behind this numerical treatment is that once crack faces contact during unloading and begin carrying appreciable compressive load, the crack tip is shielded from the strain produced by remaining portion of the fatigue cycle.

The change in the slope of the load-deflection curve is macroscopic evidence of the closure phenomenon; typically K_{cl} is taken as the stress intensity corresponding to the bend in the load-deflection curves. This bend occurs at particularly high stress intensities in L-T oriented samples of Al-Li 2090 T8E41, and K_{cl} remains appreciable in this alloy for much longer cracks and higher stress intensity ranges than other materials.

In situ high resolution x-ray computed tomography or microtomography directly probed the relationship between three-dimensional fatigue crack geometry, applied stress and physical crack closure in L-axis notched tensile samples (e.g., Fig. 2) and L-T oriented compact tension samples of Al-Li 2090 T8E41 machined from the center of 12.7 mm thick plates [19-23]. During unloading, significant fractions of the crack faces were closed at stress intensities substantially above K_{cl}, in fact at positions far from the crack tip and at stresses near the peak of the fatigue cycle. Mixed mode I-III surface contact is important in both sample geometries. Where the crack is more planar, it appears to zip shut from the crack tip during the unloading portion of a fatigue cycle. This is in contrast to large amounts of contact observed away from the crack tip and on the asperities dominating the crack face in other samples (Fig. 2). Stiffening of the samples in the lower portions of the unloading-deflection curves appears related to compressive loading of the portions of the crack which were already in contact. The da/dN vs ΔK curves for the samples studied with microtomography are consistent with others' observations; therefore, the processes in these samples appear to be representative of what occurs in larger compact tension samples.

The picture of fatigue crack growth in Al-Li 2090 T8E41 which emerges illustrates the interplay of three very different length scales. The small coherent precipitates dictate that planar slip controls deformation and subsequent cracking. The spatial correlation of grain orientations in

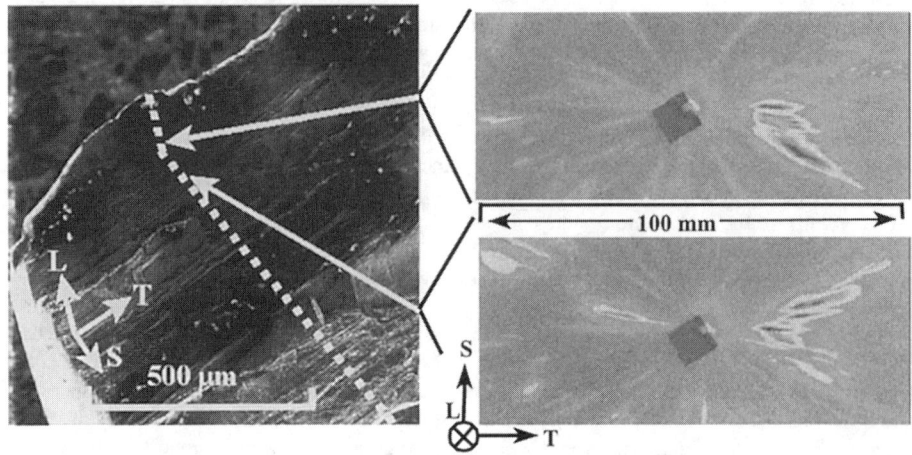

Fig.1 (left) SEM fractograph of a compact tension sample recorded at a high angle of tilt. The large asperity appears at the top, and the dashed line marks the positions where the microbeam was scanned. The central portion of diffraction patterns just within (top right) and just outside (bottom right) the asperity show 111 diffraction streaks; the arrows show where the patterns were recorded; the shift in the streaks'orientations indicates the change from one to another near single crystal volumes. The lowest intensity is behind the beam stop (pattern center, black diamond), the varying shades of gray surrounding it indicate low intensities, the white pixels show intermediate intensities and the highest intensities appear as gray and black pixels within the white envelopes. The separation between sample and image plate (2-D detector) was 245 mm. ([2], © Elsevier Sci. Ltd., Acta Metallurgica (AMI)).

in the samples (near single crystal volumes comprising a specific type of mesotexture) whose grain orientations are also highly textured (i.e., with a strong macrotexture) is a second scale midway between microscopic and macroscopic length scales. The consequences of the two shorter length scales are seen on the macroscopic scale, where crack closure and decreased fatigue crack growth rates are observed. The interaction of these different length scales and the large volume sets of data collected by x-ray microbeam diffraction and x-ray microtomography pose both an opportunity and a challenge for accurate, physically-based models of various phenomena.

ACKNOWLEDGMENTS

The microbeam diffraction experiments were performed at SSRL which is supported by the Department of Energy, Basics Energy Sciences. The work on Cu was partially supported by the Army Research Office and the Office of Naval Research. The research on Al-Li was supported by the Office of Naval Research. The following past and present Georgia Tech staff and students contributed to one or more of the x-ray studies described above: T.M. Breunig, G.C. Butler, A. Guvenilir, J.D. Haase, M.A. Langøy, D.L.McDowell, D.P. Piotrowski, T. Watt, and J.R.Witt.

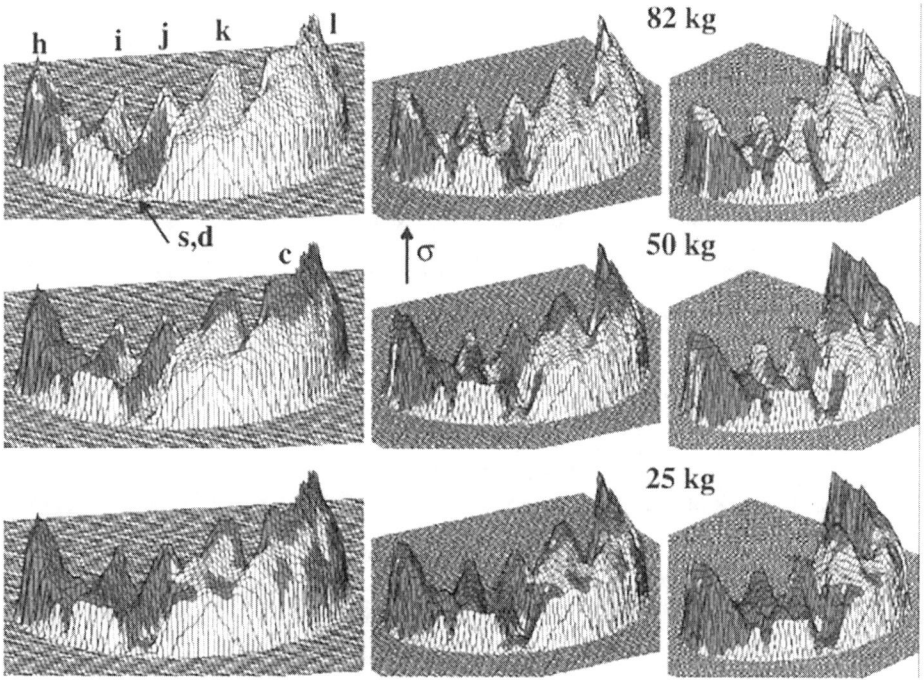

Fig. 2. Three perspectives of the mesh map representing the surface of the fatigue crack in sample NT-3. The view is from outside the sample, and the load axis is indicated by "σ" in the middle of the figure. The diameter at the notch tip was 2.1 mm for NT-3, and the crack height spanned a range of 0.85 mm. The gray pixels mark positions in contact, i.e., with openings less than 0.3 fraction of a voxel (about 1.9 μm). The top row of images is at the maximum load (82 kg, corresponding to K=7.1 MPa√m) and shows three different viewing perspectives. The middle row and bottom row of images show the same three perspectives at 50 kg and 5 kg, respectively. In the upper left image, letters 'h'-'l' mark large asperities, and 's,d' shows a position of contact away from the crack tip. In the middle left image, 'c' indicates a planar section of the crack.[2]

REFERENCES

1. J.D. Haase, "Microbeam Diffraction Mapping of Microtexture in Al-Li 2090 T8E41," MS Thesis, Georgia Institute of Technology, 1998.

2. J.D. Haase, A. Guvenilir, J.R.Witt and S.R. Stock, Acta Mater **46** (1998) 4791-4799.

3. A. Guvenilir, G.C. Butler, J.D. Haase, D.L. McDowell and S.R. Stock, Acta Mater **46** (1998) 6599-6604.

4. G.C. Butler, "Incorporation of Dislocation Substructure into Crystal Plasticity Theory," PhD Thesis, Georgia Institute of Technology, 1999.

5. G.C. Butler, S. Graham, D.L. McDowell, S.R. Stock and V.C. Ferney, J Eng Mater Tech **120** (1998) 197-205.

6. G.C. Butler, S.R. Stock and D.L. McDowell, submitted to Proceedings of AEPA 2000, June 12-16, 2000.

7. A. Godfrey and D.A. Hughes, in Proceedings of 1998 TMS Annual Meeting, pp. 191-199.

8. G.H. Campbell, S.M. Foiles, H. Huang, D.A. Hughes, W.E. King, D.H. Lassila, D.J. Nikkel, T.D. de la Rubbia, J.Y. Shu, V.P. Smyshlyaev, Mater Sci Eng **A251** (1998) 1-22.

9. K.T. Venkateswara Rao, W. Yu and R.O. Ritchie, Met Trans **19A** (1988) 549-569.

10. P.S. Pao, L.A. Cooley, M.A. Imam and G.R. Yoder, Scr Met **23** (1989) 1455-1460.

11. K.T. Venkateswara Rao and R.O. Ritchie, Int Mater Rev 37#4 (1992) 153-185.

12. G.R. Yoder, P.S. Pao, M.A. Imam and L.A. Cooley, Scr Met **22** (1988) 1241-1244.

13. E.A. Starke Jr, T.H. Sanders Jr and I.G. Palmer, J Metals **33** (1981) 24-33.

14. T.H. Sanders Jr and E.A. Starke Jr, Acta Met **30** (1982) 927-939.

15. G.R. Yoder, P.S. Pao, M.A. Imam and L.A. Cooley, in Aluminum-Lithium Alloys, Proc Fifth Aluminum-Lithium Conf, T.H. Sanders, Jr. and E.A. Starke, Jr., Eds. (1989, Mater Comp Eng Publ Ltd., Birmingham, UK) 1033-1041.

16. H.Y. Jung and S.D. Antolovich, Scr Met Mater **33** (1995) 275-281.

17. Jake D. Haase, Abbas Guvenilir, Jason R. Witt, Morten A. Langøy and Stuart R. Stock, in Mixed-Mode Crack Behavior, ASTM STP **1359** (1999) 160-173.

18. J.D. Haase, R.Morano T.Watt and S.R. Stock, in Proc Twelfth Int Conf on Textures of Mater, Vol. 1, J.A. Szpunar, Ed., (National Res Council of Canada, 1999) 123-128.

19. A. Guvenilir, T.M. Breunig, J.H. Kinney and S.R. Stock, Acta Mater **45** (1997) 1977-1987.

20. A. Guvenilir and S.R. Stock, Fatigue Fract Eng Mater Struct **21** (1998) 439-450.

21. A. Guvenilir, T.M. Breunig, J.H. Kinney and S.R. Stock, Phil Trans Roy Soc (Lon) **357** (1999) 2755-2775.

22. A. Guvenilir, S.R. Stock, M.D. Barker and R.A. Betz, in 4th International Conf. on: Aluminum Alloys Their Physical Properties and Mechanical Properties, Volume II, T.H. Sanders, Jr. and E.A. Starke, Jr., Eds.(Georgia Inst Technology, 1994) 413-419.

23. R. Morano, S.R. Stock, G.R. Davis and J.C. Elliott, in Proc Twelfth Int Conf on Textures of Mater, Vol. 2, J.A. Szpunar, Ed., (National Res Council of Canada, 1999) 1106-1111.

A MULTIPLE SLIP PLANE MODEL FOR CRACK-TIP PLASTICITY

S. J. NORONHA, S. G. ROBERTS and A. J. WILKINSON
Department of Materials, University of Oxford, Parks Road, Oxford, OX1 3PH, United Kingdom

ABSTRACT

A single slip plane dislocation dynamics based model for the brittle to ductile transition has been extended to have multiple slip planes around the crack-tip. The crack-tip plastic behaviour is studied for a variety of dislocation source configurations. The results are presented for the case of iron. The effect of modelling the plastic-zone as a single slip plane and as an array of parallel slip planes are compared.

INTRODUCTION

It is now well known that whether or not a material fails by cleavage is determined by the dislocation activity around the crack-tip. Following the work of Rice and Thomson [1], there was a surge in the literature on crack-tip plastic behaviour (for review, see [2,3]). This included several experimental observations of dislocation emission from crack-tips, notably by Ohr for a range of materials including bcc metals [3]. On the theoretical side, solutions for stress and/or displacement fields were obtained for the elastic interaction of dislocations with cracks in several different dislocation-crack configurations. The solutions were derived mainly using two approaches: either using the method of complex potentials (e.g. [4,5]) or using dislocation distributions to reduce free surface traction to zero (e.g. [6,7]). The model presented here uses solutions from the latter approach [7] for edge dislocations near a semi-infinite crack. Computer simulation studies have also been conducted to simulate the crack-tip plastic behaviour and dislocation emission criteria (e.g. [8,9]).

The single slip plane model proposed by Roberts and Hirsch [9-13] has been successful in predicting the fracture behaviour of many materials. In reality there could be many active slip planes around the crack tip. To extend the single slip plane model to describe the plastic-zone in two dimensions, we have simulated dislocation activity on several parallel slip planes around the crack tip.

THE MODEL

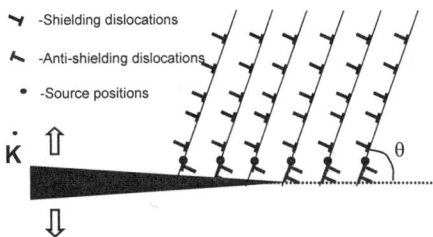 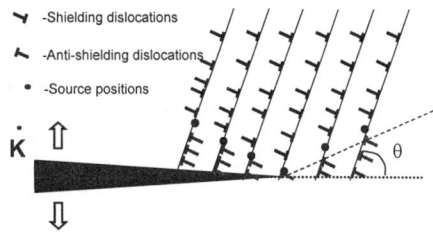

Figure 1a: Parallel slip plane model for crack-tip plastic zone, with 'parallel source' configuration.

Figure 1b: Parallel slip plane model for crack-tip plastic zone, with sources at 'maximum stressed positions'

A schematic representation of the model is shown in figure 1. We consider edge dislocations and mode-I loading. The slip planes along which the dislocations move are parallel to each other and inclined at an angle θ to the crack plane. One of the slip planes intersects the crack plane

Mat. Res. Soc. Symp. Proc. Vol. 578 © 2000 Materials Research Society

along the crack-front. Dislocations are generated at 'dipole' sources on each slip plane. When the shear stress on 'test' dislocations at the source positions exceeds a critical value (chosen as the friction stress) a pair of dislocations is generated and allowed to move away from the source according to an experimentally determined temperature-dependent velocity law. If a moving dislocation intersects the crack, it is removed from the simulation. The dislocations move against a constant (temperature-dependent) friction stress. The emitted dislocations, depending on the sign of their Burgers vector, will shield or anti-shield the crack-tip from the external load, and so the crack-tip stress intensity (k) will differ from the applied stress intensity (K). Generally, the effect of the shielding dislocations is found to dominate, and so k < K, and increasingly so as temperature increases. Hence the external (or applied) stress intensity factor for fracture, K_F (taken to be when $k = K_{Ic}$), rises with increasing temperature.

The Algorithm

For an applied stress intensity (K), the shear stress (τ_{xi}) on a dislocation at x_i is given by:

$$\tau_{xi} = \alpha \frac{K}{(2\pi x_i)^{1/2}} - \beta \frac{\mu b}{x_i} + \Sigma \tau_{ij} \tag{1}$$

where the first term represents the crack-tip stress field due to the applied stress, the second term represents the stress due to the self-image of the dislocation and the final term arises from the dislocation-dislocation interactions modified by the presence of free surfaces of the crack. Here α and β are geometrical factors, while μ and b are respectively the shear modulus and Burgers vector of the material being modelled. The relations used are those obtained in [7]. If the magnitude of the stress calculated for test dislocations at the source positions is greater than τ_f, new dislocations are generated and allowed to move away from the source. If the calculated shear stress (τ_{xi}) is above a pinning stress (τ_f), then the dislocations move over a short time step with a velocity given by:

$$v_i = \left(\frac{\tau_{xi} - \tau_f}{\tau_{xi}}\right) A \left(\frac{\tau_{xi}}{\tau_0}\right)^m \exp\left(\frac{-U}{k_B T}\right) \tag{2}$$

where A is a pre-factor, U is the activation energy, k_B the Boltzmann's constant and T the temperature. The value of the material-specific parameters was obtained from a best fit to the screw dislocation velocity data for iron [14]. The values used are $U = 1.54eV$, $\tau_0 = 1$ MPa, m = $1.251+(393/T)$. The time step is calculated as the largest time interval that allows each dislocation to move to a position no more than half-way to its nearest neighbour. It should be noted that the friction stress is a temperature dependent variable that is directly related to yield stress $(\tau_f = \sigma_y)$, which for bcc metals decreases with increasing temperature. With the above velocity law, the friction stress, rather than the $\exp(-U/k_B T)$ term is found to control the dislocation motion, since velocity increases rapidly when $\tau_{xi} > \tau_f$. Hence, for the loading rates used, the dislocations for the most part remain in quasi-static equilibrium against the friction stress. The stress intensity at the crack-tip (k) is calculated as

$$k = K - K_D \tag{3}$$

where K_D is the total dislocation induced stress intensity at the crack-tip. The stress intensity is updated according to the modelled dK/dt and the time step. In the simulations, the load (or stress intensity factor, K) is monotonically increased starting from K=0.

RESULTS AND DISCUSSIONS

The effect of source positions

In all the simulations reported here, we have used $\theta = 70.5°$, so that the slip plane intersecting the crack-front has the maximum shear stress. The Burgers vector $b = 2.54\text{Å}$ throughout. We use dipole sources of width 100 b, which generate dislocations in pairs. In the case of the single slip plane model [9], it has been found that variations in the source position within a reasonable range (1b –1000b from the crack-tip) do not strongly influence the result of the simulations. Since in this case we have many slip planes extending on either side of the crack tip, we have investigated different methods for choosing the source positions and their effect on the simulation results. Our model does not allow collision of dislocations, so we cannot use more than one source on a given plane, and the option of choosing random source positions is ruled out. Two distributions of source positions are investigated here: (a) sources at a fixed distance from the crack plane ("parallel sources", as in figure 1a), and (b) sources placed at the positions of maximum shear stress on each of the slip planes used ("maximum stressed sources", figure 1b). For $\theta = 70.5°$, the sources in the planes ahead of the crack-tip are at angle of ~60° and the sources on the planes behind the crack-tip are at an angle ~ 45° to the crack plane (schematic representation in figure 1b). It should be noted that there is an 'infertile' zone ahead of the crack at an angle 34° (dotted line in fig. 1b), for $\theta = 70.5°$, within which the shear stress acts to close the dislocation dipole and hence to oppose motion of dislocations emitted from the dipole source.

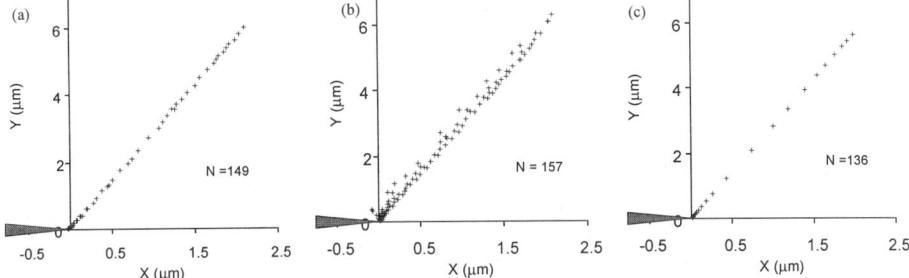

Figure 2: The dislocation configuration around crack-tip (at origin), for $\tau_f = 400$ MPa, K= 4.05MPa√m. (a) the sources are parallel to the crack plane at 52b (b) sources are at the maximum stress positions (c) single slip plane with source at 52b. In all cases the width of the dipole source used is 100b. N is the number of dislocations.

Figures 2a and 2b show the dislocation arrays around the crack-tip, from the 'parallel source' configuration and 'maximum stressed source' configurations respectively. In both cases we have used 24 slip planes each with an inter-planar spacing of 50b. In the 'parallel source' case, only 4-slip planes were activated, whereas in the 'maximum stressed sources' case, dislocations were activated on 21 of the slip planes. There is less activity in the planes ahead of crack-tip than behind it. In both cases only a single pair of dislocations are generated in the plane ahead of the crack-tip, and their net contribution is anti-shielding. In figure 2c, the dislocation configuration of single slip plane is shown. Since we remove the dislocations that intersect the crack, in all the three cases, except for one or two (closer to the crack) the crosses all represents shielding dislocations. Points towards the centre of slip planes represent more than one (3,9,27, etc.) dislocations. This is due to the 'dislocation bundling' done for computational efficiency [10]. In all cases the plastic zone generated is *very narrow* when compared to those predicted by finite-element analysis. However, such narrow plastic zones have been observed in experiment [15] where the plastic zone is contained within a single grain at the crack-tip.

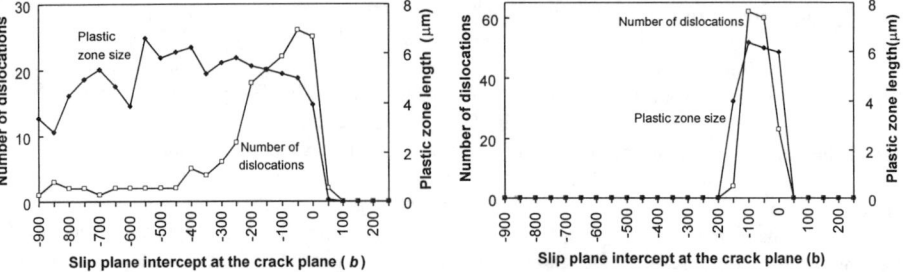

Figure 3: The number of dislocations and plastic zone length on each slip plane for $\tau_f = 400$ MPa (a) sources are at the point of maximum shear stress; (b) sources are at distance 52b above and parallel to the crack plane.

Figure 3a and 3b shows the number of dislocations in each slip-plane, and the length of the plastic zone measured along the slip planes for the two variants of the multiple slip plane model. The total number of dislocations from all the planes and the plastic zone size (length of longest active slip plane) are very similar for the two source configurations. For the maximum shear stress configuration the outermost dislocations on each of the slip planes behind the crack-tip have moved out through similar distances despite the marked reduction in the number of dislocations on planes more than ~ 200b behind the crack tip. Figure 4 shows the stress intensity at the crack tip due to dislocations (K_D) from different planes for the two source configurations. For both source configurations, the contribution to the total shielding falls rapidly beyond few slip planes behind the crack tip. However, in both cases the total shielding is roughly the same. For the case shown here at an applied K = 4.05 MPa$\sqrt{m}$, the total shielding at the crack-tip is 2.05 MPa$\sqrt{m}$ for maximum stress source configuration and 1.99 MPa$\sqrt{m}$ for the parallel source configuration. In both cases, the few slip planes closest to the crack tip are the most active and the decisive ones in the simulation. Thus the behaviour of the model is insensitive to source configuration.

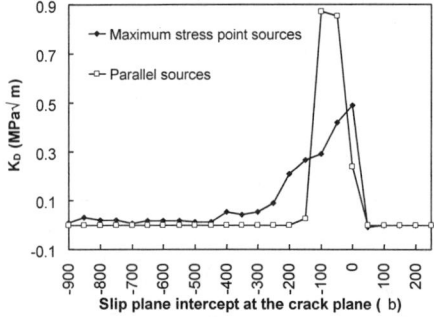

Figure 4: The negative contribution of dislocation induced crack-tip stress intensity. $\tau_f = 400$MPa.

Effect of slip plane spacing

For the case of 'maximum stress sources', we have examined the effect of varying the inter-planar distance. Figure 5 shows the plastic-zone length (measured along the slip plane) and total K_D for various inter-planar spacings. There is little change in the size of the plastic-zone or K_D with variation of inter-planar spacing. When the inter-planar spacing is reduced the dislocation

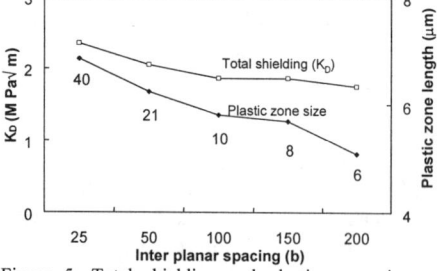

Figure 5: Total shielding and plastic zone size as a function of inter-planar spacing at k = K -K_D = 2MPa$\sqrt{m}$.

activity is seen on more slip planes. Figure 5 also shows the number of active slip planes in each case. The overall width of the plastic zone remains the same as the slip plane spacing is varied.

Comparison of single slip and multiple slip plane models

We compared the single slip plane model (source at 52b) with the multiple slip plane model with sources at the maximum stressed positions. Figure 6a shows the plastic zone size and Figure 6b shows the number of dislocations generated in the two cases. More dislocations are generated in the multiple slip plane model than in the single slip plane model.

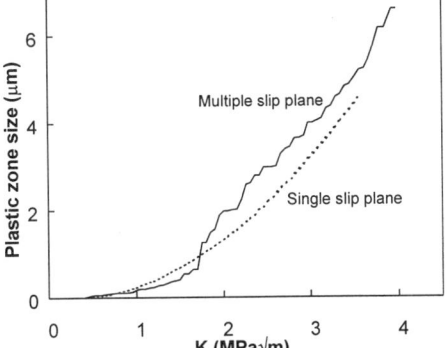

Figure 6a: The number of dislocations generated as a function of applied stress intensity for multiple slip plane and single slip plane models. (τ_f = 400 MPa).

Figure 6b: The plastic zone size as a function of applied stress intensity for multiple slip plane and single slip plane models. (τ_f = 400 MPa).

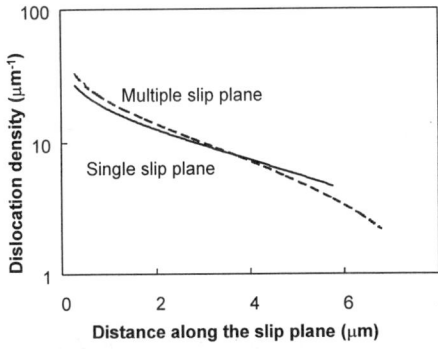

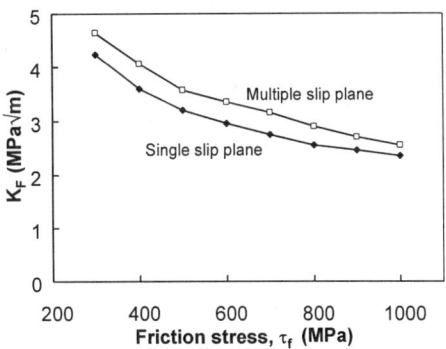

Figure 7: The dislocation density along the slip direction for the multiple slip plane and single slip plane models (K = 4MPa√m, τ_f = 400MPa).

Figure 8: The fracture stress as a function of friction stress, for single slip plane and multiple slip plane models. Fracture criterion: k = 2MPa√m.

The dislocation density measured along the slip plane for both the single slip plane and multiple slip plane models at an applied stress intensity, K = 4 MPa√m is shown in figure 7. Both have an approximately logarithmic distribution along the slip plane. In figure 8, the stress at fracture, K_F (using the fracture criterion k = K – K_D = K_{Ic} = 2MPa√m) is plotted against friction stress. We can see that the shielding caused by dislocations is around ten percent more in the case of the multiple slip plane model, and hence a higher applied stress intensity is required for fracture. The variation of K_F with friction stress (or temperature) is similar for both single slip plane and multiple slip plane models.

CONCLUSIONS

It has been verified that the multiple slip plane model for crack-tip plasticity is robust to the variation of modelling parameters, such as the number of slip planes, slip-plane spacing and source configuration. The plastic zone produced by multiple slip plane model is very narrow, compared with that obtained from finite element analysis. However, this is consistent with the experimental observations on single crystal materials [15]. The multiple slip plane model generates more shielding at the crack-tip, when compared to the single slip plane model. However, the variation of K_F with friction stress (or temperature) is similar, with K_F for the multiple slip model being ~ 10% higher. The use of the single slip plane model for single crystal materials is thus a fairly good approximation to the more complex (but more realistic) multiple slip plane models.

ACKNOWLEDGEMENTS

The work was funded by the Engineering and Physical Sciences Research Council, United Kingdom (Grant No. M05188), with support from AEA Technology and the HSE. AJW thanks the Royal Society for the provision of a University Research Fellowship.

REFERENCES

1. J. R. Rice and R. Thomson, Phil. Mag. **29**, 73 (1974).

2. R. M. Thomson in *Physical Metallurgy (4th edition)*, edited by R. W. Cahn and P. Haasen, (Elsevier Science, 1996), p. 2207-2292.

3. S. M. Ohr, Scripta Met. **20**, 1465 (1986).

4. I. H. Lin and R. Thomson, Acta, Metall. **34**, 187 (1986).

5. T. Zhang and J. C. M. Li, Acta Metall. Mater **39**, 2739 (1991).

6. S. Wang and S. Lee, Mater. Sci. Eng. A **130**, 1 (1990).

7. V. Lakshmanan and J. C. M. Li, Mat. Sci. Eng. A **104**, 95 (1988).

8. J. C. M. Li, Scripta Met. **20**, 1477 (1986).

9. S. G. Roberts, Mat. Sci. Eng. A **234-236**, 52 (1997).

10. S. G. Roberts in *Computer Simulation in Materials Science-nano/meso/macroscopic Space and Time Scales*, edited by H. O. Kirchner *et al.* (NATO ASI Series, Series E (Applied Sciences), **308**, Kluwer, Dordrecht, 1996), p. 409-433.

11. P. B. Hirsch and S. G. Roberts, Phil. Trans. R. Soc. (Lond.) A **355**, 1991 (1997).

12. P. B. Hirsch and S. G. Roberts, Phil. Mag. A **64**, 55 (1991).

13. P. B. Hirsch, S. G. Roberts and J. Samuels, Proc. R. Soc. (Lond.) A **421**, 25 (1989).

14. H. Saka, K. Noda and T. Imura, Crystal Lattice Defects **4**, 45 (1973).

15. W. Zielinski, M. J. Lii and W. W. Gerberich, Acta Metall. Mater. **40**, 2861 (1992).

DAMAGE AND CRACK PROPAGATION AT A MICROSTRUCTURAL SCALE

Elisabeth Bouchaud*, Florin Paun**, Elodie Ducourthial[+]

*DSM/DRECAM/SRSIM, CEA-SACLAY, 91191 GIF-SUR-YVETTE Cedex, FRANCE
**ONERA (DMMP), 29, Av. de la Division Leclerc, 92322 CHATILLON Cedex, FRANCE
[+] ONERA (DMSE), 29, Av. de la Division Leclerc, 92322 CHATILLON Cedex, FRANCE

ABSTRACT

A quantitative analysis of the morphology of damage cavities in metallic materials is performed. At larger length scales, the self-affine correlation length of fracture surfaces is shown to be correlated to the grain size. These observations suggest a new scenario for the origin of scaling laws observed on fracture surfaces. It is argued that they reflect strong correlations in damage created prior to fracture during crack propagation.

INTRODUCTION

Fracture surfaces of numerous materials have been shown to exhibit a self-affine structure [1-10] characterized by two roughness exponents [11,12]. One, at small length scales, is close to 0.5, and the other one, at larger length scales, has been shown to be equal to 0.78 [11-13]. These two exponents appear not to vary with the fracture mode or the material, and were conjectured to be universal. On the contrary, length scales depend quite strongly on the microstructure [9,11,12]. This is the case of the self-affine correlation length, i.e. the upper limit of the scaling domain. This is also the case of the crossover length separating the two self-affine regimes. Furthermore, this crossover length depends on the average crack velocity, and, in the cases which have been studied experimentally, was shown to decrease with the crack length [12].

In order to reproduce these observations, and model crack propagation through complex microstructures (i.e. constituted by randomly distributed elements opposing the motion of the front), we have suggested that models of lines moving through randomly distributed obstacles might be relevant [14]. In these models, the fracture surface is the *trace* left behind by the propagating line. Although the generic predictions of that category of models [9,15,16] fit qualitatively the experiments, no quantitative agreement could be obtained [12]. The central point of this paper is to offer an alternative to the line models. The latter may actually work at small enough length scales, but at intermediate length scales, one cannot ignore the presence of damage cavities ahead of the crack front, which makes it impossible to define a front properly. We argue that at these length scales, correlations between damage cavities lead to a non-trivial exponent 0.78 observed on the fracture surfaces. Our argument is based primarily on direct observations of damage cavities in two aluminum alloys (Section 1). We indeed show that in small enough cavities, only the exponent 0.5 is observed. Section 2 is devoted to the study of the fracture surfaces correlation length. We present measurements performed on various metallic alloys which show that the correlation length is of the order of the metallurgical grain size, which in all cases is the largest heterogeneity in the material. We also present the results of a bidimensional simulation in which damage is shown to have a structure controlling the morphology of the main crack. In conclusion, we offer a new scenario for the emergence of scaling laws in fracture.

Mat. Res. Soc. Symp. Proc. Vol. 578 © 2000 Materials Research Society

1. STRUCTURE OF DAMAGE CAVITIES

Damage is known to have a central role in ductile fracture : after nucleation due to cleavage of brittle second phase precipitates or to decohesion between matrix and such precipitates for example, cavities grow under the influence of the triaxiality of the local stress [17]. The main crack, which progresses through successive deflections at small length scales propagates through junctions with the cavities at intermediate length scales.

It has been shown in Molecular Dynamics simulations that this fracture mode is also the one observed in the case of Si_3N_4 and of amorphous SiO_2. In both cases, the low density regions act as nucleation sites, which concentrate the stresses, grow and coalesce with the main crack. In the case of Si_3N_4, it has been shown that the dynamic fracture surfaces exhibit, at the nanometer scale, the two self-affine regimes with exponents 0.5 and 0.8 [18,19] that other materials, fractured in a quasi-static way, exhibit at significantly larger scales [20]. All the cavities (not linked to the main crack) were found to be self-affine with an exponent 0.5 [18]. This suggested that the non trivial 0.78 could characterize the structure of the ensemble of opened cavities lying ahead of the main crack, while 0.5 characterized the growth of a single cavity. This is the idea we have tested experimentally.

In this section, we show the results of our observations of damage cavities in two different aluminum alloys broken either in monotonous tension or in fatigue. Chevron notched bar specimens of a rapidly quenched Al-Cr-Zr alloy were submitted to tension mode I loading, while fatigue cracks (R=0.1, f=10Hz) were grown in Compact Tension samples of the 7010 aluminum alloy. In both cases, crack propagation was stopped before complete failure. The samples were cut and polished within a plane containing the crack (Fig.1a).

Cavities of two types were examined in both cases, with an Atomic Force Microscope

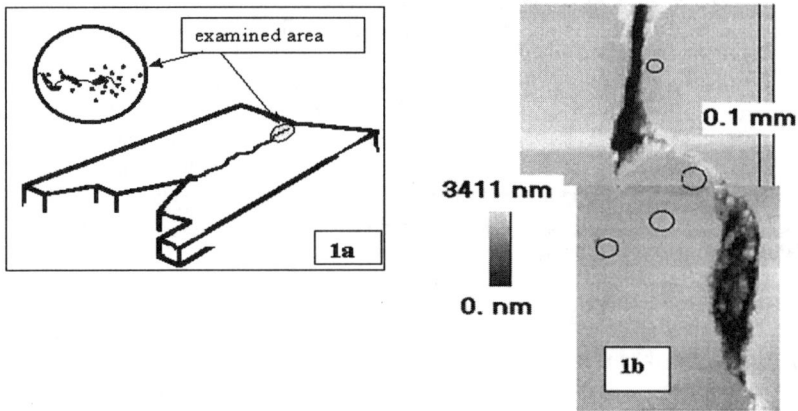

Figure 1. a) is a sketch of the region of observation valid for both cases. b) is an AFM image of the AlCrZr sample. The main crack observable on the upper part of the image is linked to a big cavity (lower part of the image). Circles indicate the presence of "secondary" cavities, of significantly smaller sizes, which are very likely independent from the main crack.

(AFM). In the case of the AlCrZr alloy, the various cavities which were explored are indicated in Fig. 1b. "Large" anisotropic cavities (5 to 15 μm wide, 30-100 μm long) visibly related to the main crack were analyzed, as well as significantly smaller cavities (3-5 μm) which appear to be more isotropic and independent of the main crack. In the large cavity shown in Fig.1b, four regions of size 10μm x10μm were explored, aligned two by two in the two directions (parallel and perpendicular to the direction of crack propagation). Profiles of 1000 points each were registered in the two directions in the four regions and pasted together in order to provide profiles of length 2000 points to be analyzed. Eight of these profiles were considered for each direction. For each of them, both the power spectrum and the Hurst method were computed and an average was made in each direction.

As shown in Fig. 2a and 2b, the roughness exponent determined from the two methods is: $\zeta \cong 0.8$ for what concerns profiles perpendicular to the direction of crack propagation. The same methods were used on parallel profiles, and no anisotropy was detected for the roughness exponent.

For what concerns small cavities (indicated by circles in Fig. 1b), the results are quite different. AFM images (Fig. 3) show different morphologies for these two types of cavities at a comparable magnification.

This qualitative difference results in differences in the roughness exponents determined in comparable domains of length scales. As a matter of fact, the roughness index measured (with both the Power Spectrum and the Hurst method [9]) in cavities smaller than 5 μm is close to 0.5, and does not exhibit the 0.8 regime.

In the case of the 7010 alloy broken in fatigue, however, cavities are anisotropic, even the small independent ones. As shown in Fig. 4b, the 0.5 regime is only visible in the direction parallel to the main crack, and at length scales smaller than 100 nm on average. This length compares fairly well to the crossover lengths determined on fracture surfaces in this case.

In the two examined cases, it indeed appears that under a certain size – which is quite small, some 100 nm, in the case of fatigued 7010 – cavities do not exhibit the 0.8 regime, but only the 0.5 one.

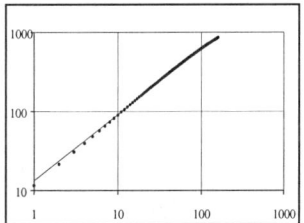
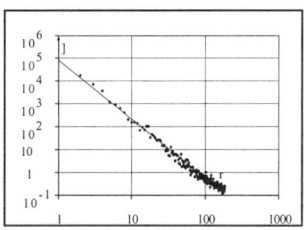

Figures 2a and 2 b show the results of the Hurst method and of the Power Spectrum method respectively. $z_{max}(r)$ is plotted as a function of r (distances are expressed in nanometers) in Fig. 2a, and is the result of an average over eight profiles. The best fit corresponds to a roughness index of 0.82. In Fig. 2b, the plotted power spectrum is also an average over the results relative to eight profiles, and the best fit corresponds to a roughness exponent of 0.79.

This is true however only in the direction of crack propagation in the case of the 7010, where the explored small cavity is quite anisotropic. Although large cavities have an anisotropic appearance in the case of the AlCrZr alloy, there is no anisotropy in the measured roughness index. It might be surprising that in both cases, the 0.5 regime observable on fracture surfaces is not visible in large cavities. As already said, it has been shown elsewhere that the extension of the 0.5 regime strongly decreases with the average crack velocity [12]. Large cavities, through which the main crack goes, are very likely located in regions of larger local stresses and/or lower toughness, where the crack velocity is higher. Small cavities probably grow more slowly. Together with observations concerning the correlation length and the structure of damage (see below), these results are an indication that, as we argue in the following, the 0.5 regime might

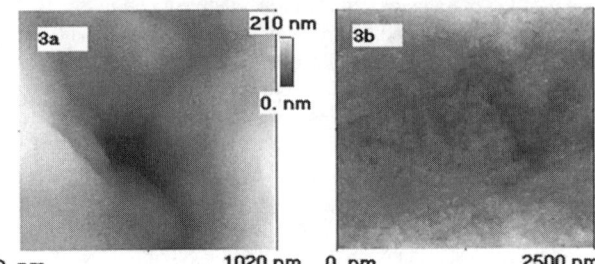

correspond to the growth of a single cavity while the 0.8 regime is due to strong correlations between these damage cavities.

2. DAMAGE CORRE-LATIONS

Figures 3a and 3b show AFM images of a large cavity and of a small cavity respectively, at comparable magnifications (the size of image 3a is 1.02 •mx1.02 •m, while the one of image 3b is 2.5 •m x2.5 •m). Image 3b exhibits more structure than image 3a, in which the relief appears to be much flatter.

As we have just seen, the 0.5 exponent might be the result of the propagation of a single crack front growing from a micro-crack or a micro-cavity nucleated from a microstructural heterogeneity in the material. On the contrary, we will see that the 0.8 regime extends to length scales at which a single front cannot be defined. In that region, fracture occurs through simultaneous growth and coalescence of these microcracks or micro-cavities, together and with the main crack. We argue that the morphology of the resulting structure – i.e. the fracture surface – reflects pre-existing correlations in damage.

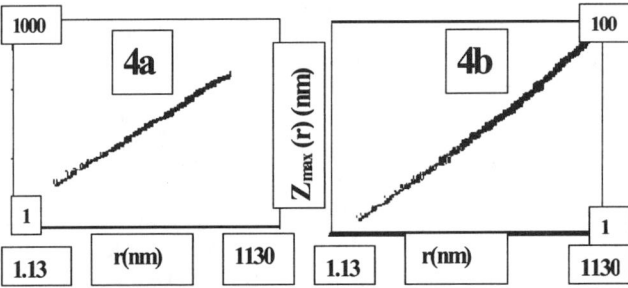

Figures 4a and 4b show $z_{max}(r)$ for the large and for the small cavities analysed respectively. The large cavity only exhibits the 0.8 regime, while the small one exhibits both the 0.8 and the 0.5 regime, but the latter only extends up to 110 nm.

Extension of the scaling domain observed in experiments

Experiments on the Ti_3Al alloy Super •$_2$ have shown that the scaling domain characterizing the self-affinity of fracture surfaces may extend over five decades of length scales [12]. In fact, it has been shown to extend up to 0.5 mm [11,12], which corresponds to the metallurgical grain size. As it can be seen in Fig. 5, it is impossible to define a single crack front in that domain of length scales, because of the very existence of damage cavities.

In the 7010 broken in fatigue, it was found again that the upper limit of the scaling domain is close to the grain size, i.e. 0.1 mm [13].

To go further, the microstructure of the N18 Ni-based superalloy was prepared in order to provide two different grain sizes, 10 and 80 •m [21]. Fatigue experiments were conducted in the same conditions, and profiles corresponding to comparable average crack velocities were analyzed in both cases. In each case, both the Hurst method and the variance method [9] were used in order to compute $z_{max}(r)$ and $h(r)$. In both cases, a saturation of $h(r)$ was observed for length scales larger than the length hence defined as the self-affine correlation length (i.e. the upper limit of the whole scaling domain). Above that distance, $z_{max}(r)$ was shown to evolve as log(r).

This correlation length was shown to be equal to the grain size in both cases. Furthermore, the behavior observed outside the scaling limit fits perfectly a random suite behavior, for which the variance is a constant while $z_{max}(r)$ grows logarithmically.

This typically "ductile fracture" behavior is, more surprisingly observed in Molecular Dynamics Simulations of ceramics [18,19] and glass [22]. In these dynamic fracture experiments, cavities are a few nanometers large, but their aspect at these small length scales is strikingly similar to the aspect of cavities in metals in the range of 10-50 •m [19].

AFM experiments performed on stress corrosion fracture surfaces of soda-lime silica glass have given results similar to the ones in metallic alloys, with a correlation length close to 100 nm [21]. Although 100 nm does not correspond to any simple structural size in glass, this ensemble of observations suggests the following picture:

- *one cannot define a single crack in most of the scaling domain*
- *within the scaling domain, for length scales larger than a single cavity size, strong correlations between damage cavities lead to the non trivial roughness observed.*
- *at length scales larger than the correlation length, cavities seem to be located at random in the material.*

Figure 5. In situ Scanning Electron Microscopy experiments on the Super •2, showing cavities opening ahead of the crack tip (top of the picture).

There is no experimental proof that the self-affine correlation length corresponds to any structure in damage. However, it is very plausible that correlations in damage cannot exceed the range of crystalline orientations.

To understand further that aspect, we have analyzed the numerical results of a bi-dimensional numerical simulation [23,24]. The material is composed of an elastic matrix containing brittle precipitates likely to break in cleavage under the stress due to the presence of a macro-crack submitted to an external mode I loading. Damage is nucleated from the cleavage micro-cracks, which may grow if the local stress intensity factor overpasses a prescribed threshold.

Extension of damage and morphology of simulated crack

A first simulation [23] was carried on without taking into account any correlation between the micro-cracks (whereas the presence of the micro-cracks, due to the main crack stress field, were actually modifying the main stress intensity factors). In this case, the simulated cracks were shown to be self-affine, with a non trivial exponent close to 0.7, up to a correlation length of the order of the average extension of the damaged zone.

In a more recent attempt to refine the model [24], average interactions between micro-cracks were introduced within the framework of a mean field analysis due to M. Kachanov [25] could be taken into account. In this case, the extension of the damaged zone was shown to be significantly smaller than the measured correlation length, and spatio-temporal damage correlations are currently being looked at in order to understand the final structure of the crack.

3. CONCLUSION

We have suggested a new scenario for fracture based on the nucleation, growth and coalescence of damage cavities. As a matter of fact, small isolated cavities do not exhibit the same structure as cavities related to the main crack. Furthermore, the correlation length observed on fracture surfaces, equal to the average metallurgical grain size in alloys, is argued to be related to correlations between damage cavities. Simple numerical models should provide a better insight into that question.

4. ACKNOWLEDGEMENTS

We are indebted to A. Morel for her technical help in AFM experiments, and to J.-P. Bouchaud and L. Van Brutzel for very interesting discussions.

References

1. B.B. Mandelbrot, D. E. Passoja, A.J. Paullay, Nature **308**, 721 (1984).
2. E. Bouchaud, G. Lapasset, J. Planès, Europhys. Lett. **13**, 73 (1990).
3. K.J. Maloy, A. Hansen, E.L. Hinrichsen, S. Roux, Phys. Rev. Lett. **68**, 213 (1992).
4. A. Imre, T. Pajkossy, L. Nyikos, Acta Metall. Mater. **40**, 1819 (1992).
5. E. Bouchaud, G. Lapasset, J. Planès, S. Navéos, Phys. Rev. B **48**, 2917 (1993).
6. J. Schmittbuhl, S. Gentier, S. Roux, Geophys. Lett. **20**, 8 (1993); ibid, 639 (1993).
7. J. Schmittbuhl, S. Roux, Y. Berthaud, Europhys. Lett. **28**, 585 (1994).
8. E. Guilloteau, H. Charrue, F. Creuzet, Europhys. Lett. **34**, 549 (1996).
9. E. Bouchaud, J. Phys. : Condens. Matter, **9**, 4319 (1997).
10. S. Morel, J. Schmittbuhl, J. Lopez, G. Valentin, Phys. Rev. E **58**, 6999 (1998).
11. P. Daguier, S. Hénaux, E. Bouchaud, F. Creuzet, Phys. Rev. E **53**, 5637 (1996)
12. P. Daguier, B. Nghiem, E. Bouchaud, F. Creuzet, Phys. Rev. Lett. **78**, 1062 (1997)
13. M. Hinojosa, E. Bouchaud, G. Marcon, in preparation.
14. J.-P. Bouchaud, E. Bouchaud, G. Lapasset, J. Planès, Phys. Rev. Lett. **71**, 2240 (1993)
15. S.F. Edwards, D. Wilkinson, Proc.Roy. Soc (London) A **381**, 17 (1982)
16. H. Leshorn, T. Nattermann, S. Stepanow, L.H. Tang, Annalen der Physik, **6**, 1 (1997).
17. D. François, A. Pineau, A. Zaoui, *Comportement mécanique des matériaux,* Ed. Hermès, (1993).
18. A. Nakano, R.K. Kalia, P. Vashishta, Phys. Rev. Lett. **73**, 2336 (1994).
19. A. Nakano, R. Kalia, P. Vashishta, Computing in Science and Engineering, Special Issue on *Dynamic Fracture Analysis*, September/October 1999, p. 39.
20. E. Bouchaud, F. Paun, Computing in Science and Engineering, Special Issue on *Dynamic Fracture Analysis*, September/October 1999, p. 32.
21. M. Hinojosa, B. Nghiem, E. Bouchaud, MRS Fall Meeting, Boston (U.S.A.) (1998).
22. L. Van Brutzel, PhD thesis (1999), available from E. Bouchaud.
23. P. Daguier, E. Bouchaud, G. Lapasset, J. Phys. IV France **8**, (1998).
24. E. Ducourthial, E. Bouchaud, J.-L. Chaboche, F. Roudolff, in preparation (1999).
25. M. Kachanov, Adv. In Appl. Mech. **30**, (1994).

CRACK PROPAGATION IN FRESHWATER AND SALINE ICE

Patrick J. Donovan *, Masahiko Arakawa **, and Victor Petrenko*
* Thayer School of Engineering, Dartmouth College, Hanover, NH 03755
**Institute of Low Temperature Science, Hokkaido University, Sapporo, Japan

ABSTRACT

Crack propagation in columnar saline and freshwater ice has been investigated with high-speed photography, acoustic emission detection and the resistance method. High-speed photography was found to be a single reliable technique. The resistance method proved effective for freshwater ice samples, but not for saline ice samples due to the presence of conductive fluid inclusions. Acoustic emissions pinpointed the moment of crack initiation, but did not correspond to the crack propagation time. Crack velocity has been characterized over a temperature range of −5°C to −30°C for freshwater and saline ice. Freshwater ice exhibited an overall average velocity of 198 m/s, and did not vary with temperature. Crack velocity in saline ice demonstrated temperature dependence, increasing from an average of 86 m/s in the −5°C to −20°C range, to 131 m/s at −30°C. The crack velocity was also shown to have a general dependence on fracture toughness K' of the material, however, the microstructural variation between samples is also shown to influence significantly the crack behavior in both saline and freshwater ice. Nonuniform crack tip advance and crack reorientation were observed as crack slowing mechanisms in freshwater ice, while in saline ice fracture crack tip blunting on voids greatly reduced average crack velocities.

INTRODUCTION

The principal objective of this study is to develop a micromechanical theoretical model by investigating the crack interaction with the microstructure of ice with high-speed photography, resistance and acoustic measurements. Previous study has demonstrated a large difference in the observed crack velocity in freshwater and saline ice. Through observation with high-speed photography, freshwater ice was found to exhibit an average crack velocity of 1063 m/s and 1050 m/s when loaded with a heavy metal wedge [1,2]. Crack velocity in freshwater ice was found to fall in the range of 100 to 1320 m/s by correlating crack extension to the change in resistance of the ice (resistance method) [3]. For saline ice, the resistance method rendered an average velocity range of 0.85 to 18 m/s [3]. Moreover, an velocity of 20 m/s was found using foil electrodes designed to tear as the crack passes to monitor crack extension [4].

Saline ice contains fluid inclusions, air bubbles, and drainage channels, whereas freshwater lacks a significant presence of these features. Therefore, the velocity difference between freshwater and saline ice may result from the differences in the microstructure. The voids of varying size in the saline ice may act to blunt the crack tip, thereby slowing crack extension. The fluid inclusions also may resist stretching induced by crack mouth opening, thereby supplying an opposite force to retard the crack [3]. Finally, it is also possible that differences in strain energy stored in the ice samples during loading in freshwater and saline ice may account for crack velocity differences.

MATERIALS AND METHODS

Columnar freshwater ice was grown with established laboratory procedure for a grain size of 5-14 mm, which is similar to the grain size of saline ice [5]. Saline ice was grown

Mat. Res. Soc. Symp. Proc. Vol. 578 © 2000 Materials Research Society

with established laboratory procedure with the same saline content as seawater [6]. Samples of both freshwater and saline ice were milled or hand planed into modified compact-type specimens 145 mm by 145 mm by 25 mm, with columns perpendicular to the plate face. A precrack of 72.5 mm was cut midway between parallel sides of the plate, and a small 10 mm minor "counter" precrack was added across from the main precrack.

The sample was installed into the testing apparatus as shown in Figure 1. Springs were used maintain constant load during the fracture event. The sample was loaded at a stress intensity factor increase of 35 kPa $m^{1/2}s^{-1}$. Acoustic emissions, load, extension and resistance signals were recorded simultaneously at an acquisition frequency of 1 MHz to 10 MHz with a Yokogawa DL-1200 oscilloscope. Acoustic emissions were used as a trigger for the oscilloscope and camera. The acoustic emissions sensor was frozen to the ice sample near the crack tip with small droplets of water. The resistance method employed foil electrodes placed 8 mm on either side of the crack as shown in Figure 1. The current running between the electrodes was measured as a potential difference across a small resistor (See Figure 1). An ULTRA NAC high-speed camera was used to capture images of the propagating crack at rates up to 10^6 frames per second. The photographs were magnified and captured with a digital camera. The crack extension was measured with NIH Image 1.62 software. Samples were tested over a temperature range of -5°C to -30°C. The average crack velocity was calculated from data points spanning a constant region among the samples. The maximum velocity was determined over all known position and time data points. K' was calculated with the sample dimensions and fracture load.

RESULTS

Comparison of Different Experimental Methods

The comprehensive tests of freshwater and saline ice are shown in Figure 2 with resistance and acoustic signals. The black section of the signals corresponds to the crack propagation time. For freshwater ice, the crack propagation time corresponded to a distinct section of the resistance signal. Despite an extended signal resulting from residual contact between the cracked halves, the crack propagation time corresponds to the steep black section in Figure 2a. As shown in

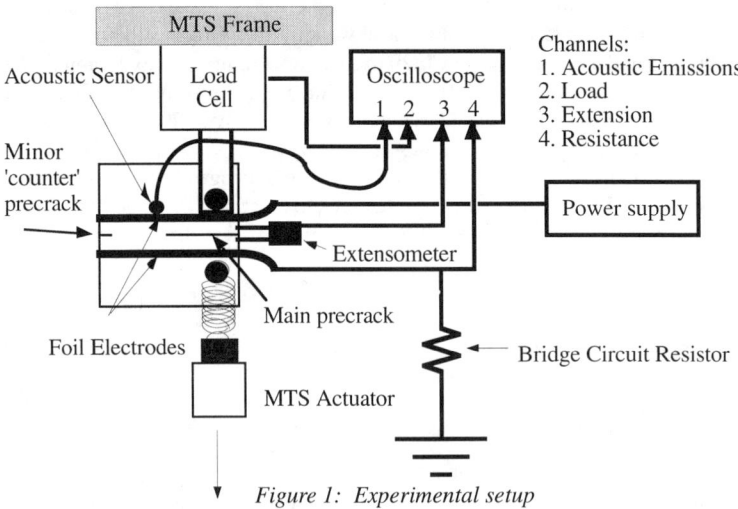

Figure 1: Experimental setup

Figure 2b, for saline ice at -10°C the resistance measurement does not correspond to the crack initiation or crack propagation time. In fact, the voltage clearly does not drop until the crack has propagated fully across the sample. The brine fluid inclusions, which are not severed during fracture, probably act as a conductive bridge across the crack, negating any change in resistance due to crack extension. Thus, the resistance method is not effective for small samples of saline ice. As shown in Figure 2, in saline and freshwater ice, the acoustic signal does not correspond to the crack propagation time, but does effectively register crack initiation.

Temperature Dependence of Crack Velocity

The temperature dependence of the average crack velocity for freshwater and saline ice is shown in Figure 3 along with the approximate brine content in saline ice at the tested temperatures [7]. The overall average crack velocity for freshwater ice was approximately 198 m/s, and remained constant over the temperature range. Freshwater ice exhibited an average crack velocity range of 98 m/s to 383 m/s, with a maximum crack velocity range of 174 m/s to 846 m/s. Saline ice exhibited no significant increase in average crack velocity between –5°C and –20°C. The overall average crack velocity over this temperature range was 86 m/s, with an average crack velocity range of 37 m/s to 156 m/s and a maximum observed crack velocity range of 83 m/s to 341 m/s. At -30°C, the overall average crack velocity increased to 131 m/s, with a range of 75 m/s to 177 m/s. The maximum crack velocity range for saline ice at –30°C was found to be 201 m/s to 429 m/s.

The saline ice crack velocities are significantly larger than those determined with the resistance method [3]. While loading configuration and other experimental factors may influence the velocities observed, the direct observation rendered by high-speed photography of the crack provides concrete evidence of the crack behavior. The velocity does not appreciably change in the -5°C to -20°C temperature range, despite a brine content decrease of half. The increase in velocity occurs below the NaCl eutectic point of -22.9°C. At -30°C, brine content is 1/4 of the brine content at -20°C. These results suggest that the drop in brine content at lower

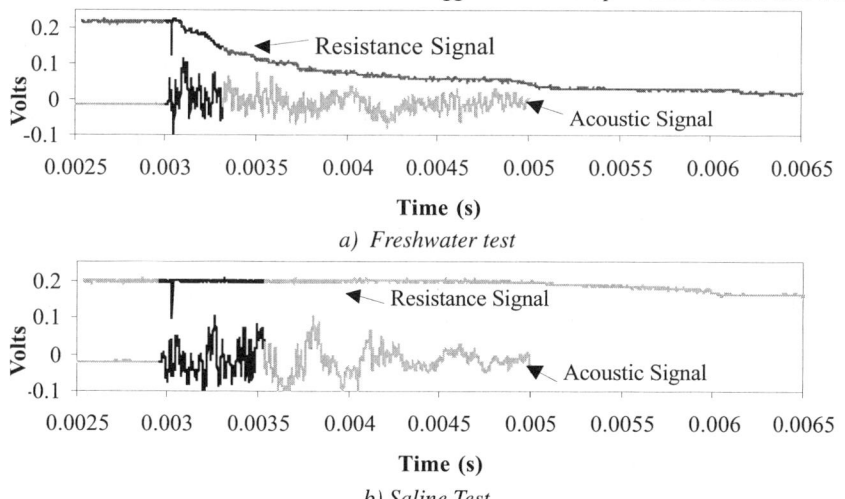

a) Freshwater test

b) Saline Test

Figure 2: Comprehensive testing results for freshwater and saline ice. The black part of the signals corresponds to the crack propagation time as determined by high-speed photography.

temperatures may influence crack propagation, although the exact mechanism is unclear. The fluid inclusions may decelerate crack mouth opening, increase fracture toughness through fluid solidification, or blunt the crack tip less as the size of the fluid inclusion decreases at lower temperatures. Since some brine drains during sample preparation, it is important to note that the theoretical fluid content may differ from the actual fluid content of the samples. Therefore, the approximate brine content is a rough guide to the relative amount of brine at a temperature but cannot be used for an absolute correlation between crack velocity and brine content. The large difference between average crack velocity and maximum crack velocity in both freshwater and saline ice implies that the crack advances at nonuniform velocity,   alternately traveling at high speed for some distance, and then slowing or arresting for a time. The variable crack speed suggests a microstructural influence on crack velocity, where the crack moves quickly before encountering an obstacle and slowing.

Fracture Toughness K' Influence on Crack Velocity

Average crack velocity is plotted against the observed fracture toughness K' in Figures 4 and 5 for freshwater and saline ice, respectively. Despite the experimental scatter in the observed average crack velocity, clear K' and crack velocity relationships are evident in these figures. As shown in Figure 4, freshwater ice exhibits a steep increasing trend with increasing K'. Likewise, as seen in Figure 5, increasing K' results in an increase of average crack velocity for saline ice, although at much slower rate than observed in freshwater ice. Therefore, for both freshwater and saline ice, an increase in the elastic energy stored in the ice results in a faster crack. The large difference in the rate of increase in average crack velocity between freshwater and saline ice suggests that the microstructural features of saline ice may be slowing down cracks due to increased energy requirements in overcoming these obstacles.   If fluid inclusions did slow the crack by resisting crack mouth opening, the K' and crack velocity trend would become more steep for -30°C tests where brine content diminishes. This trend is not evident in these results, so an active role of fluid inclusions in crack retardation cannot be concluded.

Observed Crack Slowing Mechanisms

Nonuniform crack tip advance and crack reorientation were identified as possible contributors to the observed crack velocity variability in freshwater ice. First, an example of nonuniform crack tip advance, halting all significant crack extension for 100 μs, is shown in Figure 6a. The crack front has split into a fork-like shape at the crack tip, causing the crack to arrest. Second, an

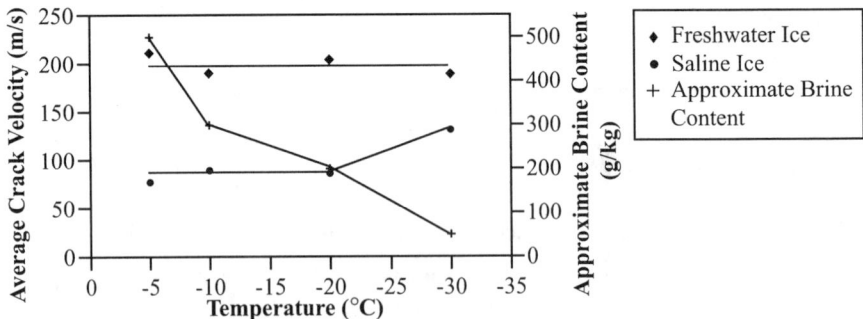

Figure 3: Temperature dependence of crack velocity in freshwater and saline ice and approximate brine content at tested temperatures [7]

example of crack reorientation impeding perceptible crack extension for 100 µs is shown in Figure 6b. The change in the thickness of the crack in the image between the section labeled 'first orientation' and 'second orientation' reflects tilting of the crack within the sample. This mechanism most likely results from the columnar structure of the ice, which may vary in vertical orientation up to 5° [5]. Disadvantageously aligned crack planes may hinder the crack from passing from one grain to another. Variation in grain size and orientation among samples may also account for the experimental scatter in observed velocity as evident in Figure 4.

While the role of fluid inclusions is unclear, crack tip blunting on voids is clearly evident in the high-speed photography of saline ice tests. In Figure 7, the tip of the crack in a saline sample has intersected a large void in the sample. This void effectively blunts the crack tip, preventing any advance for nearly 200 µs. Saline ice has a large number of smaller features such as air bubbles and fluid inclusions which cannot be seen in the high-speed photography. Since crack velocity of saline samples tended to be so low even when no crack-void interactions were evident, the population of these small features may be cumulatively slowing the crack propagation. The variability of distribution of the microstructural features of saline ice in the path of the crack may account for average crack velocity scatter relationship shown in Figure 5.

CONCLUSION

The microstructure greatly influences the speed at which crack propagation occurs. To advance

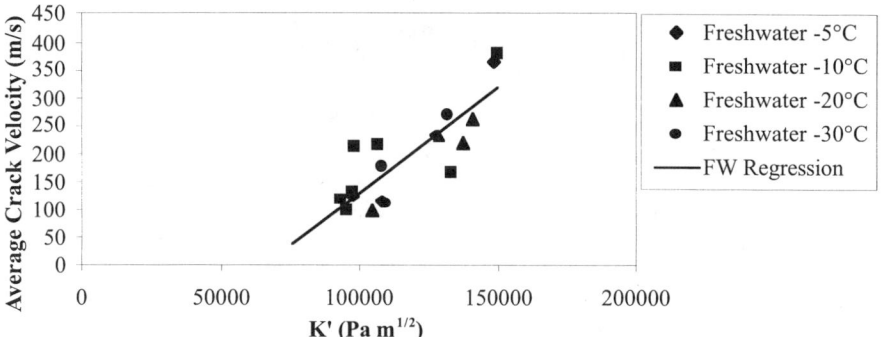

Figure 4: Crack velocity dependence on K' for freshwater ice

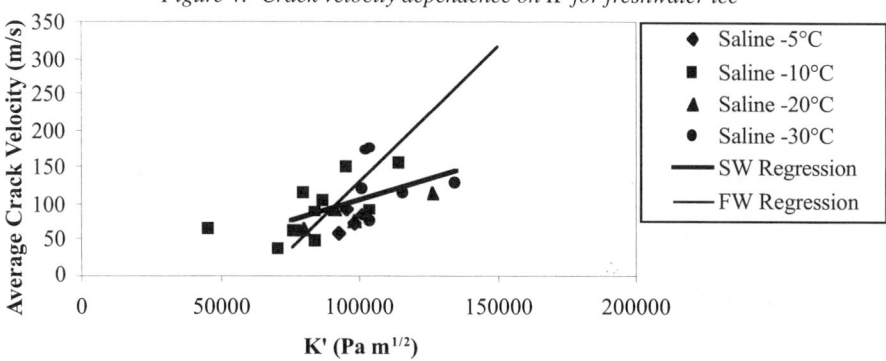

Figure 5: Crack velocity dependence on K' for saline ice. Freshwater trend is shown for comparison.

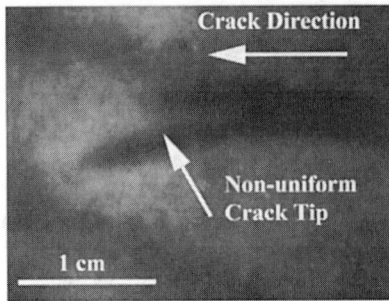 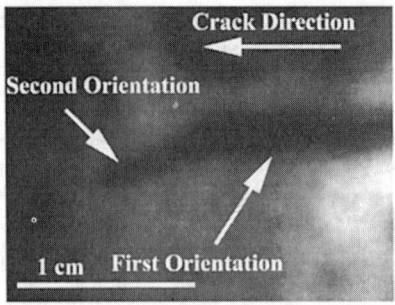

a) Nonuniform crack tip   b) Crack reorientation

Figure 6: Crack features reducing crack speed in freshwater ice

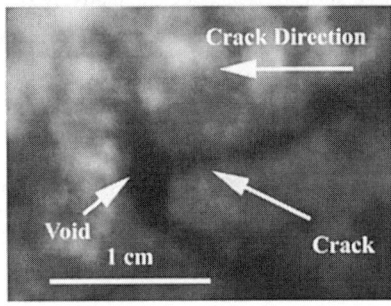

Figure 7: Crack tip blunting on a microstructural feature in saline ice

the understanding of the role of the microstructure and of the fluid inclusions in slowing the crack, research to isolate unproven retarding mechanisms should be undertaken. For example, freshwater ice with air bubbles or artificial fluid inclusions will allow small scale crack tip blunting and the role of fluid inclusions to be isolated. Moreover, observation of crack propagation with regard to specific crystallographic orientation and void distribution could account for the scatter observed in the observed average crack velocity. Once these mechanisms are elucidated, a better theoretical model of the fracture of saline ice can be formulated.

ACKNOWLEDGMENTS

The Office of Naval Research and National Science Foundation supported this research.

REFERENCES

1. A. Sato and G. Wakahama, in *Physics and Chemistry of Ice.*, edited by N. Maeno and T. Hondoh (Hokkaido Univ. Press, Tokyo, 1992) pp. 476-480
2. M. Arakawa, N. Maeno, M. Higa, J. Geophys, Res. **100** (E4), 7539 (1995)
3. V.F. Petrenko and O. Gluschenkov, J. Geophys. Res.-Sol. Ea. **101** (B5), 11541 (1996)
4. B.L. Parsons, J. B. Snellen and B. Hill. Cold Reg. Sci. Technol. **13** (3), 233 (1987)
5. N.P. Cannon, E.M. Schulson, T.R. Smith, H.J. Frost, Acta Metall. Mater. **38** (10), 1955 (1990)
6. G. A. Kuehn, R. W. Lee, Q. A. Nixon, E. M. Schulson, J. Offshore. Mech. Arctic Eng. **112**, 357, (1990)
7. A. Assur, in *Arctic Sea Ice*, U.S. Nat. Acad. Sci., Pub 598, pp. 106-138

VOID SHAPE AND DISTRIBUTION EFFECTS ON COALESCENCE IN ELASTIC-PLASTIC SOLIDS

T. PARDOEN*, J.W. HUTCHINSON
Division of Engineering and Applied Sciences, Harvard University, Pierce Hall, Cambridge, MA 02138, U.S.
*also Université catholique de Louvain, Département des Sciences des Matériaux et Procédés, PCIM, Bâtiment Réaumur, 2 Place Sainte Barbe, 1400 Louvain-la-Neuve, Belgium.

ABSTRACT

An extended Gurson model incorporating the effects of the void shape and distribution on the growth and coalescence is proposed. The emphasis is placed on void coalescence, which is modeled as a transition from diffuse plasticity around the void to transverse localized plastic yielding in the intervoid ligament. Selected results showing the importance of correctly accounting for the void coalescence stage, as well as for the void shape and distribution effects, are presented and discussed.

INTRODUCTION

Recent efforts in the development of computational models incorporating the void growth process has given rise to robust predictive methods for crack propagation in ductile solids, e.g. [1,2,3,4,5]. Most of these works employed the constitutive model initially proposed by Gurson [6], improved by Tvergaard [7], and finally extended by Needleman and Tvergaard [8]. Although good agreement with a range of experiments and void cells computations has been observed, the model as it currently stands still suffers from significant limitations:

- The transfer of experimental data obtained from non-cracked specimens for the modeling of cracked structures, and vice-versa, is not yet successful. In order to quantitatively reproduce experimental J_R curves, parameters of the model must be identified by fitting to test data taken under high stress triaxiality conditions such as from a cracked specimen (e.g. [5]). Many problems of ductile fracture in non-cracked structures occur at low to intermediate stress triaxiality, e.g. during metal forming or in structures containing notches.
- In the context of the model as it now stands, non-spherical voids can only be accounted for in an ad hoc manner by introduction of an effective porosity.
- The phenomenological criteria currently employed to signal the onset of coalescence are limited to a restricted range of conditions, which are not easily measured experimentally.

These limitations, and others, are thought to arise partly because the void coalescence stage of the ductile fracture process is not properly modeled. The objective of the present paper is to describe a realistic void coalescence model which would represent a step toward attainment of a complete model for failure due to the ductile failure mechanism of void nucleation, growth and coalescence. Selected results on void coalescence predictions, which are not captured by simpler model like critical porosity models, will be examined in order to motivate the necessity for such an enhanced model. The results presented here are all in close agreement with more exact void cell predictions [9].

EXTENDED MODEL FOR THE GROWTH AND COALESCENCE OF VOIDS

We have borrowed heavily from two contributions in the literature, and have integrated them into the enhanced model. The first contribution is the model of Gologanu-Leblond-Devaux [10], extending the Gurson model to void shape effects. Indeed, it will be shown that void shape effects must be accounted for in order to correctly predict the ductility at small stress

327

triaxiality. The second is the approach of Thomason [11] for the onset of void coalescence. Each of these has been extended heuristically to account for strain hardening. In addition, a micromechanically-based simple constitutive model for the void coalescence stage is proposed to supplement the criterion for the onset of void coalescence. Only axisymmetric stress states are considered in the present work and the solid is made of a periodic distribution of the cylindrical representative volume element (RVE) defined on Fig. 1.

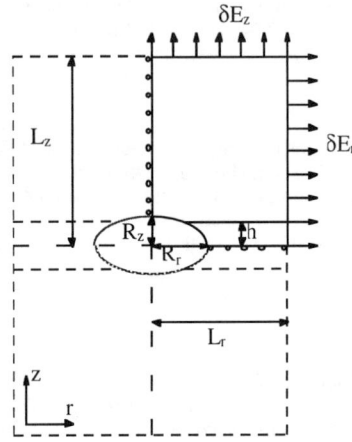

Fig. 1. Representative volume element, with the geometric parameters, symmetry lines, and boundary conditions.

Void Growth Model

The extension of the Gurson model due to Gologanu *et al.* [10], which has been adopted here to describe behavior prior to void coalescence, gives a constitutive relation for a porous elastoplastic material containing (axisymmetric) spheroidal voids. This particular model, extended for strain-hardening, contains as state variables: the components of the mesoscopic stress tensor, Σ, the porosity, f, the void aspect ratio, S, and an average yield stress for the matrix material, σ_m. The void aspect ratio is defined by $S = \ln(W)$ while $W = R_z/R_r$. The functional form of model prior to coalescence is:

$$\Phi \equiv \Phi\left(\Sigma, f, S, \sigma_m\right) = 0 , \tag{1}$$

$$\dot{f} = \left(1 - f\right)\dot{E}_{kk}^p , \tag{2}$$

$$\dot{S} \equiv \dot{S}\left(f, S, T\right) , \tag{3}$$

$$\sigma_m \dot{\varepsilon}_m^p \left(1 - f\right) = \Sigma_{ij} \dot{E}_{ij}^p , \tag{4}$$

$$\sigma_m \equiv \sigma_m\left(\varepsilon_e\right) , \tag{5}$$

$$\dot{E}_{ij}^p = \gamma \frac{d\Phi}{d\Sigma_{ij}} , \tag{6}$$

where Φ is the flow potential; E^p is the mesoscopic plastic strain tensor; (2) and (3) are the evolution laws for f and S, respectively; (4) is the Gurson [6] energy balance for the plastic

328

work allowing computation of σ_m using the effective stress-strain curve for the parent material (5); and (6) is the flow rule. The structure of the original Gurson model has been retained. The expressions for the functions such as Φ and the evolution of S are given in [9,10].

Void Coalescence Model

A criterion for the onset of void coalescence. Axisymmetric void cell computations [9,12] have shown that void coalescence consists in the localization of plastic deformation in the ligament between the voids, which, experimentally, gives rise to a flat dimpled fracture surface. (Void coalescence by shear localization is also observed in other states of stress [7], but, up to now, our model is limited to tensile localization.) Thomason [11] has studied the transition to localization for elastic-perfectly plastic solids by looking at artificially constrained localized solutions giving the load as a function of the void cell geometry. The relation between the overall tensile stress Σ_z/σ_0 and the overall strain E_z based on the full cell length is sketched qualitatively in Fig. 2. At low overall strain (small porosity), Σ_z/σ_0 required for localized yielding is far greater than the actual value from the cell. However, the actual solution peaks and falls (with the cell still deforming in a diffuse manner) until localization sets in, and then the actual solution merges with the artificially constrained localized solution which is significantly affected by the growth of the void. This is the transition point, and from this point on, the solution is localized within the ligament. For axisymmetric geometries, Thomason has proposed that the average normal stress acting on the cell at the onset of localization occurs when Σ_z attains Σ_z^{loc} where

$$\frac{\Sigma_z^{loc}}{\sigma_0} = \left[1 - \left(\frac{R_r}{L_r} \right)^2 \right] \left[\alpha \left(\frac{R_z}{L_r - R_r} \right)^{-2} + \beta \left(\frac{R_r}{L_r} \right)^{-\frac{1}{2}} \right], \tag{7}$$

where $\alpha = 0.1$ and $\beta = 1.2$. By comparing this expression with our numerical results for strain hardening materials, we also find that this expression provides an accurate estimate for the onset of localization within the cells, provided that σ_0 is replaced by an appropriate effective flow stress for the matrix, σ_m (see also [13]), and α and β incorporate a dependence on the strain hardening exponent n. The effective matrix stress, σ_m, is obtained using (4) and (5).

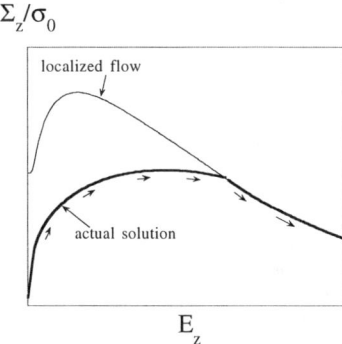

Σ_z/σ_0

localized flow

actual solution

E_z

Fig. 2. Qualitative sketch of the axial stress vs axial strain curves predicted by the constrained localized solution and by the full cell solution; the transition to localization sets in when the solution for diffuse plasticity merges with the solution for a localized plastic flow.

Thus, with attention confined to cases where Σ_z is the maximum principal stress, localization is assumed to set in when $\Sigma_z = \Sigma_z^{loc}$ where

$$\frac{\Sigma_z^{loc}}{\sigma_m} = \left[1-\left(\frac{R_r}{L_r}\right)^2\right]\left[\alpha(n)\left(\frac{R_z}{L_r-R_r}\right)^{-2}+\beta(n)\left(\frac{R_r}{L_r}\right)^{-1/2}\right]. \tag{8}$$

A fitting procedure performed on a large number of void cell results has revealed that the coefficient β is almost constant equal to 1.24 while

$$\alpha(n) = 0.1 + 0.217n + 4.83n^2 \quad (0 \le n \le 0.3), \tag{9}$$

With relation (7), a new geometrical variable related to the void spacing has entered the model. For the sake of simplicity in the formulation of the model, we have chosen to use $A = \ln(\lambda) = \ln(L_z/L_r)$. The model thus depends on all the geometric characteristics of the representative void cell: f, A (or λ), S (or W). A physical length is only required when dealing with large strain gradients. In [9], this model has proved to predict the onset of coalescence with a very high degree of accuracy for porosities ranging between 10^{-2} and 10^{-4}, stress triaxialities between 1/3 and 5, void shapes W between 1/6 and 6, and void distribution λ between 1/2 to 16.

A model for the post-localization regime. Relation (9) still pertains after the onset of coalescence and Σ_z^{loc} is replaced by Σ_z, assuming the voids do not depart significantly from a spheroidal shape. The additional equations for the evolution of the state variables during the post-localization stage are obtained under the approximation that elasticity, as well as any reversed plasticity, is neglected. In agreement with the void cell results, the half-height of the localization zone is approximated as R_z (i.e. $h = R_z$, see Fig. 1). It also follows that $R_r = L_r$. Plastic incompressibility still implies (2) for the evolution of f and the evolution of A is also elementary: $\dot{A} = \dot{E}_z$. The evolution of S can be determined by differentiating $\ln(R_z/R_r)$.

$$\dot{S} = \frac{3}{2}\left(\left(\frac{2\exp(2(A-S))}{3f}\right)^{1/3}-\frac{1}{3f}\right)\dot{E}_z \tag{10}$$

In order to evaluate the average yield stress σ_m for the material in the localized band, the average strain rate $\dot{\varepsilon}_e^{loc}$ is needed. This is obtained from the evolution of the localized band height as

$$\dot{\varepsilon}_e^{loc} = \frac{\dot{h}}{h} = \frac{\dot{R}_z}{R_z} = \left(\frac{2\exp(2(A-S))}{3f}\right)^{1/3}\dot{E}_z. \tag{11}$$

RESULTS

Void shape effect on coalescence at small stress triaxiality. Fig. 3 shows the variation of the axial overall stress as a function of the overall axial strain for three different small stress triaxialities, $T = \Sigma_h/\Sigma_e = 1/3$, 2/3 and 1 and three different initial void shapes, $W_0 = 1/6$, 1, 6. Void coalescence is detected by a sudden change of slope in the curves. At $T = 1/3$, a transition to a localized mode of yielding in the ligament is predicted to develop for the oblate ($W_0 = 1/6$) void and not for the two other void shapes. For initially spherical or prolate voids, the porosity increases slightly at the beginning of the straining and then saturates to a maximum value for further straining. The void shape effect on the ductility is also very significant at $T = 2/3$ and 1. For larger stress triaxiality ($T > 2$), ductility does not depend much on the shape.

Void distribution effects on void coalescence. The effect of the initial cell aspect ratio λ_0 on void coalescence is depicted in Fig. 4 which presents Σ_e vs E_e for $n = 0.1$, $W_0 = 1$, $T = 1$, and the void spacing, L_{r0}/R_{r0}, fixed at 3.22. The true stress - true strain curve of the matrix material is also plotted ($f_0 = 0$). The peak stresses converge to a well-defined point on the curve corresponding to $f_0 = 0$ as λ_0 increases. The limit, $\lambda_0 \to \infty$, corresponds to a single plane of voids in an infinite solid. For $\lambda_0 = 16$, the porosity is so small that there is nearly no departure from the curve $f_0 = 0$ prior to localization. The transition to a uniaxial straining mode is observed for all values of λ_0. For large λ_0, the onset of void coalescence coincides with the peak stress, which, consequently, is due to the onset of the void coalescence localization process and not due to the competition between the hardening of the matrix and the softening due to void growth. The slope of the curve after the onset of void coalescence increases with λ_0 as a result of an increasingly larger zone of elastic unloading.

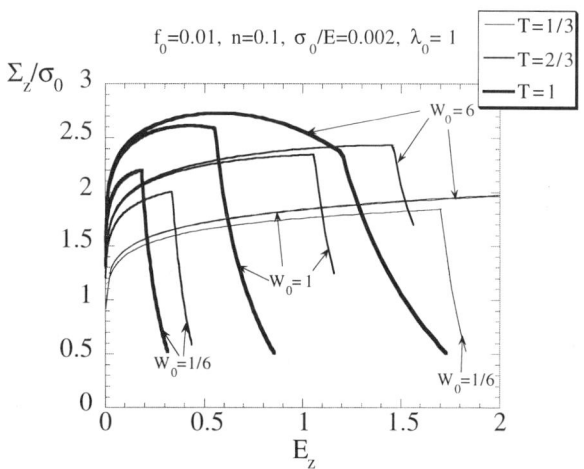

Fig. 3. Predictions obtained with the void growth - void coalescence model for $f_0 = 10^{-2}$, $\lambda_0 = 1$, $\sigma_0/E = 0.002$, $n = 0.1$ and $W_0 = 1/6, 1, 6$, at $T = 1/3, 2/3, 1$; overall axial stress vs overall axial strain.

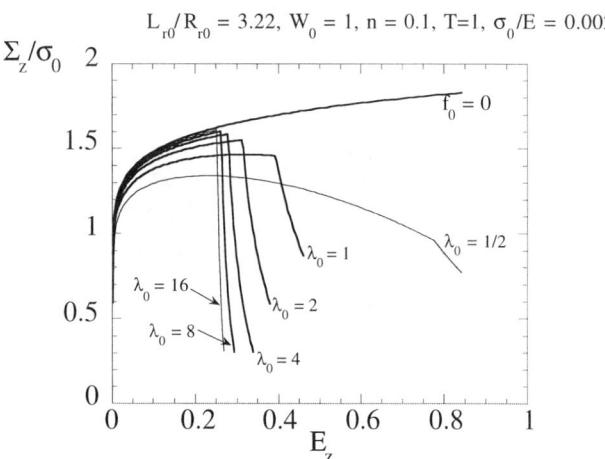

Fig. 4. Σ_z vs E_z curves for $n = 0.1$, $W_0 = 1$, a constant L_{r0}/R_{r0} ratio equal to 3.22, and $T = 1$ (a) or $T = 3$ (b), showing the effect of the cell aspect ratio.

331

CONCLUDING REMARKS

The new model only depends on the initial values of the state variable and thus avoids the use of critical porosities (for the onset of coalescence and for final separation). The two additional microstructural characteristics of the new model, S_0 and λ_0, can be obtained from the same metallographic analysis performed to ascertain f_0 and L_0. At large stress triaxiality (not discussed here), the work spent during the coalescence stage can be larger than the work spent during the void growth stage. This observation gives another motivation for the development of a constitutive model valid after the onset of localization. The comparison with the void cell simulations in ref. [9] has established that the full void growth/coalescence model is able to quantitatively account for variations of all the characteristic parameters of the representative volume element of Fig. 1: porosity, void shape, cell aspect ratio, stress triaxiality, for a wide range of matrix flow behavior. The criterion for the onset of coalescence has been shown to be very accurate for most of the cases analysed in this work. Most importantly, behavior at low and large stress triaxiality are adequately encompassed by the same model.

ACKNOWLEDGMENTS

T.P. acknowledges a fellowship from the Fonds National de la Recherche Scientifique (F.N.R.S.) Belgium, and postdoctoral fellowships from the Belgian American Educational Foundation Inc. (B.A.E.F.) and from the Université catholique de Louvain (U.C.L.). The work of JWH was supported in part by the NSF grant CMS-9634632 and in part by the Division of Engineering and Applied Sciences, Harvard University.

REFERENCES

1. Mudry F., di Rienzo F., and Pineau A., in *Non-Linear Fracture Mechanics: Volume II - Elastic-Plastic Fracture*, ASTM STP 995, edited by Landes J.D., Saxena A. and Merkle J.G. (American Society for Testing and Materials, Philadelphia, 1989) pp. 24-39.
2. Xia L., Shih C.F., and Hutchinson J.W., J. Mech. Phys. Solids **43**, 389-413 (1995).
3. Brocks W., Klingbeil D., Kunecke G., and Sun D.-Z., in *Constraint Effects in Fracture Theory and Applications: Second Volume*, ASTM STP 1244, edited by Kirk M. and Bakker A. (American Society for Testing and Materials, Philadelphia, 1995) pp. 232-252.
4. Ruggieri C., Panontin T.L., and Dodds R.H. Jr., Int. J. Fract. **82**, 67-95 (1996).
5. Gao X., Faleskog J., and Shih C.F., Int. J. Fract. **89**, 374-386, (1998).
6. Gurson A.L., J. Engng. Mater. Tech. **99**, 2-15 (1977).
7. Tvergaard V., Int J. Fract. **17**, 389-407 (1981).
8. Needleman A. and Tvergaard V., J. Mech. Phys. Solids **32**, 461-490 (1984).
9. Pardoen T. and Hutchinson J.W., Harvard Report MECH 356, 1999.
10. Gologanu M., Leblond J.-B., Perrin G., and Devaux J., in *Continuum Micromechanics*, edited by P. Suquet (Springer-Verlag, 1995).
11. Thomason P.F., *Ductile Fracture of Metals* (Pergamon Press, Oxford, 1990).
12. Koplik J. and Needleman A., Int. J. Solids Struct. **24**, 835-853 (1988) .
13. Zhang Z. L. and Niemi E., Engng. Fract. Mech. **48**, 529-540 (1994).

ATOMISTIC MODELING OF VOID GROWTH AND COALESCENCE IN Ni+H

B.P. SOMERDAY*, P.D. PATTILLO II**, M.F. HORSTEMEYER*, M.I. BASKES***
*Sandia National Laboratories, PO Box 969, MS 9402, Livermore, CA, 94550, USA
**Dept of Theoretical and Applied Mechanics, Univ of Illinois 61801, USA
***Los Alamos National Laboratory, MST-8, MS G755, Los Alamos, NM, 87545, USA

ABSTRACT

Finite strain rate atomistic simulations were conducted on Ni and Ni+H lattices containing voids to better understand the dislocation-scale mechanisms of void growth and coalescence and how hydrogen affects these damage processes. Void growth is governed by dislocations that nucleate at the void surface to transport mass away from the void as well as dislocations arriving at the void from the lattice exterior to deposit vacancies and accommodate void-surface expansion. Hydrogen can retard void growth when large local hydrogen concentrations impede dislocation nucleation and propagation at the void surface. The formation of hydrogen gas molecules in the void interior does not necessarily aid void growth. Pressure in small voids may be mitigated by the mutual interaction of hydrogen molecules and the interaction of molecules with the void surface.

INTRODUCTION

Ductile metals commonly fracture by the nucleation, growth, and coalescence of voids. Experiments have identified the mechanisms for microvoid nucleation, growth, and coalescence as well as the mechanical and microstructural variables that affect this ductile fracture mode [1-3]. Complementary studies have modeled void damage using analytical and numerical methods [2, 3]. These experimental and modeling efforts were focused primarily at the continuum level. Insight is needed on the dislocation-scale mechanisms of void damage.

Many metals are macroscopically embrittled by hydrogen (i.e., ductility is reduced), but the fracture mode remains microscopically ductile [4]. Single-crystal Ni and polycrystalline Ni containing a high volume fraction of dispersoids can be embrittled by hydrogen, but fracture progresses by void damage independent of hydrogen content [5-7]. Although hydrogen can accelerate void damage, it is uncertain how each stage of the process is affected.

The objective of this research is to better understand the dislocation-scale mechanisms of void growth and coalescence and how these stages of the void-damage process are affected by solute hydrogen. Lattices of Ni containing cylindrical voids are numerically modeled using the Embedded-Atom Method (EAM) [8, 9]. The growth and coalescence of voids are assessed as a function of applied biaxial displacement and interstitial hydrogen content using molecular dynamics simulations.

COMPUTATIONAL PROCEDURES

Modeling simulations were performed on Ni lattices that consisted of 14,000 atoms. Each cubic lattice was bounded by two sets of free surfaces that were parallel to the (100) and (010) planes. These finite surfaces had dimensions of 10.4 nm by 10.4 nm (60 lattice planes in each direction). The remaining lattice boundary was periodic with a length of 1.2 nm (8 lattice planes). As-generated lattices were constructed with one or two voids by removing atoms in a cylindrical region parallel to the periodic dimension. Single voids (1.4 nm initial diameter) were centered in the (001) planes. Double voids (1.0 nm initial diameter) were aligned in the [010] direction with a separation of one void diameter and were centered along the [100] direction. The void diameters were chosen to produce initial void area fractions (area of circular cross-

Mat. Res. Soc. Symp. Proc. Vol. 578 © 2000 Materials Research Society

section/area of (001) plane, ϕ) that were equal for the one-void and two-void configurations ($\phi=0.015$).

Hydrogen was added to the Ni lattices containing voids using a Monte Carlo (MC) method [8]. The creation of hydrogen was restricted to the region between the lattice external surfaces so that all hydrogen was contained in the lattice or at void surfaces. Since the MC method minimizes the free energy of the system, this approach ensured that hydrogen would preferentially occupy sites at the void surfaces. Two initial hydrogen concentrations were selected: 1 at% H or 10 at% H. These concentrations were achieved by selecting hydrogen chemical potentials of **–2.50 eV** and –2.1 eV, respectively, and conducting the MC simulations at 300K. These chemical potentials produce H concentrations of less than 1×10^{-4} **at%** and ~10 at% in defect-free Ni. Equilibrium was established by running the MC simulations for 3×10^{6} steps.

Void growth and coalescence were induced in Ni and Ni+H lattices by applying biaxial displacements normal to the (100) and (010) planes at the external surfaces. Molecular dynamics simulations were used to apply displacements that were uniform along the planes at a rate of 0.1 nm/s (strain rates of 1×10^{9} s^{-1} based on original lattice dimensions). The simulations were conducted at 300K for total strains up to 10%. The total void area projected through the periodic length was calculated as a function of strain for the Ni and Ni+H cases. The area occupied by the void was assessed by imposing a grid of points separated by 0.05 nm in the [100] and [010] directions in the central region of each lattice. A circular area 0.3 nm in radius was searched around each grid point for Ni atoms. The void area determined from the number of grid points at the void interior was added to the approximately 0.3 nm-wide strip of area around the void perimeter to calculate the total projected area.

Development of the EAM potentials used in this study is described in Refs. [10-12]. The functions were derived from the experimental values of solubility and diffusivity. The elastic constants predicted from the EAM functions agree with experimental results for Ni.

RESULTS AND DISCUSSION

The evolution of void growth and coalescence in Ni containing one or two voids is illustrated in Figs. 1 and 2. The lattices in Figs. 1 and 2 show changes in the void dimensions after 3%, 6%, and 10% biaxial strain. The edges of the lattices are the finite boundaries parallel to (100) and (010) planes. The views parallel to the cylindrical void axes are projections through the 8 lattice planes in the periodic length.

The voids in Figs. 1 and 2 grow and coalesce *via* the nucleation and subsequent motion of dislocations. Dislocations initially nucleate from the void surface and propagate toward the lattice exterior. This mass transport away from the voids causes the voids to grow. After about

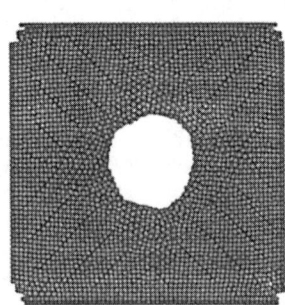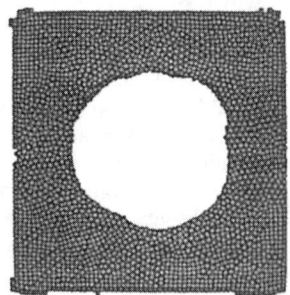

Fig. 1 Effect of biaxial strain on growth of a single void in Ni lattice: 3% strain, 6% strain, and 10% strain from left to right.

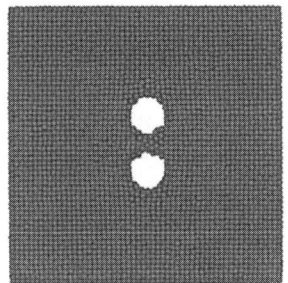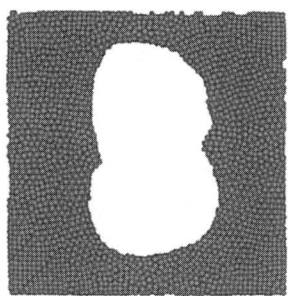

Fig. 2 Effect of biaxial strain on growth of adjacent voids in Ni lattice: 3% strain, 6% strain, and 10% strain from left to right.

4% strain, dislocations begin nucleating from the external surfaces of the lattice. This process starts at the lattice corners and is best illustrated by the lattice subjected to 6% strain in Fig. 1. Dislocations arriving at the void surface from the lattice exterior promote void growth by depositing vacancies and can accommodate the expanding surface area of the void. Thus, two sets of dislocations contribute to void growth. Voids in real materials could grow by these same dislocation mechanisms, where microstructural features such as grain boundaries serve as dislocation sources instead of the crystal free surface.

The location of two voids in close proximity to one another (1 void diameter separation) does not affect void growth under biaxial strain. Figure 3 shows plots of total void area fraction *vs* strain for the one-void and two-void configurations from Figs. 1 and 2. Prior to impingement of the two voids in Fig. 2 between 4% to 5% strain, the plots in Fig. 3 are identical. This low-strain trend demonstrates that the growth of a void is unaffected by a neighboring void, at least under biaxial strain. At higher strains, the single voids growing in the lattices of Figs. 1 and 2 have different shapes, but Fig. 3 shows that the growth rates are essentially the same. The abrupt slope-change of the plots in Fig. 3 near 4% strain is concurrent with the nucleation and propagation of dislocations from the external lattice surfaces. This suggests that void growth accelerates once dislocations arrive from the external-surface sources. Voids in real materials may not experience this abrupt change in growth rate, since the distances between voids and dislocation sources likely exceed those depicted in Figs. 1 and 2.

Hydrogen alters the evolution of void growth and coalescence in the one-void and two-

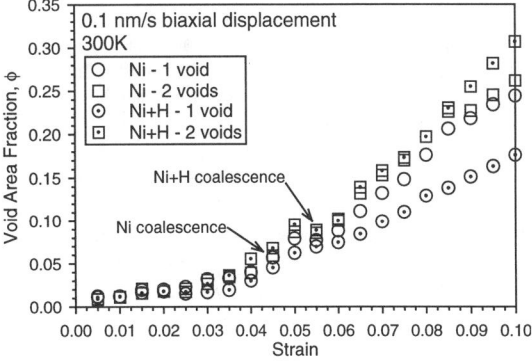

Fig. 3 Void area fraction *vs* strain for Ni and Ni+1 at% H lattices containing one or two voids.

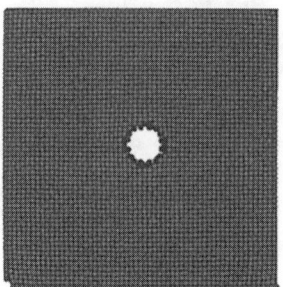

Fig. 4 Effect of biaxial strain on growth of a single void in Ni+1 at% H lattice: 3% strain, 6% strain, and 10% strain from left to right.

void Ni lattices. Figure 4 shows one-void lattices of Ni+1 at% H at 3%, 6%, and 10% biaxial strain. (The lighter-shaded atoms are Ni while the darker-shaded atoms are H.) Two features are notable in Fig. 4. First, hydrogen enters the void interior after 4% strain. This hydrogen forms molecules and remains as monatomic H (e.g., 5 hydrogen molecules and 6 monatomic H atoms are contained in the void of Fig. 4). Second, the void in Fig. 4 forms a faceted shape after 5% strain, which contrasts to the circular void in the H-free lattice of Fig. 1. The void-surface facets appear to be parallel to {110} planes. For the two-void lattice of Ni+1 at% H, hydrogen also enters the void interiors after 4% strain, but the surfaces of the growing voids do not become faceted. The voids impinge at 6% strain, which exceeds the strain for coalescence in the H-free Ni lattice. Dislocations govern void growth in the Ni+H lattices in the manner described for the H-free lattices.

Quantitative characterization of void growth in Ni+1 at% H shows that one-void and two-void lattices are affected by hydrogen differently. Figure 3 shows plots of total void area fraction *vs* biaxial strain for Ni+1 at% H containing one void and two voids. The void growth rate in the one-void Ni+1 at% H lattice is lower compared to the growth rate in the one-void H-free lattice. The void growth rates are similar for the two-void lattices until strains greater than 9%, where the growth rate is higher in the Ni+1 at% H lattice.

The results in Fig. 3 can be explained based on the void-surface coverage of hydrogen. Conflicting results suggest that H can either impede or facilitate dislocation motion [13, 14]. If H hinders the nucleation and arrival of dislocations at the void surface, then the disparate void growth behaviors of the one-void and two-void Ni+1 at% H lattices could result from different void-surface H concentrations. For equal void area fractions, ϕ, the total void surface area for a one-void configuration is 25% less than for a two-void configuration. The higher void-surface H concentration for the one-void lattice could reduce void growth. The mild acceleration of void growth in the two-void Ni+1 at% H lattice at high strain could result if the role of H reverses at lower surface concentrations and aids dislocation nucleation [15].

One additional factor that could hinder void growth in the one-void Ni+1 at% H lattice is related to the faceted void shape in Fig. 4. Void growth in this manner is analogous to precipitate nucleation and growth in solids, where precipitate shape is determined by minimizing energy through the interplay between surface energy and volume strain energy [16]. For example, precipitates with high surface energy tend to form spheres to minimize the amount of surface area. Since H lowers the surface energy of Ni, the void in Fig. 4 likely assumes a shape that minimizes volume strain energy. This restriction on void shape hinders the void growth rate.

The modeling data in Fig. 3 contrasts to experimental observations of H effects on void damage in polycrystal Ni-2 vol% ThO_2. Fractographic evidence for Ni-2 vol% ThO_2 containing

0.03 at% H suggests that void growth is enhanced compared to the H-free system [6, 17]. The proposed mechanism for accelerated void growth in Ni-2 vol% ThO$_2$ involves the pressurization of voids due to the formation of hydrogen gas. The results in Fig. 3 show that void growth is essentially unaffected by H and perhaps even suppressed despite the formation of hydrogen gas molecules in the void interiors (Fig. 4).

The discrepancy between void growth behaviors in modeling *vs* experiment suggests that high hydrogen gas pressures do not develop during the simulated void growth (e.g., Fig. 4). The gas pressure calculated assuming ideal gas behavior for the hydrogen molecules in Fig. 4 at 4% strain is 3.4 MPa, which is similar to the critical resolved shear stress of Ni (3.2 MPa [18]). This pressure should be high enough to enhance void growth but may not develop because of mutual interactions between the gas molecules and interactions between the gas molecules and void-surface atoms. The cut-off distance for these interactions is about 0.5 nm [11]. This cut-off distance is equal to about half the void radius at 4% strain, which supports the idea that hydrogen molecules interact with the void surface and with one another.

Results in Fig. 3 were augmented by conducting simulations on one-void and two-void Ni lattices containing 10 at% H for biaxial strains up to 5%. Evolution of void growth and coalescence proceeded similar to the lower-H cases (e.g., Fig. 4). Hydrogen entered the voids at lower strain and the ratio of hydrogen molecules to monatomic H atoms in the voids was considerably higher (e.g., 52 hydrogen molecules *vs* 2 monatomic H atoms for the one-void case) compared to the lower-H simulations.

Calculations of void area fraction as a function of strain in Ni+10 at% H demonstrate that higher H concentrations affect void growth. Void area fractions *vs* biaxial strain are plotted in Fig. 5 for the one-void and two-void configurations with 1 at% H and 10 at% H. Comparing results from the two-void lattice of Ni+10 at% H to those from the two-void lattice of Ni+1 at% H shows that higher H concentrations reduce void growth at strains above 3%. In contrast, growth rates in the one-void lattice of Ni+10 at% H are elevated compared to the lower-H concentration case at strains above 3%. Reduced void growth in the two-void lattice at higher H concentration is consistent with the effect of increased void-surface H hindering dislocation nucleation and propagation. Elevated void growth in Ni+10 at% H containing one void may result from internal gas pressure. The pressure calculated based on ideal gas behavior is higher for the one-void case (24.8 MPa at 4% strain) compared to pressures in the two-void case (about 17 MPa in each void at 4% strain). It is possible that gas molecules present at high densities in

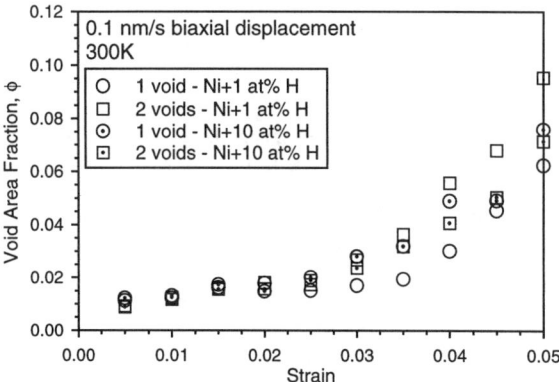

Fig. 5 Void area fraction *vs* strain for Ni+1 at% H and Ni+10 at% H lattices containing one or two voids.

small voids can overcome mutual interactions and interactions with the void surface to exert a pressure that counteracts the effect of surface H on dislocations.

CONCLUSIONS
Finite strain rate atomistic simulations on Ni and Ni+H lattices with one or two voids show that:
1. Void growth is governed by two sets of dislocations. Dislocations that nucleate at the void surface transport mass away from the void and dislocations arriving at the void from the lattice exterior deposit vacancies and can accommodate void-surface expansion.
2. Hydrogen can retard void growth. High void-surface hydrogen concentrations may impede local dislocation nucleation and propagation.
3. The formation of hydrogen gas molecules in the void interior does not necessarily aid void growth. Pressure in small voids may be mitigated by the mutual interaction of hydrogen molecules and the interaction of molecules with the void surface.

ACKNOWLEDGMENTS
This work was supported by the U.S. Department of Energy under contract # DE-AC04-94L85000.

REFERENCES
1. R. H. VanStone, T. B. Cox, J.R. Low, and J. A. Psioda, Int. Met. Rev. **30,** p. 157 (1985).
2. W.M. Garrison and N. R. Moody, J. Phys. Chem. Solids **48**, p. 1035 (1987).
3. H. G. F. Wilsdorf, Mater. Sci. Eng. **59**, p. 1 (1983).
4. A. W. Thompson and I. M. Bernstein, in *Advances in Research on the Strength and Fracture of Materials*, edited by D.M.R. Taplin (Pergamon Press **2A**, New York 1977), pp. 249-254.
5. A. W. Thompson, Met. Trans. **5**, p. 1855 (1974).
6. A. W. Thompson and B. A. Wilcox, Scripta Metall. **6**, p. 689 (1972).
7. G. C. Smith, in *Hydrogen in Metals*, edited by I.M. Bernstein and A.W. Thompson (ASM, Metals Park, OH 1974), pp. 485-513.
8. M. S. Daw, S. M. Foiles, and M. I. Baskes, Mat. Sci. Rep. **9**, p. 251 (1993).
9. M. S. Daw and M. I. Baskes, Phys. Rev. B **29**, p. 6443 (1984).
10. J. E. Angelo, N. R. Moody, and M. I. Baskes, Modelling Simul. Mat. Sci. Eng. **3**, p. 289 (1995).
11. M. I. Baskes, J. E. Angelo, and N. R. Moody, in *Hydrogen Effects in Materials*, edited by A.W. Thompson and N.R. Moody (TMS, Warrendale, PA 1996), pp. 77-90.
12. M. I. Baskes, X. Sha, J. E. Angelo, and N. R. Moody, Modelling Simul. Mat. Sci. Eng. **5**, p. 651 (1997).
13. A. H. Windle and G. C. Smith, Met. Sci. J. **2**, p. 187 (1968).
14. H.K. Birnbaum, I.M. Robertson, P. Sofronis, and D. Teter, in *Corrosion-Deformation Interactions CDI '96*, edited by T. Magnin (The Institute of Materials, London 1997), pp. 172-195.
15. S. P. Lynch, Acta Metall. **36**, p. 2639 (1988).
16. D. A. Porter and K. E. Easterling, *Phase Transformations in Metals and Alloys*, Chapman & Hall, London, 1992, pp. 142-171.
17. A. W. Thompson, in *Effect of Hydrogen on Behavior of Materials*, (Metallurgical Society of AIME, Warrendale, PA 1976), pp. 467-479.
18. T. H. Courtney, *Mechanical Behavior of Materials*. McGraw-Hill, New York, 1990, p. 83.

MICROSTRUCTURALLY INDUCED DUCTILE DEFORMATION MECHANISMS AND GRAIN-BOUNDARY EFFECTS IN POLYCRYSTALLINE AGGREGATES

W. M. ASHMAWI AND M. A. ZIKRY
Department of Mechanical and Aerospace Engineering, North Carolina State University, Raleigh, NC 27695-7910

ABSTRACT

Dislocation-density based multiple-slip constitutive formulations and specialized computational schemes are introduced to account for large-strain ductile deformation modes in polycrystalline aggregates. Furthermore, new kinematically based interfacial grain-boundary regions and formulations are introduced to account for dislocation-density transmission, absorption, and pile-ups that may occur due to grain-boundary misorientations and properties.

INTRODUCTION

Grain-boundary (GB) structure, orientation, and distribution are essential microstructural features that characterize the initiation and evolution of failure modes in crystalline metals, alloys, and intermetallics. Physically-based constitutive descriptions are needed that can account for dominant physical mechanisms that may occur at different physical scales. The primary purpose of this study is the introduction of an inelastic dislocation density-based multiple-slip crystalline constitutive formulation that can be used to obtain a detailed understanding and accurate prediction of interrelated local material mechanisms that control and affect global deformation modes in f.c.c. polycrystalline aggregates with random GB orientations and distributions. In this formulation, the length scale between multiple-slip crystalline formulations and dislocation densities is bridged by coupling evolutionary equations for the mobile and immobile dislocation densities, through the temperature dependent flow stress and slip-rates on each slip system, to a multiple-slip rate-dependent crystal plasticity formulation. The derivation of these evolutionary equations are based on accepted physical relations, and generally account for thermally activated dislocation activities such as generation, interaction, and annihilation that are generally representative of the dislocation structures in cubic crystalline metals (see for example [1]).

Most polycrystalline formulations generally do not account for GB effects such as dislocation-density and slip transmission, blockage, and absorption. These effects could result due to GB orientation, structure, or interfacial stress mismatches (see, for example, Zikry and Kao [2] for a more detailed review). In this study, GB effects are accounted for by the introduction of interfacial regions that are used to track slip and dislocation density transmissions and intersections. The width of these interfacial GB regions is taken as approximately 5% of the grain diameter. This is consistent with the experimental observations and studies of copper polycrystalline aggregates (see for example, Nes [9]). These accurate representations of overall polycrystalline aggregate behavior are needed for the prediction of failure initiation due to GBs, subgrains, and cell-walls.

MULTIPLE-SLIP CRYSTAL PLASTICITY FORMULATION

The crystal plasticity constitutive framework used in this study is based on the formulation developed in Kameda and Zikry [3]. In that formulation, it has been assumed that the deformation gradient can be decomposed into elastic and inelastic components. Starting from the decomposition of the velocity gradient, $V_{i,j}$, into its symmetric and anti-symmetric parts as

$$V_{i,j} = D_{ij} + W_{ij} \qquad (1)$$

339

where D_{ij} is the deformation rate tensor, and W_{ij} is the spin tensor. The total deformation-rate tensor, D_{ij}, and the total spin tensor, W_{ij}, are then each additively decomposed into elastic and plastic components as

$$D_{ij} = D_{ij}^P + D_{ij}^*, \qquad W_{ij} = W_{ij}^P + W_{ij}^*. \qquad (2a\text{-}b)$$

The superscript * denotes the elastic part and the superscript p denotes the plastic part; W_{ij}^* includes the rigid body spin. The inelastic parts are defined in terms of the crystallographic slip-rates as

$$D_{ij}^P = P_{ij}^{(\alpha)} \dot{\gamma}^{(\alpha)}, \qquad W_{ij}^P = \omega_{ij}^{(\alpha)} \dot{\gamma}^{(\alpha)}, \qquad (3a\text{-}b)$$

where α is summed over all slip-systems, and the tensors $P_{ij}^{(\alpha)}$ and $\omega_{ij}^{(\alpha)}$ are defined in terms of the unit normals and the unit slip vectors. For a rate-dependent inelastic formulation, the slip-rates are functions of the resolved shear and reference stresses. The power law

$$\dot{\gamma}^{(\alpha)} = \dot{\gamma}_{ref}^{(\alpha)} \left[\frac{\tau^{(\alpha)}}{\tau_{ref}^{(\alpha)}} \right] \left[\frac{\left| \tau^{(\alpha)} \right|}{\tau_{ref}^{(\alpha)}} \right]^{\frac{1}{m}-1}, \qquad (4)$$

is used here. The reference shear-strain-rate, $\dot{\gamma}_{ref}^{(\alpha)}$, corresponds to a reference shear stress, $\tau_{ref}^{(\alpha)}$. The rate sensitivity parameter, m, is material dependent, and is the ratio of the rate of change of the resolved shear stress on each slip-system to the logarithmic rate of change of the slip-rate on each slip-system. For shear slip-rates smaller than a critical value, the lattice motion is thermally activated. The rate sensitivity parameter is approximately equal to one for slip-rates greater than the critical slip-rate, and this flow is characterized by drag-controlled dislocation motion.

LOCAL DISLOCATION-DENSITY STRUCTURE

To gain a more fundamental understanding of dislocation motion, interaction, and transmission on material failure modes, the crystal plasticity constitutive formulation is coupled to internal variables that account for a local description of the dislocation structure in each crystal. Specifically, we have used the mobile and the immobile dislocation densities as the internal variables in our constitutive formulation. In inelastic deformations of ductile metals, the characteristics of the microstructure are governed by the mechanisms of dislocation production and dynamic recovery. As the material is strained, immobile dislocations are stored in each crystal, and these dislocations act as obstacles for evolving mobile dislocations. Therefore, the immobile and mobile dislocation densities can be coupled, due to the continuous immobilization of mobile dislocations.

The reference stress, on each slip-system, can be given as a function of $\rho_{im}^{(\alpha)}$, the immobile dislocation density. The reference stress that is used here is a modification of widely used classical forms (see for example, Mugharbi [8]) that relate the reference stress to a square-root dependence on the immobile dislocation density as

$$\tau_{ref}^{(\alpha)} = \tau_y^{(\alpha)} + Gb \sum_{\xi=1}^{12} a_\xi \sqrt{\rho_{im}^{(\xi)}} \qquad (5)$$

where G is the shear modulus, b is the magnitude of the Burgers vector, $\tau_y^{(\alpha)}$ is the static yield stress, and the coefficients, a_ξ $(\xi = 1,12)$ are interaction coefficients, and generally have a magnitude of unity.

Now consider a given state for a deformed material, which has a dislocation structure of total dislocation density, $\rho^{(\alpha)}$. This total dislocation density is assumed to be additively decomposed, into a mobile dislocation density, $\rho_m^{(\alpha)}$, and an immobile dislocation density $\rho_{im}^{(\alpha)}$. Furthermore, we have assumed that during an increment of strain, an immobile dislocation density rate is generated, which will be denoted by $\dot{\rho}_{im}^{(\alpha)+}$, and an immobile dislocation density rate is annihilated, which will be denoted by $\dot{\rho}_{im}^{(\alpha)-}$. We also assume that $\dot{\rho}_m^{(\alpha)+}$ corresponds to a generation of mobile dislocation densities, and $\dot{\rho}_m^{(\alpha)-}$ corresponds to an annihilation of mobile dislocation densities. Using these balance laws pertaining to the generation and annihilation of mobile and immobile dislocations (see [2] for a detailed presentation), we have derived the following coupled equations that account for the evolution of mobile and immobile dislocation densities that correspond, in an average sense, to dislocation generation, interaction, trapping, and recovery,

$$\frac{d\rho_m^{(\alpha)}}{dt} = \dot{\gamma}^{(\alpha)}\left(\frac{g_{sour}}{b^2}\left(\frac{\rho_{im}^{(\alpha)}}{\rho_m^{(\alpha)}} \right) - \frac{g_{minter}}{b^2}\exp(-\frac{H}{kT}) - \frac{g_{immob}}{b}\sqrt{\rho}_{im}^{(\alpha)} \right), \qquad (6)$$

$$\frac{d\rho_{im}^{(\alpha)}}{dt} = \dot{\gamma}^{(\alpha)}\left(\frac{g_{minter}}{b^2}\exp(-\frac{H}{kT}) + \frac{g_{immob}}{b}\sqrt{\rho}_{im}^{(\alpha)} - g_{recov}\exp(-\frac{H}{kT})\rho_{im}^{(\alpha)} \right), \qquad (7)$$

where g_{sour} is a coefficient pertaining to an increase in the mobile dislocation density due to dislocation sources, g_{minter} is a coefficient related to the trapping of mobile dislocations due to forest intersections, cross-slip around obstacles, or dislocation interactions, where g_{sour} is a coefficient pertaining to an increase in the mobile dislocation density due to dislocation sources, g_{minter} is a coefficient related to the trapping of mobile dislocations due to forest intersections, cross-slip around obstacles, or dislocation interactions, g_{recov} is a coefficient related to the rearrangement and annihilation of immobile dislocations, g_{immob} is a coefficient related to the immobilization of mobile dislocations, H is the activation enthalpy, k is Boltzmann's constant, and T is the temperature. As these evolutionary equations indicate, the dislocation activities related to recovery and trapping are coupled to thermal activation.

NUMERICAL SCHEME

The total deformation-rate tensor, D_{ij}, and the plastic deformation-rate tensor, D_{ij}^p, are needed to update the stress state of the crystalline material. A brief outline of the numerical method will be presented; for further details see [5]. An implicit finite-element analysis has been used to obtain the total deformation-rate tensor, D_{ij}. The displacements have been obtained by the quasi-Newton BFGS method. Once the displacements are obtained, the deformation tensor can be calculated. To overcome numerical problems associated with incompressible deformations, the $\bar{B}$ method has been used in the calculation of the deformation tensor. In the $\bar{B}$ method, the deformation gradient is decomposed into volumetric and deviatoric parts. The resulting volumetric deformation field eliminates spurious modes that can arise due to incompressible deformations. Once the deformation tensor is obtained from the updated nodal displacement values, the total deformation-rate tensor, D_{ij}, and the total spin tensor, W_{ij}, can be calculated at each load. To solve for the plastic deformation rate tensor, D_{ij}^p, the time derivative of the resolved shear-stress is used together with the objective stress rate, and the assumption that the elastic modulus tensor is isotropic, to obtain the following system of coupled nonlinear differential equations for each active slip-system:

$$\dot{\tau}^{(\alpha)} = 2\mu \left(P_{ij}^{(\alpha)} D_{ij} - P_{ij}^{(\alpha)} \left[\sum_{\beta=1}^{12} P_{ij}^{(\beta)} \dot{\gamma}_{ref}^{(\beta)} \left(\frac{\tau^{(\beta)}}{\tau_{ref}} \right)^{\frac{1}{m}} \right] \right). \tag{8}$$

In this derivation, it has been assumed that the lattice distortion is a function of the elastic spin in all three orthogonal directions. The solution to the system of ordinary differential equations, (8), is numerically difficult, not only due to the nonlinearity of the resolved shear stress, but also because the system of equations is numerically stiff in certain time intervals. This can lead to the growth of numerically propagated error, i.e. instability in the solution of the system of differential equations. This computational scheme is also used to update the evolutionary equations (6-7) for the immobile and mobile dislocation densities.

GRAIN BOUNDARY INTERFACIAL REGIONS

It is clear that GBs play a considerable role in controlling the mechanical and physical properties and response of polycrystalline aggregates, which in combination with other factors influence material flow and fracture. As indicated by Miller [6], most existing models treat GBs as either one-dimensional rigid walls, or only as interfacial quantities which are not accurately representative of GB morphology, structure and interfacial mismatches that may occur due to stress and strain gradients. In this study, the GB region is modeled as an interfacial region with structure and properties that are different from the bulk grain regions. This layer is assumed to be a crystalline region that has a specific orientation for its crystallographic planes. Special kinematic schemes are introduced that account for slip transmission and impedance at the GB region. These schemes are based on the identification and tracking of rotating slip systems in the interfacial region, as strain evolves, such that slip and dislocation-density compatibilities and incompatibilities are used to delineate regions of transmission and pile-ups; for a detailed presentation, see Ashmawi and Zikry [7].

RESULTS AND DISCUSSION

A polycrystalline aggregate was simulated to illustrate the effects of the presence of GB interfacial regions. The material properties that are used here are representative of polycrystalline copper [2]. Grain bulk and GB properties are given in Table (1). The initial mobile and immobile dislocation densities within GB interfacial regions were varied randomly as a function of GB misorientation (for further details, see [7]). Using the method outlined in [5], the saturated immobile dislocation density, $\bar{\rho}_{ims}$, was calculated as 10^{14} m^{-2} and the saturated mobile dislocation density, $\bar{\rho}_{ms}$, was calculated as 4.3×10^{13} m^{-2}. Using these values, the coefficient values and the enthalpy, needed for the evolution of the immobile and mobile density equations (6-7), are calculated as

$$g_{minter} = 5.53, \ g_{recov} = 6.67, \ g_{immob} = 0.0127, \ g_{sour} = 2.76 \times 10^{-5}, \ H/k = 3.289 \times 10^{3} \ ^{0}K. \tag{9}$$

All twelve-slip systems were assumed to be potentially active in each grain and GB region. Random low angle GB orientations were used with misorientations not exceeding $10°$ between adjacent grains. Fifty grains, with an average diameter of 0.1 mm, were used. An axial strain-rate of 10^{-3} /s was obtained by applying a displacement along the y-direction. This results in a plane strain deformation of the aggregate.

	Grain Bulk	GB Interfacial Region
Young's modulus, E	110 GPa	110 GPa
Static yield stress, σ_y	110 MPa	330 MPa
Poisson's ratio, ν	0.30	0.30
Rate sensitivity parameter, m	0.005	0.005
Reference strain rate, $\dot{\gamma}_{ref}$	$0.001 \ s^{-1}$	$0.001 \ s^{-1}$
Critical strain rate, $\dot{\gamma}_{critical}$	$10^4 \ s^{-1}$	$10^4 \ s^{-1}$
Burgers vector, b	$3.0 \times 10^{-10} \ m$	$3.0 \times 10^{-10} \ m$
Reference stress interaction coefficients, a_ξ ($\xi = 1, 12$)	0.50	0.50
Initial immobile dislocation density, $\rho_{imo}^{(\alpha)}$	$10^{10} \ m^{-2}$	Varies as a function of GB orientation, $10^{10} - 10^{12} \ m^{-2}$
Initial mobile dislocation density, $\rho_{mo}^{(\alpha)}$	$10^7 \ m^{-2}$	Varies as a function of GB orientation, $10^5 - 10^7 \ m^{-2}$

Table 1. Properties of grains and GB interfacial regions

The effects of the GB interfacial region on the evolution of the total dislocation density can be clearly seen in the contours shown in Fig. 1a. These contours correspond to a nominal strain of 5%. Dislocation densities, corresponding to slip system $(\bar{1}11)[01\bar{1}]$, which is one of the more active slip systems, have localized and accumulated within some of the grains and at the GB regions. This distribution indicates that slip activity in the neighborhood of these GB regions may result either in slip transmission or blockage. Furthermore, this accumulation will lead to a buildup of normal stresses [3]. If GB effects had been ignored, these dislocation density patterns would not have evolved. It can also be seen (Fig. 1b) that accumulated plastic strains, which are local deformation bands associated with all active slip systems, are concentrated in regions corresponding with high dislocation-density activity.

SUMMARY AND CONCLUSIONS

A multiple-slip crystal plasticity constitutive formulation that is coupled to the temperature dependent evolution of mobile and immobile dislocation densities has been developed for a detailed understanding and prediction of the deformation modes of polycrystalline aggregates with GB effects. The predictive capabilities and accuracy of the constitutive formulation and the specialized finite-element computational scheme have been used to investigate the effects of random GB orientations on material mechanisms in polycrystalline copper. The overall response of polycrystalline aggregates has been shown to be directly related to GB orientation, distribution, and structure. In future investigations, the constitutive formulation and the computational schemes will be used to investigate the effects of dislocation motion, interaction, and transmission on void growth and interaction in crystalline materials separated by high angle CSL and random GBs.

ACKNOWLEDGMENT

This work was partially supported by National Science Foundation grant # CMS-9713762. The computations were performed at the North Carolina Supercomputing Center. The assistance of the staff at NCSC is deeply appreciated.

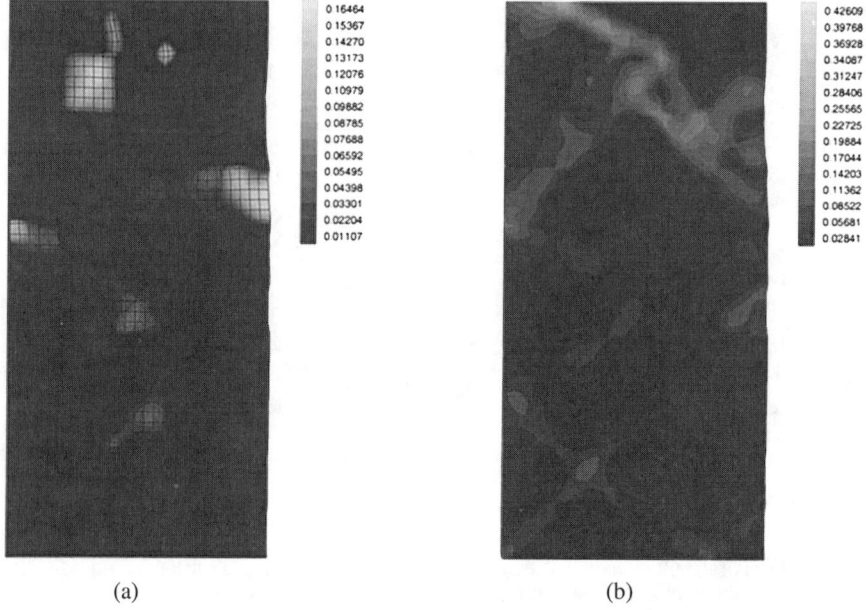

(a)   (b)

Figure 1. (a) Total dislocation densities at a nominal strain of 5% for aggregate with GB interfacial region for slip system $(\bar{1}11)[01\bar{1}]$; (b) Total accumulated plastic strain at a nominal strain of 5%.

REFERENCES

1. Bay, B., Hansen, N., Hughes, D. A. and Kuhlmann-Wilsdorf, D., Acta Metall. Mater. **40**, 205 (1992).
2. Zikry, M. A. and Kao, M., J. Mech. Phys. Solids **44**, 1765 (1996).
3. Kameda, T. and Zikry, M. A., Int. J. Plasticity **14**, 689 (1998).
4. Kubin, L. P. and Estrin, Y., Revue Phys Appl. **23**, 573 (1988).
5. Zikry, M. A., Comput. Struct. **50**, 337 (1994).
6. Miller, G. R., Int. J. Fracture **31**, 143 (1986).
7. Ashmawi, W. M. and Zikry, M. A., submitted for publication (1999).
8. Mughrabi, H., Mater. Sci. and Tech. **85**, 15 (1987).
9. Nes, E., Prog. Mater. Sci. **41**, 129 (1998).

A PARTITIONED-PROBLEM APPROACH TO MICROSTRUCTURAL MODELLING OF A GLASS-CERAMIC

A.C. FISCHER-CRIPPS
CSIRO Division of Telecommunications and Industrial Physics
Bradfield Road, West Lindfield, NSW 2070, Australia

ABSTRACT

The indentation response of a mica-containing glass-ceramic that exhibits yield in an indentation test is interpreted in terms of events occurring on the microstructural scale. This work is unique in that it links the macroscopic and microstructural properties of the specimen material in the indentation process.

INTRODUCTION

Indentation damage is a particularly important consideration in the design and application of brittle materials for structural applications. For ceramic materials with a relatively large grain size and weak grain boundaries, the characteristic Hertzian cone crack normally associated with loading of a brittle material with a spherical indenter is suppressed in favour of a region of sub-surface accumulated damage [1]. Previous work [2] has shown that the nature of the damage depends upon the microstructure. It is desirable, from the point of view of microstructural design, that the microstructural parameters that influence the indentation response be identified and quantitatively assessed. Such an analysis is demonstrated in the present work using a mica-containing glass-ceramic as a model material.

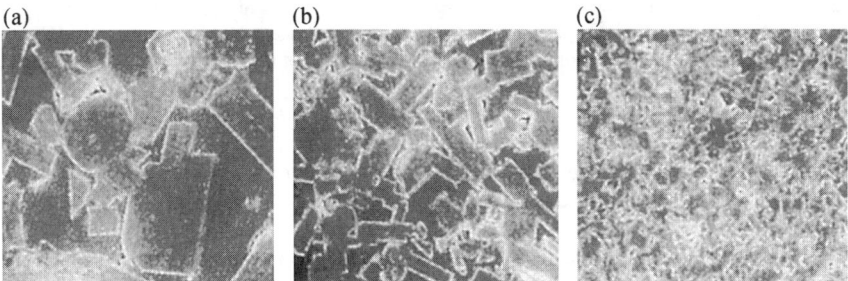

Fig. 1 Scanning electron micrographs showing the microstructure for each glass-ceramic test material. Microstructures for (a) coarse, (b) medium, and (c) fine-grained materials are shown. All surfaces polished and then etched for 10 seconds in HF.

Mica-containing glass-ceramics are known for their easy machinability; yet have a respectable long-crack toughness. As can be seen in Fig. 1, the microstructure consists of circular mica platelets in a glassy matrix and can be modified by simple heat treatment (1120°C, 1060°C 1000°C respectively). Both the average plate diameter and the aspect ratio increase with increasing crystallization temperature. The presence of the mica platelets result in an interconnecting network of weak interfaces within the material. This morphology leads to an indentation response similar to that observed in ductile materials. Identification of

345

microstructural parameters that influence the indentation response of a series of mica-containing glass-ceramics is the aim of the present work. A comparison between finite element models, experimental evidence, and theoretical considerations yields a relationship between macroscopic behaviour and events on the microstructural scale.

MACROSCOPIC BEHAVIOUR

The bonded-interface technique was used to illustrate the nature of sub-surface damage in three glass-ceramic specimens each having a different average grain size. The experimental results shown in Fig. 2 demonstrate that for the material with a relatively large grain size, the familiar Hertzian cone crack normally associated with a brittle material has been suppressed in favour of a region of distributed damaged beneath the indenter. For the relatively fine-grained material, both Hertzian cone cracks and sub-surface accumulated damage are visible.

(a)   (b)   (c)

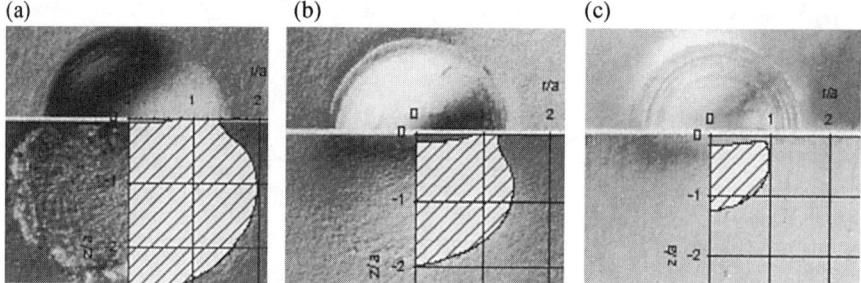

Fig. 2 Optical micrographs of indented bonded-interface specimens showing half-surface (top) and section (bottom) views. Indentations performed with WC sphere of radius R =3.18 mm at load P =2000 N for (a) coarse, (b) medium, and (c) fine-grained glass-ceramic materials. Also shown are finite element results for the size and shape of the damage zone.

MICROSCOPIC BEHAVOUR

Elastic-plastic behaviour in these nominally brittle materials occurs as a result of the relatively large level of shear in the indentation stress field resulting in shear faulting along planes of weakness in the material. Figure 3 shows an electron micrograph of the surface of a sectioned specimen for the coarse-grained specimen in the damage zone. The figure suggests that the damage consists of faults or cracks at or near the boundary between the mica platelets and the glass matrix. Since damage is observed to occur initially beneath the surface corresponding to the point at which shear stresses are a maximum and, as will be shown, the well-developed plastic zone resembles that predicted by finite-element analysis that incorporates a shear-driven failure criterion, it is postulated that such failure occurs as a result of shear in the indentation stress field.

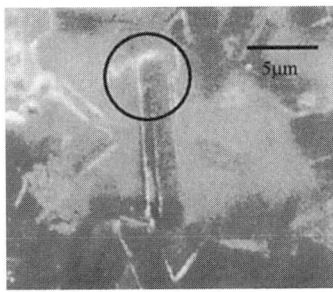

Fig. 3 Electron micrograph of surface of sectioned specimen for coarse-grained material showing detail of the sub-surface accumulated damage near a grain (circled).

THEORETICAL

Experimental evidence suggests that there is a connection between events on the microstructural scale and the macroscopic yield stress. It must be possible, therefore, to express the macroscopic shear-driven yield stress Y as a function of microstructural parameters. Let us assume that plastic behaviour occurs as sliding within the specimen material along planes of weakness as a result of shear stresses arising from the indentation stress field. On the macroscopic scale, the distribution of shear stress under indentation loading can be found by finite element modelling given a suitable choice of constitutive law for plasticity (eg. the Tresca criterion). Focussing now on the microstructure, Jaeger and Cook [3] show that slip along a plane of weakness may occur when the Coloumb criterion is satisfied:

$$|\tau| = S_o + \mu \sigma_N \tag{1}$$

In Eqn. 1, τ is the applied shear stress, μ is the coefficient of "internal" sliding friction, and σ_N is the normal stress on the plane under consideration. The parameter S_o represents the inherent shear strength of the material. The Tresca failure criterion limits the maximum shear stress within the damage zone to half that of the material's yield stress: $\tau_{max} = 0.5Y$ and is thus a special case of the Coloumb criterion, Eqn. 3b, with $\mu = 0$ and $S_o = Y/2$. In the glass-ceramic materials considered here, the planes of weakness are either the interface between the mica platelets and the glass matrix or cleavage planes within the mica itself. As a convenient starting point with a finite element analysis of the system, we may assume $\mu = 0$ and that on the microstructural scale, shear faulting is dominated by S_o, and on the macroscopic scale, shear faulting is described by the Tresca criterion.

FINITE ELEMENT MODELLING

The response of the material on the microstructural scale can be separated into that which occurs before sliding failure and that which occurs after sliding. Upon the application of increasing load, shear stresses increase until the internal shear strength of the material is reached whereupon sliding failure occurs $\tau = S_o$. At larger loads, the platelets, or grains, are debonded from the glassy matrix and become locally constrained by elastic deformations at their corners. For finite element analysis of this system, we may consider a two-dimensional "unit cell" containing a single grain as shown in Fig. 4 (a). An applied compressive load is applied to the unit cell, and the resulting deformations thus calculated. These loads and deformations then permit a uniaxial stress-strain characteristic to be drawn for each of the three test microstructures. These stress-strain curves can then the used as inputs to a macroscopic finite element model, Fig. 4 (b), which simulates the bonded-interface experiments described above.

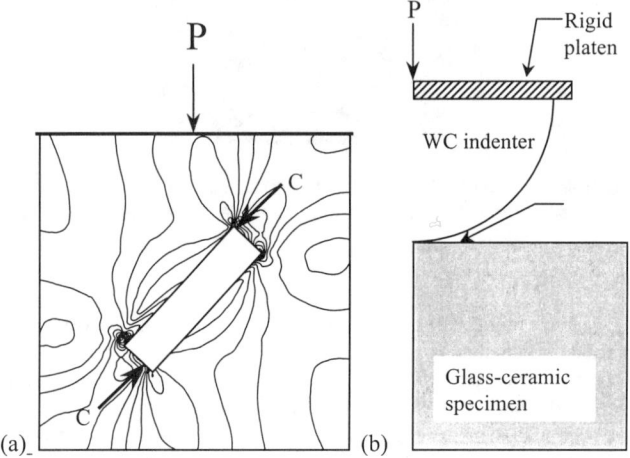

Fig. 4 Schematic of finite-element models. (a) Micro-mechanical, two-dimensional cross-section of grain embedded in an "elastic" matrix and oriented at 45° to an applied compressive stress produced by load P. Only normal compressive forces act between grain and matrix to simulate localized elastic constraint after sliding has occurred. Contours of maximum shear stress are shown (b) Macroscopic finite element model consists of axis-symmetric, frictionless loading of a half-space by a hemispherical indenter.

Microstructural Model

The central feature of the microstructural model presented here is a two-dimensional "unit cell" containing a "grain" or platelet. As shown in Fig. 4 (a), the cross-section of the grain is oriented at 45° to the direction of the application of a compressive load. The response of the structure after sliding failure is obtained by allowing only compressive normal forces to act between elements representing the grain and the surrounding matrix.

The these materials, thermal mismatch stresses arising during firing and subsequent cooling are ≈1 GPa and are comparable to the mean contact pressures in the damage region. The anisotropic nature of the thermal expansion properties of mica may be readily incorporated into the finite-element model at this level of analysis and any resulting stress distributions calculated accordingly [4]. The combined effect of grain size and aspect ratio, thermal mismatch, anisotropic material properties, etc., are thus all incorporated within the microstructural numerical analysis. Once failure occurs, no prior assumption about the nature of the localised elastic constraint need be made since all deflections are automatically computed.

The distribution of shear stress along a grain boundary for each microstructure for different load combinations was calculated and is shown in Fig. 4 (a). The most significant feature is the stress singularity which appears at the end of the grain in each case. Such a singularity is normally associated with the tip of an open crack but here is a result of a discontinuity in mechanical properties. Since shear-driven failure involves the relative sliding of the grain and matrix along the grain boundary, it is therefore appropriate that the condition for sliding of a closed fault should not be a simple uniform level of shear, but based on a fracture mechanics Mode II size effect with an inverse square root dependency on the grain diameter. In previous work [2], it has been shown that experimental evidence, in the form of a deviation from linearity in the indentation stress-strain response, is an indication of macroscopic yield. For the coarse-grained material, the macroscopic yield stress was estimated to be 750 MPa [2], and proportioning according to the inverse square root of the grain size, the corresponding yield stress for the medium and fine-grained materials are thus 1230 MPa and 2160 MPa respectively. These values of yield stress, together with the strain hardening indices to be determined in the next section, are used within the macroscopic finite-element analysis to be presented later. After internal sliding has occurred, deformation of the grain is constrained by localised elastic deformations within the grain and matrix. This leads to a macroscopic strain-hardening which can be quantified by a strain-hardening index α. Finite-element results, in the form of deflections, were obtained for a series of increasing applied mechanical loads for each of the three test material geometries while permitting only compressive normal forces between the elements representing the grain and the surrounding matrix.

Combining the effects of initiation of yield and strain hardening due to the localised constraint after yield, the resulting longitudinal stress-strain responses indicate a decrease in elastic modulus with increasing grain size with associated strain-hardening as shown in Fig. 5.

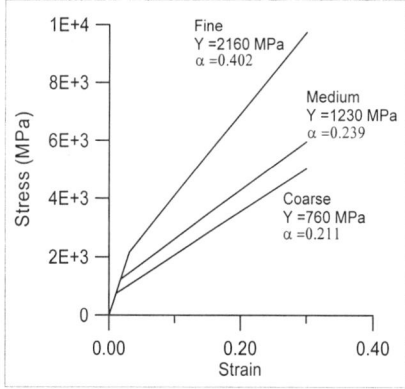

Fig. 5 Longitudinal stress-strain characteristics for the test materials obtained from microstructural finite-element analysis and used as material properties in the macroscopic analysis

Macroscopic Model

The finite-element analysis presented here, for each of the three test materials, assumes an initially spherical, elastic indenter in frictionless contact with the flat surface of an elastic-plastic specimen whose uniaxial stress-strain relationship is that obtained from the results of the previous section. These relationships embody an initial elastic response, followed by yield, and then strain-hardening. The Tresca shear-stress criterion is used as the condition of plasticity at this scale. Full details of the numerical procedure are given elsewhere [5]. The results are presented in Fig. 2 as contours of maximum shear stress which indicate the elastic-plastic boundary at $\tau_{max}/Y = 0.5$. The shapes of the plastic zone as indicated by the finite element analysis results in Fig. 2 are consistent with those observed experimentally for each of the three test materials. These comparisons verify the results of the micromechanical analysis and thus provide an explanation of the bulk properties of the materials in terms of microstructural parameters.

CONCLUSIONS

Finite element analysis at the microstructural scale provides information about macroscopic material properties. In the present case, it is concluded that yield on the microstructural scale initiates via Mode II fracture (Y = inverse square root of grain diameter) and that strain hardening index depends on aspect ratio of platelet. Values of Y and a can be used to construct a longitudinal stress-strain relationship which can be used as an input to a finite element model which simulates the actual loading experienced by the specimen on the macro scale. The size and shape of the damage zone predicted by the finite element model is consistent with the size and shape of the damage zone observed experimentally. Thus, macroscopic material properties: elastic modulus, yield stress and strain-hardening index, are described in terms of the microstructural variables: grain size and aspect ratio. This in turn leads to a connection between mechanical behaviour on the macroscopic scale and microstructural variables which may be adjusted during processing of the material.

ACKNOWLEDGMENTS

The author wishes to thank K.Chyung of Corning Inc. for supplying the glass-ceramics materials for this study. Funding for this work was provided by the United States Air Force Office of Scientific Research and the United States National Institute of Standards and Technology and is gratefully acknowledged.

REFERENCES

1. B.R.Lawn, N.P. Padture, H. Cai, and F. Guiberteau, "Making ceramics 'ductile'", Science, **263**, p1114-1116 (1994)
2. A.C. Fischer-Cripps, "Elastic-plastic behaviour in materials loaded with a spherical indenter", J.Mat.Sci., **32**, p727-736 (1997)
3. J.C.Jaeger and N.G.W.Cook, "Fundamentals of Rock Mechanics", Chapman and Hall, London, 1971.
4. H.H. Landolt and R. Bornstein, *Zahlenwerte und funktionen aus naturwissenschaften und technik, neue serie / gesamtherausgabe: k.h. hellwege,* Springer-Verlag, 1961.
5. G. Carè and A. C. Fischer-Cripps, "Elastic-plastic indentation stress fields using the finite-element method", J.Mat.Sci., **32**, p5653-5639 (1997)

LONG DISTANCE FRACTURE SURFACE ROUGHNESS ON A DENDRITIC ALUMINUM ALLOY

J. Aldaco, F.J. Garza, M. Hinojosa
Universidad Autónoma de Nuevo León, A.P. 149-F, S. Nicolás de los Garza, 66451 México.
hinojosa@gama.fime.uanl.mx

ABSTRACT

The long distance roughness of the fracture surface of a dendritic aluminum alloy is studied over a wide range of length scales. Self-affinity analysis was performed over samples broken in Charpy impact tests. Simultaneous use of Atomic Force Microscopy, SEM and stylus profilometry allowed us to cover a wide spectrum of length scales, spanning over seven decades, from a few nanometers up to one centimeter. The roughness exponent and correlation length were obtained using the variable bandwidth method. For the roughness exponent, a value of 0.8 was obtained, corresponding to the reported universal exponent. Correlation length was found to correspond well to the characteristic length of the largest heterogeneities in the complex microstructure. Our results provide information that can help to improve our understanding of the role of microstructural parameters on crack propagation mechanisms.

INTRODUCTION

It is now accepted that the fracture surfaces of heterogeneous materials are self-affine objects. Mandelbrot [1] was the first to attempt to correlate the roughness of the fracture surface of steel with its macroscopic mechanical properties. He claimed that there exists a direct correlation between the fracture toughness of the material and the roughness exponent of its fracture surface. Since then, the self-affinity analysis of fracture surfaces has been performed on a variety of materials. The use of sophisticated statistical methods [2] along with the recording of topographical data by means of STM, AFM, SEM, optical microscopy and stylus profilometry allowed us to characterize the self-affinity of the fracture surface through the determination of its roughness exponents and correlation length.

The results of many workers [3] suggest that fracture surfaces exhibit two characteristic roughness exponents. Surfaces generated by cracks propagating at high enough speeds tend to exhibit the so-called 'universal' exponent $\zeta = 0.8$ [4]. For low propagation rates, such as those obtained in fatigue tests, a smaller roughness exponent, $\zeta = 0.5$, is observed. In some cases the two exponents can be detected in the same fracture surface [5,6], at the smaller length scales, typically in the nanometer range, the $\zeta = 0.5$ exponent might be detected, with the universal exponent existing in the micrometer-millimeter range, both exponents separated at a characteristic cut-off length. For rapid kinetic conditions, the 0.8 exponent has been observed even for the smaller length scales. In any case, the role of microstructure and it relations with the self-affinity parameters are not well understood.

It seems that neither of the two roughness exponents is directly related to the microstructure. Attempts have been made to correlate the cut-off length to the size of microstructural features in some materials [7].

Fracture surfaces are self-affine for length scales up to a characteristic value called the correlation length, ξ, beyond which the surface can be considered as a flat Euclidean object.

The crack front can be imagined as a line advancing through the microstructural obstacles. It seems evident that the interactions of the crack front which the largest heterogeneities should determine the correlation length. Hinojosa [8] reported the long distance roughness of the fracture surface of a nickel superalloy, he found that the correlation length is of the order of the grain size. The results of Reyes [9] have also suggested that this correlation length corresponds to the size of the spherulites in partially crystalline polymers.

In this work we report the self-affinity analysis of the impact fracture surface of a dendritic aluminum alloy. Topographic data were recorded using AFM, SEM and stylus profilometry, a total of seven decades of length scales were covered using these techniques. The microstructure of the material is specially interesting, there are several different phases with varying sizes, the largest heterogeneities are the primary dendritic arms and the grain size, which are of the same order. We were mainly interested in finding out which of the characteristic lengths is associated with the correlation length.

EXPERIMENT

The material used in this work was an Al-Si 319-type alloy in the refined and modified condition. This alloy is employed in the automotive industry and is produced by casting. The chemical composition of the material is as follows (% wt): Si: 7.147, Cu:3.261, Fe:0.612, Zn:0.664, Mn:0.374, Ni:0.041, Ti:0.154, Mg:0.313, Sr:0.014, Al: balance. Grain refining treatment is nor ally accomplished with titanium additions and modification of the eutectic phase is obtaineu with small Sr additions. The typical microstructure is dendritic and a number of different phases are present. Figure 1 shows the six phases present in the complex microstructure: alpha phase (aluminum rich solid solution), silicon, eutectic phase, Al_2Cu, Al_5FeSi and $Al_{15}(MnFe)_3Si_2$. There is also a grain structure, not shown in Figure 1. This grain structure was analyzed on an etched specimen using Keller´s reagent. The largest heterogeneities were identified as the grain size with a value of 446 μm, and the primary dendritic arm length with an average value of 314 μm. The sizes of the other phases were much smaller, less than 40 μm.

Charpy impact specimens were machined according to ASTM standard E23 then broken to obtain the desired fracture surfaces. Fractographic observations were made by using the SEM and the AFM. Figure 2a shows a typical fracture surface as observed by SEM. A mixed intergranular and transgranular mode of propagation with evidence of some plastic deformations is observed. Figure 2b shows an AFM 3D image of this fracture surface.

Topographic data used in the self-affinity analysis were obtained by three different techniques: AFM, SEM and stylus profilometry. In each case, height profiles were obtained in a direction perpendicular to the crack propagation direction, i.e. parallel to the crack front.

In order to cover large length scales in the millimeter range, a stylus profiler was used to record profiles of a maximum length of around one centimeter. The best height resolution obtained was 0.25 microns and the tip radius was 2 μm. A typical profile consists of more than 10,000 points. SEM profiles were obtained in order to cover the scales at the micrometer level. The fracture surface was nickel plated then sectioned and polished in a plane perpendicular to the direction of crack propagation. SEM images in the BSE mode were recorded at different magnifications, from 100X up to 2000X. The profiles were extracted by image analysis, the excellent contrast between the aluminum and nickel in the BSE mode facilitated this task. The best resolution obtained in the 1024-point profiles was 0.06 microns/point. Figure 3 shows an example of the profiles obtained with the SEM technique.

AFM observations in the contact mode allowed the highest resolution profiles to be recorded and covered the nanometer length scales. Scan sizes ranging from 0.5 to 10 microns were obtained. Profiles obtained had a minimum distance between points of around one nanometer.

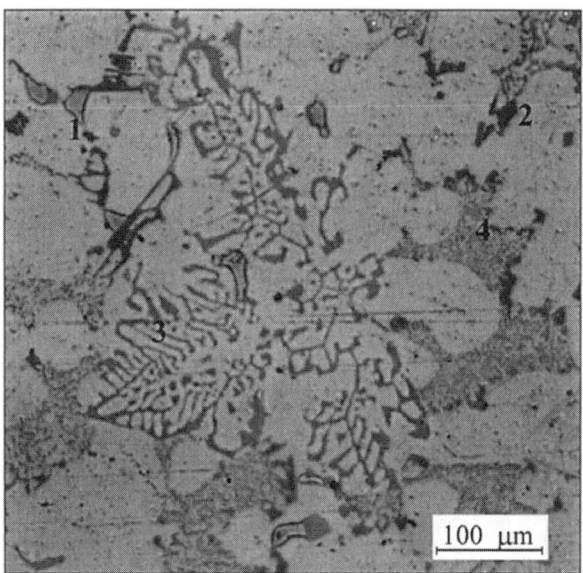

Figure 1: Microstructure of the Al-Si alloy in the as-polished condition. The different phases are the alpha matrix, Al_2Cu (1), Silicon (2), $Al_{15}(MnFe)_3Si$ (3) and the eutectic phase (4).

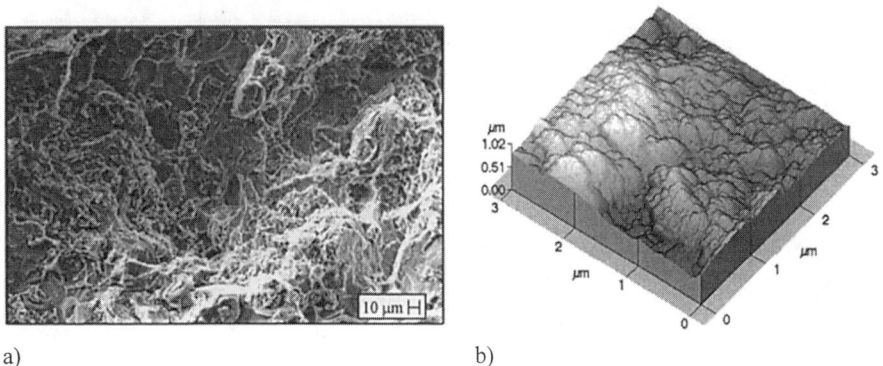

a)   b)

Figure 2: Typical fracture surface observed by SEM (2a) and AFM (2b).

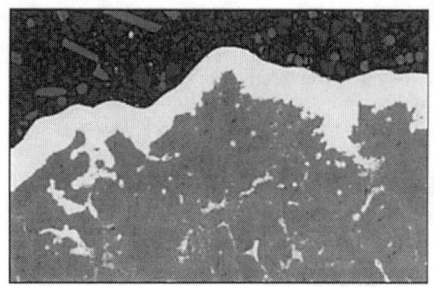 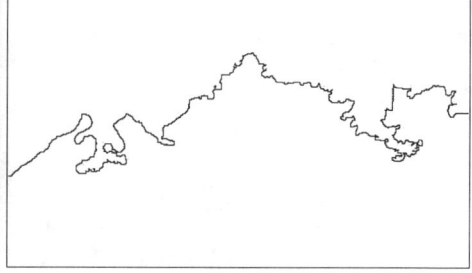

a)   b)

Figure 3: Typical SEM image of the fracture profile, (a). The respective profile extracted by image analysis is shown in (b).

Self-affinity analysis was performed using the variable bandwidth method. In this method we calculated the quantity :

$$Z_{max}(r) = < max\{z(r')\}_{x<r'<x+r} - min\{z(r')\}_{x<r'<x+r} >_x \propto r^\zeta$$

Where r is the width of the window. $Zmax(r)$ is the difference between the maximum and the minimum heights z within this window, averaged over all possible origins x of the window.

For the profilometry data, ten different profiles were averaged. For the SEM profiles, we averaged ten profiles for each magnification. In the case of the AFM profiles ten profiles were averaged for each scan size.

RESULTS

Figure 4 shows the results of the self-affinity analysis. A single averaged curve was obtained joining together the curves obtained from the different techniques. The AFM and SEM curves overlap in the region of 0.1 μm and give a continuous curve, thus showing a complete quantitative compatibility in agreement with the previous results of Daguier [7] and Hinojosa [8]. There is an small gap between the SEM and the profilometer curves but the tendency is clearly the same. The very existence of this gap manifests the inability of the SEM technique to cover the υ per region of the self-affine regime for this case, thus showing the advantage of using profilometry data to cover the largest length scales necessary to observe the limit of the regime and the determination of the correlation length. As we can see, the expected universal exponent $\zeta = 0.81$ is observed in the entire regime which spans over more than five decades. The correlation length is observed at a value of $\xi = 456$ μm.

The maximum length of the profilometer profiles were around 1 cm and the largest heterogeneities have a maximum characteristic length of less than 500 μm. This guarantees that the self-affinity analysis will permit accurate determination of the correlation length if it is of order of the largest heterogeneities. This is true in the current case, as is clearly seen in figure 4. From this result we can safely say that the correlation length corresponds to the characteristic length of the largest heterogeneities, which are the grains and the dendrites with sizes of values 314 and 446 μm, respectively. For the microstructure analyzed one can speculate that there exists a competition between the grain size and the primary dendritic arm length, the crack front advancing with characteristic step sizes which correspond to either of this features. Above this correlation limit the crack front "sees" nothing and thus the fracture surface can be regarded as flat for length scales larger than this value. Our results are in agreement with those reported by Hinojosa [8] for a nickel alloy, in which he found, figure 5, that the correlation length was of the order of the grain size in samples with a difference of one order of magnitude in grain size. Reyes [9], studying the fracture surface of a semicrystalline polymer has reported results that suggest that the correlation length is of the order of the spherulite diameter. Many doubts remain regarding the correlation between the microstructure and the self-affinity parameters, but these results should help to improve our understanding of the fracture process.

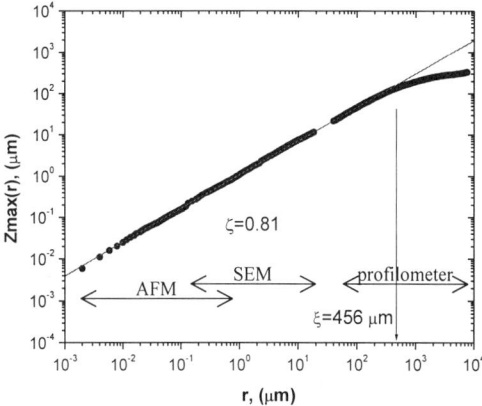

Figure 4: Self-affinity curve spanning over seven decades of length scales, the correlation length and the roughness exponent are shown.

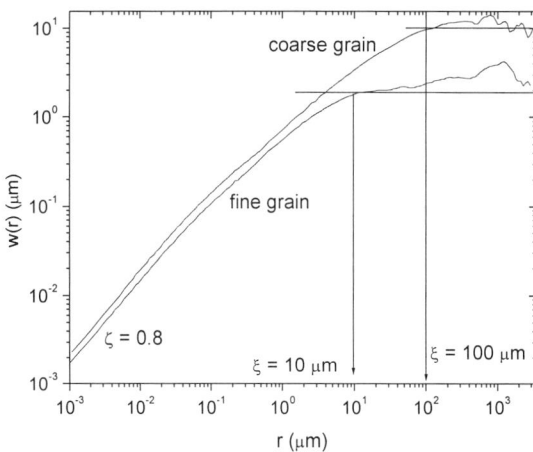

Figure 5: In nickel alloys the correlation length is of the order of the grain size [8].

CONCLUSIONS

The simultaneous use of AFM, SEM and stylus profilometry allowed us to perform self-affinity analysis of a fracture surface over seven decades of length scales, from the nanometer to one centimeter of length scale. The three techniques are shown to be quantitatively compatible. The entire self-affinity regime was studied together with its upper limit, this correlation length is of the order of the characteristic length of the largest heterogeneities, which are the grain size and the primary dendritic arm length of the Al-Si dendritic alloy. The universal exponent of 0.8 was recovered over the entire regime, in agreement with the results previously obtained for similar kinetic propagation conditions in a variety of materials.

ACKNOWLEDGMENTS

This work was supported by CONACYT and the PAICYT program of the University of Nuevo León. The authors gratefully acknowledge the invaluable help of E. Velasco, N. Cárdenas, L. Cruz and O. Garza .

REFERENCES

1. B.B. Mandelbrot, D.E. Passoja and A.J. Paullay, "Fractal Character of Fracture Surfaces of Metals", Nature, **308**, pp 721-722 (1984).
2. J. Schmittbuhl, J.P. Vilotte. S. Roux, "Reliability of self-affine measurements", Phys. Rev. E, **51** 131 (1995).
3. See the review article "Scaling Properties of Cracks", E. Bouchaud, J. Phys.:Condens. Matter **9** (1997) 4319-4344 and the abundant references therein.
4. E. Bouchaud, G. Lapasset and J. Planés, Europhys Lett. , **13**, pp 73 (1990).
5. P. Daguier, B. Nghiem, E. Bouchaud and F. Creuzet, "Pinning and Depinning of Crack Fronts in Heterogeneous Materials", Phys. Rev Lett. , **78**, pp 1062 (1997).
6. P. Daguier, S. Hénaux, E. Bouchaud, and F. Creuzet, "Quantitative Analysis of a Fracture Surface by Atomic Force Microscopy", Phys. Rev. E, **53**, 5637 (1996).
7. P. Daguier, Ph. D. thesis, Université Paris 6 (in french), November 1997.
8. M. Hinojosa, E. Bouchaud and B. Nghiem. Materials Research Society Symposium Proceedings, Volume 539, Materials Research Society, Warrendale Pennsylvania, pp. 203-208, 1999.
9. Edgar Reyes, Master Thesis, University of Nuevo Leon, Mexico (in spanish). 1999.

SELF-AFFINE MEASUREMENTS ON THE FRACTURE SURFACE OF PLASTIC MATERIALS BY AFM

E. REYES, C. GUERRERO, V. GONZÁLEZ, AND M. HINOJOSA.
Doctorado en Ingeniería de Materiales, Facultad de Ingeniería Mecánica y Eléctrica
U.A.N.L. San Nicolás de los Garza, N.L., 66450 México

ABSTRACT

The self-affine behavior of fracture surfaces of polymeric materials was qualitatively and quantitatively studied. SEM images of fracture surfaces of both polypropylene and polystyrene show Chevron marks at several magnifications. In addition, for polystyrene the mirror and Hackle zones were also observed. For quantitative analysis, the average roughness exponent, ζ, of height profiles generated by AFM images, was estimated by applying the variable bandwidth method. Values of $\zeta=0.788$ and $\zeta=0.810$ were obtained for polypropylene and polystyrene, respectively. These results are in very good agreement with the claimed universal exponent of 0.8 reported in the literature for other non-polymeric materials. By choosing the AFM appropriate operating conditions, measurements of roughness on plastic material surfaces could be performed.

INTRODUCTION

Fracture surfaces are self-affine objects characterized by a roughness exponent ζ, and exhibit scaling properties on two or three decades of length scales [1,2,3,4]. A self-affine surface is fractal up to distances of the order of a characteristic length called the correlation length, ξ; beyond this length the surface can be considered flat. In most cases, ζ is found to lie around the value 0.8, and it has been suggested that this could be an universal value, independent of the material and the fracture mode [2]. This regime is valid for high enough crack propagation speeds (uncontrolled fracture) and/or long enough length scales. For very low propagation speeds and/or small enough length scales, another self-affine regime is observed. For appropriate kinetic conditions of crack propagation, the two self-affine regimes can be observed in the same fracture surface [5].

There is still only a few experimental works reported in the literature about the fractal and self-affine behavior of polymeric material surfaces. Guerrero [6] and González [7] have been using atomic force microscopy for the measurement of the Hurst exponent on film surfaces of semi-crystalline polymers, and Tzoganakis [8], using digital image processing techniques, has obtained the fractal dimension of a linear low density polyethylene extrudate surface. Other techniques, such as scanning electron microscopy and profilometry have been successfully employed for quantitative surface analysis of both conductor and insulator materials.

The aim of this work was to determine the self-affine behavior of fracture surfaces of semi-crystalline polypropylene, PP, and amorphous polystyrene, PS, using SEM and AFM techniques. In order to accomplish this task, the roughness exponent of fracture surfaces was measured applying the variable bandwidth method.

EXPERIMENTAL

The materials used in this work were a semi-crystalline PP and an amorphous PS. For these polymers, the number average molecular weight determined by gel permeation

357

chromatography was 60,359 g/mol for PP and 76,755 g/mol for PS. The polydispersity index was 5.1 and 3.1, respectively. After differential scanning calorimetry tests, the measured PP melt temperature was 165.4 °C with a degree of crystallinity of 46.7%; the glass transition temperature was 86 °C for PS.

Samples of these materials were obtained by capillary extrusion at 190°C for PP and 180°C for PS. Filaments (about 1 mm diameter and 20 mm length) were randomly selected and cooled at room temperature. After that, the specimens were immersed in liquid nitrogen for 15 minutes. The fracture surfaces were generated by bending without control of the applied load; that means a high enough crack propagation speed without preferential direction. For SEM analysis, some samples of both polymers were gold sputtered.

The roughness was measured using an AFM in contact mode, without vacuum at room temperature. The soft nature of the polymeric surfaces is easily damaged by the tip, therefore, it is very important to determine, previous to the measurements, the appropriate operating conditions, e.g., contact force and scanning frequency. In this case, the most important restriction was to obtain sharp and clear 3-D images of the polymer surface without causing any harm to the specimens by the microscope tip. Figure 1 shows PP and PS thin film surfaces cooled from the melt to room temperature. In Fig. 1.a, the spherulitic crystallization of the PP sample can be observed; the radial growth of crystallites and the boundaries between neighboring spherulites are clear. Fig. 1.b presents a typical image of an amorphous polymer surface. This PS image shows an absence of crystalline order. For all the analyzed polymers, the best images were obtained with a contact force in the range of 8 to $15x10^{-10}$ N and a scanning frequency range from 1 to 1.5 Hz (the cantilever tip has a force constant of 0.05 ± 0.02 N/m [9]). These values are very far from the normal ones used for other rigid materials.

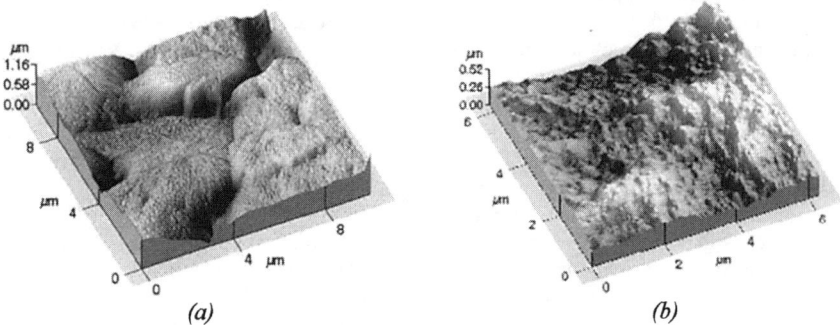

Figure 1. 3-D images of polymeric film surfaces. (a) Semi-crystalline PP sample. (b) Amorphous PS sample.

The height profiles were generated by taking 15 images from five different zones of the surface. The scan size varied from $10\mu m$ x $10\mu m$ to $2.3\mu m$ x $2.3\mu m$. For each image six height profiles (512 pixels/profile) along the direction of scanning were randomly selected. The average roughness exponent was then calculated by applying the variable bandwidth method [10].

RESULTS

The fracture surfaces were analyzed using both SEM and AFM. Qualitative SEM observations, Figure 2, show irregular curved lines that seem to depart from a point where the cracks started. These lines, named Chevron marks, give the direction of the crack propagation and also the crack origin. In Fig. 2.b, adjacent to the origin of the crack, the mirror zone and after it, the Chevron marks (Hackle zone) can be seen. This behavior corresponds to the fracture surface of amorphous materials and is the result of a change of the mechanism of crack propagation [11].

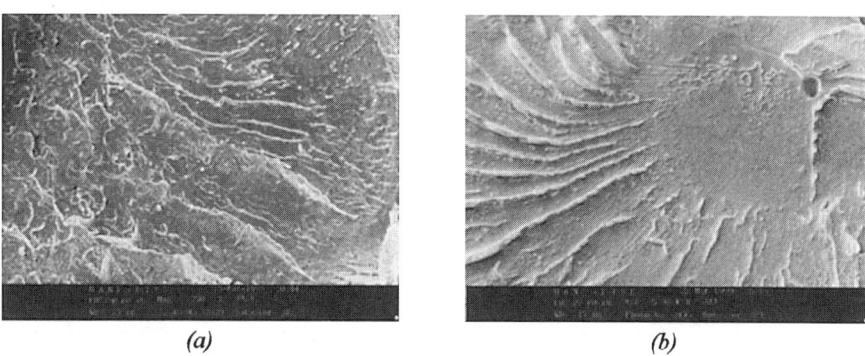

(a)   *(b)*

Figure 2. Fracture surface of polymeric materials. (a) PP sample at 250X. (b) PS sample at 5,000 X.

In all the analyzed SEM images, the fracture surfaces present similar morphologies at different magnifications. This could be considered as qualitative evidence for self-affinity.

For quantitative analysis, height profiles were generated using the AFM, Figure 3.a. 3-D images of the fracture surfaces were constructed with the corresponding height profiles, Fig 3.b.

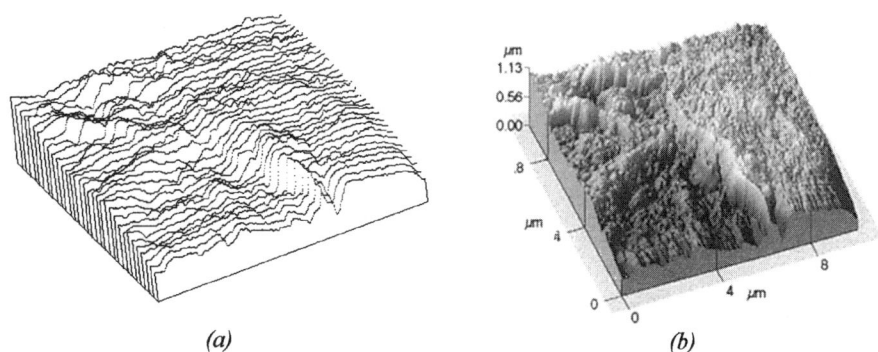

(a)   *(b)*

Figure 3. Fracture surface of a PP sample. (a) Height profiles generated by the AFM. (b) 3-D image constructed with the height profiles shown on (a).

Figure 4 shows the results of the self-affinity analysis of the fracture surfaces of both PP and PS samples. As expected, the difference between the maximum and minimum height, Z_{max}, *versus* the bandwidth size, r, on a log-log scale is a straight line spanning over at least two decades, $2x10^{-2} \mu m$ to $1x10^{0} \mu m$ for PP, Fig. 4.a, and $6x10^{-2} \mu m$ to $2x10^{0} \mu m$ for PS, Fig. 4.b. The slope of the straight line is the roughness exponent, ζ. For PP samples, ζ has a value of 0.788 ± 0.008 and for PS samples $\zeta = 0.810 \pm 0.023$. In both cases, the measured roughness exponents are in very good agreement with the claimed universal exponent of 0.8 reported for the fracture surfaces of non-polymeric materials.

The AFM scanner can obtain height profiles of $10 \mu m$ maximum lenght. Therefore, it was not possible to evaluate the correlation lenght, e.g. the limit of the self-affine regime. Future work is focusing on obtaining long distance fracture profiles.

CONCLUSIONS

The analyzed fracture surfaces of semi-crystalline polypropylene and amorphous polystyrene present a self-affine behavior with a roughness exponent very close to the claimed universal value of 0.8.

Selecting the appropriate operating conditions, it was possible to obtain sharp and clear 3-D images of plastic materials using the AFM in contact mode.

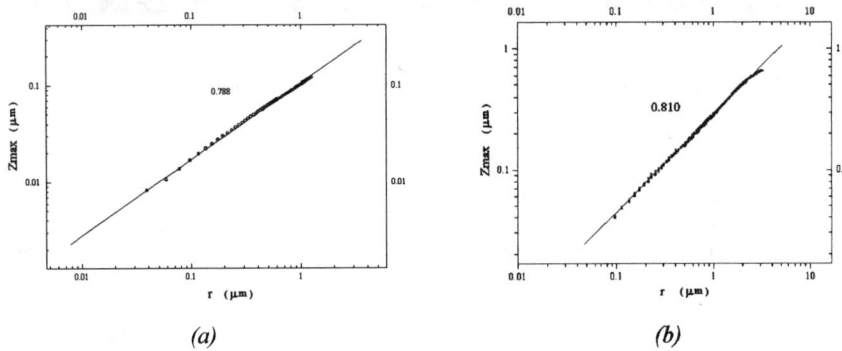

Figure 4. Z_{max} -vs- r plots. (a) PP fracture surface. (b) PS fracture surface, Hackle zone.

ACKNOWLEDGMENTS

Financial assistance from the National Science and Technology Council (CONACyT) and from the Science and Technology Research Program (PAICyT) is greatly appreciated (grants 28188-U and CA-224, respectively).

REFERENCES

1. Mandelbrot B.B., Passoja D.E., and Paullay A.J., Nature, **308**, p. 721-722 (1984).
2. Daguier P., Hénaux S., Bouchaud E., and Creuzet F., Phys. Rev. E, **53**, p. 5637-5642 (1992)
3. Bouchaud E., J. Phys.: Condens. Matter, **9**, p. 4319-4344 (1997).

4. Hinojosa M., Bouchaud E., and Nghiem B., in *Fracture and Ductile vs. Brittle Behavior-Theory, Modelling and Experiments*, edited by Beltz G.L., Selinger R.L.B., Kim K.S., Marder M.P., (Mater. Res. Soc. Proc., **539**, Boston, MA, 1998) p. 203-208
5. Daguier P., Nghiem B., Bouchaud E., and Creuzet F., Phys. Rev. Lett., **78**, p. 1062 (1997).
6. Reyes E., Guerrero C., to be published on the 2000 SPE Annual Technical Conference Proceedings, Orlando FL, May (2000).
7. González V., Hinojosa M., Guerrero C., Reyes E., to be published on the 2000 SPE Annual Technical Conference Proceedings, Orlando FL, May (2000).
8. Tzoganakis C., Price B.C., Hatzikiriakos S.G., J. of Rheology, **37** (2), p. 355-366 (1993).
9. Reyes E., M.Sc. Thesis, Universidad Autónoma de Nuevo León, Monterrey, México (1999).
10. Schmittbuhl J., Vilotte J.P., Roux S., Phys. Rev. E., **15** (1), p. 131-147 (1995).
11. Mecholsky J.J., Freiman S.W., Rice R.W., in *Fractography in Failure Analysis.* American Society for Testing and Materials, p. 363-379 (1978).

SELF-AFFINE FRACTAL CHARACTERIZATION OF A TNT FRACTURE SURFACE

L. V. Meisel*, R. D. Scanlon*, M. A. Johnson*, and Y. D. Lanzerotti**
*U. S. Army TACOM-ARDEC, Benet Laboratories, Watervliet, NY 12189
**U. S. Army TACOM-ARDEC, Picatinny Arsenal, NJ 07806-5000

ABSTRACT

A trinitrotoluene (TNT) fracture surface image is characterized in terms of a self-affine fractal structure. The fracture surface was produced by high acceleration in an ultracentrifuge when the TNT strength was exceeded. An atomic force microscope (AFM) captured the topography of a 4 micron square region on the fracture surface. The present analysis supports a self-affine description of the TNT fracture surface (wavelengths of 0.016 micron to 4.0 micron) and provides a new prespective on fracture processes in TNT. An essential step in self-affine fractal characterization of surfaces is the determination of reference surfaces. A self-affine fracture surface can be described in terms of a single-valued height function. In the TNT fracture surface, single-valued height functions, which describe surface texture can only be defined with respect to curved reference surfaces. By employing curved reference surfaces, we have demonstrated that self-affine fractal scaling can be used to characterize the TNT fracture surface. This provides important information that is not evident in the analysis of individual surface scans.

INTRODUCTION

Energetic materials, including details of the rapid chemical reactions that lead to explosive behavior, are of significant interest for scientific and practical reasons. Nevertheless, the details are still poorly characterized and understood. Energetic materials have numerous applications in propulsion systems, in ordnance, and in mining. Because of the conditions under which they are often employed, the nature of fracture processes in such materials under high accelerations is of particular interest. The characterization of fracture in trinitrotoluene (TNT) and in other energetic materials has been the subject of several investigations[1-4].

TNT FRACTURE SURFACE

The original TNT sample was melt-cast in a polycarbonate sleeve. The fracture of the sample was produced by high acceleration in an ultracentrifuge when the strength of the material was exceeded. Fracture occurred for an acceleration of 35,000 g at 25°C.

The topography of a 4 μm by 4 μm square region (projected approximately perpendicular to the z-axis of the polycarbonate sleeve) square region on the fracture surface was captured by atomic force microscope (AFM). The central 2 μm by 2 μm region (projected) was found to have been deformed by an earlier application of the AFM. The stylus force applied on the previous scan was approximately 12 nanoNewtons, and apparently it was sufficient to produce a permanent compression of the fracture surface.

A surface plot of the AFM elevation data (which is composed of 512 uniformly spaced

Mat. Res. Soc. Symp. Proc. Vol. 578 © 2000 Materials Research Society

profiles, each of which is composed of 512 uniformly spaced values) is shown in Fig. 1. The "line" visible toward the left hand side of the image was produced by omitting profiles 204 through 216 from the surface plot. This does not affect the analysis. The x-values and y-values correspond to pixel spacings (4 μm/512 = 0.0078 μm) and the z-values are in nanometers.

SELF-AFFINE FRACTAL SURFACES IN E^3 AND THE HEIGHT CORRELATION FUNCTION

Self-affine fractal surfaces are statistically invariant under anisotropic dilations (5,6). A case of particular interest pertains to surfaces in E^3, which are statistically invariant under transformations of the form

$$\{x, y, z\} \rightarrow \{\lambda x, \lambda y, \lambda^H z\}, \qquad (1)$$

where the Hurst exponent H, describes the anisotropic scaling. Equation 1 is a special case of self-affine fractal scaling in E^3. (In a more general case, the y coordinate would have a coefficient λ^K.)

An essential step in the self-affine fractal characterization of surfaces is the determination of special reference surfaces, Σ (5). Once such surfaces are determined, a self-affine fractal surface can be described in terms of a single-valued height function $h(r)$ such that the position of any point on the surface is expressable in the form $\{r, h(r)\}$, where $r \in \Sigma$.

In the usual analysis of self-affine surfaces[5,6] the reference surface, Σ, is assumed to be planar and $h(r)$ is then measured perpendicular to Σ. In many interesting cases, however, such as the topography of the earth's surface, single-valued height functions which describe surface

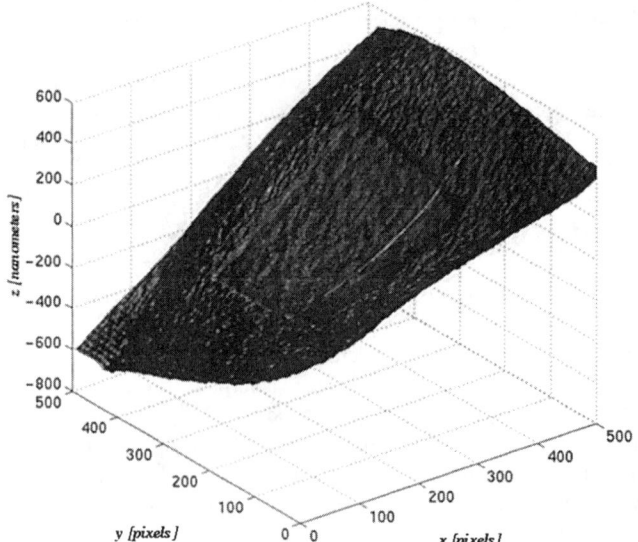

Figure 1. Surface plot of AFM elevation data for TNT fracture surface

texture can only be defined with respect to curved reference surfaces.

The invariance expressed in Eq. 1 implies that any point on a self-affine surface can be represented in the form $\{r, h(r)\}$, where the height function $h(r)$ is a single valued function of $r \equiv \{x, y\} \in \Sigma$. The scaling range is given in terms of a parallel correlation length $\chi_{//}$. The parallel correlation length $\chi_{//}$, can be regarded as giving the range at which a surface structure becomes uncorrelated. The roughness of the surface is defined in terms of a perpendicular correlation length $\chi_\perp$ (which is related to the root mean square variation of $h(r)$).

The TNT fracture surface texture can be regarded as being isotropic over Σ. In this case, the height correlation function $C_h(r)$ can be defined in terms of $h(r)$[6] as:

$$C_h(r) = \sqrt{<[h(x) - h(r + x)]^2>} \qquad (2)$$

The Hurst exponent is then defined by the small $ln(r)$ linear variation of $ln(C_h(r))$,

$$\frac{d\,ln\,(C_h(r))}{d(ln(r))} \rightarrow 2H \qquad (3)$$

for small $r \ll \chi_{//}$. The scaling range is the range of r corresponding to linearity.

For sufficiently large r, the elevations become uncorrelated and then $C_h(r)$ can be related to σ, the rms variation of $h(r)$. Thus, the large r limit of $C_h(r)$ is given by

$$C_h(r) \rightarrow 2\sigma^2 \qquad (4)$$

for $r \gg \chi_{//}$. The surface roughness, σ, is the perpendicular correlation length, $\chi_\perp$.

The parallel correction length $\chi_{//}$ is defined as the crossover from linear scaling behavior to uncorrelated elevations in $ln(C_h(r))$ versus $ln(r)$ curves. $\chi_{//}$ is the r-value corresponding to the intersection of an extrapolated straight line fit over the scaling range through the small r values, with the large r limit of $ln(C_h(r)) = ln(2\sigma^2)$.

RESULTS

Details of the determination of the reference surfaces are given in Ref. 7. Fig. 2 shows the distribution of elevations $h(r)$ that was obtained for typical 0.8 μm x 0.8 μm (100 pixel by 100 pixel) uncompressed and compressed areas on the fracture surface (Fig.1). A shift of approximately two nanometers was required to line up the centers of the distributions. The compressed elevations are plotted as 2.35 x $h(r)$ compressed in order to show the close similarities of the two distributions. Both distributions are (approximately) normally distributed.

The height correlation functions $C_h(r)$ are computed for local surface elevations data $h(r)$

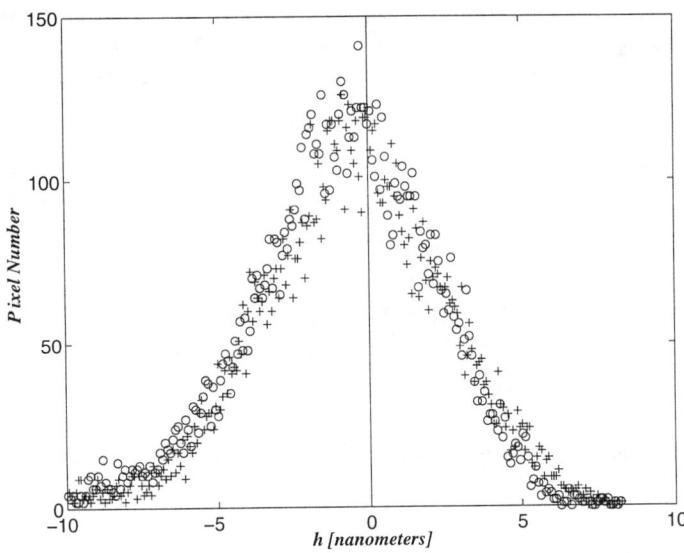

Figure 2. Distribution of elevations for uncompressed (0) and 2.35 x compressed (+) surface elevations

by means of Eq. 2. Typical log-log plots of the height correlation function $C_h(r)$ vs r for compressed and uncompressed regions are shown in Fig. 3 (from Ref. 7). The data are consistent

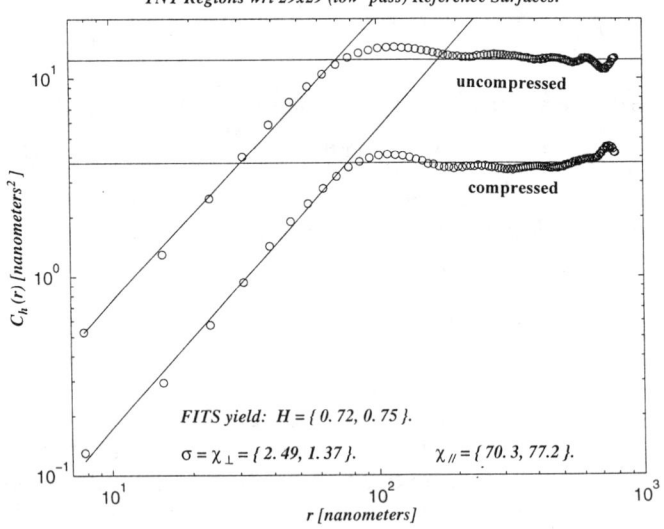

TNT Regions wrt 29x29 (low–pass) Reference Surfaces.

uncompressed

compressed

FITS yield: $H = \{ 0.72, 0.75 \}$.

$\sigma = \chi_\perp = \{ 2.49, 1.37 \}$. $\chi_{//} = \{ 70.3, 77.2 \}$.

Figure 3. Typical log-log plots of the height correlation function $C_h(r)$ versus r for uncompressed and compressed regions

with a self-affine fractal scaling model over the range from about 0.007 µm to about 0.070 µm. This was found to be approximately the range for all regions examined.

The Hurst exponent H was computed by fitting Eq. 3 over the scaling range. Analysis of a selection of randomly sited 0.78 µm by 0.78 µm (100 pixel by 100 pixel) regions yielded $H = 0.72 \pm 0.03$ and $H = 0.75 \pm 0.03$ for uncompressed and compressed regions, respectively. Fitting the large r parts of Fig. 3 (see Eq. 4) yielded the perpendicular correlation length $\chi_\perp = \sigma = 2.49 \pm 0.01$ nm for the uncompressed region and $\chi_\perp = 1.37 \pm 0.15$ nm for the compressed region.

The parallel correlation length $\chi_{//}$, given by the intersection of the small r scaling range fit line $C_h(r)$ versus r with $ln\ (2\sigma^2)$ is found to be 70.3 ± 1.7 nm for the uncompressed region and 77.2 ± 7.6 nm for the compressed region.

CONCLUSIONS

The use of curved reference surfaces enable one to perform height correlation function-based self-affine fractal analysis of textures in cases that cannot be analyzed with respect to flat reference surfaces. Employing curved local reference surfaces, we have demonstrated that self-affine fractal analysis can be employed to characterize scaling on TNT fracture surfaces. Atomic force microscopy can provide new perspectives on the fracture and deformation of energetic materials. Adapting atomic force microscopy and similar analytical analysis to the fracture surfaces of other explosives would provide important comparative data that would yield new insights into the behavior of energetic materials.

ACKNOWLEDGMENT

We thank Dr. S. A. Mogren, Columbia, MD for help in obtaining the AFM scans and converting the binary height data to ASCII height data.

REFERENCES

1. Lanzerotti, Y. D., and Sharma, J., Appl. Phys. Letters **39**, p. 455-457 (1981).
2. Lanzerotti, Y. D., Pinto, J., and Wolfe, A., in *The Ninth Symposium (International) on Detonation*, OCNR 113291-7, 1989, Volume II, p. 918-924.
3. Lanzerotti, Y. D., Pinto, J., Wolfe, A., and Thomson, D. J., in *The Tenth Symposium (International) on Detonation*, ONR 33395-12, Office of Naval Research, Arlington, VA,1993, p. 190-198.
4. Lanzerotti, Y. D., Miesel, L. V., Johnson, M. A., Wolfe, A., and Thomson, D. J., in *Atomic Resolution Microscopy of Surfaces and Interfaces*, edited by D. Smith, Materials Research Society, Pittsburgh, PA, 1997, p. 179-184.
5. Mandelbrot, B. B., *The Fractal Geometry of Nature*, Freeman, New York, 1983, p. 352.
6. Gouyet, J. F., Rosso, M., and Sapoval, B., in *Fractals and Disordered Systems*, edited by A. Bunde and S. Havlin, Springer, Berlin, 1991.
7. Meisel, L. V., Scanlon, R. D., Johnson, M. A., Lanzerotti, Y. D., Proc. APS Topical Conf. Shock Compression of Condensed Matter, June 1999 (in press).

Dislocation Interactions With
Interfaces and Grain Boundaries

THE STRUCTURE OF DEFECTS FORMED BY ABSORPTION OF CRYSTAL DISLOCATIONS IN INTERFACES IN THE HCP METALS

D.J. BACON[1], R.C. POND[1] AND A. SERRA[2]

[1]Materials Science and Engineering, Department of Engineering,
The University of Liverpool, Brownlow Hill, Liverpool L69 3GH, U.K.
[2]Department de Matematica Aplicada III, Universitat Politecnica de Catalunya,
ETSE Camins, Jordi Girona 1-3, 08034 Barcelona, Spain

ABSTRACT

Atomic-scale computer simulation has been used to investigate the interaction of crystal dislocations with two interfaces in hexagonal-close-packed (HCP) metals, namely the $\{10\bar{1}2\}$ twin boundary and a $<1\bar{2}10>/90°$ tilt boundary that is incommensurate in the direction perpendicular to the tilt axis. Crystal dislocations are absorbed in the tilt boundary with concomitant reconstruction of their cores. In the twin boundary, a broader range of interactions is observed, including defect transmission from matrix to twin and decomposition in the interface into discrete defects. The role of crystallographic features and interfacial structure is elucidated by comparing interaction processes in the two interfaces. The core structure of interfacial defects can be complex and contributes significantly to total defect energy.

INTRODUCTION

Considerable progress has been reported in determining the structure of interfaces (see, for example, [1]), but relatively few studies have revealed information about processes at the atomic level, and these are our primary concern. We have used atomic-scale computer simulation to investigate elementary processes in the hexagonal-close-packed (HCP) structure. Processes in two different interfaces have been investigated. One has been studied previously using computer simulation and transmission electron microscopy, and is the $\{10\bar{1}2\}$ twin boundary. Theoretical studies have addressed the structure of the ideal interface [2] and the nature of dislocations, which occur therein [3,4]. Transmission electron microscopy has been used to investigate both the interface structure and the character of defects in this twin boundary in zinc and α-titanium [5,6], and excellent correspondence between the theoretical and experimental studies has been found. Some preliminary simulation work has been carried out to study the response of defects to applied stress, and a novel mechanism of defect generation and movement has been revealed [7]. Perfect twin boundaries are periodic in the interface plane, but the other interface considered here is periodic in one direction and incommensurate in the orthogonal one. It is the 90°-tilt boundary formed when the (0001) basal plane of one crystal is joined to a $\{10\bar{1}0\}$ first-order prism plane of another with a $<11\bar{2}0>$ axis in common.

The interaction with the interface of various perfect dislocations from one of the crystals is considered. Three broad classes of interaction can be envisaged. Firstly, crystal dislocations may be completely transmitted through the interface into the adjacent grain. Secondly, dislocations may retain their integrity in the interface but undergo core reconstruction and/or spreading to a greater or lesser extent. Thirdly, the incident dislocation may dissociate into discrete interfacial defects, which may subsequently disperse along the interface due to their mutual repulsion. The crystallographic parameters that characterize an interface are important

371

factors in determining whether a dislocation undergoes transmission, reconstruction or dissociation. These parameters govern the topological character (i.e. Burgers vector **b** and step height h) of admissible interfacial defects. The mobility of interfacial defects in these boundaries has been studied and is reported in [8].

The topological properties of interfacial defects were set out in a recent paper using an integrated and comprehensive treatment based on the fundamental consequences of symmetry-breaking in crystalline materials [9]. The parameters of selected dislocations in the twin and incommensurate interfaces are listed in Tables 1 and 2, respectively. The notation used is as follows [10]. The upper and lower crystals are designated λ and μ, respectively. Dislocations are created in the computer model by joining λ and μ surfaces that exhibit incompatible steps of height pd(λ) and qd(μ), where p and q are integers and d represents the interplanar spacing of planes parallel to the surfaces. The smaller of the two step heights, referred to as the 'overlap' step height, is designated h. Using the present notation, either p or q equals zero in the case of crystal dislocations and both are zero for screws.

Table 1. Topological parameters of dislocations in the $(10\bar{1}2)$ twin boundary in Ti.

(Axes : $x // [\bar{1}2\bar{1}0]_\lambda$, $y // [10\bar{1}\bar{1}]_\lambda$, $z // [10\bar{1}2/\Lambda^2]$,

where $\Lambda^2 = 2(c/a)^2/3$ and $b_z = (p-q)d_{(10-12)} = (p-q)\sqrt{3}\ a\Lambda\ /\ 2(2+\Lambda^2)^{1/2}$)

$b_{p/q}$	$\|b_{p/q}\|$ [units of a]	b_x [units of a]	b_y [units of a]	b_z [units of a]	h [units of $d_{(10-12)}$]
$b_{2/2}$	0.20	0	0.20	0	2
$b_{9/9}$	0.27	0	0.27	0	9
$b_{11/11}$	0.07	0	0.07	0	11
$b_{13/13}$	0.13	0	0.13	0	13
$b_{5/6}$	0.77	0.5	0.04	-0.59	5
$b_{7/8}$	0.79	0.5	0.16	-0.59	7
$b_{10/8}$	1.31	0.5	0.26	-1.18	8
$b_{1/0}$	1.0	0.5	-0.64	-0.59	0
$b_{2/0}$	1.59	0	1.07	-1.18	0
$b_{3/0}$	1.88	0.5	0.43	-1.77	0

Table 2. Topological parameters of dislocations in the $(10\bar{1}0)_\lambda/(0001)_\mu$ interface in Ti.

b	$\|b\|$ [units of a]	b_x [units of a]	b_y [units of a]	b_z [units of a]	h [units of a]
b_y	0.14	0	0.14	0	0
$b_z^{(1)}$	0.14	0	0	0.14	1.59
$b_z^{(3)}$	0.80	0	0	0.80	0
$b_{0/0}$	1.59	0	1.59	0	0
$b_{0/0}$	1	1	0	0	0
$b_{1/0}$	1	0.5	0	0.87	0

CRYSTALLOGRAPHIC ASPECTS OF DISLOCATION DECOMPOSITION

Dislocation decomposition in the $\{10\bar{1}2\}$ twin boundary

The crystallography of twins leads to special consequences: a multiplicity of defects with finite h and **b** parallel to the interface can exist. These defects can move conservatively in principle, and have $\mathbf{b} = \mathbf{b}_{p/p}$ with p either odd or even. The set for which p is even includes the elementary twinning dislocation with Burgers vector $\mathbf{b}_{2/2}$. The relatively small magnitude of $\mathbf{b}_{2/2}$, combined with the small step height $h = 2d$, results in a stable defect having a wide core and which is mobile under the action of an applied stress [4]. The simulations show that other dislocations in this set with $\mathbf{b}_{2n/2n}$, where n is an integer >1, are unstable with respect to decomposition into n twinning dislocations, and this is consistent with experimental observations [5]. The **b** of some defects in the set with p odd are listed in Table 1. It has been found by simulation [7] and experiment [6] that, broadly speaking, defects in this set are stable in the absence of applied stress if $|\mathbf{b}_{p/p}| < |\mathbf{b}_{2/2}|$, otherwise they are unstable with respect to decompositions into products including twinning dislocations. However, even in the case of stable defects with small $|\mathbf{b}_{p/p}|$, the associated step height, h, may be large (see Table 1). As is discussed elsewhere [8], a large h tends to lead to a relatively localised dislocation core which, in turn, means that defect mobility under applied stress may be low or zero.

Three examples of dislocations with **b** inclined to the interface, i.e. $p \neq q$, are listed in Table 1. As for the previous category, it has been found experimentally and theoretically that defects which are stable in the absence of applied shear stresses have components b_y with magnitudes smaller than $|\mathbf{b}_{2/2}|$. Such defects can arise by spontaneous decomposition of crystal dislocations without the application of stress [7], for example

$$\mathbf{b}_{1/0} \rightarrow \mathbf{b}_{-(2n-1)/-2n} + n\mathbf{b}_{2/2} . \tag{1}$$

Simulations for titanium [7] show that this process occurs for $n = 3$, since the smallest value of $|\mathbf{b}_{-(2n-1)/-2n}|$ appropriate to expression (1) is $\mathbf{b}_{-5/-6}$, and the n twinning dislocations glide away along the interface leaving the $\mathbf{b}_{p/q} = \mathbf{b}_{-5/-6}$ defect isolated, as depicted schematically in fig.1(a). Reactions of this type are consistent with Frank's rule for dislocation decomposition, and we note that the sign of the component b_z for the $\mathbf{b}_{p/q}$ defect is the same as that for the initial crystal dislocation $\mathbf{b}_{1/0}$, and the sign of the twinning (or antitwinning) defects is determined by the component b_y of the crystal dislocation. The situation for a crystal dislocation with the opposite sign to that depicted in fig.1(a) is shown in fig.1(b), and the decomposition is represented by

$$\mathbf{b}_{-1/0} \rightarrow \mathbf{b}_{(2n-1)/2n} + n\mathbf{b}_{-2/-2} \tag{2}$$

with $n = 2$. However, simulations will be presented later showing that the core of the $\mathbf{b}_{3/4}$ defect reconstructs from the reentrant form depicted in fig.1(b) to a structure shown schematically in fig.1(c).

For completeness, we also indicate possibilities for decompositions for **c** and **c+a** crystal dislocations. The former is given by

$$\mathbf{b}_{2/0} \rightarrow \mathbf{b}_{(2n+2)/2n} + n\mathbf{b}_{-2/-2}, \tag{3}$$

and the latter by

$$\mathbf{b}_{3/0} \rightarrow \mathbf{b}_{(2n+3)/2n} + n\mathbf{b}_{-2/-2}. \tag{4}$$

Examples of actual decompositions are illustrated later.

Even though the dislocations characterized by $\mathbf{b}_{p/q}$ with $p \neq q$ have $b_z \neq 0$, they can, in principle, move conservatively along the interface under the action of an applied stress. This is achieved through the Serra-Bacon (S-B) mechanism found by computer simulation in [7], although core reconstructions such as that outlined above in fig.1(c) can profoundly influence mobility.

Dislocation absorption in the incommensurate $\{10\overline{1}0\}/(0001)$ interface

Unlike the case of the $\{10\overline{1}2\}$ twin boundary, no interfacial dislocations with finite h and $\mathbf{b}$ parallel to the interface can exist in the incommensurate interface. It is anticipated that at low temperature this will inhibit dislocation decomposition into discrete defects that disperse in the interface, although some limited separation might occur through a compensated-climb interaction like the S-B mechanism. At high enough temperature, decomposition into discrete defects may be assisted by climb. These considerations imply that crystal dislocations interacting with this interface at low temperature are more likely to undergo core reconstruction than to dissociate. We note that core reconstruction as depicted in figs.1(b) and (c) for the $\{10\overline{1}2\}$ twin case cannot arise in the incommensurate interface because no periodicity is present in this boundary parallel to y. Reconstruction from a reentrant to an obtuse form could occur, but would require the generation of an additional dislocation dipole.

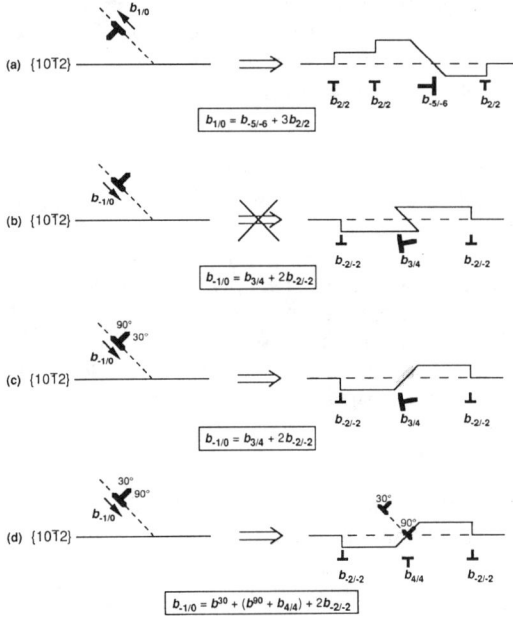

Fig.1.   Schematic illustration of the interaction with a $(10\overline{1}2)$ twin boundary of a 60°-dislocation with $\mathbf{b}$ $= \mathbf{a} = \mathbf{b}_{1/0}$. The equations show the different reactions according to the sign and position of the crystal dislocation. Notice that the reentrant core of the $\mathbf{b}_{3/4}$ defect shown in (b) reconstructs to the forms with obtuse angles shown in (c) and (d).

374

METHOD OF ATOMIC SIMULATION

The many-body potential of the Finnis-Sinclair type for α-titanium (Ti) derived by Ackland [11] has been used for the simulations. The equilibrium lattice parameter ratio, c/a, of the perfect crystal is 1.5918. The model containing the perfect $(10\bar{1}2)$ twin was created in the relaxed state using the procedure described in our previous work [3,7]. The model containing the $(0001)_\mu/(10\bar{1}0)_\lambda$ interface formed by a rotation of 90° about $[1\bar{2}10]$ was created by simply orientating the basal plane of one half-crystal parallel to the first-order prism plane of the other, with the close-packed direction $[1\bar{2}10]$ common. Since the period along the direction parallel to y is different on each side of the interface unless c/a is √3, exact periodicity was not achieved. However, c/a for Ti is such that the magnitude of $12[10\bar{1}0]$ is similar to 13[0001], i.e. 12√3a (= 20.785a) is close to 13c (= 20.693a), with a misfit of 0.4%, and it was found helpful to describe the boundary structure in terms of this pseudo-periodicity. Periodic boundary conditions were employed along the common $[1\bar{2}10]$ axis in both sets of simulations with the block thickness equal to the lattice repeat distance a. Fixed boundary conditions were imposed on the surrounding outer regions in the other two directions. The inner relaxable region contained up to 2000 atoms for the twinned model and 6000 atoms for the 90°-tilt interface simulation. (This corresponded to an inner region with long dimension 36√3a ≈ 39c in the latter case.) In both studies the inner region was relaxed by minimising the potential energy by the method of conjugate gradients.

For the simulation of the interaction of a crystal dislocation with the interface, the core of a dislocation with line direction $[1\bar{2}10]$ was created near the boundary by imposing the displacement field of isotropic elasticity on all the atoms prior to relaxation. Simulations of interfacial defects not formed directly by decomposition were initiated by introduction of these defects at a site at the relaxed interface. This method of introduction is not unique, but the final results were found to be independent of the initial configuration.

INTERACTION WITH THE $\{10\bar{1}2\}$ TWIN BOUNDARY

b = a = b$_{0/0}$ or b$_{1/0}$

Among the set with **b = a** is the screw dislocation with **b = b$_{0/0}$** parallel to the interface. Recent simulations have shown that this can be either transmitted across the interface or reflected from it by cross-slipping onto either basal or prism planes [12]. No residual dislocations are left in the boundary, which remains unchanged.

The other dislocation of this set has the Burgers vector **b$_{1/0}$** and is as depicted schematically in the left part of fig.1(a) when the positive line sense (according to the RH/FS convention) is out of the paper. The core structure of this 60° crystal dislocation before decomposition corresponds to two Shockley partials, with Burgers vectors **b^{30}** and **b^{90}** at 30° and 90° to the line direction, respectively, and a ribbon of intrinsic stacking-fault. The partial spacing is approximately 5a. For dislocations located on adjacent basal glide planes, the order of the leading and trailing partials is reversed. The decomposition of the **b$_{1/0}$** dislocation is described by expression (1), and a detailed account of the simulated structure has been given elsewhere [7]. The nature of the interfacial defect created by decomposition is independent of the leading/trailing order of the dissociated configuration.

Although the decomposition of **b$_{1/0}$** dislocations is unaffected by the partial configuration, significant differences arise in the case of **b$_{-1/0}$** dislocations. The decomposition reaction described by expression (2) shows that the signs of the interfacial defects are opposite to those in expression (1), and hence the signs of the steps are also opposite. However, as illustrated

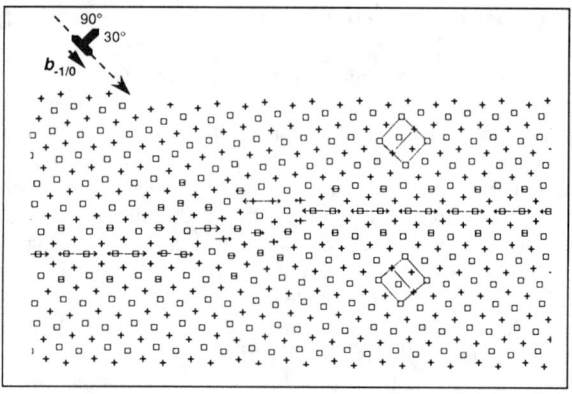

Fig.2.   Relaxed atomic positions after the $\mathbf{b}_{-1/0}$ dislocation in the upper crystal has been absorbed by the $(10\bar{1}2)$ twin boundary. The location of the interface and the dislocation core in the plots is clarified by the superimposed arrows showing atomic sites that experience a large component of hydrostatic stress (arrow to left/right = compression/tension). An outline of an hexagonal unit cell is shown in each crystal. Complete absorption of the crystal dislocation occurs when the leading partial is in the 30° orientation to its Burgers vector.

schematically in fig.1(b), this implies a reentrant step for the $\mathbf{b}_{p/q}$ defect, but such configurations were not found in our simulations. Moreover, the structures observed were found to depend on the order of the incoming partials. When the 30° partial ($\mathbf{b}^{30}$) leads, as depicted schematically in fig.1(c), complete absorption occurs according to expression (2) with n = 2, but the riser of the $\mathbf{b}_{3/4}$ dislocation is not a distinct division between the λ and μ crystals, such as that described in [7] and shown schematically in fig.1(a) for the $\mathbf{b}_{-5/-6}$ defect, and so it is represented by a shaded zone in fig.1(c). The actual atomic structure found by computer simulation is plotted in fig.2. By contrast, when the $\mathbf{b}^{90}$ partial leads, as depicted schematically in fig.1(d), it alone is absorbed and the $\mathbf{b}^{30}$ partial stands off from the boundary, separated from it by the $(0001)_\lambda$ stacking fault, and the decomposition can be described as

$$\mathbf{b}_{-1/0} = \mathbf{b}^{30} + (\mathbf{b}^{90} + \mathbf{b}_{4/4}) + 2\mathbf{b}_{-2/-2}, \tag{5}$$

b = c = $\mathbf{b}_{2/0}$

We have found in the simulations that a crystal dislocation with $\mathbf{b} = \mathbf{c}$ in the Ti model has two possible core states [12]. One, denoted 'asymmetric', is spread along a plane of $\{10\bar{1}1\}$ type: the other is 'symmetric' and is spread along two $\{10\bar{1}1\}$ planes. The $\mathbf{c}$ dislocation, like the $\mathbf{a}$, is absorbed by the boundary and the structure of the resulting interfacial defect depends on the sign of the Burgers vector and the core type. (Strictly the $\mathbf{c}$ dislocation is almost sessile and absorption was effected in the simulations by creating the dislocation right at the interface.)

Fig.3(a) shows the atomic projection of the boundary after the reaction with the 'asymmetric' dislocation, from which it can be seen that it has been absorbed by the decomposition in expression (3) with n = 4. We notice that, as with the $\mathbf{b}_{-5/-6}$ defect, the riser of the $\mathbf{b}_{10/8}$ dislocation is formed by the abutting of $(10\bar{1}0)_\lambda$ and $(0001)_\mu$ planes. This configuration allows good accommodation of the core of the $\mathbf{c}$ dislocation along the riser. A similar behaviour

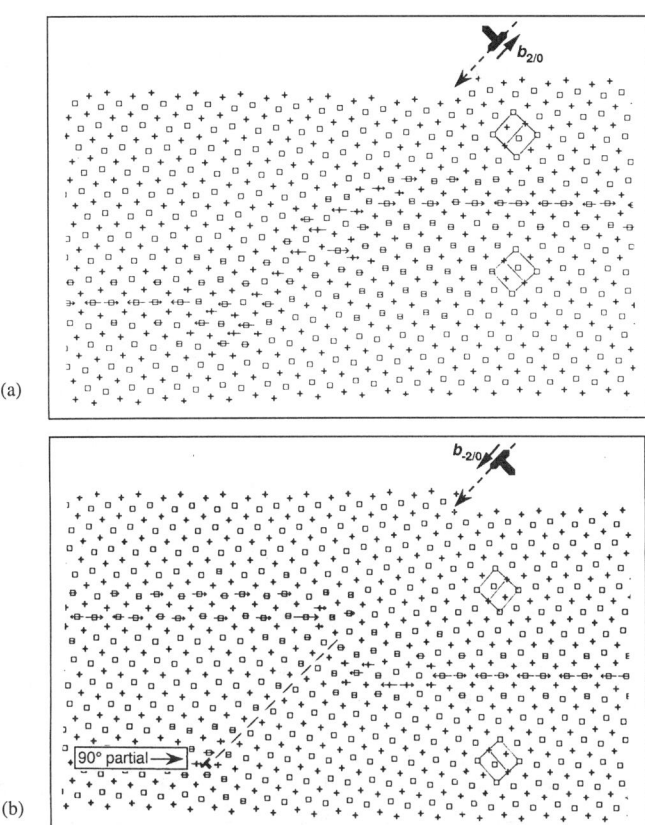

(a)

(b)

Fig.3.   Atomic plots in [1$\bar{2}$10] projection of the (10$\bar{1}$2) twin boundary after interaction by a
dislocation with **b** = **c** from the upper crystal. The arrows indicate the nature of the hydrostatic
stress at the atomic sites where it has a relatively large magnitude. (a) is the structure formed by
a dislocation with an asymmetric core. Note that the riser of the $\mathbf{b}_{10/8}$ dislocation is formed by
the abutting of (10$\bar{1}$0)$_\lambda$ and (0001)$_\mu$ planes. The structure in (b) was formed by a dislocation
with a symmetric core. Note that a 90° partial is emitted from the interface: the dashed line
indicates the basal stacking fault in the μ crystal.

is found when the **c** dislocation initially has the symmetric core, although the two risers are not
identical. This suggests that the form of a $\mathbf{b}_{10/8}$ defect depends on the form of the **c** dislocation
that created it.

    If the crystal dislocation has the opposite sign, or if it originated in the lower crystal with
Burgers vector $\mathbf{b}_{0/2}$, the reaction with the boundary could, in principle, form a reentrant
(10$\bar{1}$0)$_\lambda$/(0001)$_\mu$ riser. But such risers are not stable and relax to an obtuse configuration, as
plotted in fig.3(b) for the symmetric **c** dislocation. The steps for these reactions are smaller in
height than those for the $\mathbf{b}_{2/0}$ dislocation above, and in the case illustrated in fig.3(b), for
example, appear at first sight to be defined by the decomposition

$$\mathbf{b}_{-2/0} = \mathbf{b}_{-6/-4} + 2\mathbf{b}_{2/2} . \tag{6}$$

377

However, as a consequence of the reorientation of the $\mathbf{b}_{-6/-4}$ riser to a distorted $(0001)_\lambda/(10\overline{1}0)_\mu$ configuration, a 90° Shockley partial is emitted along a basal plane of the lower crystal, as indicated in the figure.

b = c + a

This Burgers vector has the form $\mathbf{b} = 1/3\langle 11\overline{2}3\rangle$ and is denoted as $\mathbf{b}_{3/0}$. On being inserted into the twin boundary, it decomposes according to

$$\mathbf{b}_{3/0} = \mathbf{b}_{10/8} + \mathbf{b}_{-5/-6} + \mathbf{b}_{-2/-2} \qquad (7)$$

The core of this defect is complex, but is recognisable as the superposition of a $\mathbf{b}_{10/8}$ dislocation that steps the boundary up (left part of the core) and a $\mathbf{b}_{-5/-6}$ dislocation that steps it down (right part of the core). In other words, the decomposition product of the **c+a** dislocation is not as predicted by expression (4) but is the superposition of the **c** (eq. (3)) and **a** (eq. (1)) dislocations.

DECOMPOSITION IN THE $[1\overline{2}10]/90°$-TILT BOUNDARY

Reference structure

The dislocation-free, incommensurate interface was created by abutting the $(0001)_\mu$ and $(10\overline{1}0)_\lambda$ faces of the μ and λ half crystals with coincident $[1\overline{2}10]$ axis. The relaxed interface is plotted in $[1\overline{2}10]$ projection in fig.4. The pseudo-periodicity corresponding to $12\sqrt{3}a$ in the lower crystal and $13c$ in the upper is apparent. However, the mismatch between c and $\sqrt{3}a$ is not concentrated exclusively in these zones but is distributed rather evenly along the boundary. This implies that the γ-surface [13] for this interface is relatively flat along the direction orthogonal to the common $[1\overline{2}10]$ axis, and that the easy accommodation of relative shifts of the two half-crystals along it may influence the process of dislocation absorption. Further

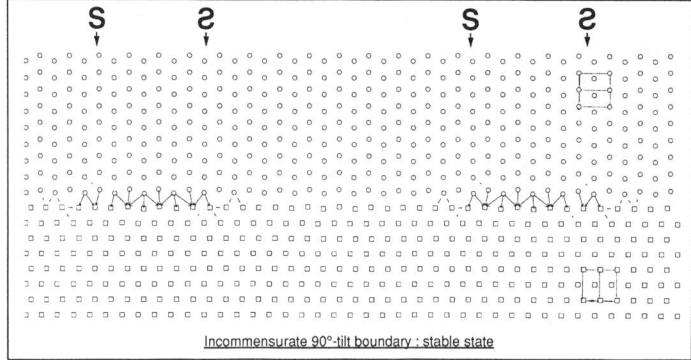

Incommensurate 90°-tilt boundary : stable state

Fig.4.   Atomic projection along the $[1\overline{2}10]$ tilt axis of the inconmensurate 90°-tilt boundary in the state of stable equilibrium, achieved by imposing small equal and opposite $[1\overline{2}10]$ shifts on the λ and μ atoms in the highly-stressed zones at the interface. The arrows show the $[1\overline{2}10]$ displacement of pairs of atoms: the longest arrows correspond to a displacement difference of 0.5a. The relaxation can be seen to be equivalent to a series of screw dipoles ... (+a/2,-a/2) ... , as indicated by the screw dislocation symbols (S) at the top edge of the figure. The outline of hexagonal unit cells in the λ and μ crystals are indicated.

378

relaxation involved $[1\bar{2}10]$ displacements of opposite sign on the μ and λ atoms in the highly-stressed zones. These relaxations can be interpreted as the introduction of pairs of closely-spaced screw dislocations of opposite sign and Burgers vector $\pm a/2$ ($= \pm 1/6[1\bar{2}10]$). The relaxed boundary is therefore equivalent to a series of screw dipoles denoted by either $(+a/2, -a/2)$ or $(-a/2, +a/2)$. Since the dipole spacing is small compared with the separation of these zones of relative displacement, there is little interaction between the dipoles. The magnitude and sign of the difference in the $[1\bar{2}10]$ displacement between neighbouring atoms is shown in fig.4 by the arrows between pairs: the longest arrow corresponds to a displacement difference of $\pm 1/6[1\bar{2}10]$. The energy of this interface is 390 mJm^{-2}, compared with 243 mJm^{-2} for the $(10\bar{1}2)$ twin. The presence of these intrinsic defects introduced by the local relaxations affects the interaction of crystal defects with the tilt boundary so that, unlike the $(10\bar{1}2)$ twin, the interaction of crystal dislocations with this interface depends on their position relative to the intrinsic screw dipoles.

Stability of interfacial defects

Dislocations with Burgers vector parallel to the interface, such as $\mathbf{b}_y$ in Table 2, were simulated but did not remain localized after their initial formation. The interface is not able to sustain shear stresses parallel to y and defects of this type were observed to spread out along the entire interface in the inner region of the computational cell during the simulation.

$\mathbf{b} = \mathbf{a} = \mathbf{b}_{0/0}$

The pure screw dislocation with $\mathbf{b} = 1/3[1\bar{2}10]$ considered in these simulations lies parallel to the common tilt axis, i.e. perpendicular to the paper in fig.4. When it was taken to lie in the lower (μ) crystal, it was created at an origin between the second and third basal planes from the interface in order to ensure that any decomposition observed was not due to an interaction at the interface-plane only. Several positions along the interface between the screw dipoles were considered. The screw in a single crystal can either dissociate on the basal plane or spread its core on the prism plane [12], and so several simulations were carried out. However, the form of the decomposition was found not to depend on this choice.

A possible decomposition product when the screw was created midway between the screw dipoles is plotted in fig.5. In this displacement difference plot, the arrows represent the difference between the $[1\bar{2}10]$ displacement of pairs of atoms relative to the relaxed stable boundary of fig.4. For fig.5, the local relaxation at the two screw dipoles prior to the introduction of the crystal dislocation corresponded to $(+a/2, -a/2)$ and $(+a/2, -a/2)$, as in fig.4. It can be seen that the screw dislocation has been attracted towards the $-a/2$ component on its left and repelled from the $+a/2$ component on its right. When the local relaxation at the screw dipoles corresponds to $(+a/2, -a/2)$ and $(-a/2, +a/2)$, the crystal dislocation is attracted towards both screw dipoles. However, in neither situation is the screw absorbed fully by the boundary and a residual distortion remains in the lower (μ) crystal. When the crystal screw dislocation is located in the upper (λ) crystal with its core spread on the prism plane, i.e. parallel to the interface, it behaves in a similar fashion to the interactions just described. When the core is spread initially on the basal plane, however, then total absorption by the boundary occurs.

$\mathbf{b} = \mathbf{c}$

When a dislocation with $\mathbf{b} = \mathbf{c}(\mu)$ is introduced into the lower (μ) crystal near the tilt boundary, the crystal dislocation decomposes by emitting a 90° Shockley partial in the upper (λ)

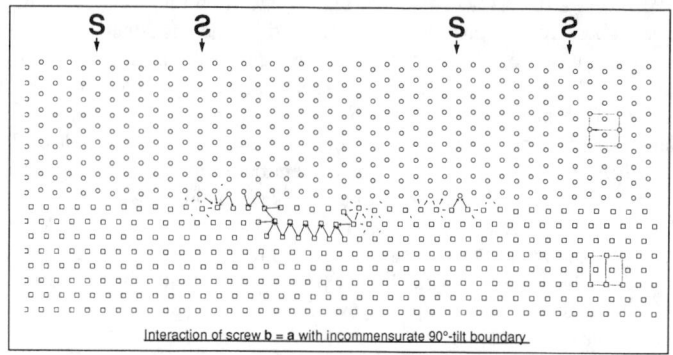

Interaction of screw **b = a** with incommensurate 90°-tilt boundary

Fig.5.   Atomic projection along the $[1\bar{2}10]$ tilt axis of the inconmensurate boundary after the interaction of crystal dislocations. The arrows show the differences in $[1\bar{2}10]$ displacement of neighbouring pairs of atoms relative to the interface in stable equilibrium: the longest arrows correspond to a displacement difference of 0.5a. A screw dislocation with **b = a** was created in the lower (μ) crystal, midway between the screw dipoles.

crystal, thereby trailing an intrinsic basal-plane stacking fault from the interface. At the interface, a dislocation with **b = c**/2 arises and a residual defect with $|\mathbf{b}| = (c/2-a/\sqrt{3})$ is localized in the λ crystal in the immediate vicinity of the interface. When an edge dislocation with **b** = **c**(λ) is near the interface, i.e. **b** parallel to the interface, it becomes delocalised by spreading its core in a manner resembling the b_y defect described above.

DISCUSSION

It is clear that the two interfaces studied in this work exhibit crystallographic and structural differences that are likely to lead to corresponding differences in their responses to incoming crystal dislocations. From the structural point of view, for example, it might be anticipated that the resistance to applied shear stress parallel to the $[1\bar{2}10]_\lambda$ direction would be similar for the two boundaries. On the other hand, the resistance to shear stress parallel to the orthogonal direction in the two interfaces is expected to be considerably smaller for the incommensurate boundary, and this was confirmed in our study [8]. A striking outcome of the present research concerns the core structure of interfacial defects exhibiting step character. In the twin boundary the most favourable configuration of such core risers is for the basal plane of one crystal to be juxtaposed with the first-order prism plane of the other. For the tilt boundary, the most favourable riser configuration corresponds to $(10\bar{1}2)$ planes of the adjacent crystals being strained into parallelism, i.e. forming a structure reminiscent of the $(10\bar{1}2)$ twin. This implies that the core energy of such defects contributes considerably to their total energy. As a consequence, decomposition of crystal dislocations into interfacial defects cannot always be reliably predicted on the basis of elastic energy considerations alone, as in the approximation leading to Frank's rule.

The interaction of $\mathbf{b}_{0/0}$ screw dislocations with the two interfaces is also different, despite the common periodicity parallel to $[1\bar{2}10]_\lambda$ in the two cases. In the instance of the twin, screw dislocations were found to be transmitted, cross-slipping onto an equivalent glide plane in the adjacent crystal, wheras they were absorbed into the incommensurate interface by reaction with the ribbons of screw-dipoles in the intrinsic structure. The nature of the interaction depends on the particular location of the line of impingement of the incoming screw dislocation with respect to the position and sign of the interfacial screw-dipoles.

ACKNOWLEDGEMENTS

This work was supported by the DGICYT project PB96-170-C03-03 of the Spanish Government, NATO Science grant CRG910900 and EEC contract CHRX-CT94-0467.

REFERENCES

1. A.P. Sutton and R.W. Balluffi, *Interfaces in Crystalline Materials*, Oxford Sci. Pub., Oxford, 1995.
2. A. Serra and D.J. Bacon, Phil. Mag. A, **54**, 793 (1986).
3. A. Serra, D.J. Bacon and R.C. Pond, Acta Metall. **36**, 3183 (1988).
4. A. Serra, R.C. Pond and D.J. Bacon, Acta Metall. Mater. **39**, 1469 (1991).
5. T. Braisaz, P. Ruterana, G. Nouet, A. Serra, Ph. Komninou, Th. Kehagias and Th. Karakostas, Phil. Mag. Lett. **74**, 331 (1996).
6. T. Braisaz, P. Ruterana, G. Nouet and R.C. Pond, Phil. Mag. A **75**, 1075 (1997).
7. A. Serra and D.J. Bacon, Phil. Mag. A **73**, 333 (1996).
8. R.C. Pond, D.J. Bacon and A. Serra, Acta Mater. **47**, 1441 (1999).
9. R.C. Pond and J.P. Hirth, Solid State Phys. **47**, 287 (1994).
10. J.P. Hirth and R.C. Pond, Acta Mater. **44**, 4749 (1996).
11. G.J. Ackland, Phil. Mag. A **66**, 917 (1992).
12. A. Serra and D.J. Bacon, Acta Metall. Mater. **43**, 4465 (1995).
13. V. Vitek, Crystal Lattice Defects **5**, 1 (1974).
14. N.E. Paton and W.A. Backofen, Metall. Trans. A **1**, 2839 (1970).
15. R.E. Reed-Hill, E.P. Dahlberg and W.A. Slippy, Trans. AIME. **233**, 1766 (1965).
16. D.A. Smith, Interface Sci. **4**, 11 (1990).
17. R. Beanland, C.J. Kiely and R.C. Pond, in *Handbook on Semiconductors,*, vol.3(a), p.1149, North-Holland, Amsterdam (1994).

DISLOCATION INTERACTIONS WITH LAMELLAR GRAIN BOUNDARIES IN TITANIUM ALUMINUM INTERMETALLICS

Jörg M. K. Wiezorek [1)], Xiao-Dong Zhang [2)] and Hamish L. Fraser [3)]
1)Department of Materials Science & Engineering, University of Pittsburgh, 848 Benedum Hall, Pittsburgh, PA 15261;
2) Reynolds Metals Company, 13203 N. Enon Church Road, Chester, VA 23836;
3) Department of Materials Science & Engineering, Ohio State University, 2041 College Road, Columbus, OH 43210;

ABSTRACT

The interactions of dislocations asscociated with deformation modes active in *hard* orientation compressed polysynthetically-twinned (PST) TiAl with lamellar boundaries have been studied by transmission electron microscopy (TEM). The deformation modes observed in the γ-phase involved *soft* superdislocation and *hard* ordinary dislocation slip, and *hard* twin systems. The transfer of these latter *hard* twin shears occurred across all types of lamellar (γ/γ)- and frequently across (γ/α_2)-boundaries. Slip of superdislocations in the minority α_2-phase lamellae has been shown to be associated with both direct transfer of piled-up twinning dislocation shears across the lamellar boundaries and the activation of interfacial sources due to pile-up stresses. The active deformation modes were consistent with macroscopic shape changes of the PST-TiAl when the various defect interactions with the lamellar boundaries and the proposed shear trasnfer processes were considered.

INTRODUCTION

Due to promising density specific mechanical properties TiAl alloys with fully-lamellar microstructure (FL-TiAl) are considered for high temperature structural applications in modern gas turbine and internal combustion engines [1]. The lamellar grains in FL-TiAl consist mainly of γ-TiAl phase lamellae with the ordered tetragonal L1$_0$-structure and a small volume fraction of α_2-Ti$_3$Al phase lamellae with the ordered hexagonal D0$_{19}$ structure [2]. The γ- and α_2-lamellae obey a crystallographic orientation relationship, $[1\bar{1}0]\gamma//[11\bar{2}0]\alpha_2$ and $(111)\gamma//(0001)\alpha_2$, and six variants of γ-lamellae, matrix variants (I$_M$ to III$_M$) and the twin variants (I$_T$ to III$_T$), coexist with the α_2-lamellae (Figure 1) [2]. The six variants of γ-lamellae are crystallographically equivalent with respect to their relation to the α_2-lamellae (Figure 1, Table I), whereas three types of possible (γ/γ)-interfaces can be categorized as true-twin or 180°-type, pseudo-twin or 60°-type, and order-rotation or 120°-type (Table 1).

The strongly anisotropic mechanical properties of FL-TiAl grains have been studied using directionally solidified material comprising only a single lamellar grain, so-called polysynthetically twinned TiAl (PST-TiAl) [1,2]. As a result of the geometric constraint imposed by the lamellar boundaries *hard* and *soft* modes of deformation and loading conditions exist for PST-TiAl. *Soft* loading conditions obtain for load axes orientations with respect to the lamellar boundaries in the approximate range of 20°$\leq \phi \leq$70°(ϕ is the angle between the boundary plane and loading direction), whereas *hard* conditions obtain for loading either parallel (ϕ=0°, A-type orientations) or normal (ϕ=90°, N-orientation) to the boundaries [2]. The *soft* deformation modes in γ-TiAl involve slip of ordinary dislocations (OD), superdislocations (SD) and ordered twinning systems (TD) associated with Burgers shears contained in the lamellar boundaries, (111). The *hard* deformation modes in γ-TiAl include those slip and twinning systems with Burgers vectors with components normal to (111). The α_2-lamellae are more difficult to deform than γ-TiAl [1-4]. The boundaries between adjacent lamellae have been proposed as efficient obstacles to the motion of *hard* mode dislocations leading to Hall-Petch effects [2,3]. The Burgers vectors associated with the *hard* slip and twinning modes must transfer across the lamellar boundaries during *hard* orientation loading [4]. This can be achieved by either direct shear transfer across the lamellar boundaries, in analogy to slip transfer processes observed across grain boundaries in polycrystalline materials [5], or by activation of dislocation sources in or near to the relevant interfaces due to the pile-up stresses generated by the incipient systems [4,6,7]. Due to the multitude of different types of possible (γ/γ)- and (γ/α_2)-interfaces it has proven difficult to assess their precise roles for the mechanical properties of lamellar

383

TiAl [3]. Furthermore, different micro-mechanisms may exist for each individual type of *hard* mode deformation system and each type of lamellar interface. Previous reports have shown that dislocations with Burgers vector $\mathbf{b}=1/3<11\bar{2}0>=<\mathbf{a}>$ may be activated in the α2-lamellae by shear transfer from the neighboring γ-lamellae, whereas the activation of **c**-component dislocations with $\mathbf{b}=1/3<11\bar{2}6>=<2\mathbf{c}+\mathbf{a}>$ is not as easy and mostly associated with operation of interfacial sources due to pile-up stresses rather than direct slip transfer [4,8,9]. Hence, dislocation interactions with the lamellar boundaries are important to understand the *hard* mode deformation behavior of lamellar two-phase TiAl.

The present paper reports transmission electron microscopy (TEM) investigations of interactions between deformation activated *hard* mode defects and the lamellar boundaries present in PST-TiAl. Implications of the findings for the strength and ductility of lamellar TiAl are discussed.

EXPERIMENTAL PROCEDURE

Coupons of PST-TiAl have been compressed at room temperature at a rate of $5.6\times10^{-4}\mathrm{s}^{-1}$ to strains between 0.01 and 0.07, parallel to $[\bar{1}10]$ (A1), along $[\bar{1}\bar{1}2]$ (A2), and also along [111] (N). Thin disks have been sectioned from the coupons with side-faces parallel to (111), $(\bar{1}10)$ and $(\bar{1}\bar{1}2)$. Standard sample preparation techniques have been employed to obtain thin foils for TEM. Fiducial marks have been rendered onto the thin foils to aid the identification of the macroscopic loading direction with respect to the electron beam direction, allowing the determination of approximate Schmid factors. The γ-variant lamellae are labeled with reference to Figure 1.

RESULTS

Compression tests

Two compression tests have been performed for each of the three hard orientation geometries A1, A2 and N respectively. The compressive yield stress at 0.002 strain, $\sigma_{0.2}$, the plastic strain, ε, and the linear strain hardening, $\delta^{*}=d\sigma/d\varepsilon$, determined from these test are listed in table 2. These mechanical properties of the PST-TiAl are consistent with previous reports [2].

TEM of the deformation microstructures

Interestingly, the deformation microstructures representative of the *hard* orientation loaded

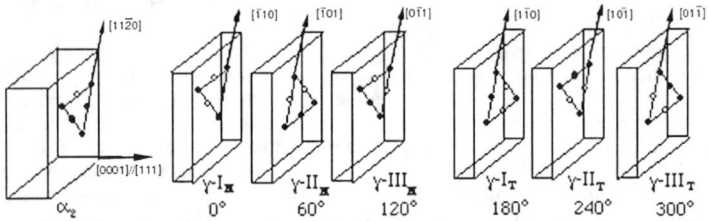

Figure 1: Schematic representation of the orientation relationships between the six γ-variant lamellae and the α2-phase lamellae in PST-TiAl and the γ-variant labeling scheme.

180°-type	60°-type	120°-type	(γ/α2)
I_M/I_T	I_M/II_M	I_M/III_M	all variants
II_M/II_T	II_M/III_M	I_M/II_T	are crystal-
III_M/III_T	III_M/I_T	II_M/I_T	lographically
--------	I_M/III_T	III_M/II_T	equivalent

Table I: The various possible types of crystallographically distinct lamellar (γ/γ)- and (γ/α2)-interfaces in PST-TiAl.

	A_1 $[\bar{1}10]$		A_2 $[\bar{1}\bar{1}2]$		N [111]	
	#1	#2	#1	#2	#1	#2
$\sigma_{0.2}$ [MPa]	533	512	513	541	847	913
ε	0.07	0.01	0.07	0.03	0.01	0.03
$\delta^{*}=d\sigma/d\varepsilon$ [GPa]	1.71	1.69	1.66	1.68	4.21	4.29

Table 2: Yield stress, strain and work hardening data from the roomtemperature tests.

samples exhibited numerous pileups of mostly *hard* mode twin systems against the lamellar interfaces, many of which appeared to transfer across these boundaries (Figure 2). TEM observations of interactions of *hard*-mode twins with the four types of lamellar interfaces in PST-TiAl after room temperature compression in the *hard* orientations are presented below.

Figure 2 depicts twinning dislocation transfer across 180°-type boundaries between γ-II$_M$ and γ-II$_T$, and across the (γ/α_2)-interface. The presence of faulted dipoles (FD) in addition to the deformation twins (T2, T3) is consistent with SD activity, the Burgers vector of which has been determined from additional tilting experiments as $\mathbf{b}=[0\bar{1}1]$. Furthermore, dislocations in the lamellar interface plane, (111), between γ-II$_M$ and γ-II$_T$ have been observed in the vicinity of the intersection of the twins T3 and T2 with the lamellar boundaries (Figure 2a). To the right-hand-side of the twin interaction with the 180°-boundary unusually large ledges in the interface have been observed (Figure 2a). These latter two types of interfacial defects, shown in more detail in Figure 2b, c, have been identified in a previous report [10] to be consistent with SD, $\mathbf{b}=[0\bar{1}1]$, and ledges with $\mathbf{b}$ parallel to $[\bar{2}11]$. Example observations of twin shear transfers across a 60°-type (γ-III$_T/\gamma$-I$_M$)-interface, and a 120°-type interface are shown in Figures 3 and 4. The interaction ot twin T3 with the 60°-boundary produces dislocations near the marker X in lamella γ-I$_M$ (Figure 3a), which have been identified as *soft* SD (Figure 3b). Figure 4 shows a twin interaction with a 120°-type boundary where transfer occurs from γ-I$_T$- to γ-III$_T$ for N-loading. These observations are consistent with the activity of the twinning systems $[\bar{1}1\bar{2}](1\bar{1}\bar{1})$, T1, and $[1\bar{1}\bar{2}](\bar{1}1\bar{1})$, T2, and transfer of T1 across the 120°-interface producing OD, $\mathbf{b}=1/2[\bar{1}\bar{1}0]$. The interaction of *hard* mode twinning in the γ-phase with the (γ/α_2)-boundaries resulted in the activation of <a>-type superdislocations, $\mathbf{b}=1/3<11\bar{2}0>$, on $\{10\bar{1}0\}$ and $\{2\bar{2}01\}$ for A-type loading (Figure 5a,b) and pyramidal slip of <c+a/2>-superpartials, $\mathbf{b}=1/6<11\bar{2}6>$, for N-loading, in agreement with other studies [4].

DISCUSSION

The TEM experiments indicated that for *hard* loading conditions primarily *hard* twinning modes active in the γ-lamellae interacted with the lamellar boundaries and twin shear transfer occured across all 4 types of lamellar boundaries, resulting in either twinning or dislocation slip in the receiving γ-lamellae (Figures 2-4), whereas either <a>-dislocations for A-type loading or <c+a/2>-superpartials for N-loading obtain as products in the α_2-lamellae (Figure 5). These dislocation-boundary interactions in TiAl are similar to dislocation-grain boundary interactions commonly observed during deformation of polycrystals [5]. The most common, succesful shear transfer across grain boundaries in polycrystals involves the boundary reaction $\mathbf{b}_1 = \mathbf{b}_2 + \Delta\mathbf{b}$, where $\mathbf{b}_1$ and $\mathbf{b}_2$ are the Burgers vectors of dislocations in grain 1 and 2 respectively, and $\Delta\mathbf{b}$ is the residual Burgers vector content left in the boundary accounting for the difference in direction and magnitude of $\mathbf{b}_1$ and $\mathbf{b}_2$. Conditions determining the most likely product slip system (slip plane and direction) in

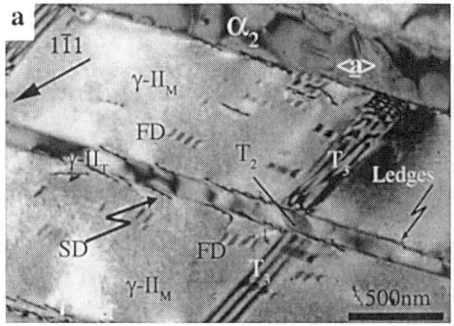
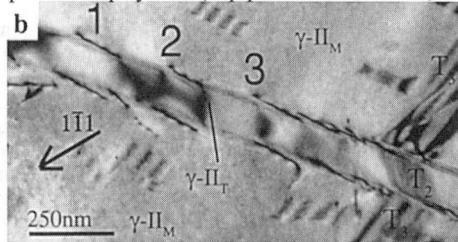
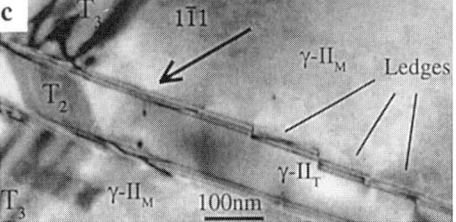

Figure 2: Twin shear transfer occurrs across lamellar boundariess; a) overview, b) and c) enlarged details of transfer across a 180°-type (γ/γ)-bounadry generating SD and ledges in the lamellar boundary (see text for details).

385

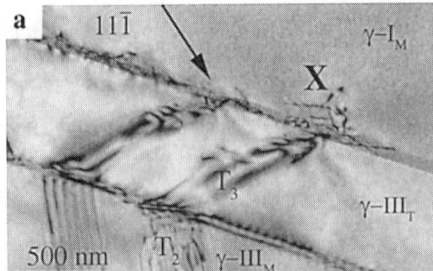 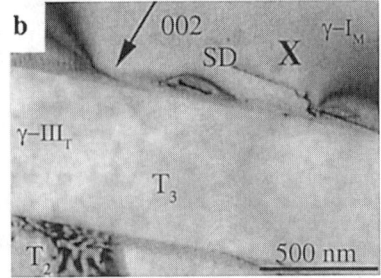

Figure 3: Examples of hard mode twin interactions with a 60°-type boundary resulting in twin shear transfer across the lamellar boundary producing soft SD.

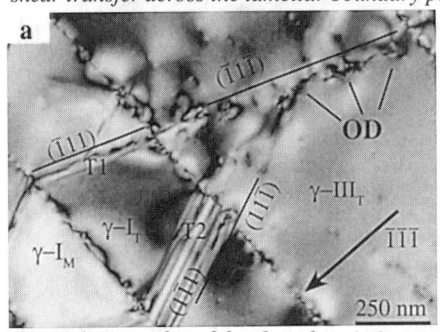 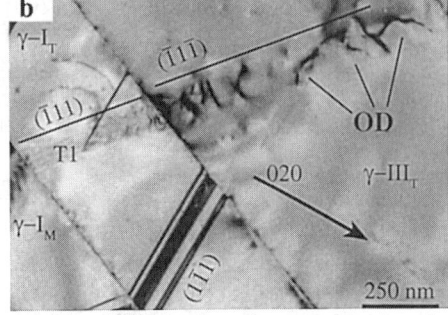

Figure 4: Examples of hard mode twin interactions with a 120°-type boundary resulting in twin shear transfer across the lamellar boundary producing hard OD.

grain 2 for a given incipient system in grain 1 have been identified empirically previously [5]:

1. Minimal rotation between the traces of the two relevant slip planes in the boundary plane;.
2. Sufficient resolved shear stress of the correct sense for the slip systems involved;
3. Zero sum of all Burgers vectors involved in a possible shear transfer reaction;
4. Reactions resulting in smaller magnitude $\Delta\mathbf{b}$ are energetically favored, at least from considerations of elastic line energy.

For all lamellar interfaces *hard* mode slip systems with common interfacial traces exist. In (γ/γ)-boundaries the close packed directions match exactly across the 180°-interfaces, whereas for 60°- and 120°-interfaces they match only approximately, as $(c/a)\approx1.02$. For 120°-interfaces the *hard*-mode octahedral slip planes are almost perfectly continuous across the boundary, whereas for 180°- and the 60°-interfaces these slip planes on either side of the lamellar boundary meet at an angle of $\approx39°$. A minimum angle of $\approx19°$ occurs between the *hard*-mode slip planes in the γ-lamellae and the potential slip planes in the α_2-lamellae that share a common trace in the interface.

Possible twin shear transfer mechanisms consistent with the experimental TEM observations (Figures 2-5) have been determined from considerations of criteria 1.-4. and crystallography. The observations of Figure 2 are consistent with *hard*-twin shear transfer across the 180°-interface according to:

$$6\times1/6[\bar{1}1\bar{2}](\bar{1}11)_{IIM} \Rightarrow 6\times1/6[\bar{1}1\bar{2}](\bar{1}11)_{IIT} + 3\times[0\bar{1}1](111) + 1/3[\bar{2}11](111).$$

A sequential dislocation reaction consistent with the observations for the 60°-interface is,

a) decomposition of the *hard* TD: $3\times1/6[1\bar{1}\bar{2}](\bar{1}11)_{IIIT} \Rightarrow [10\bar{1}](\bar{1}11)_{IIIT} + 1/2[\bar{1}\bar{1}0](\bar{1}11)_{IIIT}$,

b) transfer of the *soft* SD: $\qquad [10\bar{1}](\bar{1}\bar{1}1)_{IIIT} \Rightarrow [01\bar{1}](\bar{1}11)_{IM}$.

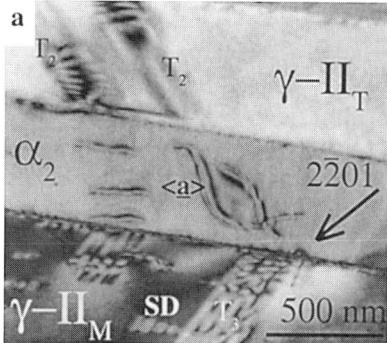
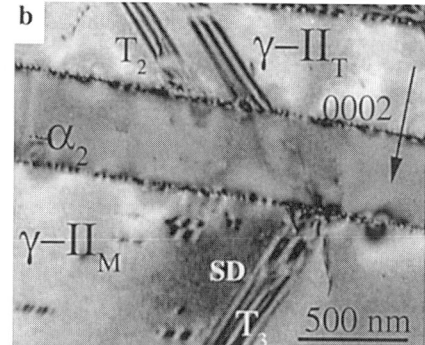
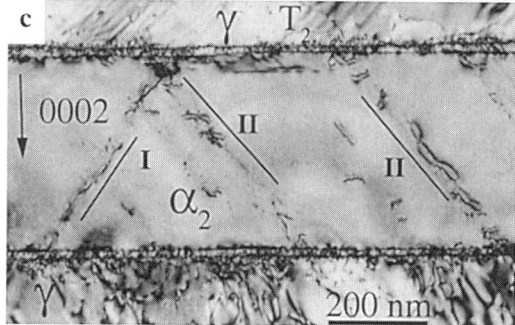

Figure 5: Examples of hard mode γ-twin interactions with (γ/α₂)-boundaries: a) and b) for A-type loading where twin shear transfer across the lamellar boundary produces <a>-dislocations; c) for N-loading γ-twin shear transfer generates pyramidal plane slip activity of <c+a/2>-dislocations on {2̄201}, label I, and {112̄1}, labels II.

For the 120°-type interface the following transfer mechanisms can be proposed:

$$3 \times 1/6[\bar{1}1\bar{2}](1\bar{1}\bar{1})_{IT} \Rightarrow 1/2[\bar{1}\bar{1}0](1\bar{1}1)_{IIIT} + 1/2[\bar{1}01](111).$$

The observations in Figure 5 are consistent with transfer mechanisms across the (γ/α2)-boundaries proposed previously [4,8], e.g. for A-type loading:

a) decomposition of the *hard* TD: $3 \times 1/6[1\bar{1}\bar{2}](1\bar{1}1)_{IIIT} \Rightarrow [10\bar{1}](1\bar{1}1)_{\gamma} + 1/2[\bar{1}\bar{1}0](1\bar{1}1)_{\gamma}$,

b) transfer of the *soft* SD: $[10\bar{1}](1\bar{1}1)_{\gamma} \Rightarrow 1/3[2\bar{1}\bar{1}0](01\bar{1}0)\alpha_2$.

The resultant <a>-dislocation can also slip on {2̄201} for certain geometries. For N-type loading more complex *hard* twin shear transfer mechanisms can generate <2c+a>{2̄201} systems (I in Figure 5c) but cannot account for <2c+a>{112̄1} systems (II in Figure 5c) [4].

All transfer mechanisms identified here are up-hill as far as elastic line energy is concerned, involve slip systems with common traces in the interface that experienced significant resolved shear stresses of the correct sign under the applied loads and the local pile-up stresses. The energy required to initiate and sustain these reactions is provided by the external stress and the local stress concentrations near the pileups. The motion of some of the residual Burgers vector content in the form of interfacial defects (Figure 2) requires the local stresses due to the *hard*-mode twin pileups, since the Schmid factors for these defects vanish for A-type and N-loading. The proposed dislocation reactions enable significant relaxations of incompatibility stresses and enhance the shear compatibility between the lamellae. For instance, for transfer across the 60°- and the (γ/α2)-interface (Figure 3 & 5) the OD system experiences a Schmid factor of opposite sign to that of the TD and SD systems, therefore dissipating stress component normal to the interface and critically enhancing interlamellar compatibility [4,10]. Incompatibility is most severe across the (γ/α2)-boundaries since the α2-phase deforms less easily than the γ-phase. Thus, transfer reactions of *hard* modes of deformation that enhance compatibility across the (γ/α2)-boundaries are especially critical.

It is interesting to note that the defects involved in the proposed hard twin shear transfer processes correlate very well with the deformation modes active in γ-lamellae for *hard* geometries [10,11]. The *hard* mode TD and OD systems have always been observed together for a given γ-variant lamella after A-type loading [10,11]. This is significant as the sum of these shears is parallel to a *soft* SD shear in the (111) interface,

e.g. $1/2[\bar{1}1\bar{2}] + 1/2[110] = [01\bar{1}]$ or $1/2[\bar{1}1\bar{2}] = 1/2[\bar{1}\bar{1}0] + [01\bar{1}]$, and may

well indicate that shear transfer as proposed here and in [4] has indeed occurred. For the N-loaded PST-TiAl only *hard* TD and OD shears have been observed in the γ-lamellae which agrees with previous studies [10,11]. Hence, the defects involved in the proposed transfer reactions are consistent with the macroscopic shape changes produced for PST-TiAl under *hard* loading conditions [10,11].

It has been shown previously for PST-TiAl that the α2-lamellae deform reasonably compatibly with the γ-lamellae for A-type loading by slip of <a>-dislocations, whereas the very difficult activation of the pyramidal <c+a/2>-systems results in a lack of compatibility between the two phases for N-orientation loading [4]. The reasonable compatibility between the deformation modes active in the γ- and α2-lamellae is consistent with the significant ductility observed for A-type loading. The strength of PST-TiAl appears to be governed for A-type loading by the characteristics of the *hard* deformation modes in the γ-phase. The vanishing ductility under tensile condition in the N-orientation may be attributed to the incompatibility between the γ- and α2-lamellae, whereas the strength of PST-TiAl for this loading geometry is governed by the characteristics of the *hard* modes in the γ-phase, and the fraction of the α2-phase. The strength of polycrystalline aggregates of FL-TiAl could be enhanced via Hall-Petch effects in the γ-phase. Conversely, a reduction in the width of the α2-lamellae would reduce the incompatibility with the γ-phase, since the pileup stresses due to *hard* modes become sufficient to drive limited lengths of <c+a/2>-dislocations across them, and improve levels of compatibility for the N-orientation. These predictions are consistent with the reports on highly refined fully lamellar TiAl [12].

CONCLUSIONS

The transfer of *hard* mode twin shears across lamellar interfaces under *hard* loading conditions has been reported and dislocation based mechanisms for the transfer have been determined. Based on the active micro-mechanism of deformation it has been concluded that the strength and the ductility of FL-TiAl may be improved by reducing the width of both γ- and α2-lamellae.

ACKNOWLEDGMENTS
The authors thank the National Science Foundation for support, grant DMR-96-22497 with Dr. B. MacDonald as program manager, and Dr. D.M. Dimiduk for providing the PST-TiAl.

REFERENCES

[1] Kim, Y.-W., Acta Metall., 1992, **42**, 1221.
[2] Fujiwara, T, Nakamura, A., Hosomi, M., Nishitani, S.R., Shirai, Y, and Yamaguchi, M., Phil. Mag. A, 1990, **61**, 591.
[3] Hazzledine, P.M., and Kad, B.K., Mat. Sci. Eng. A, 1995, **A192/193**, 340.
[4] Wiezorek, J.M.K., Zhang, X.D., Clark, W.A.T., and Fraser, H.L., Phil.   Mag. A, 1998, **78**, 217.
[5] Shen, Z., Wagoner, R.H., and Clark, W.A.T., Acta Metall., 1988, **36**, 3231.
[6] Appel, F., Clemens, H., and Wagner, R., Deformation and Fracture of Ordered Intermetallic Materials III,1996, ed. W.O. Soboyejo et al. (TMS: Warrendal, PA) 123.
[7] Smith, D.A.: Structure and Deformation of Boundaries, 1986, ed. K.N. Subramanian and M.A. Imam (TMS: Warrendale, PA), 21.
[8] Forwood, C.T., Gibson, M.A., Miller, P.R., Rossouw, C.J., and Morton, A.J., Structural Intermetallics 1997,1997, ed. M.V. Nathal et al., (TMS: Warrendal PA).
[9] Wiezorek, J.M.K., Zhang, X.D., Godfrey, A., Hu, D., Loretto, M.H.,and Fraser, H.L.,Scri. Mater., 1998, **38**, 811.
[10] Wiezorek, J.M.K., Zhang, X.D., Mills, M.J., and Fraser, H.L., MRS Symp. Proc.,1999, **552**, KK3.5.
[11] Kishida, K., Inui, H. and Yamaguchi, M., Phil. Mag. A, 1998, **78**,1.
[12] Liu, C.T., Maziasz, P.J. and Wright, J.L., MRS Symp. Proc., 1997, **460**, 83.

ATOMISTIC SIMULATIONS OF DISLOCATION-INTERFACE INTERACTIONS IN THE Cu-Ni MULTILAYER SYSTEM

SATISH I. RAO* AND PETER M. HAZZLEDINE*
Air Force Research Laboratory, Materials and Manufacturing Directorate, AFRL/MLLM, Wright-Patterson AFB, OH 45433.

*UES, Inc., Dayton, OH 45432.

ABSTRACT

Multilayered Cu-Ni has a peak yield strength four orders of magnitude higher than either Cu or Ni because the multitude of interfaces obstruct glissile dislocations. The barrier strengths of the interfaces may be traced to four mismatches across an interface: modulus, lattice parameter, chemical and slip geometry. This paper describes sample embedded atom method (EAM) simulations of dislocations crossing interfaces, designed to separate the effects of the four mismatches. The results confirm some classical calculations and emphasize the importance of three new effects (i) an interface-chemical effect in which dislocations are trapped by core spreading in the interface, (ii) a coherency-chemical effect caused by coherency strains changing effective stacking fault energies and (iii) a coherency-modulus effect in which coherency strains change elastic moduli (and hence the Koehler stress) significantly.

INTRODUCTION

Attempts to formulate a theory for the strength of multilayers fall into two categories: The Orowan approach and the Hall-Petch approach. In the Orowan approach [1,2], attention is focused on the first deviation from linear elastic behavior which occurs when a single threading dislocation can propagate along the layers of the softest phase. If the thickness of the phase is h_s, the Burgers vector $\mathbf{b}$ and the shear modulus μ, the stress required to lay down dislocations at the phase boundaries is of order $(\mu b/2\pi h_s)\ell n(h_s/b)$. As the stress rises above this point an increasing fraction of it is transferred to the harder, elastic, phase until it either fractures or yields. In the Hall-Petch approach [3,4], yield occurs when the leading dislocation in a group experiences a large enough stress τ^* to cross the strongest interface and propagate throughout the multilayer. Using the continuum approximation in a material with a propagation stress τ_0, the yield stress is $\tau_0 + (\tau^*\mu b/\pi h_s)^{1/2}$. While it is true that most multilayers become stronger as the layer thicknesses decrease, neither the $h^{-1}\ell n(h)$ dependence nor the $h^{-1/2}$ dependence is accurate over the whole range of h. Experimental measurements [5,6] of τ favor the Hall-Petch $h^{-1/2}$ dependence when h is large (>100nm) but the theory must be modified to account for two facts, first that, when h is small, the number of dislocations in a pile up is small and discrete and second, that τ^* itself depends strongly on h. The first modification progressively reduces the Hall-Petch slope as h decreases, eventually to zero at the saturation point, when dislocations cross interfaces unaided by pile ups. At this point the model is geometrically identical to the Orowan model and the saturation strength is $\tau_0 + \tau^*$. The second modification, that τ^* is not a constant but falls as h falls, creates a peak in the plot of τ against $1/\sqrt{h}$. This peak defines the 'strongest size' for the multilayer. In the Hall-Petch model the most important parameter is τ^*, the barrier strength of the interface between two layers. τ^* may be thought to have four separate components resulting from four mismatches between thick layers:
1. The modulus mismatch or Koehler effect [7]. The energy of a dislocation must rise as it passes from a layer with low μ to one with high μ. Consequently a force opposes this passage.
2. The lattice parameter mismatch [8]. Van der Merwe dislocations at the interface block a glissile dislocation crossing the interface and cause it to bow between the mismatch dislocations.
3. The chemical mismatch. The line energy of a dislocation depends on its core structure and the width of its stacking fault. An increase in this core energy from one layer to another generates a barrier at the interface.
4. Geometrical mismatch. If the slip planes or Burgers vectors do not match on the two sides of an interface, the glissile dislocation may leave a 'difference' dislocation in the interface, or acquire jogs, or, in the case of a screw crossing a twin interface, may cross-slip. All of these contribute to the barrier strength τ^* of an isolated interface.

The value of τ^* for an interface in a multilayer with very thin layers is not necessarily the same as τ^* of an identical, but isolated, interface for two reasons, the first geometrical, the second coherency. Geometrical effects become important when the dimensions of a dislocation e.g. the width of its stacking fault or its core is comparable with the layer thickness so that a dislocation straddles one or more layers. The second reason, coherency, has many effects; among the more important are (i) the glide component of the coherency stress σ_c may exert a large force on a dislocation, a force which changes sign at each interface, (ii) non-glide components of σ_c may, through the Escaig effect [9], change the effective stacking fault energy and hence the chemical component of τ^* dramatically, (iii) σ_c is so large in Cu-Ni that it significantly changes μ in both Cu and Ni (in opposite directions) thereby enhancing the Koehler component of τ^*.

In this paper we summarize a group of EAM simulations designed to calculate the various components of τ^* at both isolated interfaces and in Cu-Ni multilayers with various layer thicknesses. For this purpose several different sets of Cu-Cu, Ni-Ni, and Cu-Ni interatomic potentials have been developed.

NUMERICAL METHODS

In most of the atomistic simulations, empirical EAM potentials are developed by fitting to the properties of Cu, Ni and the disordered Cu-Ni system. Within the EAM format [10], the energy E of an ensemble of atoms is written as the sum of a pair interaction V and a local volume-dependent embedding term F as

$$E = \sum_i E_i = \sum_{i,j,i \neq j} V_{ij}(R_{ij}) + \sum_i F_i(\rho_i)$$

The argument of the embedding term ρ_i is taken to be a sum of pairwise terms, $\rho_i = \sum_{j,j \neq i} \phi_j (R_{ij})$. The form of V is a Morse potential and ϕ is an exponentially decaying function with distance. F is obtained from an exact fit to Rose's equation of state [11]. A given EAM potential is defined by varying the Morse potential parameters, as well as the decay constants in ϕ, to fit the properties of Cu and Ni. The Cu-Ni Morse pair interaction potentials are developed by fitting to the parameters of disordered fcc 0.5Cu-0.5Ni. Table I summarizes the physical properties of Ni, Cu and 0.5Ni-0.5Cu given by the various EAM potentials used in the simulations. In Table I, the potentials with subscript V are the Voter and Chen [12] potentials and these were used for determining the effects of coherency stresses on the mobility of screw dislocations. The Voter and Chen potentials were modified to isolate other particular mismatch components. The potentials labelled Cu, Ni, Cu-Ni were modifications in which the lattice parameters have no mismatch and in which the stacking fault energy mismatch is small but the elastic constant mismatch is large. These potentials were used to study the Koehler effect and the blocking strength of twinned interfaces in a (111) Cu-Ni bilayer. To study the effect of chemical mismatch, the Cu_{chm} and $Cu-Cu_{chm}$ potentials were developed to give the same lattice parameter and elastic constants as Cu, but a very different stacking fault energy.

Table I. Properties of Cu, 0.5Cu-0.5Ni, Ni given by the EAM potentials used in the simulations: Lattice parameter a_0, cohesive energy E_c, elastic constants $B = (C_{11} + 2C_{12})/3$, $C' = (C_{11} - C_{12})/2$ and C_{44}, stacking fault energy E_{sf}

Potential	a_0/nm	E_c/eV/atom	B/GPa	C'/GPa	C_{44}/GPa	E_{sf}/mJ m^{-2}
Cu_V	0.362	-3.54	142	28	81	36
Ni_V	0.352	-4.45	181	48	126	57
Cu_V-Ni_V	0.357	-4.01	159	39	103	38
Cu	0.355	-3.54	142	26	78	73
Ni	0.355	-4.45	181	51	119	92
Cu-Ni	0.355	-4.02	167	38	99	78
Cu_{chm}	0.355	-3.54	142	30	75	136
$Cu-Cu_{chm}$	0.355	-3.55	143	28	75	108

The simulation cell and procedures used in a simulation are illustrated by considering the core structures and friction stresses of screw dislocations in Cu and Ni: The simulation cell has radius 7.5 nm (of which the inner 6.3 nm is atomistic and the outer 1.2 nm elastic) and length one atom diameter along the Burgers vector. Anisotropic elastic displacements corresponding to a dissociated screw dislocation are applied to the whole cell and then the inner 6.3 nm is allowed to relax using EAM interactions with periodic boundaries along **b**. The relaxed core structures are revealed using differential displacement plots to have Shockley partial separations of 1.08 nm in Cu and 1.30 nm in Ni. A pure strain is applied parallel to **b** and on the dissociation {111} plane. The atoms are relaxed in the EAM volume and the cores are examined to see whether they have moved by as much as one lattice period. By applying gradually increasing strains, it was determined that the friction stress in both Cu and Ni is less than $5 \times 10^{-4} \mu$. In all the dislocation-interface simulations, the EAM cell was a rectangular parallelepiped, one atom thick along the dislocation line direction and typically 20 nm x 25 nm normal to the dislocation line, with the interface in the center. The EAM cell was enclosed by a 1.2 nm thick elastic shell. The dislocation (screw, 60^0 or Shockley) was placed initially in Cu and an increasing shear strain was applied, forcing the dislocation towards the interface. Examination of the displacement plots determined the strain (and stress) at which the interface barrier was overcome. In the multilayer simulations the procedure was similar but the EAM cell included several interfaces.

RESULTS
Koehler and Step Effects in a Bilayer

When a screw dislocation is introduced into Cu , it is initially repelled from the (001) Cu-Ni interface (Fig. 1,left). Only when a shear strain is applied does the dislocation move up to the interface and become blocked. Fig.1,center shows the dislocation blocked and with a slightly compressed core at an imposed strain of 0.01μ. When the strain is increased to 0.0105μ, however, first one, then immediately the second, Shockley crosses the interface into Ni, (Fig. 1,right). The Koehler component of τ^* for a screw dislocation at an isolated (001) Cu-Ni interface is close to 0.01μ. Similar simulations were carried out for screws crossing {111} and {011} interfaces, with results for τ_K^* of 0.014μ and 0.011μ respectively. The dependence of the Koehler stress on interface orientation is weak, in agreement with continuum predictions. When a screw dislocation crosses an interface, it creates no step in the interface. But any other dislocation does create a step and the energy required contributes to the barrier strength. To test the importance of the step contribution, a 60^0 dislocation was made to cross a {111} Cu-Ni interface. It did so at a stress of between 0.0125μ and 0.015μ, close to the value 0.014μ for a screw dislocation. The Koehler stress in Cu-Ni therefore appears to lie in the range 0.01μ to 0.015μ and is strongly dependent on neither the interface orientation nor the dislocation character.

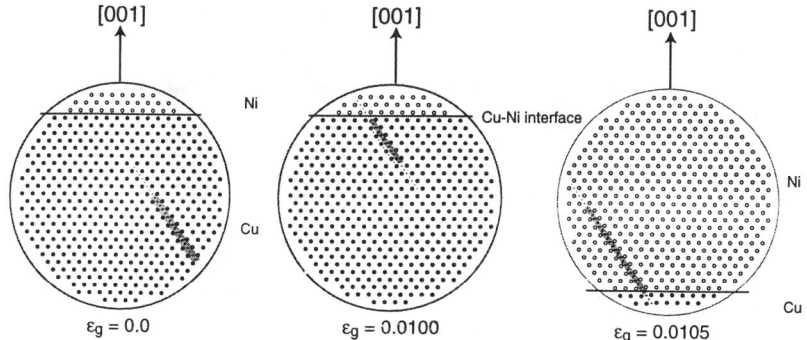

Figure 1. [110]/2 screw dislocation crosses a Koehler barrier at applied strain ~0.01

Koehler Effect in a Multilayer

The Koehler stress in a fine multilayer is not equal to τ_K^* in a bilayer because the dislocation may interact with many interfaces simultaneously. To test the importance of this effect, three simulations were run for screw dislocations crossing (001) interfaces between layers which were 16 atoms, 8 atoms and 4 atoms thick. As in the bilayer, the maximum stress was reached at the Cu->Ni interface. In the thickest multilayer (wavelength 5.68 nm), a dislocation is contained within one layer and its Koehler stress is 0.0095μ, close to the bilayer value of 0.01μ. In the thinnest multilayer (wavelength 1.42 nm) the dissociated dislocation straddles four layers and consequently experiences a smaller Koehler stress, 0.004μ. The values of τ_K^* are summarized in Fig. 2 which shows that the stress remains about constant while a single layer can accommodate a dissociated dislocation, but that the stress falls sharply as soon as the dislocation width exceeds a layer thickness.

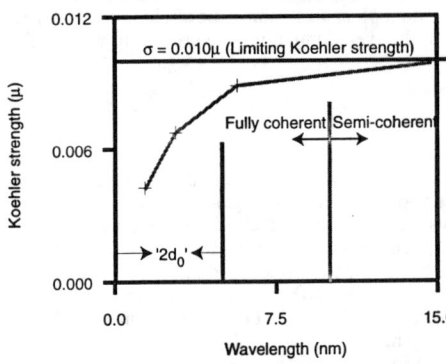

Figure 2. Koehler barrier strength as a function of the wavelength of a 50% Cu-50% Ni multilayer. Coherency effects are suppressed. The Koehler stress is low when the layer thickness is smaller than the dissociation width of a dislocation because the dislocation interacts with two or more layers simultaneously. The Koehler barrier strength is insensitive to the character of the dislocation and to the orientation of the interface.

The Chemical Interaction in a Bilayer

In this simulation of a screw dislocation approaching an (001) interface, the chemical effect is isolated from other effects by using the Cu_{chm} potential (with its attendant high stacking fault energy) as a surrogate for Ni. The value of τ^* is between 0.0025μ and 0.0038μ. After correction for the small Koehler effect in the simulation (see shear moduli in Table I), the chemical contribution to τ^* is estimated to be 0.003μ, a factor of three smaller than τ_K^*. However, see below, the chemical effect can be greatly enhanced or diminished by the presence of coherency stresses whose non-glide component may alter the effective stacking fault energies in the Cu and Ni layers in opposite directions and hence make a large contribution to the chemical stress.

Slip Plane Discontinuity in a Bilayer

The simplest slip plane discontinuity to model is that at a coherent {111} twin interface. A simulation at a Cu-Cu twin showed that a screw dislocation approaching the interface cross-slipped onto the interface plane at a low applied stress of between 0.0025μ and 0.0050μ. The screw was then firmly anchored by the fact that its stacking fault energy in the interface and hence its line energy was low. At a simulated Cu-Ni twin interface, the same process occurred at a stress only 0.0035μ to 0.006μ above the Koehler stress. 60^0 dislocations cannot cross-slip into the interface but they are strongly blocked by the slip plane discontinuity. When the 60^0 dislocation passes from Cu to Ni, it transfers from a {111} slip plane to a nearly parallel {100} plane whereas at the Ni->Cu interface, it transfers from {111} to the equivalent twinned {111} slip plane. Simulations show that τ^* is between 0.0125μ and 0.015μ at a Ni->Cu interface and between 0.05μ and 0.06μ at a Cu->Ni interface. Adding τ_K^* (0.014μ) to the former and subtracting it from the latter gives an estimate of the blocking strength τ^* of 0.03μ to 0.04μ for 60^0 dislocations.

Effects of Coherency in a Multilayer

The effects of coherency at an (001) interface between Cu and Ni were studied by using the Voter and Chen potentials with matched lattice parameters in the interface plane. The lattice parameters were made to match by inducing coherency biaxial stresses of +1.99 GPa in Ni and -2.55 GPa in Cu. When a screw dislocation was placed in Cu, the Escaig coherency stress was sufficient to separate the Shockley partials indefinitely (i.e. reduce its effective stacking fault energy to zero) so the simulation was carried out using a single 30^0 Shockley partial. In Ni, the partials of a screw would be compressed by the Escaig stress, thereby increasing the effective stacking fault energy. Apart from exerting large internal forces on a dislocation, the coherency stresses changed the elastic constants of Ni and Cu significantly, increasing the line energy of the Shockley dislocation in Ni by 6% and reducing it by 20% in Cu. In the simulation, the blocking stress of the interface was found to be 0.025μ, larger by a factor of 2.5 than the stress at which the leading Shockley crossed a Cu->Ni interface in the 'Koehler' simulation above. This large increase in the barrier strength induced by coherency stresses can be attributed in part to the enhancement in the modulus difference between Cu and Ni and partly to the enhancement in the chemical effect. It happens that in the Cu-Ni multilayer at {001} interfaces, these effects are of the same sign but this would not necessarily be true of all multilayers or all interface orientations.

DISCUSSION

The most powerful barrier investigated in this series of EAM simulations is the 'slip geometry mismatch' barrier, with a strength at Cu-Ni twin boundaries of 0.03μ to 0.04μ. In polycrystalline multilayers this is likely to be the main contributor to τ^*. In epitaxial multilayers, this effect is absent and the barrier, at long wavelengths, has two nearly equal components of $\sim0.01\mu$, one from mismatch dislocations (τ_d^*) and the other from the Koehler effect (τ_K^*) plus a smaller component, τ_c^* $\sim0.003\mu$ from the chemical effect. At Cu->Ni interfaces, all three components are positive whereas the latter two change sign at Ni->Cu interfaces. Cu->Ni interfaces are therefore stronger than Ni->Cu interfaces. In finer multilayers the layers become more coherent, the density of mismatch dislocations and τ_d^* fall and the coherency stresses rise.The rising coherency stresses enhance τ_K^* at any interface but their effect on the chemical stress τ_c^* is mixed: at {111} interfaces τ_c^* is enhanced but at {001} interfaces, it is diminished.

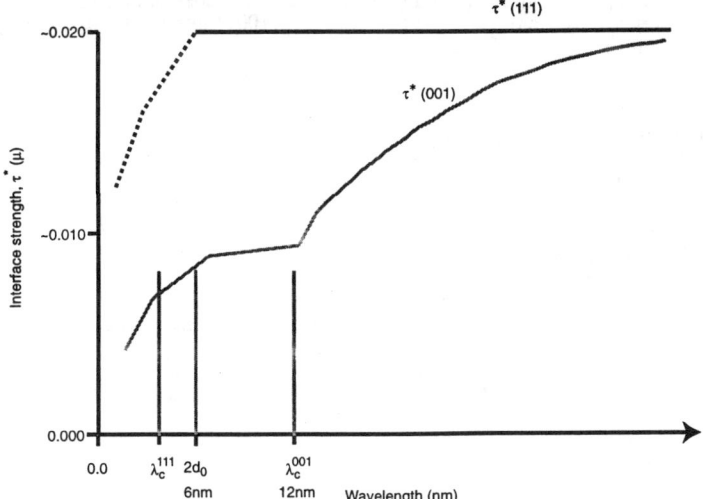

Figure 3. Schematic plot of the {111} and {001} total Cu->Ni interface barrier strengths as a function of the wavelength λ in single-crystal Cu-Ni multilayers. λ_c is the coherency limit and d_0 the width of a dissociated screw dislocation in Cu.

The resultant of all these effects is shown schematically in Fig.3: at long wavelengths, both interfaces have the same barrier strength and both weaken as λ decreases. At the $\{001\}$ interfaces, the increase in τ_K^* is roughly compensated by the decrease in τ_c^* so the main effect is the reduction in τ_d^* as coherency is approached. Below λ_c^{001}, τ_d^* is zero and τ_c^* and τ_K^* are constant. At $\{111\}$ interfaces (for which the coherence limit λ_c^{111} is considerably smaller than λ_c^{001}) the decline in τ_d^* is compensated by the changes in τ_K^* and τ_c^* so that the total τ^* is roughly constant. At very short wavelengths, comparable to the dislocation width d_0, both barriers become rapidly weaker because dislocations straddle more than one layer. At a wavelength of ~0.3 nm the barriers cease to exist as the multilayer becomes an ordered alloy.

A fuller version of this paper will be published in the Philosophical Magazine in 2000.

CONCLUSIONS

1.   The Koehler stress τ_K^* is confirmed to be 0.01μ to 0.015μ, positive at the Cu->Ni interface and negative at the Ni->Cu interface. In coherent multilayers, τ_K^* drops rapidly when the width of a dislocation is comparable with the layer width. In coherent Cu-Ni multilayers the coherency stress enhances the modulus difference between Cu and Ni and hence enhances τ_K^*.
2.   The lattice parameter mismatch stress is provided by the blocking action of van der Merwe dislocations in incoherent (long wavelength) multilayers and is of order 0.01μ, positive at all interfaces. In thinner multilayers the stress falls (to zero at the coherence limit) but its effect remains strong at all wavelengths because, as the density of van der Merwe dislocations falls, the coherency stress rises. Coherency stresses exert glide forces on mobile dislocations, modify their cores and change the elastic constants.
3.   The chemical stress itself τ_c^* is comparatively small in Cu-Ni, 0.003μ (positive at Cu->Ni, negative at Ni->Cu), but it is strongly affected by coherency Escaig stresses. At $\{001\}$ interfaces, coherency reduces τ_c^* by 0.005μ by constricting Cu dislocations and expanding Ni dislocations. At $\{111\}$ interfaces exactly the opposite occurs and τ_c^* is increased by 0.005μ.
4.   Interfaces at which the slip plane is discontinuous are powerful obstacles. 60^0 dislocations can penetrate twin interfaces only at stresses of 0.03μ to 0.04μ. Even screw dislocations are strongly blocked because they cross-slip into the interface where the stacking fault energy is low.
5.   The total barrier strength of both $\{111\}$ and $\{001\}$ interfaces is ~0.02μ at long wavelengths. As the wavelength is decreased, τ^* falls, gradually at $\{001\}$ interfaces, precipitously at $\{111\}$ interfaces, see Fig. 3.

REFERENCES

1. L. B. Freund, J. Appl. Mech. **54**, 553 (1987)
2. J. D. Embury and J. P. Hirth, Acta Met. Mat. **42**, 2051 (1994)
3. P. M. Anderson and C. Li, Nanostr. Mater. **5**, 349 (1995)
4. S. I. Rao, P. M. Hazzledine and D. M. Dimiduk, MRS Symp. Proc. **362**, 67 (1995)
5. S. L. Lehoczky, J. Appl. Phys. **49**, 5479 (1978)
6. A. Kelly, Phil. Trans. Roy. Soc. **A322**, 409 (1987)
7. J. S. Koehler, Phys. Rev. **B2**, 547 (1970)
8. F. C. Frank and J. H. van der Merwe, Proc. Roy. Soc. **A198**, 216 (1949)
9. B. Escaig, in Dislocation Dynamics edited by A. R. Rosenfield, G. T. Hahn, A. L. Bement, R. I. Jaffee (McGraw Hill, New York, 1968), p. 655.
10. M. S. Daw and M. I. Baskes, Phys. Rev. **B29**, 6443 (1984)
11. J. H. Rose, J. R. Smith, F. Guinea and J. Ferrante, Phys. Rev. **B29**, 2963 (1984)
12. A. F. Voter and S. P. Chen, MRS Symp. Proc. **82**, 175 (1987)

ACKNOWLEDGEMENTS

The authors acknowledge support from the AFRL Materials and Manufacturing Directorate Contract #F33615-96-C-5258 with UES Inc.This work was supported in part by a grant of HPC time from the CEWES DOD HPC centers: CEWES-CRAY-YMP, CRAY-C90 and ASC CRAY C90 supercomputers.

IDENTIFICATION OF 2D BOUNDARIES FROM 3D ATOM PROBE DATA, AND SPATIAL CORRELATION OF ATOMIC DISTRIBUTIONS WITH INTERFACES

O.C. HELLMAN, J.A. VANDENBROUCKE*, J. RÜSING, D. ISHEIM and D.N. SEIDMAN
Dept. of Materials Science and Engineering, Northwestern University, Evanston, IL 60208-3108
* Dept. of Physics, Stanford University, Stanford, CA 94305-4060

ABSTRACT

The Three Dimensional Atom Probe produces a real space map of the elemental identities and positions of atoms field-evaporated from a sharply pointed specimen. The analyzed volume is on the order of 20 nm x 20 nm x 100 nm. This is large enough to enclose microstructural features such as grain- or heterophase boundaries. Correlation of the measured atomic positions with such features results in an atom-by-atom description of the chemical environment of these crystallographic defects. We describe here a method for identifying these interfaces and profiling the composition in the vicinity of the interfaces without any assumptions about the interface geometry. This approach is applied to quantitative determination of interfacial segregation of Ag at a MgO/Cu(Ag) heterophase interface. We discuss the implications of our technique with respect to classical treatments of segregation at interfaces.

INTRODUCTION

Experimental techniques such as Auger Electron Spectroscopy (AES), Scanning Transmission Electron Microscopy (STEM) with Electron Energy Loss Spectroscopy (EELS), or one-dimensional Atom-Probe Field-Ion Microscopy (APFIM) can be used to measure the interfacial excess of a solute at an interface, for example grain boundaries [1] or heterophase boundaries [2]. However, these techniques are not straightforward to apply when the interface is not planar, because composition measurements are taken as a one-dimensional profile, and the interface is assumed to be planar and uniform.

In contrast, Three Dimensional Atom Probe (3DAP) [3,4] analysis results in a three-dimensional map of positions and chemical identities of the detected atoms, and therefore the three dimensional structure of interfaces can be quantified [5]. In some cases, a planar interface within this volume can be identified, and a one dimensional profile can be generated such that the data can be treated like that from a one-dimensional atom-probe analysis. However, analyses of segregation near small precipitates or of segregation to interfaces in most microstructures resulting from a phase separation are not likely to have uniform, planar interfaces. Indeed, even a basic identification of the interfaces in these structures can be complicated.

This paper outlines the data analysis procedure we use to extract a quantitative value for segregation of a solute at an interface from 3DAP data. Starting with a set of positions and identities of atoms in the sample, we first perform a *sampling*, which generates a regularly spaced three-dimensional grid of composition values. Then the grid concentrations are *interpolated* to identify the interface corresponding to a chosen composition of a given species. A *correlation* is then performed between the original set of atomic positions and this interface, wherein the distance from each atom to the interface is calculated, and a histogram of the populations of all the species with respect to proximity to the interface is tabulated. The result is that the three-dimensional atomic positions have been transformed into a single parameter representing the proximity to the interface. The normalization of this histogram results in a "proxigram", a concentration profile with respect to an interface of arbitrary geometry.

PROCEDURE

Sampling

Sampling consists of transforming the set of atom positions and chemical identities into a regular grid of concentration values. A desired grid spacing is chosen, and the composition for each grid point is defined from nearby atoms. The simplest scheme for sampling the data is to define a volume element, or *voxel*, corresponding to the grid point, and to calculate a concentration from the population within the voxel. This corresponds to applying a rectangular-shaped transfer function from the initial data onto the set of sampled data, where the width of the rectangle is the edge length of the voxel. If this width is also equal to the grid spacing, then each atom in the initial data set contributes to one voxel. Different shapes of transfer function may also be used: we commonly use a sawtooth transfer function, where each atom makes a weighted contribution to nearby grid points. The weighting is based on the distance from the position to the grid points, such that the contributions of all atoms to the grid are equal.

The choices for grid spacing and shape of the transfer function will affect the resulting concentration grid. A decreased grid spacing increases spatial resolution but also requires more memory and computation time. Wider transfer functions decrease the statistical error but also decrease resolution. The rectangular transfer function has better resolution than the sawtooth transfer function, while the sawtooth has better fidelity in reproducing a given sampling with respect to a shift in the position of the grid. For our analyses, we typically use a sawtooth transfer function with a grid spacing of 1.5 or 2.0 nm and a tooth length of double the grid spacing, so that each atom makes its weighted contribution to the eight grid points which define the cube enclosing the atom's position (note: these are not necessarily the eight closest grid points).

When the data is sampled, the atomistic nature of the data is removed entirely, and we assume that the calculated compositions represent a continuum, i.e. smoothly varying values of concentration.

Interpolation

Once the data is properly sampled, an interpolation can be performed to define an isoconcentration surface in three dimensions. This surface represents all of the points in the space where the local concentration of a species is equal to a chosen value. This is analogous to generating a contour plot in two dimensions from a two dimensional grid of data, where a contour represents a single concentration value, as in elevation contours on a terrain map. However, instead of a line, the isoconcentration surface is a two-dimensional object in three-dimensional space. We use the "marching cubes" algorithm of Lorensen and Cline [6] to generate this surface. The calculated surface is a discrete set of triangles; the vertices of each triangle lie between grid points at a point where the linear interpolation of the concentration values of the grid points is equal to the chosen value. Furthermore, each triangle has an associated direction pointing in the direction of increasing concentration.

Figures 1 and 2 illustrate the definition of an isoconcentration surface for 3DAP data from a MgO precipitate in a copper matrix. In Figure 1, the reconstructed positions of Mg atoms from the sample are shown. In Figure 2, the Mg 11 at% isoconcentration surface is shown. The value of 11 at% is chosen because it is close to the steepest concentration gradient for Mg near this particle, and thus represents the minimum of the change in position of the interface with respect to choice of the isoconcentration value. In this case, the surface points inward, towards the increasing Mg concentration in the center of the particle. A corresponding surface calculated for Cu might have a similar position but opposite sign.

Correlation

Returning to the original set of atomic positions and identities, we perform a correlation between them and an isoconcentration surface. The surface provides a point of reference for a data reduction: the 3D coordinate of each atom is reduced computationally to a single parameter: the proximity to the interface. This is computationally intensive but straightforward. Because the

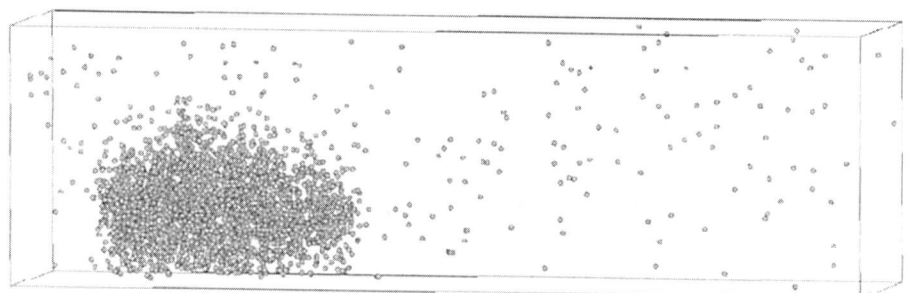

Figure 1. 3D Atom Probe reconstruction of the positions of Mg atoms in an internally oxidized Cu(Mg, Ag) alloy. The volume is approximately 17 nm x 17 nm x 57 nm

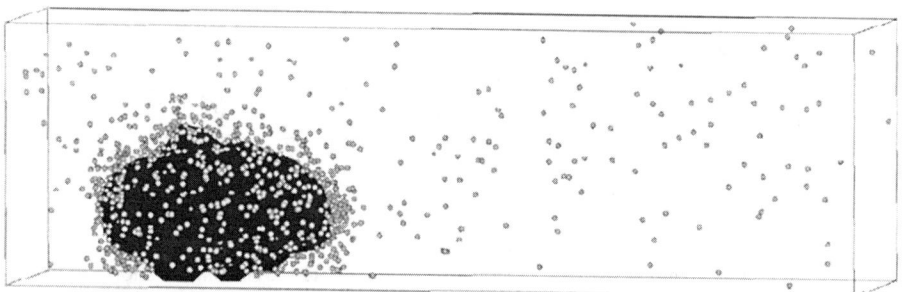

Figure 2. Representation of the Mg 11 at. % isoconcentration surface for the volume shown in Fig. 1, with Mg atoms overlay. The surface is composed of 456 triangles. The concentration grid was calculated with a grid spacing of 1.5 nm and sawtooth transfer function.

surface is just a set of triangles, the closest triangle to each atom can be identified, and its distance to the atom is calculated. If the atom lies on the side of the surface facing lower concentrations in the concentration grid, the proximity is taken to be negative. A histogram is built of number of atoms versus this proximity. It is often the case that the shortest distance to the surface is the distance to a vertex that is shared by four or more triangles. In this case, although the distance to each triangle is identical, care must be taken to choose the appropriate direction, as it is possible that the point lies on the positive side of some of the triangles, and on the negative side of others.

Normalization of the populations in each bin of the resulting histogram results in an atomic concentration value for each bin. The volume associated with each bin is not constant, as the surface can be curved and intersect with the analysis volume in topologically arbitrary ways. However, normalization removes any effect of a non-constant volume from the histogram and produces valid compositions. We note that this is accomplished without actually calculating the volume itself, or even calculating the area of the isoconcentration surface or its curvature. A normalized proxigram is presented in Fig. 3. The population of each bin of this proxigram is shown in Fig. 4. The error bars in Fig. 3 represent the one-sigma statistical error. Where the population is the highest, the corresponding error is the smallest. In Figs. 3 and 4, the best statistics are found in the two bins just below zero on the x-axis, just outside the precipitate.

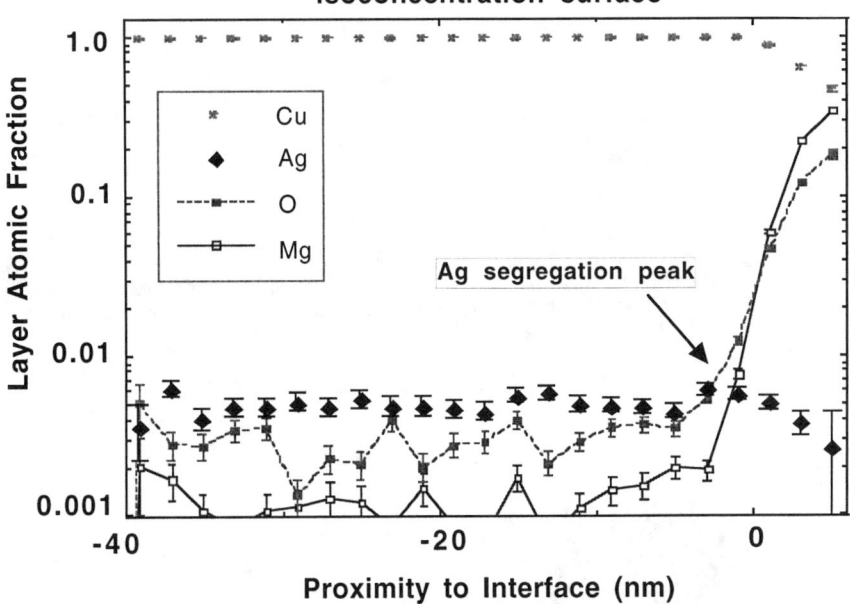

Proxigram with respect to the Mg 11% isoconcentration surface

Figure 3. Proxigram of Cu, Mg, O, and Ag with respect to the isoconcentration surface shown in Fig. 2; the bin size is 2 nm. The calculation is made over a volume approximately 18 nm x 18 nm x 119 nm, and includes two precipitates. There are no detected atoms a distance greater than 40 nm from the isoconcentration surface.

The normalization procedure also determines the upper and lower bounds of the proxigram, and the statistical error of each proxigram point. At the upper and lower bounds, the population of the bins becomes zero, because of the limited spatial extent of the analysis volume. Close to these bounds, the population becomes very small, thus increasing the statistical error at both ends. We emphasize that the x–axis of the proxigram is not a linear measure of distance, because the isoconcentration surface is not planar. In this sample, the upper bounds represent the interior of a precipitate, because the isoconcentration surface is wrapped around a small precipitate and the concentration gradient of the isoconcentration surface is pointing to the interior of the precipitate. The lower bounds of the proxigram represent the limits of the analyzed space far from the precipitate. At the region of the proxigram just outside a precipitate, where the statistics are best, each bin represents approximately 25,000 atoms. Even at this level, the statistical error for the low-concentration components is severe; the one sigma error value for a measured concentration of 0.5 % is on the order of 0.05%.

The proxigram shows a small but unambiguous Ag excess outside the MgO precipitates. The Ag concentration in the Cu matrix is observed to be 0.5 at.%. In the 5 nm outside the interface the Ag concentration increases to 0.61 at.% and the single standard deviation is well above the average concentration. The two data points of the proxigram exhibiting an interfacial excess each represent layers of approximately 25,000 atoms, and thus the measured excess in these two layers corresponds to approximately 46 Ag atoms, above the 250 Ag atoms expected in the layers for a bulk composition of 0.5 at. %. Details of this sample and its analysis appear elsewhere [7].

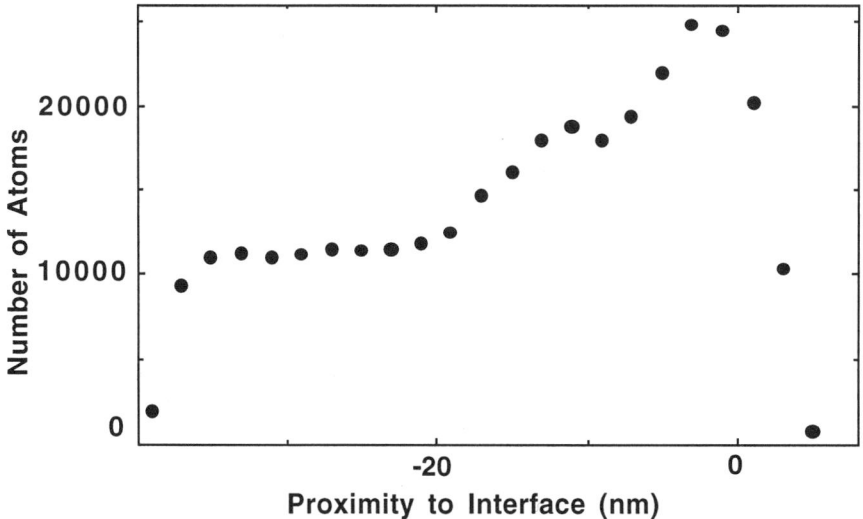

Figure 4. Distribution of atom populations in the proxigram of Fig. 3. Statistics are best in the region immediately outside the precipitates. Negative proximity indicates the distance outside of the surface.

DISCUSSION

Interfacial excess of a segregating species is commonly formulated in terms of a number of segregating atoms per unit area of interface [8-10]. The resulting proxigram does not actually have units of atoms-per-unit-area. However, assuming a constant atomic density, the y-axis can be transformed from an atomic fraction for each species to a measure of atoms-per-unit-volume. Each bin of the proxigram represents a fixed width, and thus the area under a proxigram curve represents atoms-per-area.

We note that this treatment finesses a fundamental problem in the treatment of interfacial thermodynamic quantities for non-planar interfaces. The straightforward way of calculating segregation is to count the segregant atoms and divide by the area of the precipitate's surface. As in Gibbs' treatment, this method requires the identification of the position of a dividing surface. For a non-planar interface, however, the choice of the dividing surface affects the area of the interface. If a larger precipitate radius is chosen, the measured number of segregating atoms per interface area is reduced. The Gibbsian excess, however, is a true thermodynamic quantity and should not depend on such an arbitrary choice.

One approach to this problem is to link the measured quantity to the choice of the surface. The Gibbsian excess Γ would be expressed as Γ [r], with the brackets representing the fact that the dependence of Γ on r is dependent on the arbitrary choice of the position of the dividing interface, not on an actual change in the position of the physical interface [11]. In the calculation of the proxigram, the area of the interface is never defined, but instead is a calculated result based on an expected atomic density.

CONCLUSIONS

When analyzing data from a 3D Atom-Probe for the purpose of identifying quantitative measures of segregation at interfaces, it is desired to simplify a large set of three-dimensional atomic positions and chemical identities into a one-dimensional set of atomic concentrations; where the parameter of interest is the proximity to an interface. This paper describes how such an interface is identified, and how such data deconvolution from three dimensions to one dimension is performed, while preserving as much as possible the relevant physical properties of the data set. This new approach is applied to the problem of determining the level of interfacial segregation of silver at a MgO/Cu(Ag) heterophase interface.

ACKNOWLEDGEMENTS

This research is supported by the National Science Foundation, Division of Materials Research, Grant DMR-972896, Bruce MacDonald, Grant Officer. J.A. Vandenbroucke was supported by an NSF REU grant. D. Isheim received partial support from the Deutsche Forschungsgemeingschaft (DFG), and the Alexander von Humboldt Foundation through the Max Planck research prize of DNS. J. Rüsing was supported by the U. S. Department of Energy under grant DE-FG02-96ER45597.

REFERENCES

1. B.W. Krakauer and D.N. Seidman, *Acta Mater.* **46** (1998) 6145-6161.

2. D.A. Shashkov, D.A. Muller and D.N. Seidman, *Acta Mater.* **47** (1999) 3953-3963.

3. D. Blavette, B. Deconihout, A. Bostel, J.M. Sarrau, M. Bouet and and A. Menand, *Rev Sci Instrum.* **64** (1993) 2911-2919.

4. A. Cerezo, T.J. Godfrey, S.J. Sijbrandij, G.D.W. Smith and P.J. Warren, *Rev. Sci. Instrum.* **69** (1998) 49-58.

5. L. Letellier, A. Bostel, D. Blavette, *Scripta Metall. Materialia* **30** (1994) 1503-1508.

6. W.E. Lorensen and H.E. Cline, ACM SIGGRAPH Computer Graphics **21**, No. 4 (1987), 163-169.

7. J. Rüsing, J.T. Sebastian, O.C. Hellman and D.N. Seidman , submitted to *Microscopy and Microanalysis* 1999.

8. J.W. Gibbs, *The Collected Works of J. Willard Gibbs*, Yale Univ. Press, New Haven, 1948 Vol. 1 pp. 219-252.

9. J.W. Cahn in *Interfacial Segregation*, Eds. W.C. Johnson and J.M. Blakely, Ohio, ASM, 1979, pp. 3-23.

10. A.P. Sutton and R.W. Balluffi, *Interfaces in Crystalline Solids,* Oxford Univ. Press, New York: 1995.

11. M.P.A. Fisher and M. Wortis, *Phys Rev.* **B29** (1984) 6252-6260.

ATOMISTIC MODELING OF INTERFACIAL DIFFUSION IN THE LAMELLAR $L1_0$ TiAl

M. NOMURA, D. E. LUZZI, and V. VITEK
Department of Materials Science and Engineering, University of Pennsylvania, Philadelphia, Pennsylvania, 19104-6272

ABSTRACT

Atomistic simulation employing many-body central-force potentials was performed to elucidate the diffusion mechanisms in the bulk and at lamellar interfaces assuming a vacancy mechanism. First the self- diffusion of Ti and Al in stoichiometric structures was studied. It was found that the diffusion was faster along lamellar interfaces than in the bulk; the effective activation energy for the diffusion coefficient is about ~15% lower. The simulations were then extended to investigate diffusion along lamellar boundaries with segregated Ti which is likely in Ti rich alloys. The surprising result is that diffusion remains practically unchanged when compared with the stoichiometric case. The reason is that while the path controlling the diffusion is different, the corresponding effective formation and migration energies are practically the same as in the stoichiometric case.

INTRODUCTION

In recent years lamellar TiAl with the $L1_0$ structure has been identified as a promising high-temperature structural material owing to its relatively high ductility and toughness at ambient temperatures [1-9]. At high temperatures the physical and mechanical properties of materials are commonly linked with diffusion, and thus understanding of the diffusion mechanisms in TiAl is essential for a fundamental comprehension of the high temperature properties of this compound. In general, diffusion is appreciably faster at interfaces than in the bulk because the atomic structure at the interface is less ordered than the bulk. However, lamellar interfaces consist of twin-like boundaries [10, 11]: ordinary twins (OTW), pseudotwins (PTW) and 120^0rotational faults (120RF). These possess high-symmetries and relatively well ordered structures and thus it is not a priori clear that the diffusion is faster along the lamellar interfaces. Yet, if it is, then the interfacial diffusion is likely to dominate the self-diffusion in this material because the density of interfaces in lamellar TiAl is high.

The bulk self-diffusion in TiAl and Ti_3Al was investigated experimentally by Herzig et al. [12, 13] and studied theoretically using EAM type description of atomic interactions [12-14]. The calculations show a good agreement with experiments suggesting that the EAM type potentials suffice for this study. Neither experimental nor theoretical investigations of the diffusion in lamellar interfaces have been made so far and the calculations presented in this paper are the first attempt to elucidate self-diffusion in lamellar interfaces and contrast it with bulk diffusion.

However, lamellar TiAl is Ti-rich [15] and thus it is likely that Ti will segregate to the lamellar interfaces. This has, indeed, been observed by analytical electron microscopy [16] and recent Monte Carlo simulations suggest that a strong segregation occurs at OTW and 120RF leading to the formation of a Ti_3Al-like layer at the interfaces [17]. Hence, we also investigate in this study the diffusion in such segregated interfaces in order to elucidate the effect of Ti segregation on the diffusional properties.

METHOD OF CALCULATION

The Finnis-Sinclair type central-force many-body potential for TiAl, developed by Vitek and co-workers [18, 19], was employed in this study. This potential was constructed by fitting the cohesive energies, lattice constants and elastic moduli of both $L1_0$ and DO_{19} structures. Therefore, it is expected to model both structures with reasonable accuracy.

The finite block containing the interfaces was always constructed such that the (111) planes are parallel to the interfaces; external surfaces are also parallel to these planes. The bottom half of the block remained fixed while the upper half was rotated around [111] axis by $n\pi/3$, where n=1, 2, 3, to create a pseudotwin, a 120^0rotational fault and an ordered twin, respectively. In each case the block consisted of twenty six (111) layers each composed of 120 atoms per period of the layer; the interface was positioned in the middle of the block. Periodic boundary conditions were

401

imposed parallel to the interface and free surfaces terminated the block in the directions perpendicular to the interface. The optimal structures of the interfaces studied were determined by relaxing all the atoms using a conjugate gradient method for energy minimization.

In a single element material the concentration of vacancies at a temperature T is given as

$$C_v = C_0 \exp(-E_f/k_B T) \qquad (1)$$

where E_f is the formation energy, C_0 a pre-exponential factor of the order one, related to vibrational entropy, and k_B the Boltzmann constant. In alloys the situation is more complex since formation of vacancies is necessarily accompanied by formation of anti-site defects if stoichiometry is fixed. This case was recently analyzed in detail in references [20, 14]. In stoichiometric $A_n B_m$ alloys the concentrations of vacancies at sites A and B are given by equations analogous to (1) but E_f is replaced by the following effective formation energies

$$E_{VA}^{eff} = E_{VA}^* + E_{coh} + \frac{b}{2}\left(E_{A\to B}^* - E_{B\to A}^*\right) \qquad (2a)$$

$$E_{VB}^{eff} = E_{VB}^* + E_{coh} + \frac{a}{2}\left(E_{B\to A}^* - E_{A\to B}^*\right) \qquad (2a)$$

Here a = n/(n+m), b = m/(n+m), E_{VA}^* and E_{VB}^* are the 'raw' energies of vacancies on sites A and B, respectively, equal to the differences of energies of the block with and without the vacancy; $E_{A\to B}^*$ and $E_{B\to A}^*$ have similar meaning for anti-site defects. E_{coh} is the cohesive energy per atom of the $A_n B_m$ alloy irrespective of the type of species; it should be noted that energies of individual species cannot be uniquely defined.

The migration of vacancies was studied by defining a path of the movement of an atom which when completed results in the motion of the vacancy into a nearest equivalent position. Along the diffusional path the moving atom was always confined to the plane perpendicular to the path while all the other atoms were relaxed in all directions. The vacancy migration energy, E_V^m, was then identified with the maximum energy attained during this process. Following the previous studies of the migration in the bulk TiAl [13, 14] four different paths were considered: the nearest neighbors jump in the (111) plane and in between neighboring (111) planes, a three jump cycle and an anti-structure bridge which involves formation of an anti-site defect. In this paper we only report those jumps that lead to the highest value of the diffusion coefficient defined below.

The concentration of vacancies and the vacancy migration energy determine the diffusion coefficient, D, according to the relation $D = C_v \exp(-E_V^m/k_B T)$ so that

$$D = D_0 \exp(-Q^{eff}/k_B T) \qquad (3a)$$

where the effective activation energy for diffusion is

$$Q^{eff} = E_V^{eff} + E_V^m \qquad (3b)$$

RESULTS

The calculations were first performed for the ideal $L1_0$ structure of the stoichiometric TiAl. The results are summarized in Table 1. The effective activation energies for diffusion in the bulk TiAl were measured by Herzig et al. [12] and they were found to be 2.59eV and 3.71eV for Ti and Al, respectively. It is seen that the agreement between calculations and measurements is excellent for Ti but for Al the experiment suggests that its diffusivity is lower than that of Ti while calculations lead to very similar values of Q^{eff}. The likely reason is the not entirely adequate description of Al-Al interaction in the scheme employed in this study.

The results of similar calculations for the three types of interfaces separating differently oriented lamellae in the stoichiometric TiAl are summarized in Table 2. As mentioned earlier, we only present data leading to the highest value of the diffusion coefficient, i. e. the lowest Q^{eff}. Hence, the vacancy sites and diffusional paths are generally different for different types of interfaces.

An analogous calculation was carried out for the case when Ti is in surplus and segregates to PTW and 120RF; no segregation takes place at OTW. Following the findings of the Monte Carlo simulation of Ito and Vitek [17], the central layer at the interface was replaced by the layer with the composition Ti_3Al and the effective vacancy formation energies and migration energies calculated in the same way as in the stoichiometric case. The results of this study are presented in Table 3, again for the cases that lead to the highest value of the diffusion coefficient. In this case the lowest effective vacancy formation energies have been found for the sites in the plane adjacent to Ti rich plane; these energies are very similar to those found in the stoichiometric case. The controlling diffusional jump is then from this plane into the plane with the composition Ti_3Al. The corresponding migration energy is again comparable with that found in the stoichiometric case. However, for vacancies formed in the Ti rich layer the effective formation energies and the related effective activation energy for diffusion are higher.

	E_V^{eff} (eV)	E_V^m (eV)	Q^{eff} (eV)
Titanium	1.40	1.16	2.56
Aluminum	1.62	1.01	2.63

Table 1

Calculated effective vacancy formation energies, migration energies for the nearest neighbor jumps and effective activation energies for diffusion in the bulk of stoichiometric TiAl.

	TITANIUM			ALUMINUM		
	E_V^{eff} (eV)	E_V^m (eV)	Q^{eff} (eV)	E_V^{eff} (eV)	E_V^m (eV)	Q^{eff} (eV)
OTW	1.15	1.06	2.21	1.33	0.92	2.25
PTW	1.28	0.92	2.20	1.43	0.81	2.24
120RF	1.32	0.97	2.29	1.50	0.91	2.41

Table 2

Calculated effective vacancy formation energies, migration energies and effective activation energies for diffusion in three types of interfaces encountered in the stoichiometric lamellar TiAl.

	TITANIUM			ALUMINUM		
	E_V^{eff} (eV)	E_V^m (eV)	Q^{eff} (eV)	E_V^{eff} (eV)	E_V^m (eV)	Q^{eff} (eV)
PTW	1.33	0.88	2.21	1.48	0.87	2.35
120RF	1.39	0.94	2.33	1.56	0.92	2.48

Table 3

Calculated effective vacancy formation energies, migration energies and effective activation energies for diffusion in PTW and 120RF with segregated Ti.

DISCUSSION

The results of this atomistic study suggest that in the $L1_0$ TiAl both the effective formation energies and migration energies of vacancies are lower at lamellar interfaces than in the bulk. Hence, the diffusivity will be higher along these interfaces with the corresponding effective activation energy for the diffusion about 15% lower than in the bulk. This enhancement of the diffusivity may play an important role in the high temperature properties, such as creep, in the lamellar TiAl alloys. An interesting finding is that the diffusivity is very similar in all three types of interfaces, OTW, PTW and 120RF. Without full atomistic study one would expect that owing to its highly ordered structure, the diffusivity at the OTW will be much closer to that in the ideal $L1_0$ lattice than the diffusivity in the other two interfaces.

Another interesting finding is that the segregation of Ti to PTW and 120RF affects the diffusivity in these interfaces only marginally. The Ti segregation leads to the formation of a layer with the Ti_3Al-type structure at these interfaces [17] and it could be expected that this affects significantly both the formation and migration of vacancies and thus the interfacial diffusion coefficient. However, as seen in Table 3, the effective activation energy for diffusion is practically the same as in the stoichiometric case. The reason is that the controlling process corresponds to the vacancy formation in the layer adjacent to the Ti rich layer and migration is controlled by the jump in between these layers. The corresponding effective formation and migration energies are very similar to those found in the stoichiometric case.

ACKNOWLEDGMENT

This research was supported by the National Science Foundation grant no. DMR96-15228.

REFERENCES

1. C.T. Liu, R.W. Cahn, and G. Sauthoff, editors, *Ordered Intermetallics – Physical Metallurgy and Mechanical Behaviour* (Kluwer Academic Publishers, Dodrecht, 1992).
2. Baker, R. Darolia, J.D. Whitten, and M. H. Yoo, editors, *High-Temperature Ordered Intermetallic Alloys V*, (Materials Research Society, Pittsburgh, 1993), Vol. 288.
3. R. Darolia, J. J., Lewandowski, D. T. Liu, P.L. Martin, D.B. Miracle, M. V. Nathal., editors, *Structural Intermetallics*, (TMS, Warrendale, 1993).
4. C. T. Liu, S. H. Wang, and D. P. Pope, editors, *Proc. 3rd Int. Conf. on High Temperature Intermetallics, Mat. Sci. and Eng. A* **192/193** (1995).
5. J. Horton, S. Hanada, I. Baker, R.D. Noebe and D. Schwartz,editors, *High-Temperature Ordered Intermetallic Alloys VI*, (Materials Research Society, Pittsburgh, 1995), Vol. 364.
6. J. H. Westbrook and R. L. Fleischer, editors, *Intermetallic Compounds-Principles and Practice*, (John Wiley & Sons, New York, 1995).
7. M. Yamaguchi, H. Inui, S. Yokoshima, K. Kishida and D. R. Johnson, *Mat. Sci. and Eng. A* **213**, 25 (1996).
8. S. Stoloff and V. K. Sikka, editors, *Physical Metallurgy and Processing of Intermetallic Compounds*, (Chapman & Hall, New York, 1996).
9. M. Yamaguchi, H. Inui, K. Kishida, M. Matsumoro and Y. Shirai, in *High-Temperature Ordered Intermetallic Alloys VI*, edited by J. Horton, I. Baker, S. Hanada, R. D. Noebe and D. Schwartz (Materials Research Society, Pittsburgh), Vol. 364, p. 3 (1995).
10. D. S. Schwartz and S. M. L. Sastry, *Scripta Metall.* **23**, 1621 (1989).
11. H. Inui, M. H. Oh, A. Nakamura and M. Yamaguchi, *Philos. Mag. A* **66**, 539 (1992).
12. C. Herzig, T. Przeorski and Y. Mishin, *Intermetallics* **7**, 389 (1999).
13. Y. Mishin and C. Herzig, *Mat. Sci. and Eng. A* **260**, 55 (1999).
14. C. Herzig and Y. Mishin, *Acta Mater.*, to be published, (2000).
15. C. McCullough, J. J. Valencia, H. Mateous, C. G. Levi, R. Mehrabian and K. A. Rhyne, *Scripta Metall.* **22**, 1131 (1988).
16. H. Inui, K. Kishida, M. Kobayashi, M. Yamaguchi, M. Kawasaki and K. Ibe, *Philos. Mag. A* **74**, 451 (1996).
17. K. Ito and V. Vitek, *Acta Mater.* **46**, 5435 (1998).
18. A. Girschick and V. Vitek, in *High-Temperature Ordered Intermetallic Alloys VI*, edited by J. Horton, I. Baker, S. Hanada, R. D. Noebe and D. Schwartz (Materials Research Society, Pittsburgh), Vol. 364, p. 145 (1995).
19. V. Vitek, A. Girshick, R. Siegl, H. Inui and M. Yamaguchi, in *Properties of Complex Inorganic Solids*, edited by A. Gonis, P. Turchi and A. Meike (New York, Plenum Press), p. 355 (1997).
20. M. Hagen and M. W. Finnis, *Philos. Mag. A* **77**, 447 (1998).

STRUCTURE AND STABILITY OF GRAIN BOUNDARIES IN MOLYBDENUM WITH SEGREGATED CARBON IMPURITIES

R. JANISCH, T. OCHS, A. MERKLE, C. ELSÄSSER.
Max-Planck-Institut für Metallforschung, Seestrasse 92, D-70174 Stuttgart, Germany.

ABSTRACT

The segregation of interstitial impurities to symmetrical tilt grain boundaries (STGB) in body-centered cubic transition metals is studied by means of ab-initio electronic-structure calculations based on the local density functional theory (LDFT). Segregation energies as well as changes in atomic and electronic structures at the $\Sigma 5$ (310) [001] STGB in Mo caused by segregated interstitial C atoms are investigated. The results are compared to LDFT data obtained previously for the pure $\Sigma 5$ (310) [001] STGB in Mo. Energetic stabilities and structural parameters calculated ab initio for several crystalline Molybdenum Carbide phases with cubic, tetragonal or hexagonal symmetries and different compositions, MoC_x, are reported and compared to recent high-resolution transmission electron microscopy (HRTEM) observations of MoC_x intergranular films and precipitates formed by C segregation to a $\Sigma 5$ (310) [001] STGB in a Mo bicrystal.

INTRODUCTION

Impurity atoms like Boron, Carbon, Nitrogen etc. in body-centered cubic transition metals often tend to segregate to grain boundaries. They cause modifications of the local atomic and electronic structures at the interfaces and, consequently, they are influential on various polycrystalline properties like intergranular embrittlement, corrosion or grain-boundary diffusion. For reviews of interfacial segregation see, e.g., [1,2].

For the background of the present work, for instance the following two microscopic studies are relevant: Hashimoto et al. [3] investigated theoretically the influence of substitutional P and interstitial B impurities on the atomic structure of two symmetrical tilt grain boundaries (STGB) in α-Fe by means of atomistic simulations, using pair potentials for the interatomic interactions. Pénisson et al. [4] studied experimentally the segregation of C to a $\Sigma 5(310)[001]$ STGB in a Mo bicrystal by high-resolution transmission electron microscopy (HRTEM). They observed the formation of MoC_x phases (x: local C concentration) as precipitate particles in the bulk or at the interface and as thin interfacial layers, and they determined their crystalline structure.

The topic of the present theoretical work is a microscopic analysis of the atomic structures and energetics of bulk Molybdenum Carbide phases and of interstitial C atoms segregated to $\Sigma 5(310)[001]$ STGB in Mo, employing an ab-initio electronic-structure method. In this paper, after a concise section about the theoretical ab-initio method, results for bulk MoC_x phases and for the segregation of C to the $\Sigma 5(310)[001]$ STGB in Mo are presented in two separate sections.

AB-INITIO ELECTRONIC-STRUCTURE METHOD

Our ab-initio electronic-structure method for crystalline materials is based on the local density functional theory (LDFT) [5]. Bulk phases as well as grain-boundaries are modeled by supercells with periodic boundary conditions in all three spatial directions. The influence of the atomic nuclei and the core electrons on the valence electrons is described by nonlocal, norm-conserving, nonlinear ionic pseudopotentials [6]. The Bloch wavefunctions of the valence electrons are represented by a mixed basis [7] including a finite number of plane waves, limited by a maximum kinetic energy of 16 Rydberg (1 Rydberg = 13.6 eV), five atom-centered, localized functions with angular d symmetry per Mo atom and three functions with p symmetry per C atom. Brillouin-zone integrals are calculated by sufficiently dense discrete sampling and Gaussian broadening. With this mixed-basis pseudopotential method total energies and atomic forces can be calculated efficiently and accurately, allowing for structural optimizations and for quantitative determinations of phase stabilities (formation energies) and grain-boundary energetics (interface energies and segregation energies).

STRUCTURES AND ENERGETICS OF CRYSTALLINE MoC_x PHASES

For comparison with the Molybdenum Carbide precipitates and interfacial layers observed in the HRTEM experiment [4], and to provide bulk references for the energetics of the grain-boundary segregation addressed in the next section, the formation energetics of bulk MoC_x phases was determined by ab-initio LDFT calculations. For the concentrations $x=0$, 0.5, 1.0, and "∞" (pure Mo, Mo hemicarbide and monocarbide, pure C), the following lattice structures of Mo were considered: body-centered tetragonal with the special cases body-centered cubic (bcc=bct with $c/a=1$) and face-centered cubic (fcc=bct with $c/a=\sqrt{2}$), simple cubic (sc), hexagonal close-packed (hcp) and hexagonal primitive (hpr). As interstitial sites for C atoms, regular (fcc, hcp) or tetragonally distorted (bcc) octahedra, regular tetrahedra (fcc) and cubes (sc), and trigonal prisms (hpr) of Mo were considered. All structural parameters (a and c/a) and internal atomic displacements (for hcp and hpr phases) were optimized by total-energy minimization. The resulting energy-volume relations per Mo atom were well represented by the "universal" equation of states (EOS) [8].

Formation energies of all considered MoC_x phases were determined with respect to the minimum energies of pure bcc Mo and pure C. For C, both the diamond and the graphite structure were considered. With respect to structural energy differences of 10 kJ/mol and more, which are relevant here, diamond and graphite are energetically degenerate. Therefore, for C only the diamond results are used here. Equilibrium lattice parameters a_0 and bulk moduli B_0 extracted from the EOS for diamond C and bcc Mo are given in table 1. Compared to the experimental data taken from [9], the theoretical data show slight underestimation of a_0 and overestimation of B_0, which are typical for computationally accurate ab-initio LDFT results. The EOS of diamond C and of Mo in five different structures are displayed in figure 1. For Mo, as expected, the bcc structure is most stable, followed by the two close-packed structures (fcc and hcp), which are almost degenerate. The two primitive structures (hpr and sc) have much higher energies.

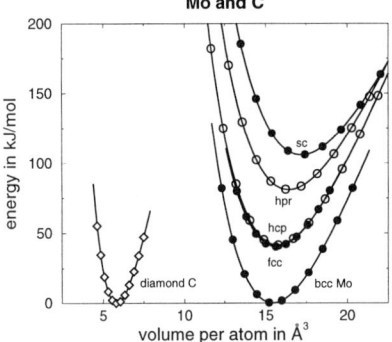

Figure 1: EOS E(V) for diamond C, cubic and hexagonal Mo phases.

		a_0 in Å	B_0 in GPa
bcc Mo	theo.	3.14	277
	exp.	3.15	273
diamond C	theo.	3.60	505
	exp.	3.57	545

Table 1: Equilibrium lattice parameters a_0 and bulk moduli B_0 of bcc Mo and diamond C. The theoretical values were determined by fitting the universal equation of state [8] to a set of ab-initio data of total energies versus unit-cell volumes. The experimental data are taken from [9].

The calculated EOS for $MoC_{0.5}$ phases are shown in figure 2. C located on octahedral sites in bcc Mo with expanded volume is very unfavourable because of a very short separation between C and two of the six surrounding Mo atoms that form an irregular, tetragonally compressed octahedron. The bcc $MoC_{0.5}$ phase ($c/a=1$) is unstable against a tetragonal distortion (Bain's displacive deformation). The energy minimum with respect to variation of the tetragonal c/a and V appears very close to $c/a=\sqrt{2}$, which corresponds to fcc Mo and regular interstitial octahedra for C (fcc_o). Compared to pure Mo and C, fcc $MoC_{0.5}$ is still unstable. Among the other considered phases $MoC_{0.5}$, fcc Mo with C on tetrahedral sites (fcc_t) is very high in energy, and hpr Mo with C in trigonal prisms of Mo is also less favourable than fcc Mo with C on octahedral sites. Only hcp $MoC_{0.5}$, with C coordinated by trigonally distorted octahedra, is energetically slightly more stable than the pure materials. This is in accordance with the experimentally determined most stable bulk phase of MoC_x for $x \approx 0.5$ (see, e.g., [10]). It also corresponds to the intergranular hexagonal Mo_2C precipitates observed in [4]. The intragranular tetragonal Mo_2C precipitates reported in [4], on the other hand, have a tetragonal structure close to the bulk fcc $MoC_{0.5}$ phase. Although being unstable as a bulk phase at ambient conditions, the HRTEM experiments [4] indicate that it can be stabilized as precipitates embedded into pure Mo or as a thin interfacial layer. This issue deserves further theoretical effort.

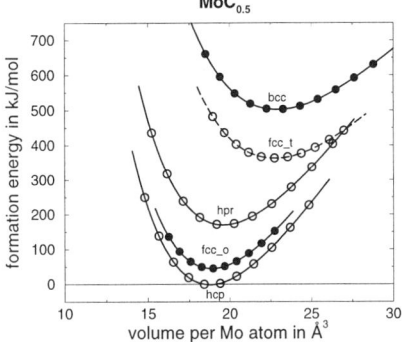 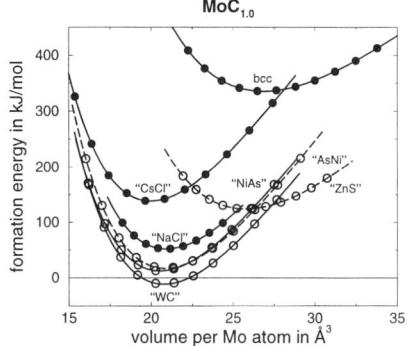

Figure 2: EOS E(V) for cubic and hexagonal $MoC_{0.5}$ phases.

Figure 3: EOS E(V) for cubic and hexagonal $MoC_{1.0}$ phases.

Figure 3 displays the EOS of $MoC_{1.0}$ phases. Again, the coordination of C in bcc Mo is highly unfavourable, and bcc $MoC_{1.0}$ is unstable against a tetragonal distorsion. Also the eightfold cubic coordination of C in sc Mo ("CsCl" structure) and the fourfold tetrahedral coordination of C in fcc Mo ("ZnS" structure) lead to highly negative formation energies. With sixfold coordinations, octahedron and trigonal prism are competitive. For $x=1.0$, the fcc phase ("NaCl" structure with C in Mo octahedra) is least stable, and the hpr phase ("WC" structure with C in Mo trigonal prisms) is most stable. The latter phase is also found experimentally as the most stable phase for $x \approx 1$ (see, e.g., [10]). This illustrates a close relationship of MoC to tungsten carbide. Energetically intermediate to the "NaCl" and "WC" phases is the "NiAs" phase with Mo forming a hcp lattice and C filling octahedral sites on a hpr sublattice. In the phase named "AsNi" the Mo and C sublattices of the "NiAs" phase are mutually exchanged. These two phases are energetically very similar, emphasizing the competition of trigonal-prism and octahedron interstitial sites for C.

SEGREGATION ENERGETICS FOR C ATOMS AT THE $\Sigma5$ STGB OF Mo

For the microscopic study of the segregation of impurities to grain boundaries in transition metals by means of ab-initio electronic-structure calculations, we selected one particular model case: the segregation of C atoms, which occupy bulk interstitial sites, to the $\Sigma5(310)[001]$ STGB in Mo. This choice is motivated by the following reasons: The atomic structures of pure $\Sigma5$ STGB in bcc metals, namely Nb and Mo, have already been studied extensively, both experimentally via HRTEM and theoretically, using empirical interatomic potentials in various levels of sophistication as well as ab-initio electronic-structure techniques. More informations about such studies of the pure Σ5 STGB in Nb and Mo are given for instance in our recent papers [11]. Furthermore, some thermodynamic data of segregation enthalpies derived from experiment are available at least for C segregated to grain boundaries in Mo polycrystals (cf. [2], table 3) and to the Σ5 STGB in α-Fe (cf. [2], table 4; for various material properties α-Fe can be considered to be similar to Mo).

In [11] optimized atomic structures of the pure Σ5 STGB in Mo with respect to relative axial and lateral grain displacements were determined by ab-initio LDFT calculations, employing the same technique as in the present work. It was found that a grain-boundary configuration having a mirror symmetry with respect to the interface plane, which had been determined earlier by HRTEM as the equilibrium configuration for Nb [12], was unstable for Mo with respect to a symmetry-breaking displacement of the grains parallel to the [001] tilt axis. This theoretical result was supported by a recent HRTEM experiment with a Mo bicrystal [13]. The optimized, symmetry-broken configuration (denoted by (7) in [11] and here as well) was energetically preferred with respect to the mirror-symmetric configuration (denoted by (4) as in [11]). The calculated interface-energy difference was $E_{GB}(7) - E_{GB}(4) = -106$ mJ/m^2 in favour of configuration (7).

To investigate the energetics of C to the Σ5 STGB in Mo, we took from [11] the two supercells of the optimized configurations (4) and (7), each containing twenty relaxed Mo atoms. The atomic arrangement in the supercells in a projection parallel to the [001] tilt axis are shown schematically in figure 4. In this projection the symmetry-breaking grain displacement which distinguishes (7) from (4) is not visible. The twenty Mo atoms on a (002) plane contained in the base-centered orthorhombic supercells are marked by large grey spheres. The Mo atoms on the neighboring (002) planes one half bcc lattice parameter a_0 above or below are marked by small black spheres.

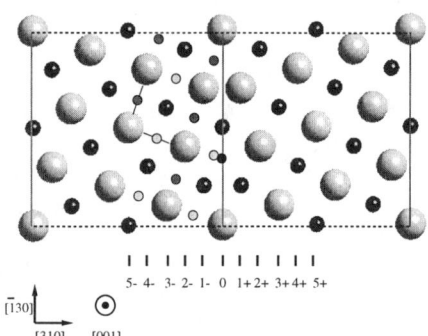

Figure 4: A two-dimensional projection of the Σ5 STGB supercell. The large filled circles mark the twenty basis atoms and their periodic images in a (002) plane, the small black spheres mark the atoms in adjacent (002) planes. The (310) grain-boundary planes are located vertically in the center and at the left and right borders of the supercell. The three locations of octahedral sites are (a) below the big grey spheres, and marked with small circles, (b) dark grey and and (c) light grey. The C position in case (d) is marked by the small black circle on layer 0.

Next, two C atoms, which form one (310) monolayer per each Mo grain because of the periodic boundary conditions, were inserted into the supercells of configurations (4) and (7). Relaxations of the Mo atoms due to the C addition were neglected. The C atoms were put on octahedral sites in different distances from the grain boundaries. Because the octahedra in the bcc lattice are tetragonally compressed, they can be oriented in three different directions. The resulting segregation energies, referred to the energy of C in the most bulk-like layers 5+ and 5−, are plotted in figure 5 as function of the distance from the interface layer 0. The energies for configuration (4) (configuration (7)) are marked by filled (empty) circles connected by solid (dashed) lines.

In the case (a), the tetragonal axis of an octahedron filled with C is parallel to [001]. The C positions are one half lattice parameter a_0 above or below the large grey spheres and also, because of the translational symmetry of the base-centered orthorhombic supercell, above or below the small black spheres. The segregation energy for this octahedron orientation is very weakly dependent on the axial position with respect to the interface, because in this case the C atoms are in sites which are highly constrained by symmetry, like in bulk bcc Mo without any volume expansion.

408

Much stronger energy changes appear for the cases (b) and (c) (marked by small dark and light grey circles in figure 4 in one of the two grains) where the octahedra are oriented under angles of about 18.5° and 71.5° (the orientations of the tetragonal axes are marked by small black lines in two octahedra) with respect to the interface and less constrained by the supercell symmetry. Although still no atomic relaxations are allowed, the electronic structure can react more flexibly to the presence of the interface. Close to the interface layer 0 there are considerable energy gains compared to the bulk-like positions on the layers 5+ and 5−. Nevertheless, because of the neglect of any relaxations, supercell symmetry constraints or rather high forces acting on the Mo atoms surrounding the octahedral C atoms the energy data (a)-(c) shown in figure 5 do not provide more than a purely qualitative behaviour of the segregation energetics. It just illustrates that it is more favourable for C to be located close to a grain boundary than in the bulk volume of bcc Mo if no Mo lattice relaxations are allowed.

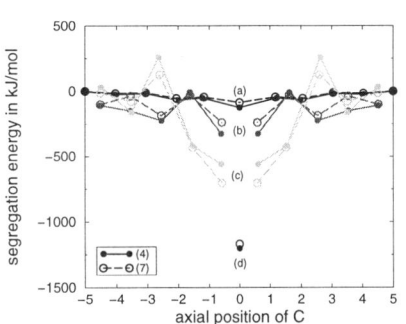

Figure 5: Energetics of segregation of C on octahedral sites in the bcc Mo grains without relaxation.

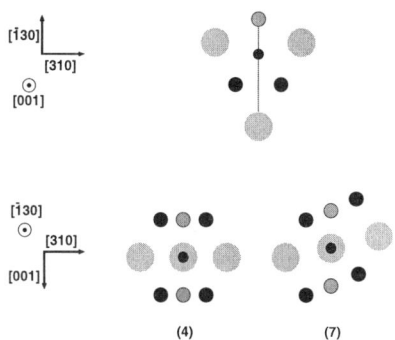

Figure 6: Local Mo clusters around a C atom on the interface layer 0, (marked by a central small black circle), displayed in [001] (top) and in [$\bar{1}$30] (bottom) projections for configurations (4) (left) and (7) (right).

The highest energy gain is achieved for the case (d) with the C atoms located directly on the interface layer 0, marked in figure 4 by a small black circle surrounded next by an acute "black" Mo triangle. The local atomic arrangements are shown enlarged in figure 6, again in the same [001] projection (top) and in the [$\bar{1}$30] projection (bottom). For configuration (4) (left), the C atom is surrounded by a capped, acute triangular prism of Mo atoms. In configuration (7) (right), this prism is sheared in the direction of the tilt axis. The C atom positions in the prisms were varied to minimize the total energies, and the resulting forces on the surrounding, fixed Mo atoms appeared rather small. Therefore, the energies of these arrangements are useful for quantitative estimates of segregation energies.

First, as figure 5 shows, for configuration (4) the energy is lower than for configuration (7). The related forces on Mo atoms in (7) are pointing towards positions in (4). This means that the relative stability of (4) and (7) is changed by the presence of C. The resulting calculated interface-energy difference is $E_{GB}(7) - E_{GB}(4) = +335 \text{mJ/m}^2$ in favour of (4), predicting a most stable configuration of the $\Sigma 5$ STGB in Mo with a monolayer of segregated C. An important reason, why (4) is favoured over (7), seems to be related to the local triangular Mo prism surrounding the C atoms. For (4) it is similar to the C environment in the most stable bulk phase, hpr $MoC_{1.0}$ ("WC"). For (7), on the other hand, it is sheared and therefore the Mo-C distances are considerably shortened for two corners and elongated for two others. Therefore, the local sixfold coordination of C is highly distorted and costs a certain amount of energy. This energy change can overcome the interface-energy difference of (4) and (7) of the pure grain boundary and stabilize the mirror-symmetry.

Second, by relating the total energies per C atom of (4) and (7) to the total energy per C atom of the most stable bulk phase $MoC_{1.0}$, the segregation energies were estimated quantitatively: $E_S(4) \approx -50$ kJ/mol is close to experimental enthalpy values of -70 kJ/mol for C in polycrystalline Mo (cf. [2], table 3), or between -40 and -60 kJ/mol for C at different tilt grain boundaries in bicrystals of α-Fe (cf. [2], table 4). $E_S(7) \approx +250$ kJ/mol, on the other hand, indicates an instability of the symmetry-broken grain boundary induced by the segregated C. This is in accordance with experimental experience: For quantitative determinations of geometric translation states and grain-boundary symmetries, for instance by HRTEM, it is important to use transition-metal bicrystals with high purity, because segregated impurities can affect the results. Calculations for B and N impurities are in progress.

SUMMARY

The energetics of bulk MoC_x phases and of C segregated to $\Sigma5(310)[001]$ STGB in Mo was investigated theoretically by means of ab-initio electronic-structure calculations, which are based on the local density functional theory. The calculated most stable bulk phases for $x=0.5$ and 1.0 are hexagonal close-packed $MoC_{0.5}$ with C located in Mo octahedra, and hexagonal primitive $MoC_{1.0}$ ("WC") with C in Mo trigonal prisms. For the grain-boundary segregation of C in Mo, using super-cells with and without mirror symmetry that where taken from [11], qualitative segregation energy profiles were calculated, which illustrate the tendency for C to segregate to the $\Sigma5(310)[001]$ STGB. Quantitative estimates of segregation energies were possible for C monolayers located directly on the interface planes, which are well compatible with measured segregation enthalpies. Furthermore, the energetics indicate that the relative stability of different grain-boundary translation states can be affected by the segregated impurities.

REFERENCES

[1] A. P. Sutton and R. W. Balluffi, *Interfaces in Crystalline Materials* (Clarendon Press, Oxford, 1995), ch. 7.
[2] P. Lejček and S. Hofmann, Critical Reviews in Solid State and Materials Sciences **20**, 1 (1995).
[3] M. Hashimoto, Y. Ishida, R. Yamamoto and M. Doyama, Acta metall. **32**, 1 (1984).
[4] J. M. Pénisson, M. Bacia and M. Biscondi, Philos. Mag. A **73**, 859 (1996).
[5] P. Hohenberg and W. Kohn, Phys. Rev. **136**, B864 (1964); W. Kohn and L. J. Sham, Phys. Rev. **140**, A1133 (1965).
[6] D. Vanderbilt, Phys. Rev. B **32**, 8412 (1985); S. G. Louie, S. Froyen and M. L. Cohen, Phys. Rev. B **26**, 1738 (1982).
[7] C. Elsässer, N. Takeuchi, K. M. Ho, C. T. Chan, P. Braun and M. Fähnle, J. Phys.: Condens. Matter **2**, 4371 (1990); K. M. Ho, C. Elsässer, C. T. Chan and M. Fähnle, J. Phys.: Condens. Matter **4**, 5189 (1992); B. Meyer, C. Elsässer and M. Fähnle, *Fortran90 Program for Mixed-Basis Pseudopotential Calculations for Crystals* (MPI für Metallforschung Stuttgart, unpublished).
[8] J. H. Rose, J. R. Smith, F. Guinea and J. Ferrante, Phys. Rev. B **29**, 2963 (1984).
[9] C. Kittel, *Introduction to Solid State Physics* (Wiley, New York, 1986).
[10] L. E. Toth, *The Transition Metal Carbides and Nitrides* (Academic Press, New York, 1971).
[11] C. Elsässer, O. Beck, T. Ochs and B. Meyer, Mat. Res. Soc. Symp. Proc. Vol. **492**, 121 (1998); T. Ochs, O. Beck, C. Elsässer and B. Meyer, Philos. Mag. A (in press, 1999); T. Ochs, C. Elsässer, M. Mrovec, V. Vitek, J. Belak and J. A. Moriarty, Philos. Mag. A (in press, 1999).
[12] G. H. Campbell, S. M. Foiles, P. Gumbsch, M. Rühle and W. E. King, Phys. Rev. Lett. **70**, 449 (1993).
[13] G. H. Campbell, J. Belak and J. A. Moriarty, Acta mater. **47**, 3977 (1999).

INFLUENCE OF GRAIN BOUNDARY STRUCTURE
ON LIQUID METAL PENETRATION BEHAVIOR

Liping Ren*, D.F. Bahr* and R.G. Hoagland**
*MME Department, Washington State University, Pullman, WA 99164
**Los Alamos National Laboratory, Los Alamos, NM 87545

ABSTRACT

The penetration of Ga along Al grain boundaries under stress-free conditions is investigated in the present study. In-situ SEM observations indicate that the penetration rate of Ga along Al grain boundaries at room temperature ranged from 6.4 to 9.2 μm/s, which is similar to the rate of diffusion in the liquid state. For a specific high energy grain boundary, the grain boundary misorientation is determined from the TEM diffraction Kikuchi pattern, and a molecular statics simulation method was employed to investigate grain boundary structure. A comparison of the structure of this high energy boundary is made with the $\Sigma 11(131)[101]$ tilt grain boundary that is not penetrated by Ga in the absence of the applied stress. The results indicate that the grain boundary plane void structure in the high energy grain boundary may provide void channels for Ga monolayer penetration. In addition, penetration behavior investigated under different length scales supports this model.

INTRODUCTION

Liquid metal embrittlement (LME) is the phenomenon by which normally ductile metals or alloys exhibit brittle fracture behavior while in contact with certain liquid metals. A wide range of materials have been found to be embrittled by liquid metals. For example, industrially important alloys such as brasses and bronzes, and carbon and stainless steels, may be embrittled by molten metals with both low and moderate melting points including mercury, gallium, solders, and even copper based alloys [1]. The detailed causes and mechanisms of LME have still not reached yet but a few promising hypotheses have been proposed. The most popular ones are tensile decohesion [2], indicating adsorbed atoms lower the cohesive strength of nearby metallic bonds, and enhanced shear [3] claiming that plastic deformation is easily reached locally.

Generally, LME failure could be interganular or transganular and requires an applied stress sufficient to produce plastic deformation. But in severe cases, when the LME failure occurs under no applied stress, it is usually time-dependent and interganular, e.g. Ga embrittles Al in stress-free conditions. From energetic point of view, this process is generally considered to be the reduction in interfacial energy which results when a high energy grain boundary is replaced with two lower energy aluminum-gallium interfaces [4]. Unfortunately, this description does not permit quantitative assessment of the time dependent embrittlement nor does it provide insight into the detailed atomic mechanism.

In the present study, a computer simulation using a molecular statics method was applied to grain boundary structure studies. The void distribution maps on a specific high energy grain boundary and the special $\Sigma 11(131)[101]$ tilt grain boundary were constructed after relaxation. By linking grain boundary atomic structures and in-situ experimental observation, mechanisms of Ga penetration on high energy Al grain boundaries at test conditions are suggested and some experimental evidence is provided.

Mat. Res. Soc. Symp. Proc. Vol. 578 © 2000 Materials Research Society

EXPERIMENT

Samples for SEM observations are 25mm × 10mm sheets cut from a 99.99% pure aluminum casting ingot which has a grain size of about 5 mm. Samples were cut with a diamond wheel and mechanically thinned. After annealing at 300°C for 1 hour, samples were electropolished to a thickness of ~1 mm and then etched. Etched grains are examined on both sides of the sample to make sure the grain boundaries go through the sample thickness for in-situ penetration observations. A small droplet of 99.9999% gallium was placed onto the surface of the sample, and the aluminum surface was scratched through the gallium droplet to break the surface oxides and promote wetting. The samples were then placed in a JEOL JSM-6400 scanning electron microscopy with a CCD camera system to record the Ga penetration process.

The experimental detail for in-situ TEM observations can be found elsewhere [4]. To establish all five angular variables necessary to determine the grain boundary misorientation, Kikuchi patterns and bright-field images were recorded parallel to the optic axis. The penetration processes were also recorded with a CCD camera.

Computer models were constructed for the $\Sigma11(131)[101]$ tilt grain boundary and the specific grain boundary where the misorientation data was obtained from the TEM diffraction Kikuchi pattern. There are two regions for each model. In region I, the inner region, atoms are allowed to move individually, region II provides necessary neighbors for energy and force calculations in region I. The interaction potential used in the present study was developed by F. Ercolessi and J.B. Adams [5] using the Embedded Atom Method (EAM).

RESULTS

(1) In-situ TEM observations

In –situ TEM observations indicate that the penetration front is wedge shaped, as shown by the dotted line parallel to the penetration front in Fig. 1 (b) compared to (a) before penetration. The fiduciary marks in Fig.1 are from the CCD camera recording through the TEM screen. The penetration distance at different times was measured every 1/6 second. The results are plotted in Fig. 3(a).

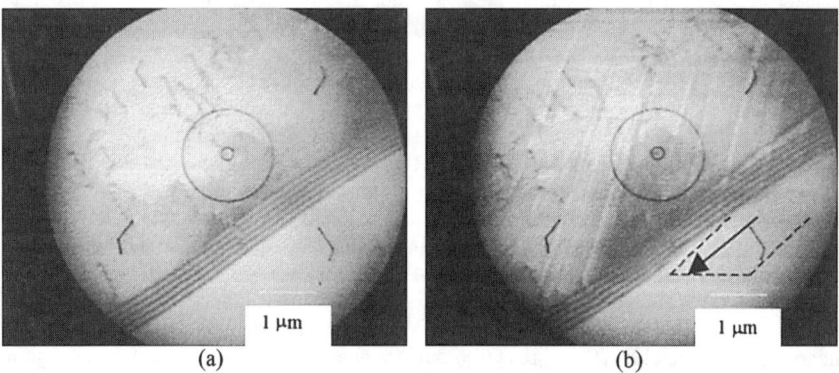

| (a) | (b) |

Fig. 1 TEM observation of Ga penetration on Al grain boundary. (a) grain boundary before penetration and (b) during penetration, the arrow in (b) shows the penetration direction, the penetration front is next to the arrowhead and dotted line show the penetration front shape.

(2) In-situ SEM observations

In in-situ SEM observations, the general measured length scale is about 1 millimeter. Since the grain boundary plane angle may change along the gain boundary, each observation can be considered as the average behavior on different grain boundary planes under same misorientation grain boundary. Since the Ga is a higher atomic number than the Al, the penetration front appears as light spots under backscattered electron (BSE) image as shown in Fig. 2(a), (b) and (c). From the Fig. 2, It is noted that the penetration front is discontinuous when the penetration proceeds and this part can roughly be as long as a few millimeters. The results of measured penetration rates are plotted as shown in Fig. 3(a) for TEM and Fig. 3 (b) for different grain boundaries in SEM observations.

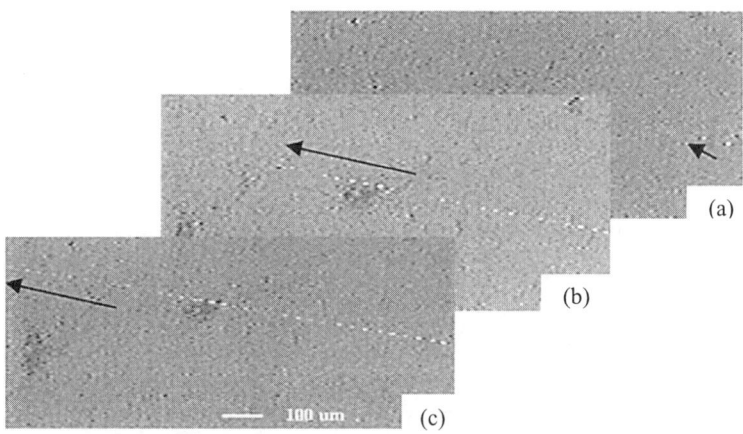

Fig. 2 BSE image of Ga penetration on an Al grain boundary with the penetration rate of 7.16 μm/s, arrows in the figures show the penetration direction, the penetration front is next to the arrowhead. (a) penetration front at t = 0 s, (b) penetration front at t = 73 s, (c) penetration front at t = 118 s. Light spots are Ga.

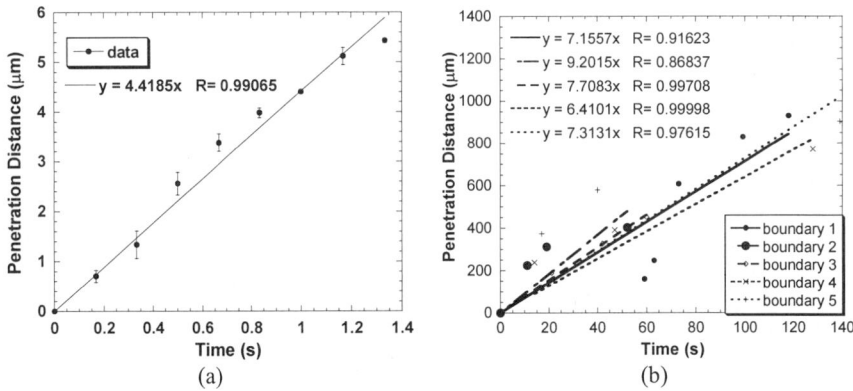

Fig.3 Plot of penetration distance as the function of penetration time in (a) in-situ TEM observations, and (b) in-situ SEM observations of 5 different grain boundaries.

The penetration rates vary from 6.4 to 9.2 μm/s for in-situ SEM observations. And also, the penetration rates for in-situ TEM observations are in agreement with other results which show penetration rates in the range of 0 to 12 μm/s for differently oriented grain boundaries [4]. For the SEM observations, it seems that there is not much difference for the Ga penetration on randomly oriented grain boundaries.

(3) Computer models

The grain boundary misorientation was determined from the TEM diffraction Kikuchi pattern on bicrystal samples. The laboratory coordinates system was constructed in such a way that both of the grain boundary misorientation and grain boundary plane can be easily determined [4]. The direction cosine of each grain orientation to laboratory coordinates was measured in the form of matrix for two grains, named grain A and grain B, respectively, as follows.

$$T_A = \begin{pmatrix} 0.720 & 0.683 & -0.122 \\ -0.440 & 0.314 & -0.841 \\ -0.536 & 0.660 & 0.527 \end{pmatrix} \qquad T_B = \begin{pmatrix} 0.928 & -0.218 & 0.300 \\ 0.354 & 0.760 & -0.544 \\ -0.110 & 0.612 & 0.783 \end{pmatrix}$$

According to the matrix T_A and T_B, rotation axes and angle was determined and listed in Table 1 with estimated Σ value and grain boundary planes.

Table 1 Grain boundary misorientation and grain boundary plane

Grain boundary planes (B/A)	(153)/(212)
Σ Value	$\Sigma 29$
Rotation angle (θ)	146.7°
Rotation Axes (in B)	<773>

Computer models were constructed using the data from the above calculation, and as a comparison, were also constructed for the special grain boundary $\Sigma 11(131)[101]$ using the method mentioned in the experimental section. After models were constructed, atoms within region I were relaxed to their low energy states via molecular statics, using conjugate gradient minimization. After an equilibrium state was reached, the grain boundary energy can be calculated as the excess energy of the atomic set with the boundary in comparison with the energy of the same amount of atoms in ideal lattice. The calculated grain boundary energy for the $\Sigma 29(153)[773]$ grain boundary is about 773 mJ/m^2 which is much higher than that of the $\Sigma 11(131)[101]$ grain boundary, 146 mJ/m^2.

To locate the voids in grain boundary plane, a grid of points, 4nm × 1nm × 1nm with grid spacing of 0.02 nm, was first superimposed onto the grain boundary plane of the computer model. All grid points closer than 0.17 nm to any atom center were then deleted. The remaining grid points defined the void distribution map. The void distribution maps for the $\Sigma 11(131)[101]$ grain boundary and $\Sigma 29(153)[773]$ grain boundary are shown in Fig. 4 (a) and (b), respectively. Note that 0.17 nm was chosen for defining the void. The idea here is that the distance between the point and atoms is slightly larger than 0.165 nm, the distance from the center of 3 atoms in close packed plane (111) to each atom. This way, in a perfect crystal, there are still some 'void' by this definition. In order to reduce this effect, a reference map was constructed within a perfect crystal and the points in the same positions as the reference map will be removed before plotting void maps for the investigated grain boundaries.

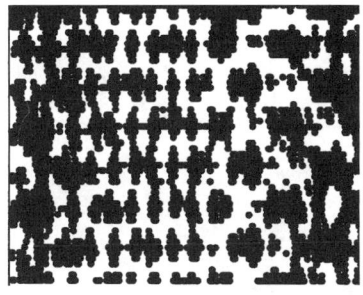 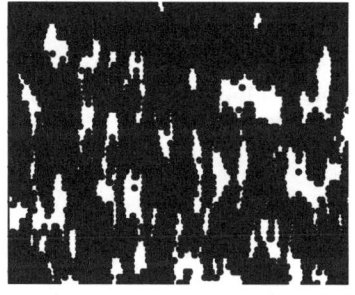

(a) Void distribution map for Σ11(131) [101] grain boundary, 4223 points

(b) Void distribution map for Σ29(153) [773] grain boundary, 7121 points

Fig.4 Void distribution maps for different grain boundaries. The void points in the same positions as the reference map were removed. (Voids are dark)

DISCUSSION

From the results of in-situ SEM observations, there is little difference in penetration between differently oriented grain boundaries. The measured penetration rates under these test conditions ranged from 6.4 to 9.2 μm/s. The penetration rate obtained by in-situ TEM on the Σ29 (153)[773] grain boundary is 4.4 μm/s which is in agreement with other results [4], and on the same order as the SEM observations. The reason that the penetration rates do not vary greatly with differently oriented grain boundaries under SEM observations may be because the grain boundary plane changes under the investigated length scale so that the penetration behavior exhibits combined behavior on different grain boundary planes, although the grain boundary misorientation is same for each observation.

All observations so far indicate that over dimensions of the order of the measured grain boundaries, the velocity is independent of time. However, due to the difficulty in performing longer time experiments, it is conceivable that the motion of Ga in Al may follow more of a Fick's law behavior. In this case, a first order approximation using experimental diffusivity in solid or grain boundary data [6], the estimated velocity should be a few orders or magnitude lower than the penetration rates obtained in the present study. This may indicate that the controlled penetration process proceeds within liquid channels in grain boundary planes.

The driving force for Ga penetration on Al grain boundaries is the energy reduction resulting from the replacement of the Al grain boundary with two Ga-Al interfaces. The surface energy of Ga on the (110) plane of Al is about 689 mJ/m^2 [7], and the energy of the Σ11(131)[101] grain boundary is 146 mJ/m^2. Therefore, Ga penetration on the Σ11(131) [101] grain boundary is not energetically favorable, as the total energy of the system would not lower by the formation of two Al-Ga interfaces. On the other hand, the Σ29(153)[773] grain boundary energy is 773 mJ/m^2, the penetration process will be energetically favorable if the energy of Ga on the (153) plane of Al is less than 386 mJ/m^2, which seems reasonable based on the first-principles calculations of Ga adsorption on different Al crystal surfaces [7].

When the penetration is energetically favorable, the penetration process could be associated with the void channel width. By quantitatively analyzing void distribution maps, the average channel width of the Σ11[101](131) grain boundary is approximately one half of the void width of the Σ29 (153)[773] grain boundary. If this void channel is large enough for a few monolayers Ga to penetrate, a contiguous network of voids would allow Ga penetration on the Σ29(153)[773] grain boundary, as it has been shown experimentally on the other high energy grain boundaries [4,9]. This mechanism would result in penetration fronts shown schematically in Fig. 5(a). Experimentally, by placing the Ga on the polycrystal Al surface for hours, and observing the samples on the other end, pin holes were found along the grain boundaries as shown in Fig. 5(b), which can be considered as the special path formed as above mentioned process. The discontinuous Ga front shown in Fig.2 could be further evidence for this network path model.

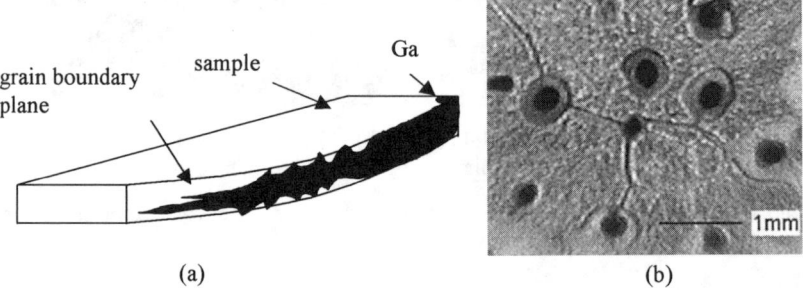

(a)   (b)

Fig. 5 (a) Schematic of Ga penetration grain boundary by penetrating special path along grain boundary plane, (b) pin holes along the grain boundaries. Some grain boundaries are not shown due to the short etching time.

CONCLUSIONS

It is suggested that the void channels of high energy grain boundaries provide a preferred path for the penetration of Ga monolayers into high energy grain boundaries. The penetration behavior investigated by SEM observations experimentally demonstrates that there may exist some contiguous network of voids that could be preferably penetrated by Ga. The rate of penetration appears similar to that of diffusion of liquid Ga rather than motion in the solid state.

ACKNOWLEDGMENTS

This work was supported by the US Department of Energy under contract DE-FG06-87ER45287. The authors are also very grateful for the helpful discussion and input from our coworkers R.C. Hugo and Boxiong Ding.

REFERENCES

1. M.G. Nicholas and C.F. Old, J. Mater. Sci. **14**, 1 (1979).
2. N.S. Stoloff and T.L. Johnston, Acta Met. **11**, 251 (1963).
3. S.P. Lynch, Embrittled by Liquid and Solid Metals, ed. M.H. Kamdar, 105, Proc. Met. Soc. AIME (1982).
4. R.C. Hugo and R.G. Hoagland, Scripta Mater., **38**, 523 (1998).
5. F. Ercolessi and J.B. Adams, Materials Theory and Modeling Symposium, Master. Res. Soc., Pittsburgh, PA, 31, (1993).
6. I. Kanr, and W. Gust, Handbook of Grain and Interphase Boundary Diffusion Data, **1**, Ziegler Press, Stuttgart (1989).
7. R. Stumf and P.S. Feibelman, Phys, Rev. **B54**, 5145 (1996).
8. R.C Hugo, Ph.D. Thesis, Washington State University (1998).
9. T. Watanabe, S. Shima, and S. Karashia, Embrittled by Liquid and Solid Metals, ed. M.H. Kamdar, 105, Proc. Met. Soc. AIME (1982).

TWINNING DEFORMATION IN MARTENSITE MICROSTRUCTURE

G.J.ACKLAND and U.PINSOOK

Department of Physics, The University of Edinburgh, James Clerk Maxwell Building, Kings Buildings, Mayfield Road, Edinburgh EH9 3JZ

ABSTRACT

Molecular dynamics is used to study twinning deformation in a martensite microstructure obtained from rapid cooling β zirconium through the bcc-hcp transition. The microstructure is composed of $(10\overline{1}1)$ twin boundaries and boundary dislocations which sometimes spread across the twins to form stacking faults. A series of such equilibrium microstructures subjected to discrete, increasing $< 10\overline{1}2 > (10\overline{1}1)$ shear strain. The stress-strain curve has stick-slip behaviour with yield stress of ≈ 5.0Kbar and yield strain of $\approx 3.8\%$. Deformation occurs through movement of twin boundaries in segments between boundary dislocations. Straight perfect twin boundaries do not move.

INTRODUCTION

Named after the prototype in steels, martensitic phase transformations are extremely common in metals. Typically the change from a high temperature (austenite) phase to a low temperature (martensite) phase with lower symmetry is such that each atom moves to a specific site in the martensite defined by its original position in the austenite. Usually there is also a shear deformation of the unit cell and which the sample cannot accommodate so an austenite single crystal will transform into a strain-accommodating microstructure comprising a number of twins: crystallites of martensite with different orientation of the symmetry-breaking crystal axes.

The behaviour of these twinned microstructures gives rise to some extraordinary phenomena such as superelasticity and the shape memory effect. The cause is that martensitic microstructures deform by movement of twin boundaries rather than dislocations. Here, we present a molecular dynamics computer simulational model of a system which exhibits the phase transition and associated twinning deformation in order to understand the essential processes on the atomic level which govern plasticity in martensite.

Specifically, we use a model for a single element, zirconium [1,2], which exhibits a martensitic phase transition from bcc to hcp via a soft phonon mechanism [3,4]. Real zirconium does not exhibit shape memory effect or superelasticity because it can deform by a dislocation mechanism, so in our simulations we will suppress the formation of dislocations by our choice of boundary conditions. To bridge the lengthscale gap between the atomistic and the twinned microstructure, we require large molecular dynamics calculations of up to 200000 atoms. We also want to achieve reasonable statistics for defect formation, so we repeat each calculation several times with different initial conditions. 200000 atoms is still small relative to the scale of typical microstructures, and tends to favour formation of high dimensional defects at the cost of bulk strain.

Mat. Res. Soc. Symp. Proc. Vol. 578 © 2000 Materials Research Society

CALCULATIONS

We use two approaches to produce representative martensitic microstructures: since symmetry requires that the self accommodating twin microstructure of hcp zirconium derived from a bcc crystal comprises $(10\bar{1}1)$ twin boundaries [5], we could set up a supercell containing stripes of $(10\bar{1}1)$ twins. Alternately, we could simulate the phase transition with molecular dynamics and apply strain to the resultant microstructure. The simulated transition forms vicinal [6] twins oriented at 60°, their boundaries have the structure of $61.5°$ $(10\bar{1}1)$ twin boundaries with dislocations to accommodate the mismatch [7].

We carry out our simulations using the Parrinello-Rahman Lagrangian in which the vectors defining the simulation supercell are treated as dynamical variables and the atomic positions are defined by fractional coordinates within the supercell [8]. This allows the samples to be deformed by applying a finite strain to the supercell vectors while the relative coordinates of the particles remain fixed. Once the strain has been applied we allow the structure to evolve under molecular dynamics for several picoseconds, until the calculated stress and appearance of the microstructure have stabilised. This is taken as the equilibrium elastically strained structure.

For analysis, the elastically strained microstructure is quenched to a local minimum of the energy with respect to atomic positions. Each atom is coloured according to the arrangement of its near-neighbours as uncharacterised (lightest circles), fcc (grey) or one of the possible hcp variants (dark greys and black) compatible with the (111) growth plane of the martensite [7]. Once the appropriate growth plane has been determined for a given sample, a slice through the sample is taken: parallel slices tend to be similar. These slices enable us to identify the essential features of the microstructure. The post-transition microstructure exhibits a number of objects which are thermodynamically stable for very fine microstructures but whose energy scales unfavourably with system size, and might be expected to vanish in coarser structures.

Partial twin boundary dislocations

In close packed metals, dislocations with integer Burgers vector can split into two partial dislocations separated by an area of stacking fault. The partial dislocations have smaller strain fields than a full dislocation, and so the splitting reduces the energy of the total defect. Against this, the energy of the stacking fault increases the energy. Since the stacking fault energy cost is proportional to the separation, while the splitting energy gain falls as $1/r^2$, there is an optimal separation which minimises the defect energy [9].

Twin boundary dislocations, such as those appearing in the vicinal twins of the current system, have a related mechanism. Partial twinning dislocations form on adjacent twin boundaries, linked by a strip of stacking fault. The two partials have lower core energy than a single full boundary dislocation and this, not their strain field, is the energy gain. Against that must be balanced the cost of forming the stacking fault: as before this is proportional to its length, but now the length is determined by the size of the twin. Consequently, twin boundary dislocations will form as partials in fine microstructures, while full dislocations will be favoured in coarser microstructures. The crossover is determined by the ratio between stacking fault energy and the energy difference between full and partial twin boundary dislocations. However, the E1 stacking fault cannot be removed by a simple shear: in practice this makes them very difficult to eliminate once formed

Metastable bulk phases

Small domains of metastable crystal structures can also be formed. In the present case fcc is metastable at all temperatures. However, the basal plane interface between hcp and fcc (which preserves close packing) is very low in energy compared with the vicinal $(1\bar{1}01)$ twin boundary. Moreover, fcc can form basal plane interfaces with all three hcp variants simultaneously. Consequently, small regions of fcc can be thermodynamically stable in the microstructure, the energy cost of the unfavourable crystal structure being compensated for by the energy gain from the low energy interfaces. Such stabilisation of small regions is similar to epitaxial growth of metastable crystal structures on surfaces of, or as inclusions in, a host material.

Both partial boundary dislocations and metastable phases are consequences of finite twin size: the higher dimension defect will ultimately become unstable.

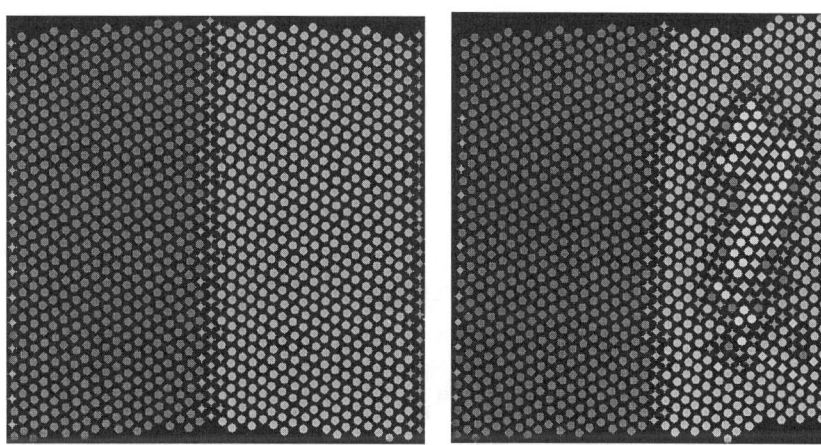

Apply shear strain

Perfect Twins $(1\bar{1}01)$ Boundary is sessile, new twin forms

FIG. 1. Deformation by applying shear to a straight twin boundary: the material responds by nucleating another variant. Circles indicate local coordination corresponding to three hcp variants, diamonds to fcc and stars to boundary atoms.

Shear deformation of twinned microstructure

For samples containing only straight $(10\bar{1}1)$ twin boundaries, very large elastic stresses were developed before plastic deformation occurred, well in excess of the shear strength of real materials. Moreover, when the material finally yielded it did so by nucleation of another twin within an existing one, in preference to motion of the boundary. From this we conclude that perfect twin boundaries have higher strength than the bulk material, and play no role in deformation of the real material.

Figure 1 shows a typical straight twin boundary simulation, before and after application of shear. The inclusion in the second figure appears to have numerous stacking faults, and may be a boundary-stabilised ABCAB stacked metastable structure.

419

The behaviour of microstructure derived from the transition is more interesting, and we discuss it with respect to two typical runs. For rapid cooling of larger systems we find it possible to obtain three variants, and the evolution of one such simulation is shown in figure 2. Given enough time such microstructure will ripen to a laminate structure [7], but the applied shear enhances the process, favouring the development of the one variant (coloured black) which best accommodates the strain.

The stacking faults between twin dislocations are remarkably robust (they are an integral part of the vicinal twin microstructure, not a separate defect). Even when the sample has fully transformed into a single martensitic variant, the stacking faults remain, terminated by true crystal dislocations, and when we continued to apply shear to the single domain sample it was found to yield by dislocation motion.

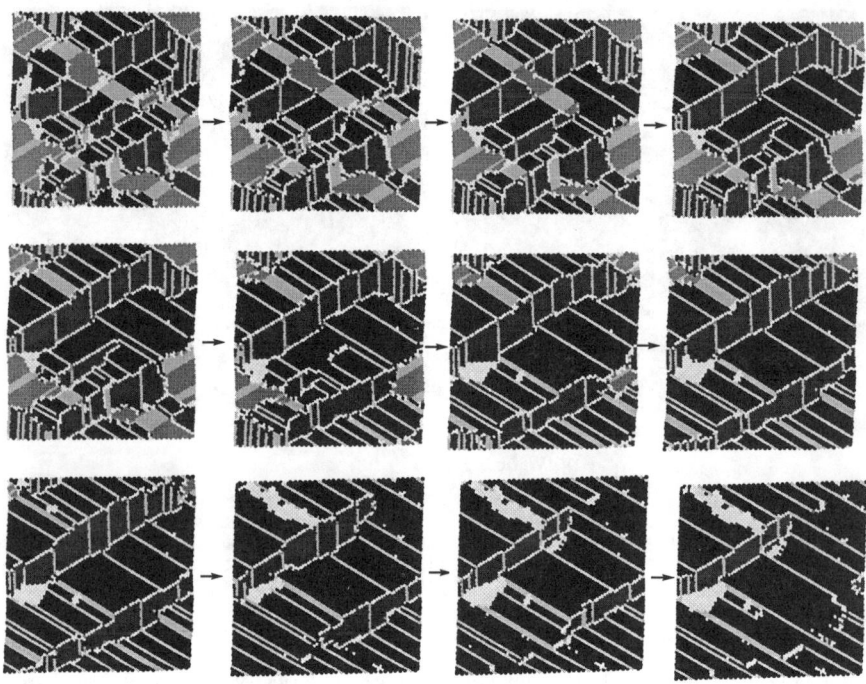

FIG. 2. Effect of shear on a large, rapidly cooled microstructure - the supercell is much thinner in the perpendicular direction (5 layers). Initially three hcp variants are present (darker greys) and some boundary stabilised fcc (lighter grey: notice the lack of 'undefined' atoms at the very sharp, low energy hcp-fcc interfaces). As strain is increased, the twelve snapshots show the elimination of one variant (and with it the need for high energy boundaries which previously stabilised fcc), and then the steady growth of the black variant. This simulation is large enough to show the elliptical shape of the last grey twin. Note the presence throughout of stacking faults (light grey atoms).

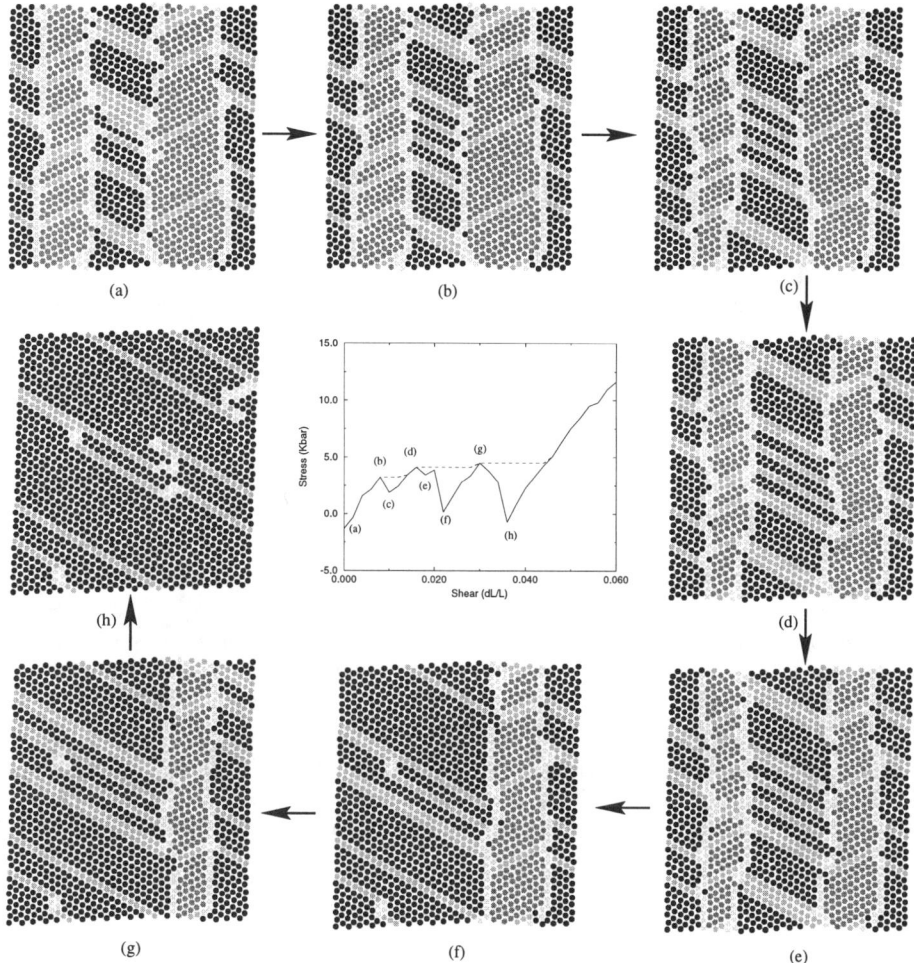

FIG. 3. Effect of shear on a laminate microstructure. One set of twins grows at the expense of the others, ultimately eliminating them. Growth occurs in sections between the stacking faults, which tend not to move. Central figure shows the stress-strain relationship for this simulation, figures (a) to (h), clockwise from top left, show quenched microstructure for strains and denoted in the figure. The dotted line shows the implied response to an applied stress. Beyond 4% strain, deformation is primarily elastic with some partial dislocation motion.

For a slow-cooled structure from a relatively small simulation in a cubic supercell which produced four twins, the applied strain had the effect of making one grow at the expense of the other. However, the twin boundary dislocations do not move. Twin growth occurs by rapid motion of jogs between sessile partial twin boundary dislocations. We presume that the jog moves between the two twin boundaries to enable the stacking fault to climb, but we have not observed this object.

The preferential growth of a particular twin variant is shown in figure 3 for a small, laminate microstructure. The stress-strain relation for the simulation is also shown - the initial elastic region is followed by a stick-slip type of deformation corresponding to yielding of different regions of the twin boundary. Obviously, in a macroscopic sample this region would flatten out by averaging over yielding in many environments, giving the characteristic superelastic behaviour. Once the whole sample has converted to one variant, the stress-strain curve rises steeply.

CONCLUSIONS

Simulations of formation and shear of a model martensitic material have shown the expected formation of a twinned microstructure which responds to shear by preferential growth of one twin. This growth occurs by motion between *partial twin boundary dislocations*, defects which exist in pairs on opposite sides of a twin connected by a stacking fault. Twin boundaries without these dislocations are more rigid than the bulk and do not move.

The stacking faults associated with these dislocations do not move, and can survive even after the unfavoured variants are eliminated by annihilation of twin boundaries. The elimination of a twin leaving behind its boundary dislocations results in true dislocations which can facilitate further plastic deformation.

While the results of these simulations suggest some interesting new phenomena, it is prudent to add a few caveats. Firstly, although the potential is fitted to data for zirconium it does not give a reliable description of shear deformation in that material, however we believe that several generic features of twinning deformation are revealed. Secondly, the objects responsible for deformation in our simulation may be stabilised by the finite size of the system. This is not as serious a constraint as it first appears, since martensites grow from extremely fine tips at the austenite/martensite boundary, and the partial twin boundary dislocations will be stable here. Subsequently, as we have seen, they are remarkably robust, outlasting even the twinned microstructure itself in shear calculations.

REFERENCES

[1] G.J.Ackland, S.J.Wooding and D.J.Bacon, *Phil. Mag.* **A 71**, 553 (1995).
[2] M.W.Finnis and J.F.Sinclair, *Phil. Mag.* **A 50**, 45 (1984).
[3] Z.Nishiyama, *Martensitic transformations* (Academic, New York, 1978).
[4] U.Pinsook and G.J.Ackland Phys Rev.B 59, 13642-13649 (1999)
[5] A.Serra and D.J.Bacon, *Mater. Sci. Forum* **126-128**, 69 (1993).
[6] A. P. Sutton and R. W. Balluffi Interfaces in Crystalline Materials, Oxford (1995)
[7] U.Pinsook and G.J.Ackland, Phys.Rev.B, 58, 11252-57 (1998)
[8] G.J.Ackland, D.Phil thesis, Oxford University (1987).
[9] J. P. Hirth and J. Lothe: Theory of Dislocations, McGraw-Hill. (1982)

Microstructural Modeling for
Industrial Metals Processing

MODELLING OF GRAIN REFINEMENT IN ALUMINIUM ALLOYS

A. L. GREER, A. TRONCHE and M. VANDYOUSSEFI
University of Cambridge, Department of Materials Science & Metallurgy, Pembroke Street,
Cambridge CB2 3QZ, UK.

ABSTRACT

Commercial grain refiners for aluminium solidification are so potent that the barrier for grain initiation is that for free growth rather than for nucleation itself. In this case quantitative prediction of grain size is possible. For small melt volumes a successful isothermal-melt model is presented. This is extended to directional solidification in a temperature gradient using cellular-automaton modelling of grain growth with finite-element heat-flow calculations.

INTRODUCTION

Industrial solidification of aluminium alloys often involves grain refinement (inoculation) to ensure uniform, fine, equiaxed grain structures [1]. Refinement facilitates faster casting and brings improved properties; it is achieved by adding to the melt particles (within a master alloy) which can act as substrates for heterogeneous nucleation. For common Al-Ti-B refiners these particles are TiB_2. Problems with existing refining practice include: variability in performance, 'poisoning' of refinement in the presence of some solutes, and refiner efficiencies (no. of grains per added particle) of at best 1%. It would be desirable to be able to predict grain size as a function of the key parameters (refiner addition level, cooling rate, solute content of alloy), but this is expected to be difficult. While microstructure prediction can be quantitative when growth dominates (e.g. [2]), prediction of nucleation rates remains elusive. Nucleation rates depend very sensitively on the solid-liquid interfacial energy which is rarely well known. For heterogeneous nucleation, the contact angle may show a distribution of values and is also not well known. In any case, for potent nuclei the classical approach based on spherical-cap nuclei with a defined contact angle may be invalid; an alternative adsorption model has been proposed [3], but a full kinetic treatment is still lacking.

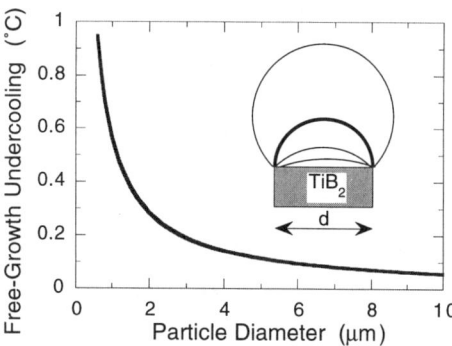

<u>Fig. 1.</u> The undercooling necessary to initiate free growth ΔT_{fg} from the assumed-circular (0001) face of a TiB_2 disc of diameter d. This undercooling arises from the Gibbs-Thomson shift of the solid-liquid equilibrium and is given by eq. (1). The critical condition for free growth is when the solid-liquid interface has the minimum radius of curvature, i.e. is hemispherical as shown in the inset.

Existing refiners appear to be very potent and may readily form nuclei of solid aluminium, either as very flat spherical caps or by adsorption [3]. For TiB_2 particles, the solid appears from microscopical studies to form on the basal (0001) faces [4]. For such nuclei to grow into grains, the solid-liquid interface must be able to break free from the particle without its radius of curvature becoming less than that of the critical homogeneous nucleus at the given undercooling. The critical undercooling for 'free growth' is given by

$$\Delta T_{fg} = 4\,\sigma\,/\,\Delta S_V\,d,\qquad(1)$$

425

where σ is the solid-liquid interfacial energy (taken in the present work to be 158 mJ m^{-2} [5]), ΔS_V is the entropy of fusion per unit volume (1.112 MJ °C^{-1} m^{-3}), and d is the average diameter of the major cross-section of the particle. Figure 1 shows that ΔT_{fg} is comparable with overall measured undercoolings ($\leq$ 0.5 °C [6]) for likely particle diameters of ~1 µm. If the barrier to free growth, rather than true nucleation, is rate-controlling, quantitative prediction of grain initiation becomes possible as d is readily measurable.

ISOTHERMAL-MELT MODEL

Maxwell and Hellawell [7] pointed out that growing grains interact thermally rather than through their solute fields and that recalescence can limit the number of nucleation events. As detailed elsewhere [8, 9] their model has been adapted to use the free-growth criterion rather than nucleation. The inputs to the model (independently determined) are σ, ΔS_V, the enthalpy of fusion per unit volume (9.5×10^8 J m^{-3}), the heat capacity of the melt per unit volume (2.58×10^6 J °C^{-1} m^{-3}), the solute diffusivity in the melt (for all solutes taken to be 2.52×10^{-9} m^2 s^{-1}, the value for titanium [9]), and the measured particle size distribution (Fig. 2). It is also necessary to know the liquidus slope m (°C wt.%$^{-1}$), partition coefficient k and content C_0 (wt.%) for each solute present in the melt. The growth of grains (diffusion-limited growth of spheres, modelled as in [8, 9]) is assumed to start at the relevant ΔT_{fg} for each particle, the latent heat release is computed, and the numerical model calculates the cooling curve for a given rate of external heat extraction. As the thermal diffusion length is typically twice the dimension of the sample (few cm) in the tests being modelled (standard grain-refining 'TP-1' tests [10]), the approximation is made (following [7]) that the melt is spatially isothermal. Both measurements on actual TP-1 tests and the model show recalescence due to latent heat release, with maximum undercoolings in good agreement at ~0.2 °C [9]. As the melt is undercooled, according to eq. (1) grains will grow from the largest particles first, and with $\Delta T \leq 0.2$ °C only the shaded region in Fig. 2 initiates new grains.

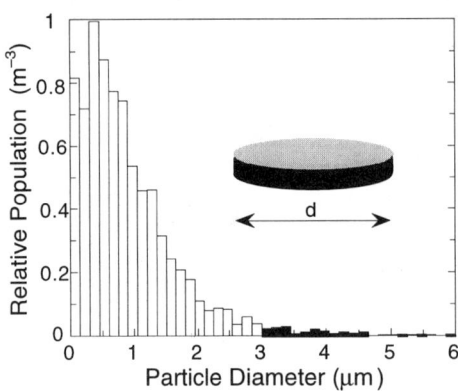

Fig. 2. The measured diameter distribution of TiB$_2$ particles in the commercial Al-5Ti-1B (wt.%) refiner used here. The particles are hexagonal platelets [4] approximated as discs (inset). The relative populations (shown) are estimated from intersections in SEM micrographs of polished surfaces, corrected for sectioning effects. Absolute populations are then estimated from the known volume fraction of TiB$_2$ phase. The refiner is very inefficient as under normal conditions only the largest particles (shaded area) initiate grains.

Figure 3 shows that without adjustable parameters, the model gives quantitative predictions of grain size of commercial purity (CP) aluminium as a function of the amount of added refiner. These data can be replotted as the numbers of grains and of refiner particles per unit volume (inset in Fig. 3), making clear the low efficiency of commercial refiners. The model also gives good predictions for the variation of grain size with cooling rate [9]. Spittle and Sadli [11] have measured grain size in refined melts with many different solutes and solute addition levels. The model has been used with relevant m, k and C_0 values for each of the cases measured in [11], providing the comparison in Fig. 4. For equiaxed structures (grain size < 400 µm) the agreement between model and experiment is good; at larger sizes the measured structures are likely to be columnar and beyond the applicability of the model.   This weakness of the model is fundamentally because it ignores temperature gradients (crucial in the columnar-equiaxed transition [12]). The model has another weakness in that it assumes essentially perfect nucleation. When nucleation is impaired, as in badly made refiners [13] or with solute poisoning (e.g. [14]), the model breaks down.

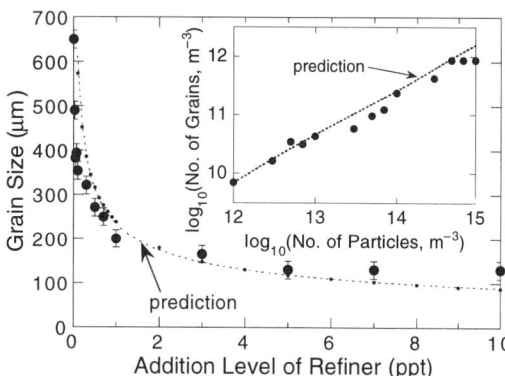

Fig. 3. Comparison of measured (TP-1 test) and calculated (isothermal-melt model) grain sizes in CP-aluminium as a function of the addition level of refiner.   The inset shows the same data plotted as the numbers of grains and of refiner particles per unit volume. The efficiency of the refiner (the ratio of these populations) decreases at higher refiner addition levels.

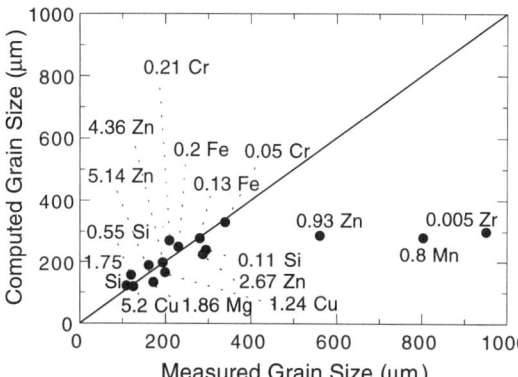

Fig. 4.   Comparison of measured (TP-1 test data from [11]) and calculated (isothermal-melt model) grain sizes in binary Al alloys with a wide variety of solute types and levels. Measured grain sizes > 400 μm correspond to columnar structures not treated by the modelling. The solute contents (wt.%) of the alloys are marked.

However, with standard refiners and no poisoning solutes the model makes usefully quantitative predictions of the grain size in small-scale grain-refining tests. The match with experiment validates: the limitation of refiner efficiency (Fig. 2) based on recalescence [7], the free-growth model (eq. (1)), and the parameters used. With the fundamental grain-initiation behaviour on a known size distribution of refiner particles thus established, the extension of the modelling to more complex, non-isothermal cases is considered next.

CELLULAR-AUTOMATON MODEL

Industrial-scale processes such as DC casting are on a much larger scale than the few cm represented in the TP-1 test [10]. In such processes there are significant temperature gradients without overall undercooling of the bulk. Locally the solidification is directional with defined interface velocity V and temperature gradient G. Laboratory-scale Bridgman experiments have been widely used to study steady-state directional solidification under controlled conditions (e.g. [15]).   On the other hand, solidification modelling, particularly coupled thermal and microstructural simulation, can provide valuable details about the evolution of solidification grain structure in a temperature gradient in combination with experiment.

Here commercial software *calcoMOS*™ [16] is used to model directional solidification in a Bridgman sample. The simulation is physically-based and stochastic. A cellular-automaton (CA) grid is defined, in which solidification (dendritic in the present case) proceeds from cell to cell. The CA approach takes account of randomly dispersed heterogeneous nuclei, random grain orientation, preferred dendrite growth directions ($\langle 100 \rangle$) and the growth kinetics. The growth law for the dendrite tips is calculated from the model of [17], and approximated as a power-law dependence of velocity on local undercooling. The period before the morphological breakdown into dendritic growth is ignored. Nucleation is assumed to be on sites which become active at a spectrum of undercoolings. As it happens, this approach is quite compatible with the behaviour illustrated in Fig. 2. In calcoMOS™, however, the distribution of nucleation events as a function of undercooling is taken to be Gaussian. The nucleation can be both at the surface and internal, but only the latter is of immediate interest. Both the nucleation and growth are simulated in the same CA grid, within which the cell size must be somewhat less than the spacing between active dendrite tips for reliable results. An enthalpy-based finite-element (FE) algorithm is used to calculate temperature distributions and heat flow. The FE mesh is much coarser than the CA grid, but the thermal computation is fully coupled with the latent heat released in the CA cells and with changes in specific heat. Details of the two-dimensional (2-D) coupled CA-FE model (as used here) are given in [18], and further developments are described in [19].

In the present work the CA-FE model is applied to model solidification in a particular Bridgman furnace with a well-characterised temperature profile. To facilitate comparison with experiment (to be reported elsewhere), the modelling is for Al - 4.15 wt.% Mg. For the Gaussian distribution of nucleation undercoolings in the model, the median undercooling (peak of the distribution) was set at $\Delta T_m = 3$ °C, and the standard deviation at $\Delta T_\sigma = 0.5$ °C, while the maximum number of nucleation events N_{max} (i.e., the number of inoculant particles) was used as an adjustable parameter. In future, the undercooling distribution in the CA-FE model will be matched to that determined using the isothermal-melt model (previous section), but in the present preliminary work, the values of ΔT_m and ΔT_σ were selected only to facilitate the computation. In the thermal calculation, it is assumed that there is no heat exchange through the sample sides in the non-cooled part of the sample.

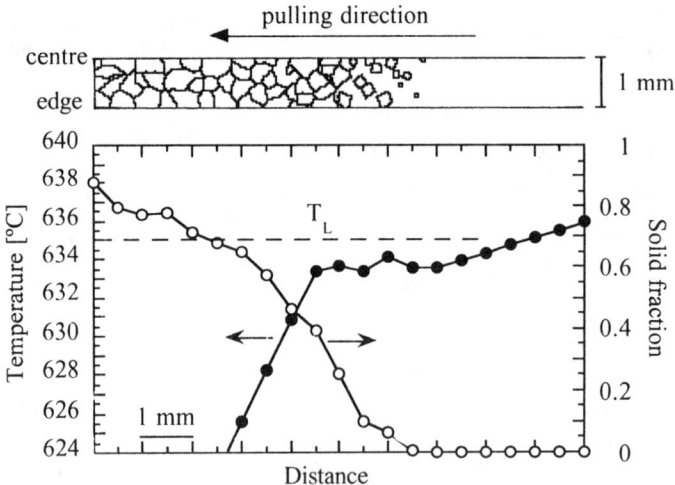

Fig. 5. A grain structure (in 2 mm × 10 mm area, half shown) developing in the CA-FE model during directional solidification of Al - 4.15 wt.% Mg with $V = 0.1$ mm s^{-1} and $N_{max} = 10^{11}$ m^{-3}: 2-D microstructure and the corresponding profiles of temperature and solid fraction.

Figure 5 shows a simulated microstructure, with corresponding temperature and solid fraction (f_s) profiles (the average of 3 profiles parallel to the pulling direction). The temperature gradient in the liquid is lower than that in the solid, because of the temperature profile of the furnace. In considering the f_s profile, it should be noted that the material within the dendritic envelopes is not fully solid (the residual liquid fraction is calculated using a Scheil approach), and that Al - 4.15 Mg shows eutectic solidification at 450 °C outside the region shown.

In the solidifying (mushy) zone, the growing grains are warmer than the surrounding liquid by ≤ 2°C (because of latent heat release). Despite averaging over 3 positions across the width of the sample, the snapshot profile in Fig. 5 still shows such local temperature fluctuations. Apart from these, the mushy zone appears to be isothermal and undercooled by $\Delta T_{mush} \approx 1.5$ °C. It is clear that, through the release of latent heat, the solidification significantly perturbs the temperature profile in the Bridgman sample, a factor normally discounted [12]. In the isothermal-melt model (previous section), the balance between the external heat extraction and the latent heat release determines the maximum undercooling and leads to recalescence. In the present case, the same balance determines ΔT_{mush}, limiting the number of grain-initiation events. The uniformity of ΔT_{mush} in the mushy zone ensures that all the initiation events occur early in the mushy zone (i.e. at low f_s), as is consistent with the simulated microstructure. The length of the mushy zone is affected by the processing parameters (pulling velocity V and temperature gradient G) as well as by N_{max}. For a given V, increasing G or decreasing N_{max} leads to a shorter mushy zone, then to elongated grains, and finally to columnar growth. These results are consistent with the trends reported in [18], and will be explored more quantitatively in future.

Figure 6 shows how the population of grains N_g varies with the population of inoculant particles (N_{max}). For given solidification conditions (V and cooling rate $V \times G$ measured in the liquid just ahead of the mushy zone), the efficiency of grain refinement decreases with refiner addition level, analogous to the behaviour shown in the inset in Fig. 3. While the efficiencies indicated in Fig. 6 are unrealistically high, the gradients of the lines (i.e. y in $N_g \propto N_{max}^y$), up to 0.65, are comparable with the $y = 0.71$ for the measured data in Fig. 3. In Fig. 6 it is clear that for a given N_{max} increasing the cooling rate increases N_g. This is illustrated more directly in Fig. 7. The measured data are fitted with an N_{max} of somewhat over 10^{13} m^{-3}, a reasonable value. The trends shown in Fig. 7 not only reproduce the measured data (from Bridgman samples) but also are very similar to measured and calculated data for TP-1 test samples [10].

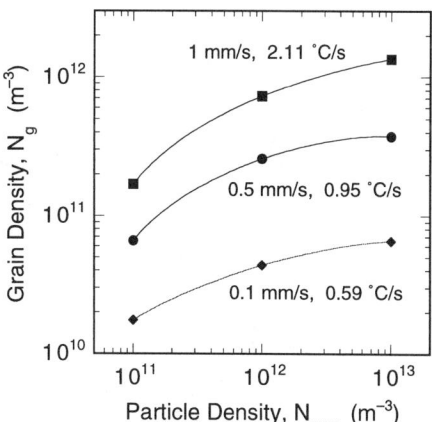

Fig. 6. Calculated (CA-FE model) grain density N_g as a function of the density of refiner particles for selected V and cooling rate. The density N_g (per unit volume) is estimated from the 2-D simulated microstructure by applying standard stereological corrections.

CONCLUSIONS

Recognition that the significant barrier to grain initiation is that for free growth, rather than for nucleation per se, permits quantitative modelling of solidification and prediction of grain size in inoculated aluminium melts. A simple model assuming a spatially isothermal melt, and with measured refiner particle size distribution as input, has been extensively validated against measured effects of refiner addition level, cooling rate and melt composition in small-sample grain-refining tests. In these tests, melt recalescence limits refiner efficiency. The fundamental behaviour so determined can be applied in more general models of microstructural development in solidification, with finite-element solutions for the temperature distribution and cellular-automaton modelling of grain growth. While this CA-FE modelling has not yet been tested

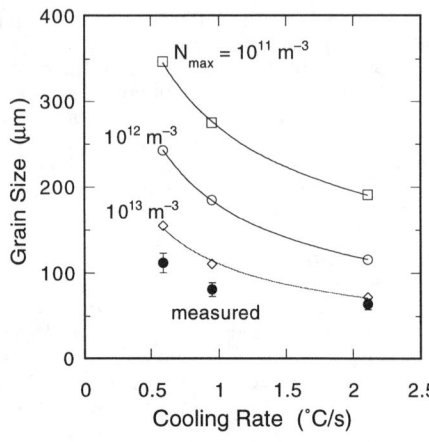

quantitatively against experiment, it successfully predicts the trends of grain size with processingconditions in situations such as Bridgman solidification in which there are significant temperature gradients. The CA-FE modelling suggests that in the presence of a temperature gradient, grain refinement is still limited, as in the isothermal-melt model, by the latent heat release during solidification. Modelling of the kinds described may permit optimization of refining practice and design of improved refiners.

Fig. 7. Comparison of measured (Bridgman) and calculated (CA-FE model) grain size as a function of cooling rate for selected N_{max}. The best fit is with a nucleant particle density N_{max} slightly greater than 10^{13} m^{-3}.

ACKNOWLEDGEMENTS

This work is funded mainly by the Engineering and Physical Sciences Research Council (UK). Further support from and interactions with Alcan International Limited Banbury Laboratory, London and Scandinavian Metallurgical Co. Limited, and Pechiney Centre de Recherche de Voreppe are gratefully acknowledged.

REFERENCES

1. D. G. McCartney, Int. Mater. Rev. **34**, 247 (1989).
2. W. Kurz and R. Trivedi, Mater. Sci. Eng. **A179-180**, 46 (1994).
3. W. T. Kim and B. Cantor, Acta Metall. Mater. **42**, 3115 (1994).
4. P. Schumacher, A. L. Greer, J. Worth, P. V. Evans, M. A. Kearns, P. Fisher and A. H. Green, Mater. Sci. Technol. **14**, 394 (1998).
5. N. Eustathopoulos, L. Coudurier, J. C. Joud and P. Desré, J. Cryst. Growth, **33**, 105 (1976).
6. M. Johnsson, L. Bäckerud and G. K. Sigworth, Metall. Trans. A, **24**, 481 (1993).
7. I. Maxwell and A. Hellawell, Acta Metall. **23**, 229 (1975).
8. A. M. Bunn, P. V. Evans, D. J. Bristow and A. L. Greer, in *Light Metals*, edited by B. Welch (TMS, Warrendale, PA, 1998) pp. 963-968.
9. A. L. Greer, A. M. Bunn, A. Tronche, P. V. Evans and D. J. Bristow, Acta Mater., submitted.
10. *Standard Test Procedure for Aluminum Alloy Grain Refiners: TP-1* (The Aluminum Association, Washington, DC, 1987).
11. J. A. Spittle and S. B. Sadli, Mater. Sci. Technol. **11**, 533 (1995).
12. J. D. Hunt, Mater. Sci. Eng. **65**, 75 (1984).
13. A. M. Bunn, A. L. Greer, A. H. Green and M. A. Kearns, in *Solidification Processing 1997*, edited by J. Beech and H. Jones (Univ. of Sheffield, Sheffield, 1997) pp. 264-267.
14. A. M. Bunn, P. Schumacher, M. A. Kearns, C. B. Boothroyd and A. L. Greer, Mater. Sci. Technol. **15**, 1115 (1999).
15. P. V. Evans, J. Worth, A. Bosland and S. C. Flood, in *Solidification Processing 1997*, edited by J. Beech and H. Jones (Univ. of Sheffield, Sheffield, 1997) pp. 531-535.
16. Calcom SA, Lausanne, Switzerland.
17. W. Kurz, B. Giovanola and R. Trivedi, Acta Metall. **34**, 823 (1986).
18. Ch.-A. Gandin and M. Rappaz, Acta Metall. Mater. **42**, 2233 (1994).
19. Ch.-A. Gandin, Ch. Charbon and M. Rappaz, ISIJ Int. **35**, 651 (1995).

A MODEL FOR PREDICTING WELD METAL GRAIN REFINEMENT IN *G-V* SPACE

Ø. GRONG* and C.E. CROSS**
* Dept. of Materials Technology and Electrochemistry, Norwegian University of Science and Technology, N-7491 Trondheim, Norway
** Metallurgical Eng. Dept., Montana Tech of the University of Montana, Butte, MT 59701

ABSTRACT

A model has been developed which allows for the graphical representation of undercooling and nucleation site density in *G-V* space (i.e. temperature gradient G versus growth rate V). This model is used to explain a unique grain structure in weldments where equiaxed grains may be found both at the fusion boundary and along the center of the weld metal. Details of this model will be given and discussed in relation to grain refinement predictions in *G-V* space.

INTRODUCTION

The grain structure of weld metal typically consists of columnar grains that nucleate off of base metal grains (i.e. epitaxial nucleation) [1]. In some instances, however, equiaxed grains have been observed along the weld centerline [2-5] and, more recently, they have also been observed along the fusion zone boundary [6-14]. Thus, there appears to be the possibility for a dual transition in weld metal grain structure, first from equiaxed to columnar, then from columnar to equiaxed.

The occurrence of centerline equiaxed grains has been attributed to the high growth rate V and low temperature gradient G characteristic of the weld centerline as depicted in Figure 1. The corresponding low G/V (or $G/V^{\frac{1}{2}}$) ratio associated with this region has traditionally been taken to represent high undercooling and the possibility for nucleation of new grains [15]. However, this rationale fails to account for the presence of equiaxed grains at the fusion zone boundary, a region characterized as having a high G/V ratio.

Use of the G/V ratio as an indicator for undercooling originates from Chalmers' analysis of constitutional undercooling [16], where a critical G/V value was shown to denote the transition from planar to cellular growth. Large values of G/V, greater than the critical value, represent conditions for stable planar growth and the absence of undercooling. This criterion has since been extended to predict growth transitions from cellular to cellular-dendritic, cellular-dendritic to dendritic, and dendritic to equiaxed-dendritic. The problem with using the Chalmers criterion to predict these other modes of solidification is that this model does not account for undercooling due to curvature.

A more rigorous analysis of undercooling by Hunt [17], which accounts for undercooling due to curvature, has been applied to welding and has demonstrated the possibility for fusion line equiaxed grains [18]. The purpose of this work has been to use a modified form of Hunt's analysis to demonstrate graphically, in *G-V* space, how it is possible to have a dual transition (equiaxed-columnar-equiaxed) in weld metal grain structure.

Mat. Res. Soc. Symp. Proc. Vol. 578 © 2000 Materials Research Society

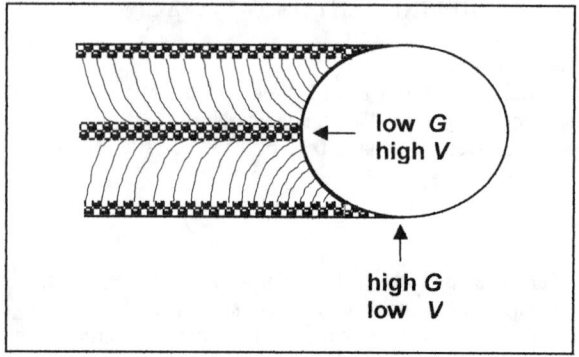

Figure 1: Schematic diagram showing top view of weld
pool and weld metal with both equiaxed and columnar grains.

MATHEMATICAL MODELING

In the following, a simple analytical model for the nucleation and growth of equiaxed
grains ahead of the advancing interface will be presented, based on a modification of the
original Hunt theory [17]. The main objective of this modeling exercise is to document
that the equiaxed grain formation during solidification is either favored by the
combination of a high thermal gradient G and a low growth rate V or a low G and a high
V. The latter condition is met in the center of castings and fusion welds, where equiaxed
grains are most frequently observed. Moreover, the solidification conditions existing in
welding imply that the combination of a high G and a low V is attained close the fusion
line, which, in turn, may lead to the formation of a narrow band of fine equiaxed grains in
the adjacent weld metal. Both incidents can be rationalised by means of simple
mechanism maps showing contours of constant undercooling ΔT and nucleation site
density N_0 in G-V space.

COMPONENTS OF THE MODEL

The model consists of two components, i.e. a model for the undercooling ahead of the
advancing columnar grains and a simple model for the equiaxed grain formation within
this region. The principle behind this model is illustrated in Figure 2. The
transformation in grain structure, from columnar to equiaxed, will occur only if the
heterogeneous nucleation sites, located ahead of the advancing columnar grains, become
activated and grow to impingement to obstruct the columnar grains. This goal can be
accomplished with a combination of either a large number of nucleation sites and a slow
growth rate, or a small number of nucleation sites and a fast growth rate.

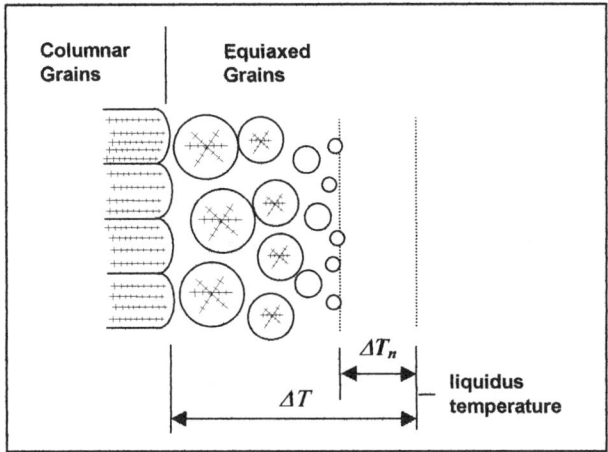

Figure 2: Schematic diagram showing heterogeneous nucleation and growth of equiaxed grains ahead of columnar grains.

Undercooling During Cellular and Dendritic Growth

As a starting point, the model of Burden and Hunt [19] is used to obtain an analytical expression for the undercooling ΔT as a function of the growth rate V of the columnar grains and the actual thermal gradient G in the liquid:

$$\Delta T = \frac{DG}{V} + AV^{1/2} \tag{1}$$

The first term in Equation 1 represents the undercooling associated with a flat solid/liquid interface, whereas the second term includes the effects of curvature through the parameter:

$$A = 2\left(-\frac{2m(1-k_0)C_0\Gamma}{D} \right)^{1/2} \tag{2}$$

where D is the solute diffusivity in the liquid, m is the slope of the liquidus line in the phase diagram, k_0 is the solute partitioning coefficient and Γ is the Gibbs-Thompson parameter.

Equiaxed Grain Growth

The next step is to invoke a simple growth model for the equiaxed grains that form within the undercooled region ahead of the advancing columnar grains. If local site

433

saturation is assumed (i.e. all grains nucleate instantaneously at time $t = 0$), the mean diameter of these grains d can be calculated from their growth rate V_e [17]:

$$d = 2\int_0^t V_e dt = \frac{2}{A^2}\int_0^t \Delta T^2 dt \qquad (3)$$

At steady state, the cooling rate $C.R. = d(\Delta T)/dt$ given by the product VG. By utilizing this relationship, it is possible to substitute for time and integrate Equation 3 in temperature. Taking the lower integration limit equal to the undercooling at the actual heterogeneous nucleation temperature ΔT_n, we obtain:

$$d = \left(\frac{2}{A^2 VG}\right)\int_{\Delta T_n}^{\Delta T}(\Delta T)^2 d(\Delta T) \qquad (4)$$

from which

$$d = \frac{2}{A^2}\frac{\left[(\Delta T)^3 - (\Delta T_n)^3\right]}{3VG} \qquad (5)$$

In order to develop a fully equiaxed grain structure, growth must proceed until the grains physically impinge on each other. If there is only one nucleation event per particle, the total number of heterogeneous nucleation sites per unit volume N_0 necessary to reach complete impingement is approximately given by the relationship:

$$N_0 = \left(\frac{1}{d}\right)^3 \qquad (6)$$

Construction of Diagrams

Based on the above set of equations, it is possible to construct so-called ΔT and N_0 contour plots in G–V space, provided that relevant input data for A and D are available. For many commercial aluminium alloys within the 2xxx, the 6xxx and the 7xxx series, a typical value of A would be $2\ ^\circ C\ s^{1/2}\ mm^{-1/2}$, while D is close to $3\times10^{-3}\ mm^2\ s^{-1/2}$ [20].

Figure 3 shows an example of ΔT contour plot, as calculated from Equation 1. As expected, the combination of a high thermal gradient G and a low growth rate V or a low G and a high V favor high undercoolings. In practice, this means that the undercooling passes through a local minimum as we move from the upper left corner to the lower right corner of the diagram along a line representing a constant cooling rate. Thus, there exist combinations of G and V that makes nucleation and growth of new equiaxed grains ahead of the advancing columnar grains rather difficult during welding and casting.

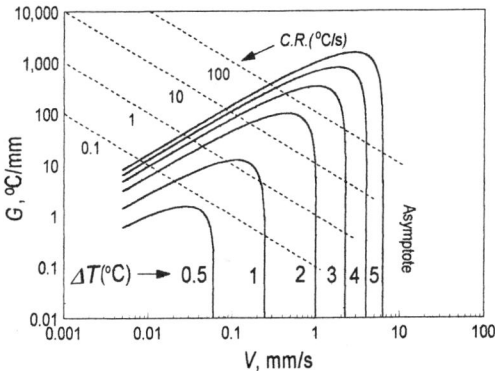

Figure 3: Graph in *G-V* space using Equation 1 to define contour lines of equal undercooling. Dotted lines denote constant cooling rate.

Figures 4 and 5 show corresponding contour plots of N_0 in *G–V* space for two different nucleation conditions, i.e. no energy barrier against nucleation ($\Delta T_n = 0$) and intermediate energy barrier against nucleation ($\Delta T_n = 1°C$), respectively. From these diagrams it is obvious that grain refining in regions of low undercooling requires the use of extremely efficient inoculants. At the same time the nucleating particles must be finely dispersed within the liquid prior to solidification. Note in Figure 5 how the addition of a nucleation barrier results in large region in *G-V* space where nucleation is inhibited.

DISCUSSION

The important difference between grain refinement of castings and fusion welds can be seen from the diagrams of Figures 3-5. If we for example consider an alloy containing a fixed number density of a given type of particles N_0, the solidification conditions existing in casting imply that the chances of achieving grain refining will increase as the columnar grains approach the centre of the mould (i.e. low *G* and high *V*).

In contrast, in fusion welds grain refining should be most difficult in an intermediate position along the periphery of the weld pool because the undercooling ahead of the advancing columnar grains passes through a local minimum as they grow from the fusion line towards the centre of the weld. In reality, the picture is more complicated due to fading, as N_o is likely to decrease with time after addition of the grain refiner due to coarsening or dissolution of nucleating particles. The fading problem is particularly important in fusion welding because the liquid metal becomes superheated in contact with the arc. This reduces the chances for the nucleating particles to survive in the weld pool, and the resulting drop in N_0 will finally determine the degree of grain refinement that is achieved during solidification.

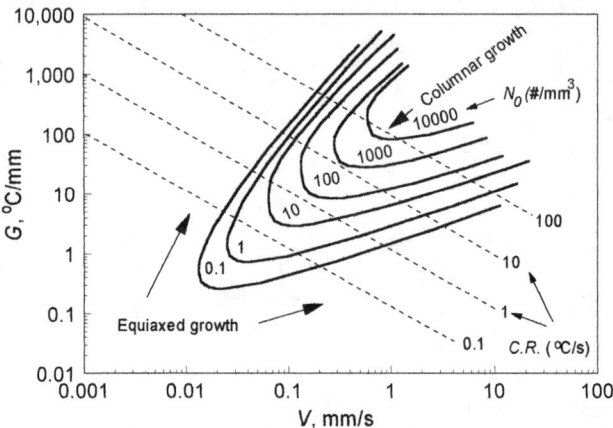

Figure 4: Graph in G-V space using Equations 5 and 6 to generate nucleation density contours. Assume $\Delta T_n = 0$. Dotted lines denote constant cooling rates.

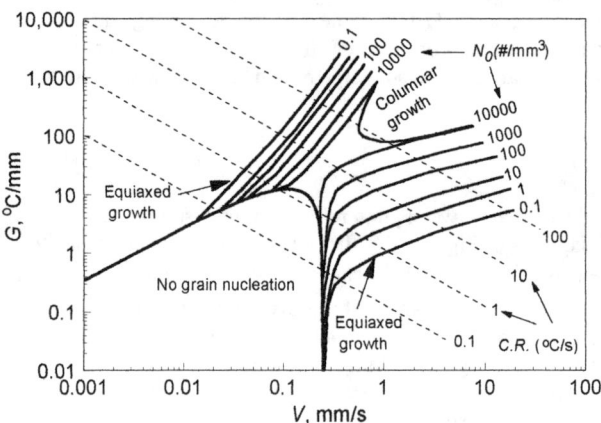

Figure 5: Graph in G-V space using Equations 5 and 6 to generate nucleation density contours. Assume $\Delta T_n = 1°C$. Dotted lines denote constant cooling rates.

CONCLUSION

The model proposed in this paper has demonstrated how it is possible to define regions in $G\text{-}V$ space where it is most likely to form equiaxed grains. Specifically, there are two divided regions predicted by this model where high undercooling may promote grain refinement: one region at high G and low V, and another region at low G and high V. In between, there is a region of low undercooling where columnar grains are more likely to exist. This coincides precisely with observations made in weldments, where fine grains are found both at the fusion boundary and at the weld center, with columnar grains in between (Figure 1).

REFERENCES

1. W.F. Savage and A.H. Aronson, Weld. J., **45**, 85s (1996).

2. M. Kato, F. Matsuda and T. Senda, Trans. Jpn. Weld. Soc., **3**, 69 (1972).

3. Y. Arata, F. Matsuda and A. Matsui, Trans. JWRI, **3**, 89 (1974).

4. T. Ganaha, B.P. Pearce and H.W. Kerr, Met. Trans., **11A**, 1351 (1980).

5. S. Kou and Y. Le, Met. Trans., **19A**, 1075 (1988).

6. C.E. Cross, L.W. Loechel and G.F. Braun, in *Proc. 6th Int. Al-Li Conf.* (Deutsche Gessellschaft fur Materialkunde, Oberusel, Germany, 1991), p. 1165.

7. S.R. Shah, J.E. Wittig and G.T. Hahn, in *Int. Trends in Welding Sci. and Tech.* (ASM Int., Materials Park, Ohio, 1992), p. 281.

8. Y. He, D. Gao, L. Wu and L. Ming, in *Proc. 3rd Int. Conf. Al Alloys* (SINTEF, Trondhiem, Norway, 1992), p. 385.

9. C.E. Cross and W.T. Tack, in *ASM Handbook-Vol. 6* (ASM Int., Materials Park, Ohio, 1993), p. 549.

10. A. Gutierrez, J.C. Lippold and W. Lin, Mat. Sci. Forum, **217-222**, 1691 (1996).

11. M.F. Lee, J.C. Huan and N.J. Ho, J. Mat. Sci., **31**, 1455 (1996).

12. K.K. Soni, R. Levi-Setti, S. Shah and S. Getz, Adv. Mat. Process., **149**, 35 (1996).

13. A. Gutierrez and J.C. Lippold, Weld. J., **77**, 123s (1998).

14. G.M. Reddy, A.A. Gokhale, K.S. Prasad and K Prasad Rao, Sci. Tech. Weld. Join., **3**, 208 (1998).

15. W.F. Savage, C.D. Lundin and A.H. Aronson, Weld. J., **44**, 175s (1965).

16. W.A. Tiller, K.A. Jackson, J.W. Rutter and B. Chalmers, Acta Met., **1**, 428 (1953).

17. J.D. Hunt, Mat. Sci. Eng., **65**, 75 (1984).

18. C.E. Cross, Ø. Grong and M. Mousavi, Scripta Mat., **40**, 1139 (1999).

19. M.H. Burden and J.D. Hunt, J. Crystal Growth, **22**, 99 (1974).

20. W. Kurz and D.J. Fisher, *Fundamentals of Solidification* (Trans. Tech. Pub., Switzerland, 1986).

CRYSTALLOGRAPHIC TEXTURE AND YIELD BEHAVIOR OF
Al-Cu-Li (2195) PLATE

K.E. CROSBY*, R.A. MIRSHAMS**, S.S. PANG*
*Mech. Engr. Dept., Louisiana State University, Baton Rouge, LA 70803, kcrosby@lsu.edu
**Mech. Engr. Dept., Southern University, Baton Rouge, LA 70813

ABSTRACT

Existing experimental texture analysis capabilities allow testing of theories on plasticity using polycrystal models. Aluminum-lithium alloys, which are particularly suited for aerospace applications due to excellent strength to weight ratio, typically possess pronounced texture. Al-Cu-Li (2195) thick plates were deformed by cold rolling to various thickness reductions. The plates exhibit a texture gradient through the plate thickness accompanied by a variation in yield strength values. The difference in yield strength values is related to the texture variation in terms of the texture components (ideal crystallographic orientations) identified from experimental measurements. A modified Taylor-based polycrystal plasticity model developed by Kocks and his colleagues is used to predict yield surfaces by incorporating texture data. Results of this investigation show that the texture intensities measured at certain ideal orientations increase or decrease with increasing deformation. These textural changes influence the value and anisotropy of the yield strength of the alloy.

INTRODUCTION

Studies of the various Al-Li alloys [1-3] conclude that the various particles that form during aging provide barriers to dislocation motion during slip. The major strengthening precipitate, T_1 (Al_2CuLi), forms with artificial aging only on {111} habit planes, preferentially at dislocation sites. This impedance provides these materials with improved strength properties over conventional aluminum alloys while their chemical compositions lead to lower densities than their conventional counterparts. For this reason, Al-Li alloys are of significant interest to the aerospace industries. Al-Li alloys are typically anisotropic and they exhibit strong textures.

Anisotropy of mechanical properties in different directions of measurement is a concern in the forming of metals into shapes and parts. It is tied into considerations of the yield locus. Various factors cause anisotropy in metals, including elongated grains [4] and the presence of second-phase precipitates [1,5]. Researchers agree that crystallographic textures or preferred orientations resulting from thermomechanical treatment such as hot or cold rolling or stretching are most directly responsible for anisotropy in metal alloys [6]. For Al-Li alloys, crystallographic texture may also have an indirect effect on anisotropy resulting from the heterogeneous distribution of the primary strengthening precipitates on specific habit planes [7]. For these reasons, texture analysis is important for characterization of Al-Li materials.

Textures are usually presented as pole figures or orientation distribution function (ODF) plots. However they may also be represented in terms of components, thus reducing the representation of the orientation distribution into a small set of specific orientations which describe a large number of crystallites present in the specimen [8]. This is useful for relating the presence of certain texture components to material behavior. In this paper, such discussion is made, relating the measured texture and idealized texture components to uniaxial yield strength and the biaxial yield surface predicted by polycrystal plasticity modeling for artificially aged Al-Cu-Li (2195) rolled plates.

Mat. Res. Soc. Symp. Proc. Vol. 578 © 2000 Materials Research Society

EXPERIMENTAL

Material processing and all texture measurements were performed at Los Alamos National Laboratory (LANL), Los Alamos, New Mexico, within the Materials Science & Technology (MST) and Center for Materials Research (CMR) divisions. Aging treatments were performed at Louisiana State University's Materials Characterization Laboratory.

Material

The Al-Cu-Li 2195 received material was received from Reynolds Metals Co. as 3.81 cm.-thick plates in the T3 condition (heat treated and cold rolled); the weight percentages of the aluminum alloying elements are given in Table 1. Sections of the as-received plate were solutionized in air for one hour at 540°C then quenched in water. Immediately following, the sections were additionally cold rolled to various thickness reduction percentages (10, 20, and 30% reductions) to induce various amounts of deformation. Artificial aging was performed at 180°C for 15 h. The artificial aging parameters were chosen based on an aging study [9] on 2195 plate showing that peak aging (optimum T_1 precipitation) is achieved near the defined time and temperature.

Table I. Al-Cu-Li 2195 Alloy Composition

Al	Cu	Li	Fe	Mg	Mn	Si	Ti	Zn	Zr	Ag
Balance	3.9	.9	.04	.33	<.01	.02	.02	.01	.14	.32

Selecting Through-Thickness Location for Samples

As shown in Figure 1, hardness measured at seven positions through the thickness of the plate shows a profile symmetric about the center of the plate. It is therefore assumed that there is agreement of measured values from one side of the plate to the equivalent position on the other side. A notation for through-thickness position may be assumed such that the position is described as a percentage of the total thickness, t, as measured from the plate surface. In this fashion, the following designation is given: position $1=0.05t$, position $2=0.2t$, position $3=0.35t$, and the center position $4=0.5t$. Samples were cut normal to the rolling plane; through-thickness position for samples was chosen based on the hardness profile. Samples were chosen from the region where maximum hardness is measured, between positions $0.2t$-$0.35t$. Specimens obtained from this region are designated as $0.2t$. Other samples were also obtained from the center of the plate, $0.5t$ (where hardness profiles showed the minimum values), for comparison.

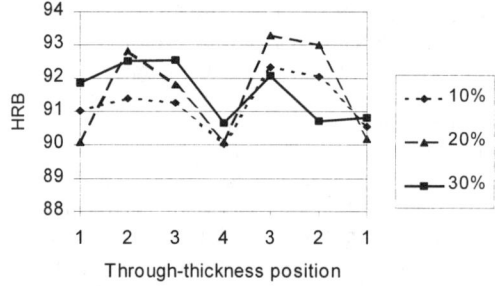

Figure 1. Through-thickness hardness profile of Al-Cu-Li plate

Texture Measurement

For texture measurement, slices approximately 3 mm thick and 25 mm × 25 mm were cut from the bulk material. Polishing and etching of the surface was performed to remove the damage layer resulting from cutting. A Scintag Inc. X-ray goniometer with a Fe-Kα radiation source was used for X-ray diffraction for the measurement of diffraction intensities. Because the Al-

Cu-Li 2195 has a very large grain size, texture was measured for three samples and then averaged to achieve more statistical data.

Mechanical Testing

Uniaxial tension tests have been performed in different directions of samples from material rolled to different percentages. Tensile specimens were extracted from the bulk material using electric discharge machining. An Instron™ Model 1125 (Serial No. 6615) screw-driven mechanical testing machine was used to pull the tensile specimens at a strain rate of .001 s^{-1} at room temperature.

RESULTS

Texture

The computer software, *popLA* [10], developed at LANL was used to convert the recorded intensities for (111), (200), and (220) pole figures. These pole figures are in turn used for the calculation of the ODF using *popLA*. All of the Al-Cu-Li 2195 samples exhibit typical rolling textures. While there is a transition observed between the 0.2t and 0.5t positions, visual comparison of pole figures and ODFs among samples of different thickness reduction does not reveal noticeable change. Another method of viewing textural differences is to plot the intensity change along the "fiber" that connects ideal orientations. Fiber representations enable the comparison of measured intensities or other texture "volume" data between materials with different processing histories. The α- and β-fibers are usually identified in FCC alloys. The α-fiber is defined in Bunge notation by the variation of the Euler angle ϕ_1 at constant $\Phi = 45°$ and $\phi_2 = 0°$. It runs from the Goss orientation through the brass orientation to {011}<011>. The β-fiber position runs from the Cu orientation, through the S orientation and connects to the α-fiber at the brass orientation.

The α-fiber is shown in Figure 2 as the variation of measured intensities along the fiber. For all specimens, the measured intensities are negligible at the Goss recrystallization orientation. The intensity values begin to increase at the Bs orientation for the specimens from the 0.5t position, reaching a maximum between $\phi_1 = 45°$ and $\phi_1 = 65°$. Around $\phi_1 = 70°$ the intensity of the 0.5t position samples tapers off to zero while there is a peak observed for the 0.2t samples.

Figure 3 shows a plot of the intensities along the ideal β-fiber. Locations of the copper, S, and brass components along the fiber are indicated. The intensities at the 0.2t position are significant at the copper and S orientations as shown in Figure 3. The measured intensity at the copper orientation is least for the as-received specimen. The measured intensity with 10% rolling at 0.2t increases by approximately 315 pct. Again, the intensity at the copper orientation increases slightly with another rolling pass (20% 0.2t). However, a third rolling pass (30% 0.2t) appears to degrade the intensity at the copper orientation, resulting in an intensity decrease of approximately 45 pct. from the maximum value at 20% 0.2t. At the S orientation, only the as-received specimen's measured intensity has value. At the brass orientation, all measured intensities are negligible for the 0.2t specimens of Figure 3.

Comparatively, for the specimens from the 0.5t position shown in Figure 3, the measured intensities are negligible at the copper orientation. The measured intensities at the S orientation are on the same scale as those of the 0.2t specimens in Figure 3. The intensity of the center position increases rapidly along the β-fiber from the S orientation to the brass orientation for the 10% 0.5t specimen. However, with a second (20% 0.5t) and third (30% 0.5t) rolling pass, the measured intensity at the brass orientation on the β-fiber is significantly reduced.

441

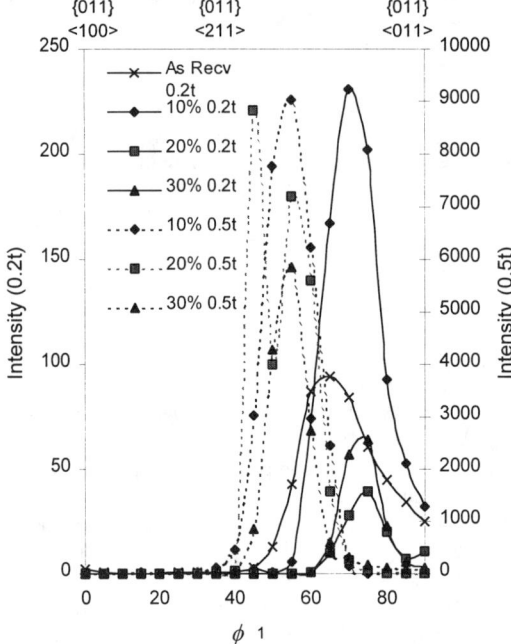

Figure 2. Texture intensities along the ideal α-fiber. Locations of ideal orientations along the fiber are indicated.

Mechanical Testing

Figure 4 compares the variation of 0.2% offset yield strengths with specimen orientation, determined from stress-strain curves. Each data point represents the average of at least three tests (except 30% in the rolling direction–only one data point). The yield values follow the same trend regardless of material processing, displaying a maximum in the rolling direction. The strength of the material increases significantly with artificial aging when the major strengthening precipitate, T_1, forms.

Yield Simulation

The software, *popLA*, contains a component, *LApp* (Los Alamos polycrystal plasticity), which employs a modified Taylor model to predict properties such as deformation textures and R-values of polycrystal materials. It is used here to predict the yield surfaces of the Al-Cu-Li 2195 samples. *popLA* converts the measured textures (continuous distributions) into a weighted grains file (discrete distributions) describing the texture as if the grains have been individually measured or if they are the result of a simulation. *LApp* uses the weighted grains file to calculate the yield surface in 3-D or as a 2-D projection onto any set of axes. The 2-D biaxial projection is presented in Figure 5 for the 1st quadrant of Cauchy stress space (σ_1-σ_2). The uniaxial tensile yield stresses (reference Figure 4) are presented in Figure 5 as single points on the axes.

CONCLUSIONS

Rolling textures tend to develop particularly around the copper, brass, and S orientations in aluminum alloys [11]. Figure 3 shows the intensity at the copper orientation for the 0.2t specimens to increase up to 20% deformation then decrease after 30%. After a small amount of deformation, the intensity at the S orientation diminishes. The intensity of the 0.5t specimens at the brass orientation is significantly higher with 10% deformation than with additional rolling. Increasing deformation may sharpen the texture by increasing the intensity [12]. After a certain amount of deformation occurs, shear bands may form which may contribute to scattering around the main texture components [8], thus decreasing the sharpness and intensity at the ideal orientation. Shear bands are observed at 0.2t in 30% samples using a light microscope.

The marked through-thickness texture intensity gradient is illustrated in Figure 2 where the intensities at the 0.5t position are much greater than at 0.2t. The through-thickness texture variation in the plate is likely due to the differing amounts or modes of deformation experienced

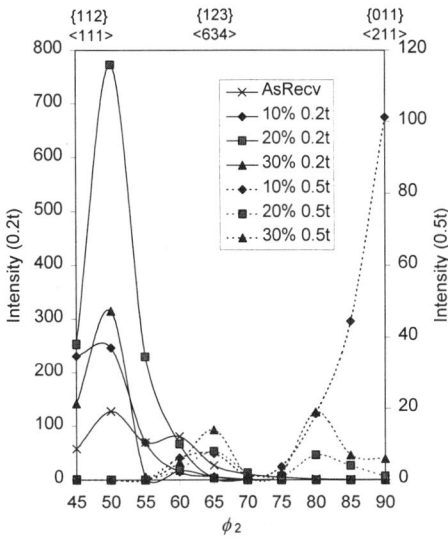

Figure 3. Texture intensities along the ideal β-fiber. Locations of ideal orientations along the fiber are indicated.

at different positions though the plate. Most of the rolling process is concentrated near the surface of the plate whereas the center of the plate experiences the least amount of the total deformation. Thus the center of the plate may not experience processes such as breaking of grains that cause texture heterogeneities which can degrade the sharpness of a texture.

Uniaxial tensile yield increases significantly with artificial aging once the T_1 precipitate forms. The yield values exhibit slight anisotropy because the values are consistently higher in the rolling direction than in the transverse direction. This occurs regardless of aging procedure. Therefore it can be concluded that the texture plays the most prevalent role in anisotropy of yield. However, as stated previously, the T_1 precipitates may have an indirect effect on anisotropy due to heterogeneous distribution of the precipitates on {111} habit planes [7].

Yield loci predicted for ideal orientations including copper, brass, and Goss textures show copper orientations producing more symmetrical yield loci with respect to the line of equibiaxial tension ($\sigma_1=\sigma_2$) compared to brass and Goss, whereas the latter orientations produce distortion of the loci in the direction of σ_2 [11]. Changes in texture intensity and through-thickness texture gradients can affect yield locus shape and size such that textures of higher intensity may exhibit sharper corners near the balanced biaxial range [13]. The predicted biaxial yield loci of the 0.2t specimens in Figure 5 are less anisotropic in the uniaxial directions than the 0.5t loci. Yet, the loci of the 0.2t specimens are less symmetric than the 0.5t loci. The vertex near the balanced biaxial range is sharper for the 0.5t than for 0.2t. Although particular components can be isolated to describe the textures of these materials, the actual measured textures vary from ideal orientation and the yield surface model is calculated based on the experimental texture data. There is good agreement of experimental yield results and predicted yield loci in the uniaxial directions.

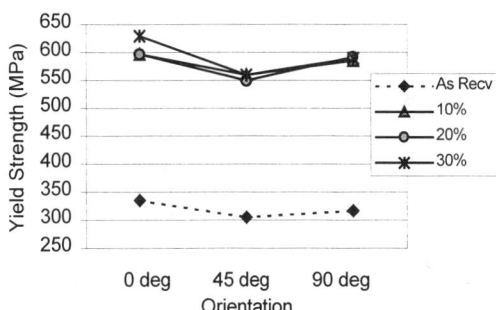

Figure 4. Variation of yield strength (uniaxial tension) with specimen orientation to the rolling direction.

ACKNOWLEDGMENTS

This research was partially supported by the Louisiana Board of Regents with Lockheed Martin Manned Space Systems under the contract LEQSF (1994-97)-ENH-PLEx-05 and through the Louisiana Board of Regents fellowship support for Ms. Crosby. Partial support was also provided by the NASA/LaSPACE grant under account

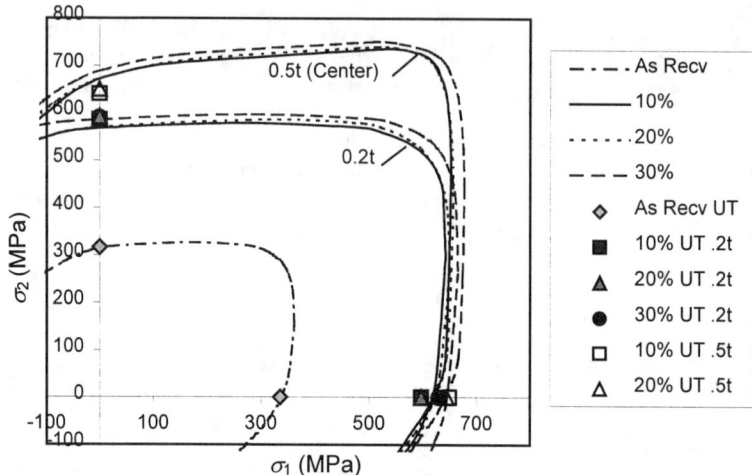

Figure 5. Predicted σ_1-σ_2 biaxial yield surface (Quadrant I). Uniaxial tensile (UT) test results indicated by single points as denoted by the legend.

number 127-40-5102 and the EPSCoR-BoR fellowship. The authors would like to thank LANL (MST-8) for technical and financial support to complete this research. Special thanks are due to S.R. Chen, J. Bingert, C. Necker, M. Lopez, and M. Lovato of LANL for their invaluable help.

REFERENCES

1. A.K. Vasudevan, W.G. Fricke, Jr., M.A. Przystupa, and S. Panchaadeeswaran, in *Proc. of the 8th Int. Conf. on Textures of Materials (ICOTOM 8)*, edited by J.S. Kallend and G. Gottstein (The Metallurgical Society, Warrendale, PA, 1988), p.1,071-1,077.

2. V. Gerold, H.J. Gudladt, and J. Lendvai, Phys. Stat. Sol. **A131**, p. 1509 (1992).

3. E.W. Lee in *Light Materials for Transportation Systems*, edited by N.J. Kim, Center for Advanced Aerospace Materials, 1993, pp. 79-92.

4. U.F. Kocks and H. Chandra, Acta Metall., **30**, p. 695 (1982).

5. J. Mizera, J. Driver, E. Jezierska, and K. Kurzydlowski, Mater. Sci. Eng., **212A**, p. 94 (1996).

6. A. Fjeldly and H.J. Roven, Acta Mater., **44**, p. 3,497 (1996).

7. N.J. Kim and E.W. Lee, Acta Metall., **41**, p. 941-948 (1993).

8. U.F. Kocks, C.N. Tomé, and H.R. Wenk, *Texture and Anisotropy: Preferred Orientations in Polycrystals and their Effect on Materials Properties*, Cambridge University Press, Cambridge, UK, 1998, p. 185, 227.

9. B. Skrotki, G.J. Shiflet, and E.A. Starke, Jr., in SEAS Report No. UVA/538865/MSE95/101, (Dept. of Materials Science and Engineering, University of Virginia, Charlottesville, VA).

10. J.S. Kallend, U.F. Kocks, A.D. Rollet, and H.R. Wenk, Mater. Sci. Eng., **132A**, p. 1 (1991).

11. F. Barlat and O. Richmond, Mater. Sci. Eng., **95**, p. 15 (1987).

12. Zeng, X.H, Ahmad, M. and Engler, O., Mater. Sci. Technol., **10**, p. 581 (1994).

13. X.H. Zeng and F. Barlat, Met. Trans., **A25**, p.2,783 (1994).

INFLUENCE OF GRAIN SIZE ON RECRYSTALLISATION DURING HOT WORKING OF AUSTENITIC STAINLESS STEELS

J. A. WHITEMAN*, Y. CHOI**, and C.M.SELLARS*
*IMMPETUS (Institute for Microstructural and Mechanical Process Engineering:The University of Sheffield),Mappin Street,Sheffield, S1 3JD, UK.
** now POSCO, PO Box 35, 790-600, Pohang City, Kyungbuk, Korea.

ABSTRACT

During the hot rolling of austenitic stainless steels, complete static recrystallisation is expected between passes unless finishing temperatures are low. Typically progressive refinement takes place to grain sizes in the range 20-50μm. However, most experimental studies of the effect of strain, strain rate, temperature and initial grain size on recrystallisation kinetics and recrystallised grain size under hot working conditions have been carried out on initial grain sizes greater than 50μm. Empirical relationships from these data and from more limited results of C-Mn steels have been extrapolated to smaller grain sizes for use in models of microstructural evolution during rolling.

Recent development of a physically based model for the effects of initial grain size, assuming that site saturated nucleation occurs at grain corners, grain edges, grain faces and at intragranular sites leads to interdependence of the effects of strain and grain size on nucleation density and hence on recrystallised grain size and recrystallisation rate. Experimental evidence available in the literature and some new results on finer grained Type 316 stainless steel are reviewed and compared with the expectations from the model.

INTRODUCTION

Empirical relationships that describe the flow stress of steels during hot working and show the influence of strain, strain rate, initial grain size and temperature on the kinetics of static recrystallisation, recrystallised grain size and grain growth rate have been combined to give an overall model for the development of flow stress and microstructure during hot rolling [1-4]. The differences between such models have been reviewed [5]. These models have limitations in that they do not allow the specific deformation conditions that characterise particular parts of a rolled section to be modelled throughout the rolling process. The advent of finite element (FEM) models that can define the specific conditions at any point in the stock and the development of physically based models for the development of microstructure during rolling both require the influence of the important variables to be re-examined. The aim of this paper is to review the influence of grain size on the flow stress and on the recrystallisation process, and to present new data for a Type 316 stainless steel, which covers smaller grain sizes than much previous work.

FLOW STRESS

The stress strain curves for an austenitic stainless steel, a C-Mn steel in the austenite range and nickel have a characteristic shape. The initial work hardening rate gradually decreases with increase in strain until a peak stress is reached. Beyond the peak, the stress is reduced by dynamic recrystallisation until a steady state value is attained. Figure 1 shows data obtained from torsion testing of nickel. The influence of strain rate and initial grain size can be seen. At all strain rates smaller grain size gives peak stress at a lower value of strain. There is also some evidence for a small increase in work hardening rate with decrease in grain size.

Mat. Res. Soc. Symp. Proc. Vol. 578 © 2000 Materials Research Society

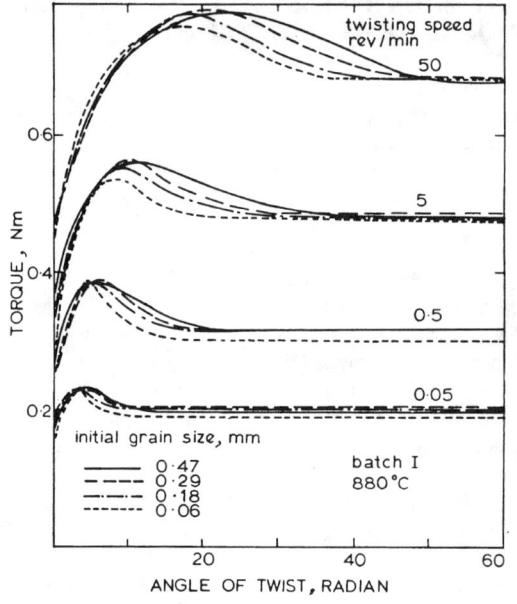

Figure 1
Typical torque/twist curves
for nickel with different
starting grain sizes [6]

There is a critical strain level for the onset of dynamic recrystallisation. This is typically about 0.8 of the peak strain for a material like stainless steel, which dynamically recrystallises rather sluggishly.

Recent work[7] on type 316 stainless steel, using plane strain compression testing, shows clearly both the decrease in strain to peak stress as the grain size decreases, and the increase in work hardening rate associated with the decrease in grain size, as shown in Figure 2. The increase in slope of the work-hardening and recovery part of the stress strain curve reflects the higher resistance to deformation of fine grain-sized material.

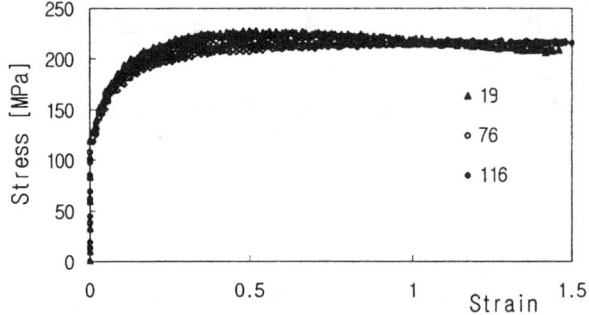

Figure 2
Stress strain curves
for type 316 stainless
steel tested in plane
strain compression at
1000°C and 1s^{-1},
showing the effect of
initial grain sizes
given in microns in
the figure.

The critical strain for dynamic recrystallisation decreases and the recrystallisation rate increases with decreasing grain size in accord with the work on nickel [6]. This is reflected in the stress strain curve corresponding to the small grain size deviating from the smooth work-hardening and recovery region earlier and falling to a steady state at a lower strain than the large grain size material.

The data for all grain sizes and deformation conditions, reported by Sah et al. [6], indicate that the dynamically recrystallised grain size depends only on the equivalent tensile stress, and may be described by the relationship.

$$\sigma = \sigma_0 + Kd^{-m} \tag{1}$$

Where σ is the steady state flow stress, σ_0 is the initial flow stress and K and m are constants. m has a value between 0.5 and 0.75, and d is the measured grain size. The smaller peak strain for recrystallisation, with small initial grain size, is consistent with a process of recrystallisation that has rapid nucleation and very limited growth. The limited growth is due to the dynamically recrystallised grains becoming rapidly work hardened and losing their driving force for further growth, while other regions achieve enough strain to become recrystallisation nuclei.

STATIC RECRYSTALLISATION

Static recrystallisation between individual passes, is the main cause of microstructural change and complete softening, during hot working. The kinetics of recrystallisation in relation to the time available during multi-pass deformation are critical in influencing subsequent microstructural evolution. Typically the fraction recrystallised X follows an Avrami equation with time t.

$$X = 1 - \exp(-0.693 (t / t_{0.5}))^k \tag{2}$$

The kinetics can therefore generally be described by the value of the exponent k and the time to 50% recrystallisation, $t_{0.5}$. Experimental work on Type 304 stainless steel in torsion [8] and Type 316 stainless steel in plane strain compression [9] has resulted in equations for the effect of hot working variables on the kinetics of static recrystallisation. In both cases a range of grain sizes was investigated, but the smallest grain size was about 60 microns.

The value of the exponent k was found to be 2 for the 304 stainless [8], although some lower values were obtained. In the 316 stainless steel [9] the values were close to 1. In both cases the fraction recrystallised was obtained by quantitative metallography and the position in which the measurements were made was carefully chosen. In the 304 stainless work the measurements had to be taken at a depth that was associated with a particular strain. In the 316 stainless work measurements taken on longitudinal specimens gave values of less than 1 for k at strains less than 1. Measurements taken from transverse sections along lines of constant strain for the lower strain specimens gave a value of 1.

FRACTION RECRYSTALLISED

For modelling of the microstructural development during hot working it is essential to know the factors that influence the time to 50% recrystallisation and the recrystallised grain size, This latter is normally smaller than the initial grain size. These two variables are shown as a function of equivalent strain for 316 stainless steel [9] in Figure 3. The $t_{0.5}$ measurements are from longitudinal sections.

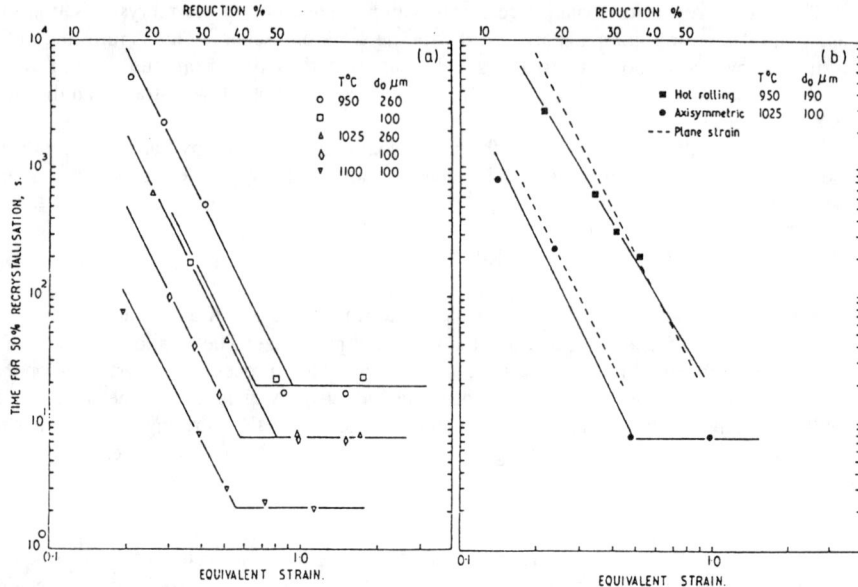

Figure 3 Time to 50% static recrystallisation as a function of applied equivalent strain [9], (a) from plane strain compression   (b) from rolling and axisymmetric compression

Results for plane strain compression in Figure 3a can be compared with those from rolling and axisymmetric compression in Figure 3b and good agreement between data from different working processes is found. As has been previously found in C-Mn steels[1] in the low strain region $t_{0.5}$ shows a dependence on both strain and grain size, but above a critical strain, at which dynamic recrystallisation starts, the recrystallisation time is independent of both strain and initial grain size. The strain effect in the lower strain region in figure 3 is satisfactorily represented by a power law, independent of the original grain size in the range of grain sizes investigated. The dependence of recrystallisation time on initial grain size in the low strain region has been established for 304 stainless steel [8]. The other factor that influences the recrystallisation time is the deformation temperature. All these factors can be combined in a relationship [9].

$$t_{0.5} = 4.0 \times 10^{-15} \, \varepsilon^{-3.6} \, d_o^{1.33} \, Z^{-0.38} \, \exp(475000 / R \, T) \qquad (3)$$
$$\varepsilon \leq \varepsilon_c$$

where d_o has the units of μm.
Assuming the same activation energy for recrystallisation in the high strain regime gives

$$t_{0.5} = 1.8 \times 10^{-7} \, Z^{-0.6} \, \exp(475000 / R \, T) \qquad (4)$$
$$\varepsilon \geq \varepsilon_c$$

where
$$\varepsilon_c = 7.5 \times 10^{-3} \, Z^{0.06} \, d_o^{0.37} \qquad (5)$$

The above relationships give reasonable correlation with data obtained by rolling, axisymmetric compression and plane strain compression, and with other published data on type 316 stainless steel.

RECRYSTALLISED GRAIN SIZE

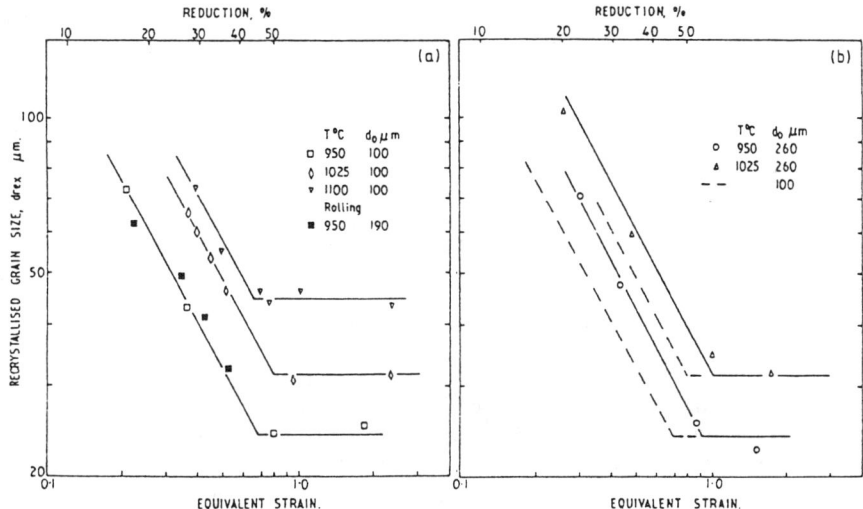

Figure 4   Recrystallised grain size as a function of applied equivalent strain [9]
showing   (a) effect of temperature
(b) effect of initial grain size

Grain size (d_{rex}) after complete recrystallisation for type 316 stainless steel [9] is plotted as a function of strain in Figure 4. This shows the influence of the same variables, strain, temperature and initial grain size in the low strain region, but the influence of grain size is reduced. The equivalent derived relationships are

$$d_{rex} = 4.7 \times 10^2 \, \varepsilon^{-1} \, d_o^{0.3} \, Z^{-0.1} \qquad (6)$$
$$\varepsilon \leq \varepsilon_*$$

$$d_{rex} = 2.65 \times 10^3 \, Z^{-0.1} \qquad (7)$$
$$\varepsilon \geq \varepsilon_*$$

where $\varepsilon_* = 1.8 \times 10^{-1} \, d_o^{0.3} \qquad (8)$

These equations give reasonable correlation with other published results on type 316 stainless steel.

One of the issues from the prior work on type 304 stainless steel is why the recrystallisation curves change shape and become less steep as the grain size increases, as is seen in Figure 5. In an attempt to address this problem Sellars[10] developed a simple model to describe the minimum recrystallised grain size that could be obtained.

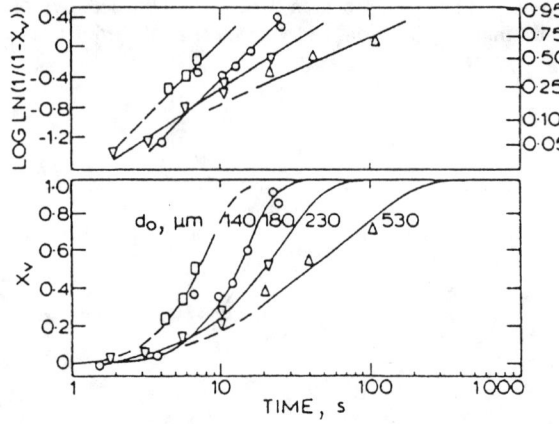

Figure 5.
Effect of original
grain size on static
recrystallisation
curves for type 304
stainless steel
deformed at 1050 °C
and 1s^{-1} to a strain of
0.5 [8].

MODEL FOR STATIC RECRYSTALLISATION

The model is shown schematically in Figure 6. It is based on two general observations.
(a) the nucleation of recrystallisation occurs preferentially at grain boundaries. (b) recrystallised
grains are nearly equiaxed unless their growth is constrained by particles. As illustrated in Figure
6 ,when the recrystallised grain size d_{rex} is relatively coarse compared with the original grain
size d_o, recrystallisation can occur completely from preferential sites on grain boundaries.
However if grains nucleated at grain boundaries stop growing when they are nearly equiaxed for
small recrystallised grains recrystallisation must be completed by nucleation at intragranular
sites.

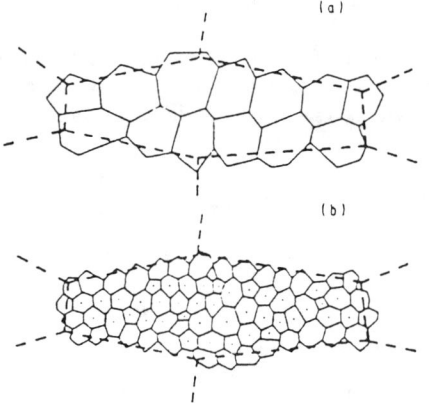

Figure 6
Schematic drawing of recrystallised
grains formed in a deformed
original grain. (a) entirely by grain
boundary nucleation and (b) by
grain boundary and intragranular
nucleation. Intragranularly nucleated
grains marked with a dot.

For plane strain deformation, eg rolling of slab or strip, the lower limit of recrystallised
grain size (d^*_{rex}) for grain boundary nucleation only occurs when

$$d^{*}_{rex} \cong \frac{d_o}{2} \cdot \frac{h_f}{h_o} \qquad (9)$$

where h_o and h_f are the initial and final thickness of the slab or strip and are related to the equivalent tensile strain ε through the Levy-von Mises relationship as

$$\varepsilon = \frac{2}{\sqrt{3}} \ln \frac{h_o}{h_f} \qquad (10)$$

When $d_{rex} > d_{rex}^{*}$, recrystallised grain size depends on the original grain size through its effect on the effective density of nucleation sites per unit volume and the exponent of grain size in equation (6) is expected to be 1/3 for nucleation on grain surfaces and 2/3 for nucleation on grain edges. Data on recrystallised grain size for specimens strained less than the critical strain for dynamic recrystallisation [8,9,11] are shown in Figure 7. There is a clear change in slope in the relationship between d_{rex} and d_o when $d_o/(d_o)_{lim} = 1$. Below this value $d_{rex} < d_{rex}^{*}$ and above it $d_{rex} > d_{rex}^{*}$, giving values of the exponent of ~ 0.25 and 0.5 respectively.

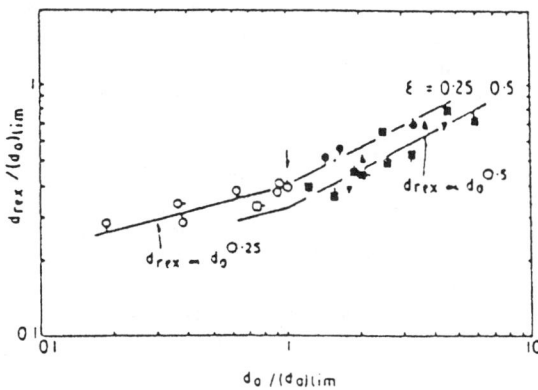

Figure 7.
Experimental data for recrystallised grain size and original grain size in stainless steels.[10]

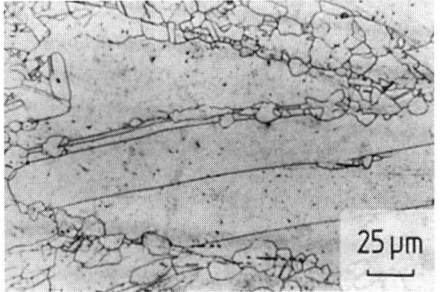

Figure 8
Micrograph showing evidence of intragranular nucleation of static recrystallisation on twin boundaries in type 316 stainless steel with a large grain size [12].

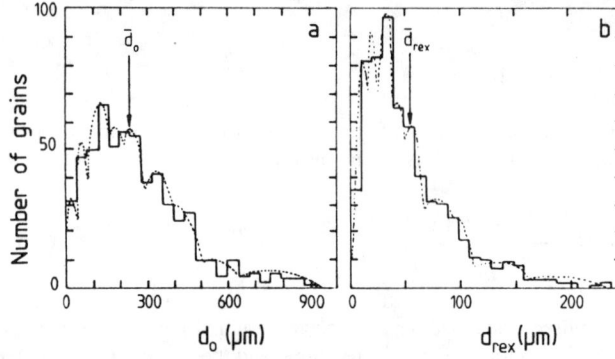

Figure 9
Measured initial and recrystallised grain sizes in type 316 stainless steel [12]

Experimental work on the initial and recrystallised grain size distributions in type 316 stainless steel by Wadsworth and Sellars[12] showed clear evidence for the nucleation of recrystallised grains on twin boundaries, see Figure 8. Quantitative measurements show that the recrystallised grain size is much smaller than the initial grain size under most conditions of working. An example of these distributions is shown in Figure 9. In this case an initial mean grain size of 250μm produces a recrystallised grain size of 70μm. Both the initial and recrystallised grain sizes closely approximate to a log normal distribution. A simple model for the fraction recrystallised was developed using the distribution data and assuming that the recrystallised regions were spherical. Two limiting assumptions were made about the growth of the spherical regions : (a) the continuous line assumes homogeneous nucleation and that all grains grow at a constant rate until recrystallisation is complete. (b) the broken line is for grain boundary nucleation with a grain size distribution in Figure 9a. Recrystallising grains grow at the same rate until impingement occurs at the centre of the original grain, when growth ceases.

Figure 10 shows the experimental observations lie between the two limits. At small fractions transformed the first approximation appears to be closely followed. There is no significant effect of original grain size, although for coarser initial grain sizes some intragranular nucleation would be required. This probably occurs by the nucleation of recrystallisation at twin boundaries.

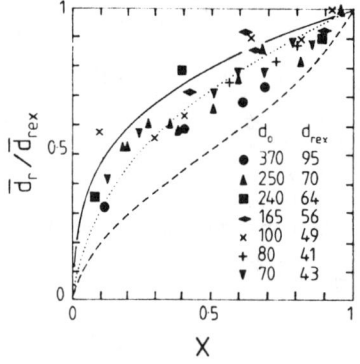

Figure 10
Calculated ratio of recrystallising /final mean linear intercept grain size for different fractions recrystallised. Limits given by two models for growth of recrystallised regions shown by solid and dotted lines [12]. Observations from type 316 stainless steel, deformed by 25% reduction.

452

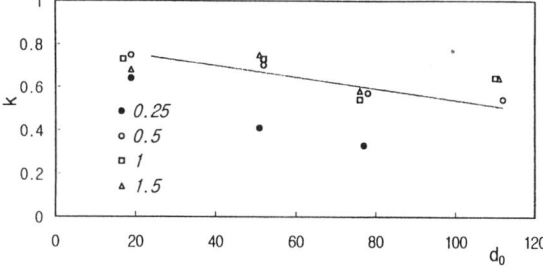

Figure 11.
Avrami exponent k as a function of initial grain size in type 316 stainless steel. Recrystallisation kinetics measured by stress relaxation in PSC test to the equivalent strain indicated[7].

Recent work by Choi, on type 316 stainless steel used the technique of stress relaxation to measure the kinetics of static recrystallisation[7]. Most of the tests were performed at 1000°C, with deformation carried out at 10s⁻¹. The work covered a wider range of grain size than previously, with grain sizes as small as 13μm being obtained after annealing. These small grain sizes grew during reheating for testing and the minimum effective grain size was 19μm. For most grain sizes, k was found to be between 0.8 and 0.6 with a slight trend for decrease in k with increase in initial grain size. Some anomalously low values of k were obtained at the lowest strain of 0.25. This was thought to be due to the inhomogeneous nature of the strain, which is confined to slip line fields and means that local regions experience strains much higher than the mean strain at low levels of strain. The variation of k with initial grain size is seen in Figure 11.

One other feature of this work is seen in the relationship between t_{50} and strain as shown in Figure 12. With the smaller grain sizes, there is a clear dependence of the slope of the strain dependent part of the curve on initial grain size. The value of this slope decreases from $\varepsilon^{-4.3}$ for grain sizes above 75 μm, which is in good agreement with previous work, to $\varepsilon^{-1.4}$ for the smallest grain size used. A 50μm grain size gives an exponent of $\varepsilon^{-2.3}$. A similar variation in the strain exponent with grain size has been reported by Hodgson for C-Mn steels [13]. There is also some suggestion in Figure 12 that the t_{50} in the high strain region, where dynamic recrystallisation occurs may have a grain size dependence.

The dependence of t_{50} on grain size was found to have an exponent of 1.3 [7], which is

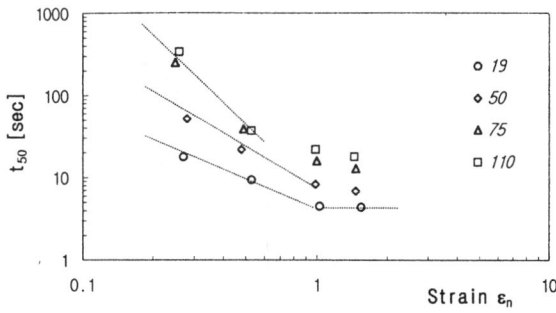

Figure 12
t_{50} as a function of strain for type 316 stainless steel. Recrystallisation kinetics measured by stress relaxation in PSC test [7].

lower than the value of 2 found for type 304 stainless steel by Barraclough and Sellars [8], although a lower value of 0.8 had been reported for the material with the smallest grain size of 60μm and discounted as untypical. A value of 1.33 for the grain size exponent had been obtained by Barbosa and Sellars[9] for type 316 stainless steel.

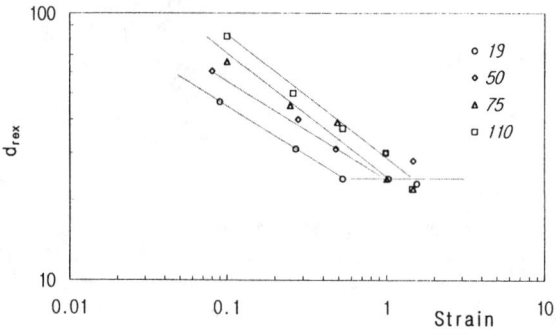

Figure 13
Recrystallised grain size as a function of strain for type 316 stainless steel [7].

Figure 13 shows the recrystallised grain sizes [7] as a function of strain in type 316 stainless steel. In the low strain region the recrystallised grain size depends on strain to the power $\varepsilon^{-0.5}$ for the coarser initial grain sizes and $\varepsilon^{-0.4}$ for the finer initial grain sizes below about 50μm, although there are only limited data at these low grain sizes. An exponent of this magnitude has been reported for 304 stainless [11], but more generally an exponent of the order of 1 is found.

The dependence of recrystallised grain size on initial grain size in the type 316 stainless steel [7] is seen in Figure 14 and is similar to that reported by Barbosa and Sellars[9]. The exponent of d_0 is 0.3 but shows no clear evidence of a change in dependence as shown in Figure 7. Another feature seen in Figure 14 is the limiting d_{rex} found as the strain is increased. This indicates that the grain size dependence changes, as dynamic recrystallisation becomes dominant.

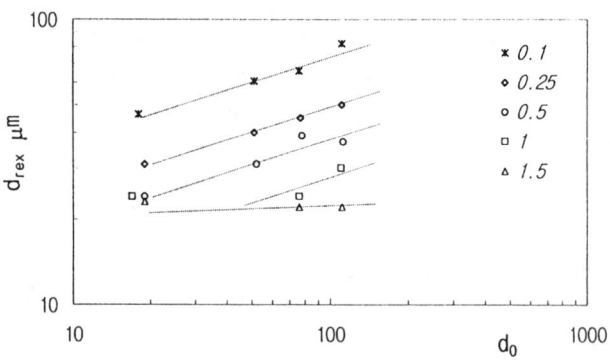

Figure 14
Recrystallised grain size as a function of initial grain size for type 316 stainless steel [7]

DISCUSSION

Data for type 316 stainless steel with a small initial grain size differ in a number of significant ways from results with larger initial grain size, and several of the parameters used in an empirical model for fraction recrystallised and recrystallised grain size depend on initial grain size. This has important implications for modelling of structural changes that result from hot working. The type of empirical equations discussed so far are inadequate for modelling the detailed changes in local strain, temperature and strain rate that may be calculated using finite element techniques to simulate real working processes.

In order to model this new situation, physically based models with functions of internal state variables are required. A model of recrystallisation in aluminium alloys has been developed [15], which calculates the flow stress, the nucleation density and the growth rate of statically recrystallised regions in terms of variables that can be measured experimentally. The nucleation density per unit volume (N_v) is a function of the probability of the different boundary sites: grain corners, grain edges, grain faces and intragranular sites with mobile boundaries eg deformation boundaries with a misorientation of greater than $10°$ within an original grain. Constants for each type of boundary are a function of strain. The resultant N_v versus strain is shown in Figure 15

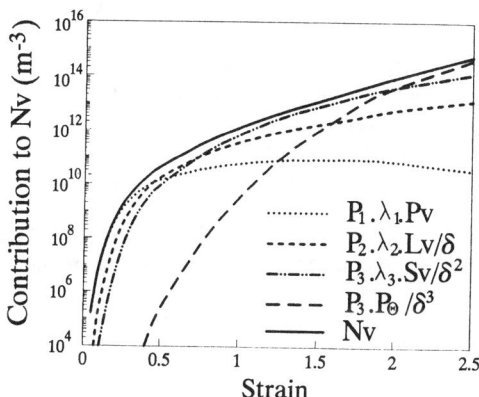

Figure 15
Contribution to N_v of different nucleation sites as a function of strain [15].

At small strains nucleation of static recrystallisation is totally dominated by grain corners and as the strain increases other sites begin to be important. Figure 15 was calculated for an initial grain size of $117\mu m$ and the strain at which other sites become dominant is dependent on the initial grain size. This results in a predicted continuous decrease in the dependence of d_{rex} on d_o with increasing strain from an exponent of about 0.7 to about 0.25. In order to calculate curves equivalent to Figure 15 for stainless steels, measurements of subgrain size and misorientation and dislocation density will have to be obtained. Because of the intrinsically fine subgrain size in stainless steels, the measurement of misorientation is very difficult and it may only be possible to obtain adequate information by using a FEG SEM using technique.

CONCLUSIONS

1) The flow stress in stainless steel is influenced by the initial grain size. Small grain size gives peak stress at lower strains and increases the initial work hardening rate.

2) The empirical relationships describing fraction recrystallised and recrystallised grain size have different values for the parameters as the grain size is varied in stainless steels. They do not provide an adequate model to describe the details of hot deformation processes.

3) Physically based models for structural change in stainless steels can in principle be developed, as has been done for aluminium. There is a need to obtain information about structural variables such as subgrain size and misorientation and dislocation density. This will require a considerable amount of experimental effort.

REFERENCES

1. C.M.Sellars, *Hot Working and Forming Processes,* edited by C.M.Sellars and G.J.Davies, Metals Society (London) 1980, pp. 3-15.
2. P.Choquet, A.Le Bon and Ch.Perdrix, *Strength of Metals and Alloys,* edited by H.J. McQueen et al, Proc. ICSMA7, Montreal, Canada,1985, **12**, pp.1025-1030.
3. W.Roberts, A.Sandberg, T. Siwecki and T.Werlefors, *HSLA Steels:Technology and Applications,* Proc. Conf. Philadelphia, ASM 1983, pp.67-84.
4. T.Senuma and H.Yada, *Annealing Processes-Recovery, Recrystallisation and Grain Growth,* edited by N. Hansen et al., Proc. 7th Riso International Symposium on Metallurgy and Materials Science, Riso, Roskilde, Denmark, 1986, pp.547-552.
5. C.M.Sellars, *Modelling- an Interdisciplinary Activity,* edited by S.Yue, Proc. Int. Symp. Mathematical Modelling of Hot Rolling of Steel, The Canadian Institute of Mining and Metallurgy, Montreal, Canada, 1990, pp 1-18.
6. J.P.Sah, G.J.Richardson and C.M.Sellars, Metal Science, **8,** pp 325-331, (1974).
7. Y.Choi, M.Phil. thesis, University of Sheffield, 1999.
8. R.Barraclough and C.M.Sellars, Metal Science, **13**, pp 257-266, (1979).
9. R.A.N.M.Barbosa, and C.M.Sellars, Materials Science Forum, **113-115**, pp 461-466, (1993).
10. C.M.Sellars, *Modelling of Structural Evolution during Hot Working Processes,* Riso Int. Symp. on Annealing Processes- Recovery, Recrystallisation and Grain Growth, Roskilde , Denmark, 1986, pp 167- 187.
11. D.J.Towle and T. Gladman, Metal Science, **13**, pp 246-256, (1979).
12. J.E.J,Wadsworth and C.M.Sellars, *Effect of Grain size distributions on the Observed Microstructure during Static Recrystallisation of Type 316 Stainless Steel,* edited by T Chandra, Recrystallisation 90, The Minerals, Metals and Materials Society, Warrendale Pa, USA, 1990, pp 417- 422.
13. P.D.Hodgson, Ph.D. thesis, University of Queensland,(1993).
14. A.Sandberg and R.Sandstrom, Mat. Sci. and Tech., **2**, pp 917-925, (1986).
15. C.M.Sellars, *Microstructure Modelling in Hot Deformation,*edited by B Hutchinson et al., Proc. Int. Conf. On Thermomechanical Processing: in Theory, Modelling and Practice [TMP]², The Swedish Soc. for Materials Technology, Stockholm, Sweden, 1997, pp 35 – 51.

THE CELLULAR AUTOMATON SIMULATION OF MICROSTRUCTURAL EVOLUTION DURING DEFORMATION PROCESSING OF METALS

CHRIS H.J. DAVIES

Department of Materials Engineering, Monash University, Clayton, Victoria, Australia, 3800.
E-mail: Chris.Davies@eng.monash.edu.au

ABSTRACT

The computer simulation of the evolution of microstructure during deformation processing is a desirable but elusive goal. In order to be effective, models must be tied to deformation parameters (temperature, strain rate, strain), and grain size distributions and recrystallisation kinetics must be predicted with high accuracy if the evolution of microstructure is to be tracked through several passes. This paper examines the cellular automaton (CA) simulation technique which has the potential to enable the modeller to accomplish these goals.

Aspects of the CA approach examined are the three dimensional representation of microstructure and its evolution and the incorporation of texture representation into simulations. Also investigated will be the limitations of the technique, in terms of the error that can be expected when different boundary conditions are imposed on a simulation.

Although at a rudimentary stage of investigation, the interfacing of CA simulations to deformation simulations (eg, finite element) will be examined. The paper concludes with a discussion of the challenges facing the implementation of CA simulations in industrial process models.

INTRODUCTION

Cellular automata are routinely used to simulate various aspects of isothermal static recrystallisation [1-7], delivering simulations of microstructural evolution from which the kinetics of the process can be derived. However, they are not unique in this; there are several simulation techniques able to represent microstructural evolution in a realistic manner, but of these, CA simulations would appear to be the best suited to modelling industrial processing, when strain and temperature gradients exist in a material.

The so-called 'Avrami machine' models and related geometrical models [8-11] effectively employ a mean-field approach to growing recrystallised grains. Spherical grains grow into a homogeneous medium, and it is one of the deficiencies of these types of models that they do not appear able to account for growth into a strain energy gradient (although the effect on nucleation can be incorporated); the model does not see beyond the boundary of a growing grain. Such models are, however, able to reproduce grain size distributions and kinetics for any number of defined texture classes and are readily coupled to experimental data. None-the-less the limitation would seem to preclude such models from being used to simulate industrial processing, during which temperature and strain gradients are prevalent. Monte-Carlo simulations [12-14] are a probabilistic technique most commonly associated with grain growth simulations rather than recrystallisation. However, whereas the CA method looks only at the sites with the potential to take part in the transformation (ie, those sites at the interface between recrystallised and unrecrystallised regions), Monte-Carlo simulations perform an interrogation at each site in each time step, and thus involve many computationally-redundant steps. Furthermore, and in contrast to the geometrical Avrami-machine simulations, there is some

457

question as to how readily the simulated microstructural evolution can be related to experimental conditions [15].

I choose to employ CA simulations because they can be directly and easily related to the microstructure and kinetics of a real system through calibration of the rules that govern the CA[7], and because of the richness of information storage afforded by their cellular structure. Cellular automaton simulations, then, combine the attractive probabilistic nature of the Monte-Carlo method with the practical advantage associated with the geometrical approach of being easily coupled to experiments.

This paper explores CA simulations of static recrystallisation of deformed structures. In particular, the errors that can be expected of a CA simulation are quantified according to CA size and boundary conditions for isothermal conditions. These simulations are conducted as a precursor to simulations of recrystallisation in materials with deformation phase-field gradients. The simulations are also used in an inverse manner- to inform the experimenter not only which experiments are required but also the level of precision required of the experimental data.

STRUCTURE OF THE CELLULAR AUTOMATON

A cellular automaton is a matrix of contiguous cells in one, two, or three dimensions. Each cell is a finite-state machine, its state being controlled by interaction with neighbouring cells and a set of rules describing the transition from one state to another. The CA is thus defined by the neighbourhood type (the number and location of neighbours a cell can interrogate during the simulation), a nucleation rule, and transition rules. In addition, the cells in a CA have *attributes* which determine how the cell interprets the nucleation and transition rules. In a metallurgical process, the attributes are descriptors of the deformation phase field and the microstructure.

The basic automaton used in this work was a three dimensional CA, with up to eight million cells. The cells were defined as cubes, although any tessellating shape may be employed. The neighbourhood used was a Moore neighbourhood, rendered in three dimensions and modified to give realistic grain shapes. Inspecting a single cell of the CA, a simple Moore neighbourhood in three dimensions has 26 neighbour cells (a cube of side-length three cells, minus the central cell which is the one under inspection). The modified Moore neighbourhood allows the central cell to interact with cells along each of the diagonals of the Moore cube in turn (Figure 1)[1]. Each neighbourhood contains 12 cells. The CA used in this work differs from other CAs [4, 16] in that an unrecrystallised cell under investigation at the boundary of a recrystallised grain interrogates the grain, and, based on its attributes and the transition rules, determines whether it will change state. The alternative methodology [4, 16] has recrystallised cells at the boundary interrogate the unrecrystallised cells. The shape of the resultant grains is unaffected by changes to the inspection methodology. At the boundary of the CA (the faces of the automaton) either periodic or mirror boundary conditions are employed; the effect of boundary conditions is one of the parameters under investigation, and will be discussed in detail later.

The nucleation rule is a probability for nucleation at any site is given by the relation:

$$P_N = \frac{\dot{N}dt}{N_{CA}} \qquad (1)$$

---

[1] The modified Moore neighbourhood is only required to render realistic microstructures for cubic cells. Were the cells arranged such that their centres were equidistant, the simple Moore neighbourhood could be used; however, visualisation of the data would be more complicated.

where $\dot{N}$ is the nucleation rate (determined by experiment), and N_{CA} is the number of cells present in the CA. For each cell in the CA, a random number is generated and compared to the calculated probability. Nucleation may be either site saturated or continuous.

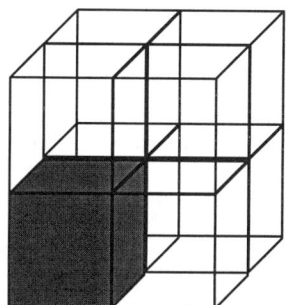

<u>Figure 1: Half of one of the four modified Moore neighbourhoods. In any one time step the central shaded cell interrogates the six hollow cells and their reflection through the centre.</u>

The transition rule can be automatic (every cell with a previously-transformed cell in its neighbourhood will also transform), or controlled by a probability of growth:

$$P_G = \frac{1}{s_{CA}} \int_{t_1}^{t_2} G dt \qquad (2)$$

where G is the interface migration rate, and s_{CA} is the cell size. In this work, we shall treat only the condition under which growth is automatic once the calculated interface has entrapped the next unrecrystallised cell. In this case in effect P_G is set to one, and only when the relation is satisfied (by incrementing the numerator) is growth allowed to proceed. The microstructures and kinetics resulting from each of the two updating procedures are to all intents indistinguishable.

The progress of the recrystallised interface is calculated for each texture class rather than for each grain prior to entering the CA routine proper. In effect, the position of a 'virtual interface' is incremented, and the CA rules allowed to determine whether or not the new position is acceptable for each grain in turn. This procedure has little effect on the efficiency of a simulation under isothermal conditions, but was implemented to improve the speed of simulations when nucleation and growth rates are varied through the CA.

For each texture class the interface migration rate is calculated according to the Cahn-Hagel method [17], and fitted to the relation

$$G_i = \alpha_i' t^{n_i'} \qquad (3)$$

which upon integration yields

$$r_i = \alpha_i t^{n_i} \qquad (4)$$

where r_i is the radius of a grain of the ith texture class (using the same notation as Juul Jensen [11] to represents the texture class under consideration) and α_i and n_i are constants related to the constants in Equation (3).

Knowing the constants in Equation (3), the nucleation rate is deduced for each texture class [7]:

$$N_v^i = \frac{X_{Vex}^i 3\left(1 - n_i'\right)}{4\pi\alpha_i' t^{3\left(1-n_i'\right)}} \qquad (5)$$

where X_{vex} is the extended volume fraction.

SIMULATIONS

Several simulation scenarios were implemented in order to quantify the errors that can be expected of a CA. This type of analysis is essential if CAs are to be used in a predictive manner. Errors can arise from two sources: errors which are inherent to the methodology- herein referred to as CA errors- and errors arising from inaccurate or inadequate experimental data- herein referred to as sensitivity errors.

CA Errors

Cellular automaton simulations of static recrystallisation under isothermal conditions are often conducted with periodic boundary conditions at each edge of a 2D CA [1] or face of a 3D CA [4, 7] Periodic boundary conditions are not, however, essential: they are employed in an attempt to minimise edge or face errors when examining a small area or volume of material. When simulating microstructural evolution in a material subject to temperature and strain gradients, periodic boundary conditions are clearly inappropriate in all axes. In addition to boundary errors, there is also the potential of introducing errors related to the size of the CA, and in particular to the number of recrystallised grains. Two simulation scenarios were run to investigate CA errors; both examined isothermal static recrystallisation.

The first scenario looked at the effect of CA size and boundary conditions for ideal JMAK conditions- a random distribution of nuclei and constant growth rate- with site-saturated nucleation, under which conditions, we should expect a JMAK exponent of 3. Automata of 100^3, 150^3, and 200^3 cells were set up with either all periodic or all mirror boundary conditions. The probability of nucleation was set at 2.4×10^{-6} [6], and in Equation (4), $\alpha = 0.36$, and $n = 1$; the time step was 0.001s. The nucleation probability was chosen to deliberately expose any effect of a small number of nuclei. Simulations were compared to the JMAK relation for N grains:

$$X = 1 - \exp\left(- N\theta\right) \qquad (6)$$

where

$$\theta = \frac{4\pi r^3}{3V_0}$$

and r is the average radius of one growing grain, and V_0 the total volume of the CA.

Because the JMAK relation is strictly only valid in the limit as the number of grains tends to infinity [18], the simulations were also compared to the equation to which the JMAK relation is an approximation, which is valid for all N:

$$X = 1 - \left[1 - \frac{\theta}{N}\right]^N \qquad (7)$$

The simulation results were fitted to the form of the JMAK equation more common in the metallurgical literature:

$$X = 1 - \exp\left(- bt^n\right) \qquad (8)$$

and the constants b and n determined.

The second scenario quantified the effect of changes to boundary conditions on a single face of the CA, a scenario which is more akin to hot rolling: essentially constant temperature along and across the width of a plate, with variation through the thickness. Representation of texture was included and the effect of CA size was restricted by using larger nucleation probabilities. The simulations were compared to experimental data, obtained for three texture classes using electron back-scattered diffraction (EBSD) in the SEM (Table I), under conditions in which site saturated nucleation was found to prevail. The effect of CA shape was also evaluated; if strip rolling is to be simulated, the microstructure will vary most in the thickness direction and the CA should be able to represent the microstructure from surface to centre-line. Simulations using CA of size 64x64x1920 and 32x32x7808 were compared to those produced with the 200x200x200 cube.

Table I: Nucleation and growth parameters for each texture class [7].

Texture class	Nuclei density (grains/μm^3)	α' (μms^{n-1})	n'
Cube	7.42×10^{-5}	0.668	0.689
Rolling	9.18×10^{-5}	0.319	0.684
'Random'	9.05×10^{-5}	0.224	0.619

Sensitivity Errors

Sensitivity errors were investigated in an attempt to establish some boundaries for the accuracy required of the experiments which lead to the CA rules. Using the second scenario above as a base, and for a CA of 200^3 cells, the α' term in the growth rate of the cube grains (the fastest growing grains) was increased or decreased by one percent, and the nucleation density was altered according to the dependence in Equation (8). Because nucleation rate calculations can vary independently of growth rate (through measurements of fraction recrystallised), the density of nuclei of cube grains was also independently increased by six percent- which equates to an increase of only one percent in measured fraction of recrystallised grains.

RESULTS

CA Errors

The effect of boundary conditions on the kinetics of the simulation is marked for the two smaller simulation sizes (Figure 2): periodic boundaries lead to simulations whose results come much closer to the expected curve than do those simulations employing all mirror boundaries. However, this difference is only due to differences in the value of 'b'; in all cases mirror boundary conditions predict a JMAK exponent closer to the expected value of 3 (Table II). None-the-less the integrated absolute errors resulting from simulations employing mirror boundaries are substantially larger than for periodic boundaries (Figure 3). These integrated errors scale with CA size for both types of boundary. The predicted volume fraction recrystallised deviates most from the analytical equations in the mid-range of volume fractions (Figure 3) for all simulation conditions except the smallest CA sizes. Of interest is the increasing deviation of the JMAK curve from the precise solution as the CA size- and, consequently, the number of grains- decreases (Figure 2).

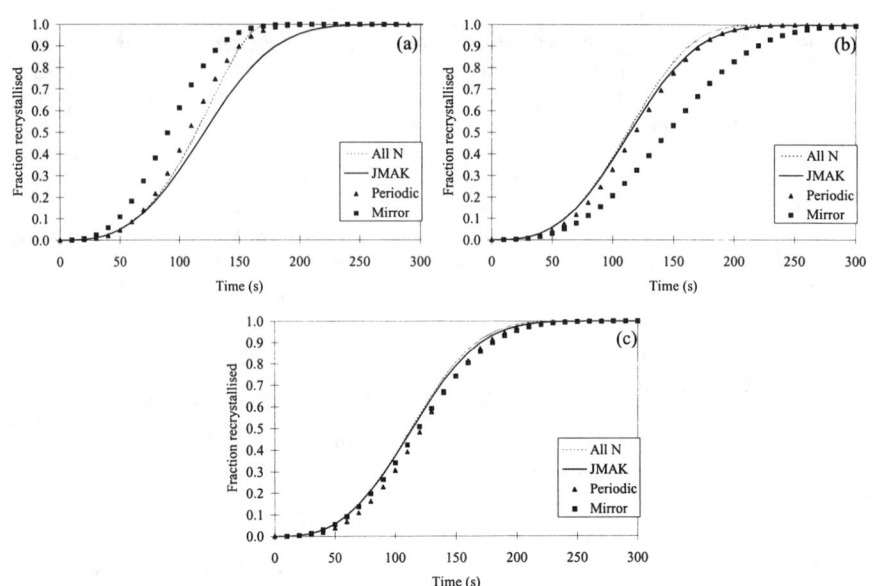

Figure 2: Comparison of simulated kinetics with Equation (6) - JMAK- and Equation (7) - All N- for two boundary conditions and simulations of size (a) 100; (b) 150; (c) 200.

Table II: Calculated JMAK parameters for simulations run under the first scenario

Property	CA size and boundary type					
	100 mirror	150 mirror	200 mirror	100 periodic	150 periodic	200 periodic
n	3.04	2.95	2.91	3.56	3.25	3.25
b	7.68×10^{-7}	2.95×10^{-7}	6.42×10^{-7}	4.07×10^{-8}	1.28×10^{-7}	1.14×10^{-7}
SD n	0.42	0.20	0.22	0.22	0.15	0.13
SD b	1.55×10^{-6}	2.08×10^{-7}	9.96×10^{-7}	4.23×10^{-8}	1.26×10^{-7}	7.08×10^{-8}

At a CA size of 200, the two types of boundary lead to predictions which are to all intents the same (Figure 2), although simulations employing mirror boundaries lead to a JMAK exponent closer to the theoretical value (Table II). The integrated errors are approximately equal (Figure 4), although the maximum error is lower for mirror boundaries than for periodic.

A feature of the simulations with mirror boundary conditions was that the time predicted for complete recrystallisation was roughly double the time predicted using periodic boundary conditions and calculated using Equations (6) and (7), although all methods yielded times to achieve a fraction of 0.95 within 10s of one another. The run-time for the mirror boundary simulations was correspondingly longer.

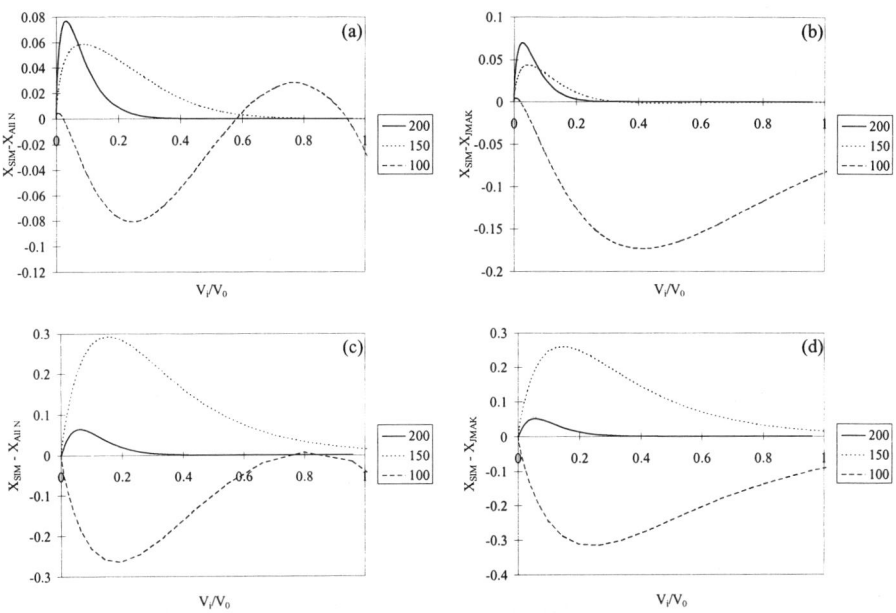

Figure 3: Errors resulting from simulations for two boundary conditions and of size shown: (a) periodic boundaries compared to Equation (7); (b) periodic boundaries compared to Equation (6); (c) mirror boundaries compared to Equation (7); (d) mirror boundaries compared to Equation (6). V_i/V_0 is the extended volume fraction of one grain.

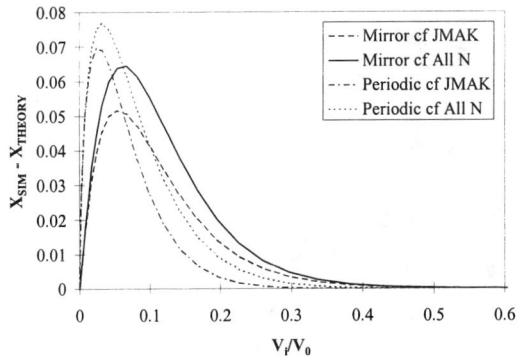

Figure 4: Comparison of errors for simulations of size 200.

The second scenario- comparing simulations with experimental data- showed no significant difference between predictions using the two CA boundaries (Figure 5). In each case at low fractions recrystallised the simulations predicted a lower fraction than was found experimentally. However, it was again found when a CA using a mirror boundary was the medium for simulation that the time predictions exhibited a long tail above a recrystallised fraction of 0.99. The predicted grain size distribution was found to contain a higher fraction of small grains than was observed experimentally (Figure 6)- leading to a mean grain size for the simulations lower than the experimental mean grain size- although the shapes of the two distributions were similar.

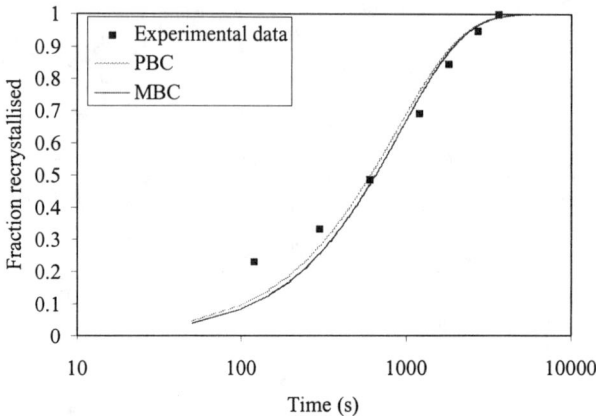

Figure 5: Comparison of simulated kinetics with experimental data [7]- determined by hardness measurement- for all periodic boundaries (PBC), and two periodic boundaries and one mirror boundary (MBC).

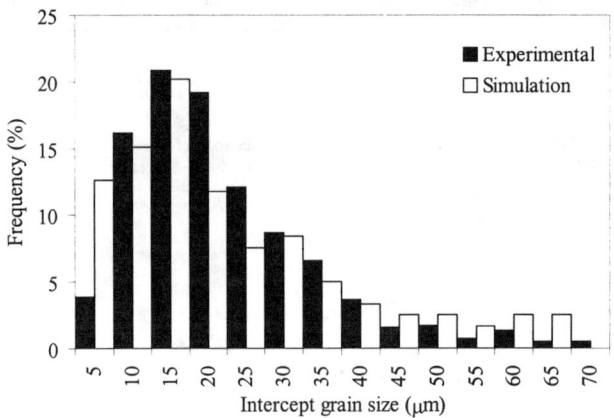

Figure 6: Aggregate grain size distribution resulting from simulations with periodic boundary conditions [7]

The shape of the CA had a negligible effect on simulation kinetics (Figure 7) but the predicted times for complete recrystallisation increased as the aspect ratio of the CA was

464

increased. Clearly the study of CA shape needs to include investigation of the resultant grain size and shape distributions.

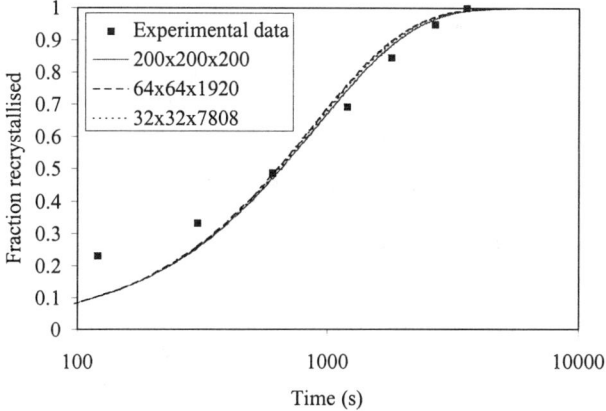

Figure 7: Effect of CA shape on the simulated kinetics.

Sensitivity Errors

Variations in the α' term for the growth of cube grains coupled with the corresponding variation to the nucleation density had little effect on the kinetics of the simulations (Figure 8a). Increasing the nucleation density independently of the interface migration rate led to a small decrease in recrystallisation times (Figure 8b), as might be expected.

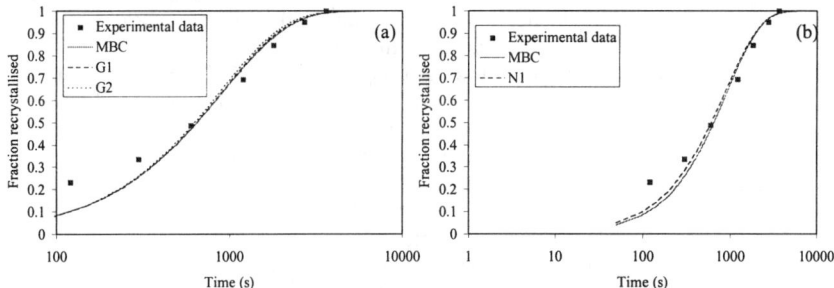

Figure 8: (a) Effect of varying the α' term for growth of cube grains with a dependent change to the nucleation density; G1- α' increased by 1%; G2- α' decreased by 1%; (b) Density of cube grains increased by 6% (N1), for constant α' term.

DISCUSSION

Figures 5 and 6 are the pivotal figures in this paper, and such under-predictions are not uncommon in other simulation methodologies (eg, [10, 11]: what is it about the simulations that leads to an under-prediction of volume fraction recrystallised at low fractions, and to a simulated grain size distribution that is skewed towards smaller grain sizes compared to the experimental distribution? Clearly these two facets are related, and the discussion revolves around the CA methodology, and the experimental information used to establish the CA rules.

Errors inherent to the CA methodology were studied by comparing simulations to exact analytical solutions. Errors are large for small automata, and are still significant for CA of size 200: a difference of 0.05 fraction recrystallised between simulation and exact solutions is the best that the CA under investigation can achieve. However, whereas in Figure 5 the greatest difference between simulation and experiment are found at fractions below 0.3, the errors are largest in the mid range of fractions recrystallised (Figures 3 and 4). Moreover, the maximum difference between simulation and experiment (Figure 5) is 0.12, and we can expect the intrinsic error for the same size of CA to be lower as the number of nuclei increases. Thus we cannot explain the under-prediction illustrated in Figure 5 solely by errors resulting from the methodology.

Of importance for simulations of hot rolling is that all mirror boundaries lead to a *smaller* deviation from the analytical solution than do all periodic boundaries for the largest CA. When only one mirror boundary is used in conjunction with two periodic boundaries (Figure 5), the deviation from experimental data is slightly larger than for simulations using three periodic boundary conditions, but the simulations are still within an acceptable margin of one another. However, coupled to this apparent advantage is the attenuating effect of mirror boundaries on the final stages of recrystallisation, and the similar effect of CA shape. The effect at mirror boundaries is thought to arise from the stagnation of grains and for this reason it is difficult to envisage how a simulation might circumvent the problem. That larger aspect ratio CAs enhance this effect is probably due to growth in the short-axis direction concluding well before growth in the long axis.

The simulations are quite resistant to realistic changes in the experimentally-determined parameters. The changes made to parameters in the equation describing the growth of cube grains and to the measured volume fraction of the cube class are within the bounds of differences such as might arise from examination of the data by two independent researchers. These changes do not significantly affect the simulations, and we are thus left with the question of how to explain the differences between simulation and experiment.

It is clear from the foregoing that it is the rules employed which are inadequate for a precise simulation of the microstructural evolution. The power of the CA method is that each cell may hold a number of attributes- descriptors of the microstructure- and that the microstructural evolution is controlled locally. However, the rules used to run CAs have been generally *average* in nature- mean interface migration rates for texture classes, mean (random) location of nucleation sites, and mean values of stored energy- although some aspects of local effects have been addressed qualitatively in simulations [2, 4].

In contrast, our understanding of the physical systems which CAs presume to simulate is far richer. It is accepted that recrystallisation will consume areas of high strain energy early in the process, and we can infer from this that nucleation rates and growth rates may be larger than average in the early stages of the transformation. The CA does not take this into account, and indeed it is only relatively recently that the experimental work needed to provide the rules has emerged [20, 21]. Although it is well known that interface velocity can be represented by the equation $v = MP$ - where M is the mobility of the boundary, and P the driving pressure (stored energy) - experimental characterisation of these factors for recrystallisation is very recent. Using EBSD, Huang and Humphreys [20] determined velocity-pressure relations for recrystallising grains in aluminium annealed at several temperatures. Critical to the experiments was the mapping of the misorientation across subgrains in the deformed material. Similarly, strain localisation at various grain boundary sites has been revealed to have a large effect on nucleation of static recrystallisation [21]. These observations are *local* and *microstructurally-dependent* and I speculate that rules based on such information would lead to a better match of the simulations to experiment, both in terms of kinetics and grain size distribution.

Currently then, cellular automata lacks the right experimental descriptions to be a truly predictive simulation methodology for static recrystallisation. Although it is a straightforward matter to provide finite element predictions as input to the CA, without the necessary precision in the CA rules the coupling of CAs and finite element can be assessed only qualitatively. As an example of through-thickness variations in microstructure (surface to centre-line of a plate), an arbitrary 10% difference in growth and nucleation parameters was imposed either side of a plane through the centre of a CA and the microstructure inspected qualitatively. The outcome of this scenario was little observable difference in the form of the microstructure away from the dividing plane (Figure 9). However, in some sections at the plane a clear dividing line was observed, possibly where grains had nucleated close to the dividing plane.

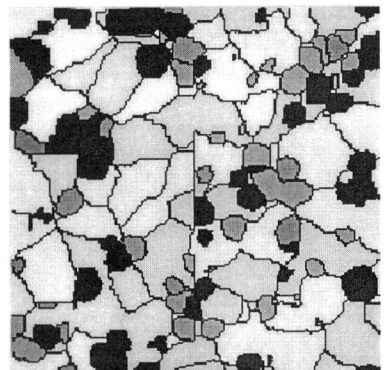 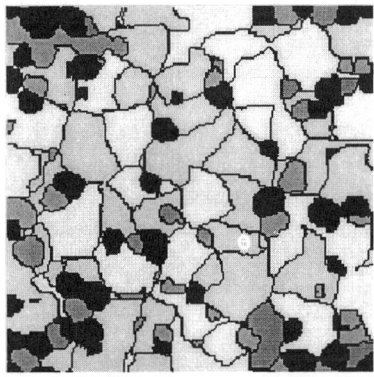

Figure 9: Microstructures resulting from an arbitrary variation in the nucleation and growth rates imposed on either side of a plane through the centre of the CA; the slices shown were taken at different depths in the CA.

In attempting to simulate industrial processing we must then consider the level of detail required of the experiments that supply the rules: what, for example, is reasonable spacing of temperatures at which experiments must be performed. Similarly, because of the marked variation in strains in any deformation test, and the strong effect of strain on recrystallisation kinetics [22] finite element simulations of experiments may be required in conjunction with measurements of the local microstructural variations if the CA rules are to operate with sufficient precision.

CONCLUSIONS

Assessing the errors inherent in CA simulations of microstructural evolution leads to the conclusion that most of the difference between simulations and experimental data is due to the inadequacy of the rules employed to drive the CA. Even with low nucleation rates, for a CA of eight million cells, errors of only 0.05 in the fraction recrystallised are observed, whereas differences of up to 0.12 are observed between experiment and simulation.

Encouragingly for simulations requiring gradients across a CA, mirror boundaries result in lower errors than periodic boundaries. However, such boundaries also lead to an attenuation of the time to complete recrystallisation. In conjunction with the shapes of CA likely to be required for through-thickness simulations, the attenuation is further accentuated.

If truly predictive simulations of industrial processes are to be achieved, detailed descriptions of the spatio-temporal variation nucleation and growth with prior deformation and temperature are prerequisite. It is likely that the local deformation will have to be evaluated by simulation to achieve the required accuracy in the experimental descriptions.

ACKNOWLEDGEMENTS

This work was supported under the auspices of the Australian Research Council Large Grant scheme; some of the research was conducted at IMMPETUS- the Institute for Microstructural and Mechanical Process Engineering: The University of Sheffield- in the UK, whilst the author was the holder of an Australian Academy of Science travel fellowship. The contribution of each organisation is gratefully acknowledged. The Engineering Faculty Research Committee of Monash University are thanked for their travel grant to attend the MRS meeting.

REFERENCES

1. H.W. Hesselbarth and I.R. Gobel, *Acta Metall. Mater.*, **39**, 2135-2143 (1991).
2. R.L. Goetz and V. Seetharaman, *Metall. Mater. Trans A*, **29A**, 2307-2321 (1998).
3. V. Marx, D. Raabe, O. Engler, and G. Gottstein, *Textures and Microstructures*, **28**, 211-218 (1997)
4. V. Marx, F.H. Reher, and G. Gottstein, *Acta Mater.*, **47**, 1219-1230 (1999).
5. C.H.J. Davies, *Scripta Metall. Mater.*, **33**, 1139-1143 (1995).
6. C.H.J. Davies, *Scripta Mater.*, **36**, 35-40 (1997).
7. C.H.J. Davies and L. Hong, *Scripta Mater.*, **40**, 1145-1150 (1999).
8. Mahin, K.W. Hanson, K., and Morris, J.W., *Acta Metall.*, **28**, 443-453 (1980).
9. Marthinsen, K., Lohne, O., and Nes, E., *Acta Metall.*, **37**, 135-145 (1989).
10. Furu, T., Marthinsen, K., and Nes, E., *Mater. Sci. Tech.*, **6**, 1093-1102 (1990).
11. D.Juul Jensen, *Scripta Metall. Mater.*, **27**, 1551-1556 (1992).
12. Srolovitz, D.J., Grest, G.S., and Anderson M.P., *Acta Metall.*, **34**, 1833-1845 (1986).
13. Srolovitz, D.J., Grest, G.S., M.P. Anderson, and Rollett, A.D., *Acta Metall.*, **37**, 2115-2128 (1989).
14. M. Miodownik, A.W. Godfrey, E.A. Holm, and D.A. Hughes, *Acta Mater.*, **47**, 2661-2668 (1999).
15. A.D. Rollet, *Progr. Mat. Sci.*, **42**, 79-99 (1997).
16. M. Rappaz, and Ch.-A. Gandin, *Acta Metall. Mater.*, **41**, 345-360 (1993).
17. Gokhale, A.M., and DeHoff, R.T., *Metall. Trans.*, **16A**, 559-564 (1985).
18. S. Fletcher, M. Thomson, and T. Tran, *J. Electroanal. Chem.*, **199**, 241-247 (1986).
19. D.Juul Jensen, *Metall. and Mater. Trans.*, **28A**, 15-25 (1997).
20. Y. Huang and F.J. Humphreys, *Acta Mater.*, **47**, 2259-2268 (1999).
21. C.M. Sellars, in *Proc. Intl. Conf. on Thermomechanical Processing: in Theory, Modelling and Practice*, ed B. Hutchinson *et al.* (Swedish Soc. for Materials Technology, Stockholm, Sweden 1997), p. 35 - 51.
22. P.L. Orsetti Rossi and C.M. Sellars, *Mater. Sci. Technol.*, **15**, 193-201 (1999).

MODELING THE EFFECT OF NOISE ON MULTISCALE DISLOCATION CELL STRUCTURES

M. A. MIODOWNIK[1] , E. A. HOLM[2] AND D.J.BROWNE[1]
[1] University College Dublin, Mechanical Eng. Dept., Belfield, Dublin, Ireland.
[2] Sandia National Labs, Department of Materials Modeling & Simulation, NM, USA.

ABSTRACT

Plastic flow in metals occurs by a dislocation mechanism in which strain is accommodated by a dislocation flux.   This results in tangled networks of dislocations which self-organize into stable cell structures with two characteristic length scales. These structures have a number of remarkable properties not least of which is that they exhibit scaling. We examine the idea that the evolution of geometrically necessary boundaries is dominated by random fluctuations, construct a simple model of cell orientations and show that scaling of misorientation distributions emerges a direct result of Gaussian white noise. We investigate whether hierachical microstructures containing both incidental dislocation boundaries and geometrically necessary boundaries can be studied using the same model. The results are compared with the experimental measurements and show good agreement.

INTRODUCTION

The dislocations produced during plastic deformation often form characteristic patterns known as dislocation cell structures [1]. The cell structures of medium to high stacking fault energy fcc materials have been studied in detail by Hughes et al. [2] who note the existence of boundaries that subdivide an original grain at two length scales. The larger scale is defined by long, continuous dislocation boundaries which they call geometrically necessary boundaries (GNBs). Within the blocks defined by these GNBs smaller scale cell boundaries are formed, termed incidental dislocation boundaries (IDBs). The patterns can be characterised by average cell spacing, cell spacing distributions, average misorientation angle, and misorientation angle distributions. Hughes et al. [2,3] have shown that for a wide range of fcc materials, the misorientation angle distributions of the IDBs and the GNBs exhibit a scaling property. Thus the misorientation distribution functions are dependent only on the average misorientation, θ_{av}, and so the varying distributions collapse onto to a single curve when scaled by θ_{av}.

Pantleon [4,5] has hypothesized that dislocation boundaries form as a result of statistical trapping of glide dislocations. Arguing that a Gaussian white noise leads to a bias of dislocation fluxes he shows that there will be a temporary formation of boundaries which exhibit scaling. Miodownik et al. [6,7] have taken a different approach and examined the hypothesis that the evolution of dislocation cells is dominated by random fluctuations of the cell orientations. This reposes the problem in terms of cell orientations rather than boundary misorientations. The Gaussian white noise can then thought of as allowing cell orientations to random walk in Orientation space. This is a diffusion problem, a direct analog of the 'thin film' problem in metallurgy [8]. Miodownik et al.[7] solve diffusion equations to get a time dependent solution for the orientation distributions and from this derive the misorientation distributions and the strain dependence of the average misorientation angle. The misorientation distributions scale with θ_{av} and the misorientation distribution function agrees very well with the experimental data for IDBs. In this paper we use the result from [7] that deformation-induced noise creates a Gaussian spread of cell orientations and extend the approach to systems which contain both IDBs and GNBs.

COMPUTER SIMULATION

GNB Structures

First we consider a small region of a deformed crystal that has split up in a idealised array of NxM (N=100, M=100) square blocks, see figure 1. Due to noise we assume the distribution of orientations is a Gaussian distribution about a mean orientation O_{ref} [6]. To simulate this a set of rotation angles r_{xy} is defined which has a Gaussian distribution, and a set of axes $[uvw]_{xy}$ are defined which are randomly distributed with a uniform probability on a unit sphere. The 3x3 transformation matrix R_{xy} of a rotation of r_{xy} about $[uvw]_{xy}$ is calculated and the orientation of each block, O_{xy} is then given by:

$$O_{xy} = R_{xy} O_{ref} \qquad (1)$$

The distribution of misorientations is calculated by cataloging the list of pairs of domains which define the boundaries present in the system. A misorientation matrix $M_{xy,x'y'}$ between each pair is calculated such that:

$$M_{xy,x'y'} = O_{xy} O_{x'y'}^{-1} \qquad (2)$$

where O_{xy} and $O_{x'y'}$ are the orientations of neighbouring blocks. This misorientation is then expressed in angle:axis format, taking the solution with the minimum misorientation angle from the equivalent descriptions allowed by cubic symmetry [9-11]. This defines the misorientation-angle $\theta_{xy,x'y'}$.

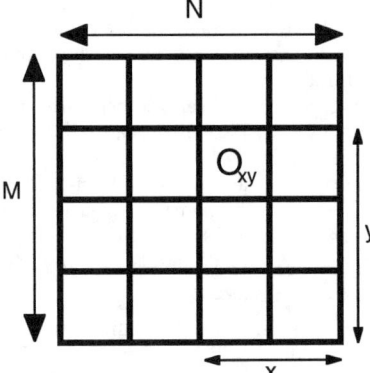

Fig 1. 2D square block structure (NxM) used to simulate idealised GNB network. Each block is identified by block coordinates (x,y), the orientation of each block is O_{xy}.

The distribution of misorientation angles as a function of average misorientation is shown in figure 2. Note the average misorientation is determined the standard deviation, σ_{block}, of the Gaussian noise. As expected the spread of misorientation increases with average misorientation. However the data exhibit the property of scaling; when scaled by the average misorientation, the data collapse onto a single curve. This scaling curve shows excellent agreement with the experimental data for GNBs by Hughes et al. [2], see figure 3.

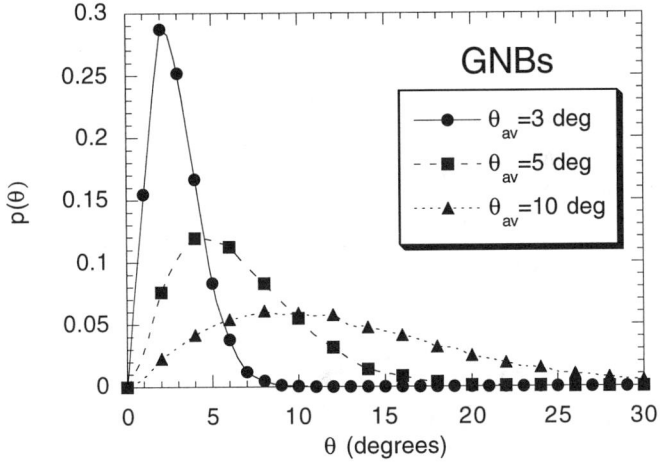

Fig 2. Misorientation distribution functions for GNBs generated by Gaussian Noise.

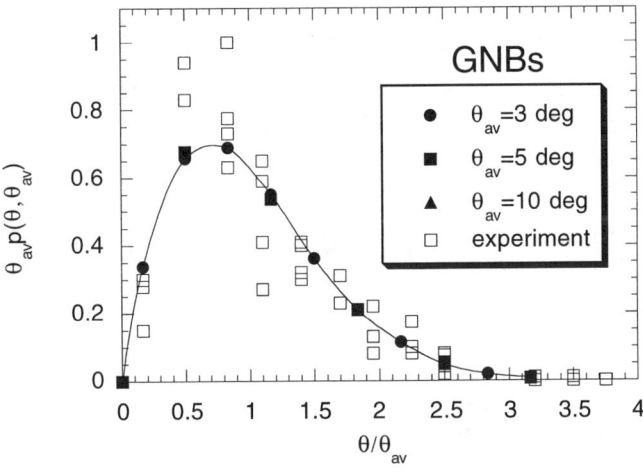

Fig 3. Scaled misorientation distribution functions for the GNBs generated by Gaussian noise. Experimental data from Hughes et al. [2] is shown as comparison. The experimental measurements were taken on rolled pure aluminium (99.996%) with strains from 5-50% cold reduction or ε_{VM} = 0.06-0.80.

GNB/IDB Structures

Now we consider an idealised 2D structure which contains both cells and blocks. Each of the MxN (M=100, N=100) blocks of the previous simulation is broken up into a square array of qxq cells, see figure 4. It is assumed that the cell orientations are created by a noise term and thus have a Gaussian distribution with standard deviation, σ_{cell}, about the orientation of the block. Thus each block orientation is replaced by a distribution of cell orientations allocated by the method described in the previous section. When $\sigma_{cell} \ll \sigma_{block}$ this procedure naturally produces a heirachal microstructure containing low angle and high angle cell boundaries, which can be thought of as IDBs and GNBs. The IDBs in the system are defined as boundaries between cells in the same block and the GNBs are between cells in different blocks.

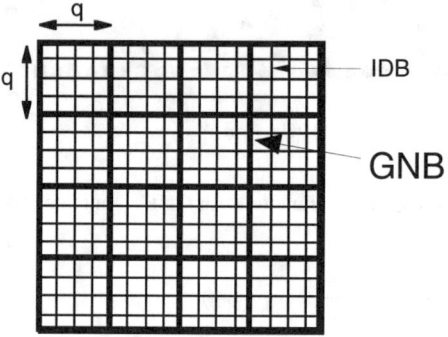

Fig 4. 2D square structure used to simulate idealised GNB- IDB network. Each block in figure 1 has been further subdivided into qxq square cells. The orientations of the cells are generated by a Gaussian Noise.

The standard deviations σ_{block} and σ_{cell} were chosen to give an average misorientation 5 deg for GNBs and 1 deg for IDBs. These were chosen because they were representative of experimental structures measured by Hughes et al.[2]. Although clearly they are arbitrary and there is no physical basis for differentiating between the noise terms for different length scales. This is a severe drawback of the current approach. That said, the scaled distributions show excellent agreement with the experimental data. Figure 5 shows the IDB distributions and is compared to the large sample of IDB experimental measurements of many different fcc metals studied by Hughes et al.[3]. It is difficult not to conclude that noise is responsible for these structures. Figure 6 shows the GNB scaled distributions as a function of q. Again there is good agreement with experiment although it should be mentioned that the data set is much smaller and from only a single material. The inclusion of IDBs into the structure has very little effect on the GNB distributions. The shape of the distributions is also relatively insensitive to q, confirming the analysis by Hughes et al. that the average misorientation is the most important parameter.

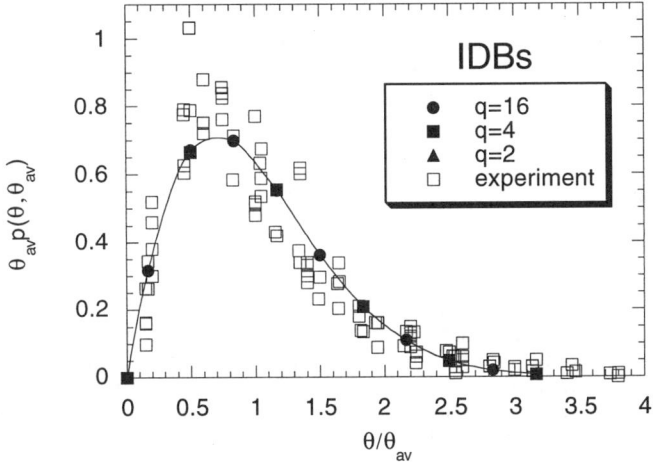

Figure 5. Scaled misorientation distribution functions for the IDBs generated by Gaussian noise. Experimental data for IDBs collated for a number of deformed fcc metals by Hughes et al. [3] is shown as comparison.

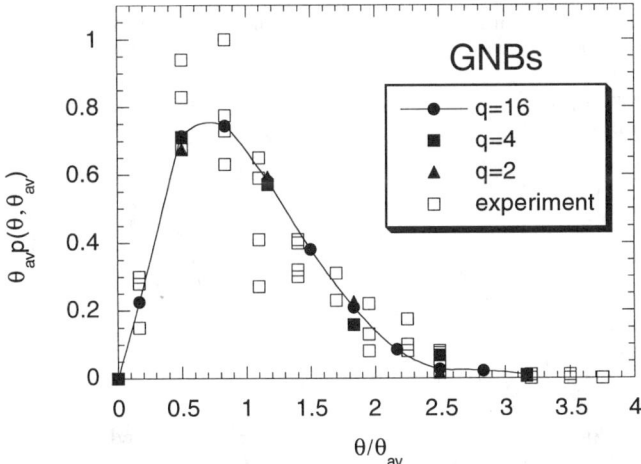

Figure 6. Scaled misorientation distribution functions for the GNBs generated by Gaussian noise. Experimental data for GNBs from Hughes et al [2] is shown as comparison. The experimental measurements were taken on rolled pure aluminium (99.996%) with strains from 5-50% cold reduction or $\varepsilon_{VM} = 0.06$-0.80.

DISCUSSION AND CONCLUSIONS

The model presented assumes the existence of a multiscale cell structure. It is the presence of the IDB and GNB boundaries and their absorption of glide dislocations that allows noise to be reflected in misorientation distributions. The boundaries are, in a sense, vehicles for the expression of Gaussian noise in the system. The model says nothing about how these boundaries are initially formed. It does not simulate deformation only the effects of noise generated by deformation. It is especially important to note that GNBs are discussed only in terms of their ability to absorb statistical dislocations. The model should be thought of purely as a tool to investigate the effect a Gaussian white noise on the θ_{av} and the scaling of misorientation distributions.

Pantelon [4,5] and Miodownik et al. [6,7] have shown with different approaches that the scaling of IDB misorientation distributions can be explained by the presence of white noise during deformation. This begs the question as to whether noise is responsible for the scaling of other features of dislocation cell structures? We have constructed a simple model to treat GNBs as another hierachical level of the essentially the same phenomneon. If the GNBs are affected by noise in the same way as IDBs, the scaled misorientation distribution of both IDBs and GNBs match the experimental distributions very well. Scaling arises naturally from noise.

ACKNOWLEDGMENTS

The Authors are grateful to Prof.D.Chzran, Dr.D.A.Hughes and Dr.A.W.Godfrey for useful discussions. Part of this work was performed at Sandia National Laboratories, supported by the U.S. Department of Energy under contract number DE-AC04-94AL85000. Dr.M.A.Miodownik is grateful for support through a Newman Scholarship at University College Dublin, Ireland.

REFERENCES

1. Kuhlmann-Wilsdorf D. and Hansen N. (1991) *Scripta metall. mater.* **25**, p.1557
2. Hughes D.A., Liu Q., Chrzan D.C. & Hansen N. (1997) *Acta mat.* **45**, p.105.
3. Hughes D.A., Chrzan D.C., Liu Q., & Hansen N. (1998) *Phys. Rev Lett.* **81**, p.4664.
4. Pantleon W. (1996) *Scripta metall. mater.* **35**, p.511
5. Pantleon W. (1998) *Acta mat.* **46**, p.451.
6. Miodownik M.A., Holm.E.A., Godfrey A.W., Hughes D.A., LeSar R. (1999) Mat. Res. Soc. Symp. Proc. Vol 538.
7. Miodownik M.A., Srolovitz D.J., Smereka P. and Holm E.A. submitted to*Acta mat.*
8.P.G.Shewmon,(1983) "Diffusion in Solids" McGraw-Hill Book Company.
9. Gertsman V.Y, Zhilyaev A.P., Pshenichnyuk & Valiev R.Z. (1992) *Acta metall.mater.* **40**, p.1433.
10. Garbacz A. & Grabski M.W. (1989) *Scripta metall.mater.* **23**, p.1369.
11. Haraze J. & Shimizu R. (1990) *Acta metall.* **38**, p.1395.
12. Bay B., Hansen N. Hughes D. and Kuhlmann-Wilsdorf D. (1992) Acta metall. mater. **40**, p.205.

THIXOTROPY OF SEMISOLID METALS

Andreas N. Alexandrou*, Gilmer R. Burgos*, Vladimir M. Entov**
* Semisolids Processing Laboratory, Metals Processing Institute, WPI, Worcester, MA. 01609
** Institute for Problems in Mechanics of Russian Academy of Science, pr. Vernadskogo, 101, 117526, Moscow, Russia

ABSTRACT

Understanding the time-dependent flow behavior of metal alloys in semisolid state is essential for the further development of the process. In the present investigation, the thixotropic behavior of semisolid slurries is modeled using conservation equations and the Herschel-Bulkley fluid model. The rheological parameters are assumed to be functions of the solid volume fraction, and of a structural parameter that changes with processing history. The evolution of the structural parameter is described by a first order kinetic differential equation that relates the rate of build-up and break-down of the solid skeleton. The model is implemented into a computer code to predict die filling.

INTRODUCTION

The processing of metal alloys in semisolid state can be divided into: (a) preprocessing, and (b) processing stage. During preprocessing, the raw material is melted and allowed to cool while growing dendrites are broken up using mechanical or electromagnetic stirring. The resultant slurry has an equiaxed microstructure made up of round, rosette-like crystals mixed in eutectic liquid. The specially preprocessed material is either immediately injected into a die in a process named rheocasting or solidified in billet forms for later processing (thixoforming). In the last process, the billets are reheated to a temperature in the mushy zone, and then injected into a die (thixocasting) or shaped between closed dies (thixoforging).

The process offers distinct advantages over similar methods for near-net shaping. The advantages are derived from the characteristic rheology of the material whose state lies between that of a pure solid and a pure liquid. The material can stand freely when it is not sheared and consequently, it can be easily manipulated. Furthermore, due to its higher than liquid viscosity, the flow remains mostly laminar minimizing the possibility for gas entrapment, thus making possible the use of heat treatment to obtain superior mechanical properties. Due to the fine uniform microstructure, the process can be used to produce parts with high strength and integrity, complicated geometries and very thin walls. Moreover, the process is performed at a temperature lower than in conventional casting, resulting in shorter solidification time, less possibility of shrinkage, increased die life, and lower energy requirement [3].

Despite the attractive features of the process, its implementation to industrial applications is hampered by technical problems primarily due to the complex rheology of the material. The material behavior during filling of dies is not well known and therefore, the process is difficult to control. The theoretical understanding of semisolid metals (SSM) during shape making operations is still under development. In this study, a constitutive theory based on the Herschel-Bulkley fluid model is presented to contribute to the understanding of the material and to individualize the most important properties that govern the material behavior. The model is implemented into a finite element code to simulate the filling process in a simple two-dimensional cavity.

Mat. Res. Soc. Symp. Proc. Vol. 578 © 2000 Materials Research Society

RHEOLOGY OF SEMISOLID METALS

Most of the early studies in semisolid metals were devoted to the rheological behavior of the slurry. In equilibrium steady-state shear flow experiments, SSM samples were cooled continuously to a given average solid fraction while sheared at constant shear-rate. Under these conditions, SSM behave as shear thinning (pseudoplastic) fluids with effective viscosity decreasing with increasing shear rate [5], [6], [7], [13]. However, experimental data on transient flows show a shear-thickening behavior, i.e., increasing effective viscosity with increasing shear rate [7], [9]. This apparent contradictory behavior is explained considering that in constant shear-rate experiments, the structure of the material is allowed to evolve to a new steady state corresponding to the imposed shear field whereas in rapid transients, the structure of the material does not have enough time to adjust to the new conditions. Experimental results also show that these materials resist finite shear stresses before deformation begins [7], thus behaving like Bingham fluids. Additionally, using hysterisis-loop tests for Sn-15%Pb SSM [5] and shear-rate step experiments [12], it was also found that SSM are thixotropic materials.

Since SSM are concentrated suspensions of alpha phase particles in eutectic liquid, the most plausible explanation for the rheology described above is that in the solid fraction range for thixoforming (0.4-0.5), the solid particles form a skeleton of interconnected particles that defines the apparent mechanical behavior of the system. This was evidenced in isothermal shear-rate step-up/step-down experiments wherein it was found that the break-down is faster than the restructuring of the network formed by the solid metal particles [7], [8], [12]. Therefore, it can be inferred that the evolution of the structure of the material during processing is governed by a number of kinetic phenomena of different characteristic time-scales. As a result of these kinetic processes, the rheological properties of the material, such as effective viscosity and yield stress, decrease with structure breakdown and increase with its build-up.

THIXOTROPY OF SEMISOLID METALS

The isothermal thixotropic behavior of SSM is modeled using the conservation of mass and momentum equations for an incompressible fluid, a rate-equation that governs the evolution of the material structure, and a constitutive relation based on the Herschel-Bulkley fluid model.

STRUCTURAL KINETICS

To characterize the time-dependent rheological behavior described in the previous section, a non-dimensional structural parameter λ is introduced to define the state of the structure. In a fully structured state, i.e. when all the particles are connected, λ is assumed to be unity. In a fully broken state, λ is assumed to be zero. In a partially broken state, λ assumes a value between zero and unity. The evolution λ is defined by a first-order rate equation, similar to those used to describe chemical reaction kinetics [10]. It is assumed that the rate of break-down depends on the fraction of links existing at any instant and on the deformation rate. Similarly, the rate of build-up is assumed to be proportional to the fraction of links remaining to be formed,

$$\frac{\partial \lambda}{\partial t} + \mathbf{u} \cdot \nabla \lambda = a(1 - \lambda) - b\lambda \left[\frac{D_{II}}{2}\right]^{1/2} exp\left(c\left[\frac{D_{II}}{2}\right]^{1/2}\right) \tag{1}$$

where the recovery parameter a, and the break-down parameters b and c are empirical constants. D_{II} is the second invariant of the rate of strain tensor defined as $D = \nabla\mathbf{u} + (\nabla\mathbf{u})^T$. The exponential dependence on the deformation rate, in the rate of break-down term of Equation (1), is included in order to account for the fact that the shear stress evolution for the shear-rate step-up experiment is faster than for the step-down case [8], [12].

CONSTITUTIVE RELATION

In agreement with the experimental evidence, the constitutive relation that relates the stress tensor with the kinematics of the flow is based on the Herschel-Bulkley fluid. This model is a combination of Bingham, and power-law fluid models. The Bingham fluid model accounts for the finite yield stress of the material while the power-law model describes the shear-thickening or shear-thinning behavior of SSM [1]. For the Bingham part we used the continuous model introduced by Papanastasiou [11].

$$\tau = \left\{ K(s,\lambda) \left[\frac{D_{II}}{2}\right]^{\frac{n(s,\lambda)-1}{2}} + \frac{\tau_y(s,\lambda)\left[1 - exp(-m\left|\sqrt{D_{II}/2}\right|)\right]}{\sqrt{D_{II}/2}} \right\} D \qquad (2)$$

where $K(s,\lambda)$ is the consistency index, $n(s,\lambda)$ the power-law index, $\tau_y(s,\lambda)$ the yield stress, and τ the viscous stress tensor. The parameter m is the stress growth exponent that controls the exponential growth of stress.

The material parameters a, b, and c in Equation 1, and $K(s,\lambda)$, $n(s,\lambda)$, and $\tau_y(s,\lambda)$ in Equation 2 are identified using appropriate experimental data. In the present investigation, we used the values obtained from the analysis of the experimental results by Modigell et al. [9] for Sn-15%Pb with $s = 0.45$ [2]. These values are: $a = 0.035$, $b = 0.00015$, and $c = 0.001$. Figure 1 shows the yield stress, power-law index, and consistency index as function of the structural parameter. The predictions indicate that the yield stress follows a "S" type curve behavior with the structural parameter. The yield stress is zero for a fully broken structure ($\lambda = 0$) and a finite value τ_o for a fully structured state ($\lambda = 1$). The power-law index n decreases quadratically with the structural parameter, while the consistency index K increases exponentially.

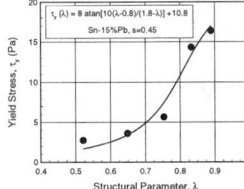 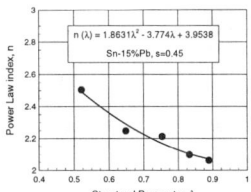 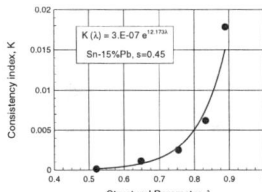

Figure 1: Yield Stress, τ_y, Power-Law index, n, and Consistency index, K, as function of the structural parameter, λ: ● Data obtained from the analysis of the experimental results of Modigell et al. [9], — Fitted [2].

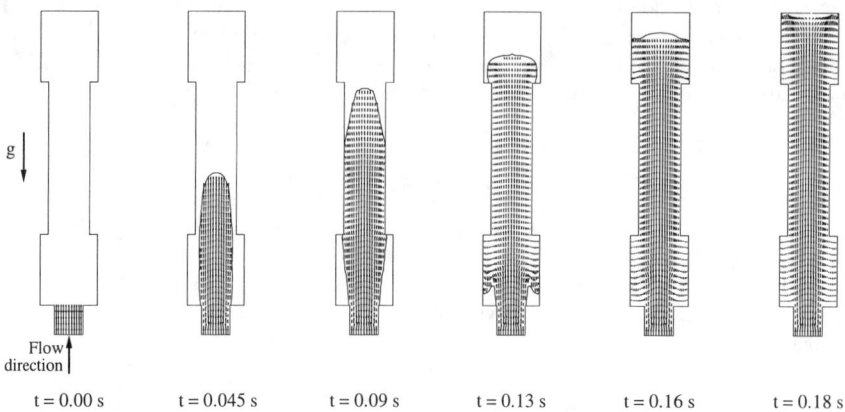

Figure 2: Filling of a tensile bar cavity against gravity, $Re = 195$, $Bi = 0.12$.

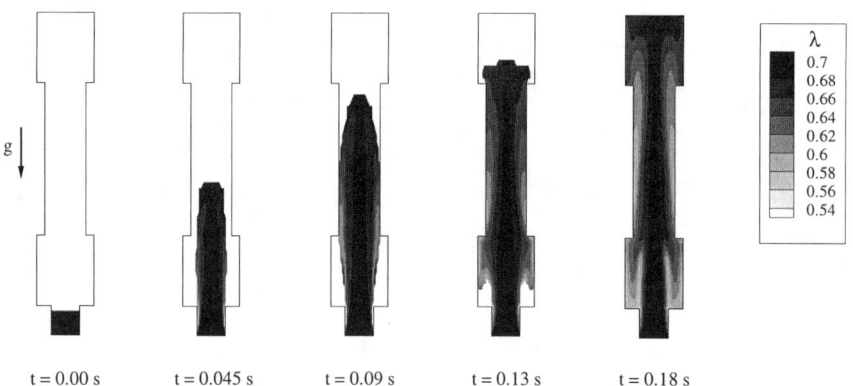

Figure 3: Evolution of the structural parameter λ in the filling of a tensile bar cavity against gravity ($Re = 195$, $Bi = 0.12$).

FILLING OF A SIMPLE CAVITY

In the present study, the constitutive model presented in the previous section is used to simulate the filling against gravity of a two-dimensional tensile bar cavity. This simple cavity is chosen in order to show the importance of the thixotropic behavior of SSM slurries. As initial condition, it is assumed that the material is moving with a uniform velocity equal to 1 m/s and a structural parameter $\lambda = 0.7$. The same conditions are assumed as boundary conditions at the inlet section of the die. No-slip boundary condition is imposed at the channel walls. We also assumed that the air exits the mold without constraint so that the pressure on the interface remains constant and equal to atmospheric pressure. Since the density and dynamic viscosity of the air are much smaller than those of the metal, interface interactions are neglected.

The filling front is tracked using the Volume of Fluid (VOF) technique introduced by Hirt and Nichols [4]. The solution to the conservation of mass and momentum equations along with the constitutive relations and front-tracking equation is obtained using the Streamline Upwind Petrov-Galerkin (SUPG) and the Pressure Stabilized Petrov-Galerkin (PSPG) finite element formulations [14]. The non-linear system of equations is linearized by a Newton-Raphson procedure and solved using an iterative solver based on the BiConjugate Gradient Stabilized Method [15]. Element-by-element storage is used in order to avoid assembly of element matrices to form a global coefficient matrix and the Jacobi preconditioning to speed up convergence.

Figure 2 shows the evolution in time of the filling front for the conditions stated above. Initially, a straight jet fills partially the lower-end and the middle section of the cavity. Then, the filling of the two-end sections are completed progressively. A back-flow pattern is generated in the lower-end section due to the stepped change in cross section. Flow defects may be formed in this region if appropriate welding does not happen. Moreover, the high shear rates generated in this region and close to the walls result in breakdown of the structure as depicted in Figure 3. Only a small core region penetrating the bar remains almost undeformed at the end of the filling process.

CONCLUSIONS

The thixotropic behavior of semisolid metals during processing has been modeled using a Herschel-Bulkley fluid model. The time dependent rheological parameters are function of the alpha phase volume fraction and and an internal variable that changes with the processing history. Results are presented for the filling of a tensile bar cavity with Sn-15%Pb SSM with $s = 0.45$. Although, this is a simple cavity, the results show that the model is capable of predicting the evolution of the structure of the material during the filling process.

References

[1] G. Burgos and A. Alexandrou, *J. of Rheol.*, **43**(3), 485-498 (1999); Proceedings of the 1998 ASME Fluids Engineering Division Summer Meeting, ASME, Washington D. C., June 21-25 (1998).

[2] G. Burgos, A. Alexandrou and V. Entov, Symposium *Synthesis of Light Metals III*, edited by F. Froes, C. Ward-Close, P. McCormick, and D. Eliezer, San Diego, Califor-

nia, Feb 28-March 4, 1999; *Light Metals 1999*, Proceedings of the 128th TMS Annual Meeting, edited by C. Eckert, San Diego, California, Feb 28-March 4 (1999)

[3] G. Chiarmetta, in *4th International Conference on Semi-Solid Processing of Alloys and Composites*, edited by D. H. Kirkwood and P. Kapranos, The University of Sheffield, England, 204-207 (1996).

[4] C. Hirt and B. Nichols, J. of Comp. Physics, **39**, 201-225 (1981)

[5] P. Joly, Ph. D. Dissertation, MIT, 1974.

[6] P. Joly and R. Mehrabian, J. Mat. Sci., **11**, 1393-1418 (1976).

[7] P. Kumar, C. Martin, and S. Brown, Acta Met. et Mater., **42**(11), 3595-3602 (1994); **42**(11) 3603-3614 (1994).

[8] M. Mada and F. Ajersch, *Metal & Ceramic Matrix Composites: Processing Modeling & Mechanical Behavior*, edited by R. Bhagat, TMS, 337-350 (1990).

[9] M. Modigell, K. Johannes, J. Petera (private communication).

[10] F. Moore, Trans. Brit. Ceramic Soc. **58**, 470-494 (1959).

[11] T. C. Papanastasiou, J. of Rheol, **31** 385-404, 1987.

[12] H. Peng and K. Wang, in *4th International Conference on Semi-Solid Processing of Alloys and Composites*, edited by D. H. Kirkwood and P. Kapranos, The University of Sheffield, England, 2-9 (1996).

[13] D. Spencer, Ph. D. Dissertation, MIT, 1971.

[14] T. E. Tezduyar, Advances in Applied Mechanics, **28**, 1-44 (1992).

[15] H. Van Der Vorst, SIAM J. Sci. Statist. Comp.., **13**, 631-644 (1992).

CHARACTERIZATION AND MODELING OF PRECIPITATION KINETICS IN ALUMINIUM 7000 ALLOYS

J.C. WERENSKIOLD, A. DESCHAMPS
LTPCM / ENSEEG, Domaine Universitaire de Grenoble, 38402 St. Martin d'Heres, France

ABSTRACT

The precipitation kinetics of 7108.70 aluminum alloy has been investigated in a wide range of temperatures by *in situ* Small Angle X-ray Scattering (SAXS) and Transmission Electron Microscopy (TEM), and computer modeled by use of an internal-state variable model which predicts the evolution of microstructural parameters. The modeling and experiments were done for isothermal heat treatment at 120, 140, 150, 160 and 170°C. The industrial T6 and T7 treatments have also been investigated.

INTRODUCTION

The 7108.70 alloy is widely used in automotive applications, mainly in car bumpers, due to its high mechanical properties. The process route for these alloys includes a solution treatment, a stretch forming operation followed by some natural aging and a two step artificial aging heat treatment. The first step ranges from 100°C to 120°C which is within the stability range of GP-zones [1], and the second step from 140°C to 170°C which is the temperature range for η' and η precipitation [2].

In order to improve and control the heat treatments in these alloys it is important to know both which phases are present and the kinetics of the precipitation reaction at different times and temperatures.

The first step to reach this goal is to measure, in a quantitative manner the precipitation kinetics at the different temperatures involved and then to apply a modeling approach to the experimental results.

In the present work the precipitation kinetics for several isothermal heat treatments between 100 and 170°C has been investigated experimentally and modeled by an internal state variable model concerning two regimes, *nucleation and growth* and *growth and coarsening*.

EXPERIMENTS

The 7108.70 alloy was provided by Hydro-Raufoss Automotive Research Center as extruded plates. Alloy composition is 5.5%Zn, 1.2%Mg, 0.16%Zr and 0.15%Fe (all in wt%). The solution treatment was 30 minutes at 480°C followed by water quenching, and resulted in a fully fibrous structure, with an average sub-grain size of approximately 3.5 μm.

The SAXS experiments were performed *in situ* during heat treatment, measuring scattering vectors in the range from 0.02 to 0.5Å^{-1} from 80μm thick samples with Cu K$_\alpha$ radiation. Scattering curves were corrected for background, fluorescence and absorption effects. Isothermal heat treatments were performed at 100, 120, 140, 150, 160 and 170°C with holding times up to 24 hours. A fast heating ramp of 320°C/hr was used, close to industrial heating ramps. The industrial T6 treatment was also investigated by SAXS: 6 hours at 100°C and 6 hours at 150°C with the same heating ramps, 320°C/hr.

Mat. Res. Soc. Symp. Proc. Vol. 578 © 2000 Materials Research Society

The particle dimensions were calculated using the Guinier approximation, which gives the gyration radius of the particles [3].

The nature and distribution of precipitates were investigated by Transmission Electron Microscopy (TEM) for the industrial T6 and T7 treatments. The T7 heat treatment is similar to the T6, but the last step is preformed at 170°C for 6 hours. Samples were prepared by ultrasonic cutting of 3mm discs and electropolished in a double-jet Tenupol by a 33% Nitric acid solution in methanol maintained at -20°C and 15 V. Samples were observed on a Jeol 3010 microscope.

TEM Observations

Samples from the as-quenched, the first step, the T6 and the T7 states were investigated by TEM. In the as-quenched state, one could observe numerous Al_3Zr dispersoids, mainly on dislocations and sub-grain boundaries, showing a very effective pinning.

The T6 sample showed a uniform distribution of precipitates in matrix, except from a precipitate free zone of approximately 40nm center to edge. Their size was measured by image analysis to be 22Å in radius for a sphere of equivalent size. Their shape is mostly disc shaped. The nature of the precipitates is determined from the $<111>_{Al}$ zone axis diffraction pattern. Most precipitates appear to be η', but some η, mainly in the orientations η_1 and η_2 on dislocations and grain boundaries, can be detected.

The T7 sample showed coarser precipitates. Their size was measured to 43Å in radius. Their shape is similar to the ones in the T6 sample. The precipitates are mainly η in the orientations η_1, but some η' can still be detected.

Precipitation Kinetics

The evolution of precipitate size with aging time at the different temperatures investigated is shown in figure 1 as the Guinier radius vs. time.

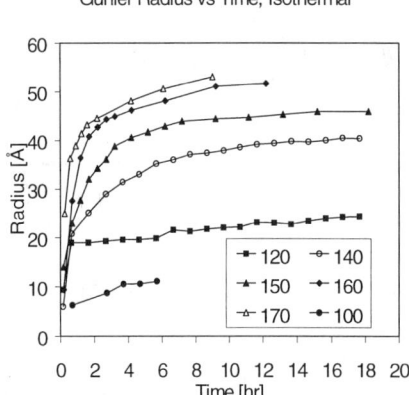

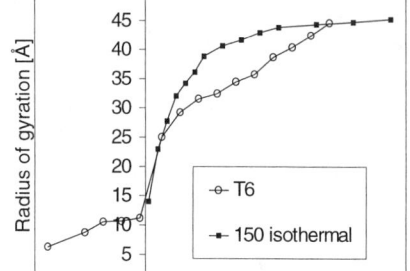

Figure 1. Experimental results. Guinier radius evolution versus time for isothermal aging.

Figure 2. Comparison between industrial T6 and isothermal aging at 150°C

The activation energy for nucleation:

$$\Delta G^{*} = \Delta G_{0} / \ln^{2}(C/C_{eq})\qquad(3)$$

In these expressions T is the absolute temperature, k is Boltzman's constant, v_{at} is the average atomic volume (supposed identical in the precipitates and the matrix), C is the solute concentration in the matrix at time t, C_{eq} is the equilibrium solute concentration, and γ is the interfacial energy between precipitate and matrix. ΔG_{0} is an adjustable parameter reflecting the heterogeneous nature of the nucleation mechanism, and is expected to depend on the nucleation conditions. The details of the model can be found in [6].

The parameters included in the model all have a physical meaning. The diffusion constant follows an Arrhenius law and was adjusted to describe the diffusion over the entire temperature range. The equilibrium concentration was adjusted according to an exponential law in order to obtain $C_{eq} = 1\%$ at 160°C and 5,4% at 400°C [6]. Interfacial energy was adjusted to the experimental results and taken as 280mJm^{-2}.

The differences in behavior at various temperatures resulting from different nucleation mechanisms will thus appear in the temperature dependence of ΔG_{0}.

Results from the modeling

The best fits for the evolution of Guinier radius with time at the various temperatures between 120 and 170°C are shown in Figure 4a to e. For each temperature, the evolution of the solute concentration predicted by the model is represented in Figure 3. As expected, the higher precipitate growth rate as the temperature increases is associated with a faster decrease of the solute concentration. The model is able to predict correctly the precipitation kinetics at all temperatures with a single adjustable parameter ΔG_{0}.

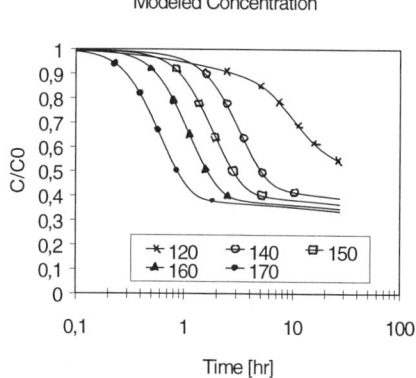

Figure 3. The variation in concentration as given from the model.

After reviewing the results, it is interesting to discuss the values for the activation energies chosen at the different temperatures, ΔG_{0} as a function of temperature. First it is useful to recall the theoretical value for ΔG_{0} if nucleation was homogeneous :

$$\Delta G_{0}^{\,hom} = 16/3 \times \pi \gamma^{3}(v_{at}/kT)^{2}\quad \Rightarrow \quad \Delta G_{0}^{\,hom} \propto T^{-2}\qquad(4)$$

Thus, one can expect that at constant nucleation mechanism (in terms of heterogeneous nucleation, this would correspond to a constant wetting angle), ΔG_{0} should change with the temperature proportionally to T^{-2} since the T dependence of C/C_{eq} is very weak. Figure 5 shows

The 100°C curve shows precipitate radius between 6 and 11Å. At this temperature, GP-zones are the only precipitates present. The curves from 120 to 170°C are within the stability range of η' precipitates. All curves in this range shows a first measured point around 8Å after 10 minutes. These are probably GP-zones which have survived the heating ramp. The next measured point is ranging from 19Å for 120°C to 24Å for 170°C. This is probably the nucleation radius for η' precipitates. The increase in nucleation radius with time indicates a more difficult nucleation at higher temperatures.

Figure 2 shows a comparison between the isothermal 150°C and the T6 heat treatment. The T6 sample shows at all measured times a lower precipitate size, thus indicating that the first step at 100°C has a refining effect by providing more nucleation sites for η' precipitates.

DISCUSSION AND MODELING OF PRECIPITATION KINETICS

The precipitation reaction in these alloys starts with GP-zones which may follow some sort of clusters [4]. The GP-zone solvus depends on the composition of the alloys and is around 120°C for the 7108 [1]. At this temperature the GP-zones are replaced by the metastable η' phase which heterogeneously nucleates on the dissolving GP-zones [5]. The η' phase can appear up to more than 180°C, however, it is also replaced by the equilibrium η $MgZn_2$ phase during isothermal heat treatment. The η' to η transition does not seem to be of great importance for practical purposes. It has never been shown that the transition has any influence on either the precipitation kinetics or the hardening potential of the alloy.

Modeling Approach

The model used in this work was developed by one of the authors [6] for a different 7000 alloy containing higher levels of solute and for isothermal heat treatments at 160°C. One aim of the present modeling work is to determine the flexibility of the precipitation model, and in particular in terms of changing the heat treatment temperature. Changing the temperature will influence many parameters such as diffusion, equilibrium concentration in solute, and nucleation mechanisms.

The primary assumptions of the model are that the alloy is considered pseudo binary with an equivalent solute, having its own equilibrium concentration and diffusion constant, thus the ternary nature of the alloy is not taken into account. The precipitates are assumed to be pure solute. The complex precipitation sequence, $\alpha \to \alpha + GP\text{-zones} \to \alpha + \eta' \to \alpha + \eta$ is replaced by a single precipitation process which is assumed to be η'. The GP-zones do not enter the model directly but as nucleation sites for η' precipitates and are thus included in the activation energy for nucleation, ΔG_0.

The main thermodynamic equations entering the model are (1), the driving force for phase separation from an ideal solid solution model:

$$\Delta g = -kT \ln(C/C_{eq}) / v_{at} \qquad (1)$$

The critical radius for precipitate dissolution in solid solution at a concentration C (which stands both for nucleation and for coarsening) :

$$R^* = 2\gamma v_{at} / kT \ln(C/C_{eq}) \qquad (2)$$

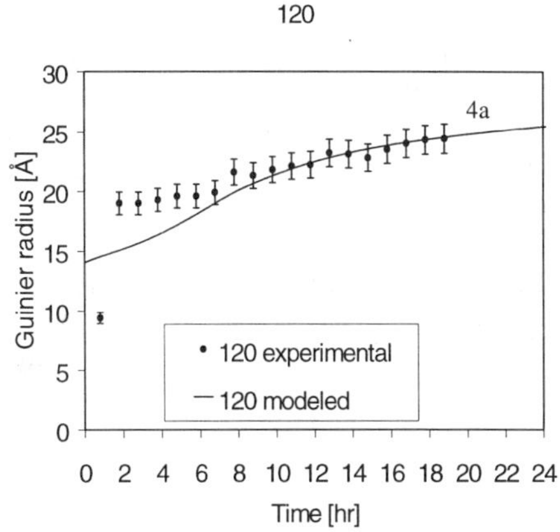

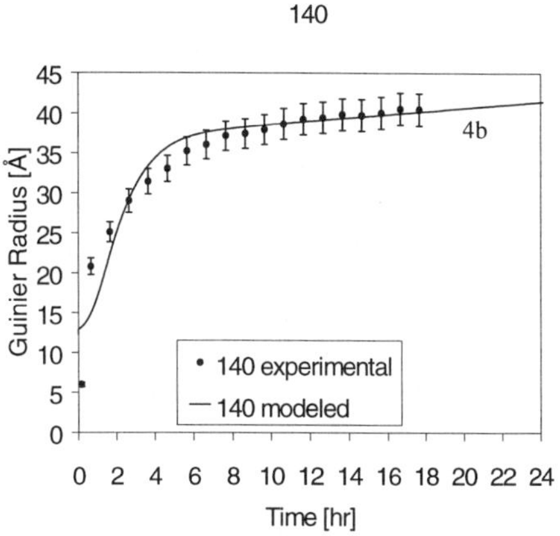

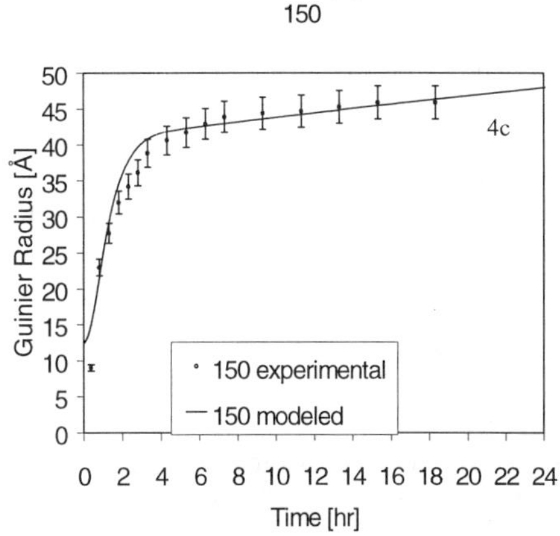

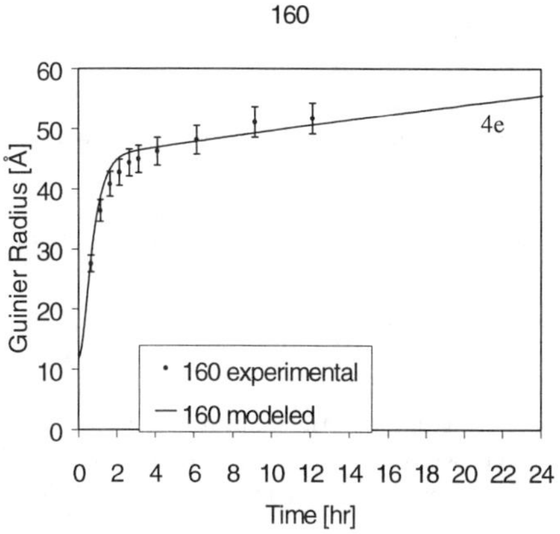

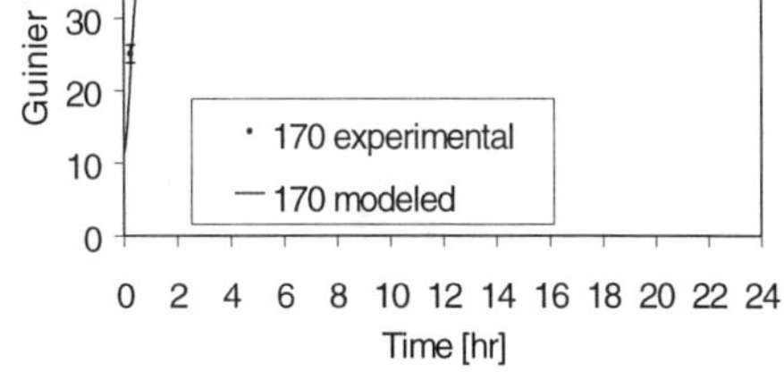

Figure 4a to e. Results of the modeling: experimental results of the Guinier radius along with the best fit from the model.

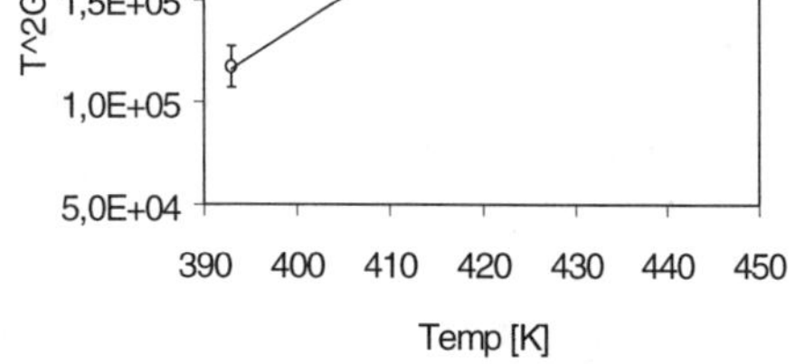

Figure 5. A plot of $T^2\Delta G^0$ vs. T, showing a straight line, indicating that nucleation is more difficult at higher temperatures.

that the evolution of $\Delta G_0 T^2$ with temperature is perfectly linear and not constant. The interesting outcome of this result is that it reflects different nucleation mechanisms acting over this temperature range. However, the detailed evolution of the nucleation barrier, namely $\Delta G_{0'} T^2 \propto T$, is a mere coincidence. This is in agreement with the general accepted nucleation sequence in this system: it is known that nucleation of η' is very sensitive to the degree of clustering of the structure, in terms of vacancy rich clusters (VRC) or GP zones [2, 7, 8].

In the thermal cycle that we have used in this study, the degree of clustering before the rapid heating to the aging temperature is very limited, and thus these clusters are likely to dissolve partially before any nucleation may occur. Therefore, it is expected that at higher aging temperature the nucleation of η' will be more difficult, thus due to larger dissolution of the potential nucleation sites in the ramping stages.

CONCLUSION

In the present paper we have characterized quantitatively over a range of temperatures the precipitation kinetics. The simple model proposed in ref[6] allows us to describe the kinetics for both precipitate radius and precipitated volume fraction with a limited number of adjustable parameters whose physical meaning and order of magnitude are known. The only really adjustable parameter ΔG_0 is evolving with T in a way which clearly reflects that the change in nucleation mechanism depends on the nature of the phase which precipitates (η', η).

In future work, non isothermal heat treatments and different alloys will be studied with the simple model in order to come closer to industrial heat treatment.

ACKNOWLEDGMENTS

The authors would like to thank Hydro Raufoss Automotive Technology Center for providing the materials and for economical support.

We would also thank senior scientist Frederic Livet at LTPCM for helping with the experiments and Prof. Yves Brechet for stimulating discussions.

REFERENCES

1. G. Groma and E. Kovacs-Csetenyi, Phil. Mag. *1975*, 869

2. T. Ungar, J. Lenvai and I. Kovacs, Aluminium **55**, 663 (1979)

3. O. Glatter and O. Kratky, *Small Angle X-ray Scattering* (Academic press 1982)

4. H. Löffler, I. Kovacs and J. Lendvai, J .Mater. Sci. **18**, 2215 (1983)

5. A.K. Mukhopadhyay, Q.B. Yang and S.R. Singh, Acta Metall. **26**, 267 (1994)

6. A. Deschamps and Y. Brechet, Acta mater. **47**, 293-305 (1999),

7. J. Lendvai, G. Honyek and I. Kovacs, Scripta Metall. **13**, 593 (1979)

8. Loffler, Kabisch, Gueffroy, Radomsky, Honyek and Ungar, Kristall und Technik **14**, 721 (1979)

AUTHOR INDEX

SUBJECT INDEX

.